Handbook of Research on the Internet of Things Applications in Robotics and Automation

Rajesh Singh
Lovely Professional University, India

Anita Gehlot
Lovely Professional University, India

Vishal Jain
Bharati Vidyapeeth's Institute of Computer Applications and Management (BVICAM), New Delhi, India

Praveen Kumar Malik
Lovely Professional University, India

IGI Global
DISSEMINATOR OF KNOWLEDGE

A volume in the Advances in Computational Intelligence and Robotics (ACIR) Book Series

Published in the United States of America by
IGI Global
Engineering Science Reference (an imprint of IGI Global)
701 E. Chocolate Avenue
Hershey PA, USA 17033
Tel: 717-533-8845
Fax: 717-533-8661
E-mail: cust@igi-global.com
Web site: http://www.igi-global.com

Library of Congress Cataloging-in-Publication Data

Names: Singh, Rajesh (Electrical engineer), editor. | Gehlot, Anita, editor. | Jain, Vishal, 1983- editor. | Malik, Praveen Kumar, editor.
Title: Handbook of research on the Internet of Things applications in robotics and automation / Rajesh Singh, Anita Gehlot, Vishal Jain, and Praveen Kumar Malik, editors.
Description: Hershey, PA : Engineering Science Reference, an imprint of IGI Global, [2020] | Includes bibliographical references and index.
Identifiers: LCCN 2019003699| ISBN 9781522595748 (hardcover) | ISBN 9781522595762 (ebook)
Subjects: LCSH: Internet of things--Industrial applications. | Automation. | Robotics.
Classification: LCC T59.5 .I455 2020 | DDC 629.8/954678--dc23 LC record available at https://lccn.loc.gov/2019003699

This book is published in the IGI Global book series Advances in Computational Intelligence and Robotics (ACIR) (ISSN: 2327-0411; eISSN: 2327-042X)

British Cataloguing in Publication Data
A Cataloguing in Publication record for this book is available from the British Library.

All work contributed to this book is new, previously-unpublished material. The views expressed in this book are those of the authors, but not necessarily of the publisher.

For electronic access to this publication, please contact: eresources@igi-global.com.

Advances in Computational Intelligence and Robotics (ACIR) Book Series

Ivan Giannoccaro
University of Salento, Italy

ISSN:2327-0411
EISSN:2327-042X

Mission

While intelligence is traditionally a term applied to humans and human cognition, technology has progressed in such a way to allow for the development of intelligent systems able to simulate many human traits. With this new era of simulated and artificial intelligence, much research is needed in order to continue to advance the field and also to evaluate the ethical and societal concerns of the existence of artificial life and machine learning.

The **Advances in Computational Intelligence and Robotics (ACIR) Book Series** encourages scholarly discourse on all topics pertaining to evolutionary computing, artificial life, computational intelligence, machine learning, and robotics. ACIR presents the latest research being conducted on diverse topics in intelligence technologies with the goal of advancing knowledge and applications in this rapidly evolving field.

Coverage

- Robotics
- Computational Intelligence
- Cyborgs
- Artificial Intelligence
- Evolutionary Computing
- Machine Learning
- Cognitive Informatics
- Agent technologies
- Computer Vision
- Neural Networks

IGI Global is currently accepting manuscripts for publication within this series. To submit a proposal for a volume in this series, please contact our Acquisition Editors at Acquisitions@igi-global.com or visit: http://www.igi-global.com/publish/.

The Advances in Computational Intelligence and Robotics (ACIR) Book Series (ISSN 2327-0411) is published by IGI Global, 701 E. Chocolate Avenue, Hershey, PA 17033-1240, USA, www.igi-global.com. This series is composed of titles available for purchase individually; each title is edited to be contextually exclusive from any other title within the series. For pricing and ordering information please visit http://www.igi-global.com/book-series/advances-computational-intelligence-robotics/73674. Postmaster: Send all address changes to above address.

Titles in this Series

For a list of additional titles in this series, please visit:
https://www.igi-global.com/book-series/advances-computational-intelligence-robotics/73674

Handbook of Research on Advanced Mechatronic Systems and Intelligent Robotics
Maki K. Habib (The American University in Cairo, Egypt)
Engineering Science Reference • copyright 2020 • 400pp • H/C (ISBN: 9781799801375) • US $295.00 (our price)

Edge Computing and Computational Intelligence Paradigms for the IoT
G. Nagarajan (Sathyabama Institute of Science and Technology, India) and R.I. Minu (SRM Institute of Science and Technology, India)
Engineering Science Reference • copyright 2019 • 347pp • H/C (ISBN: 9781522585558) • US $285.00 (our price)

Semiotic Perspectives in Evolutionary Psychology, Artificial Intelligence, and the Study of Mind Emerging Research and Opportunities
Marcel Danesi (University of Toronto, Canada)
Information Science Reference • copyright 2019 • 205pp • H/C (ISBN: 9781522589242) • US $175.00 (our price)

Handbook of Research on Human-Computer Interfaces and New Modes of Interactivity
Katherine Blashki (Victorian Institute of Technology, Australia) and Pedro Isaías (The University of Queensland, Australia)
Engineering Science Reference • copyright 2019 • 488pp • H/C (ISBN: 9781522590699) • US $275.00 (our price)

Machine Learning and Cognitive Science Applications in Cyber Security
Muhammad Salman Khan (University of Manitoba, Canada)
Information Science Reference • copyright 2019 • 321pp • H/C (ISBN: 9781522581000) • US $235.00 (our price)

Multi-Criteria Decision-Making Models for Website Evaluation
Kemal Vatansever (Alanya Alaaddin Keykubat University, Turkey) and Yakup Akgül (Alanya Alaaddin Keykubat University, Turkey)
Engineering Science Reference • copyright 2019 • 254pp • H/C (ISBN: 9781522582380) • US $185.00 (our price)

Handbook of Research on Deep Learning Innovations and Trends
Aboul Ella Hassanien (Cairo University, Egypt) Ashraf Darwish (Helwan University, Egypt) and Chiranji Lal Chowdhary (VIT University, India)
Engineering Science Reference • copyright 2019 • 355pp • H/C (ISBN: 9781522578628) • US $295.00 (our price)

IGI Global
DISSEMINATOR OF KNOWLEDGE

701 East Chocolate Avenue, Hershey, PA 17033, USA
Tel: 717-533-8845 x100 • Fax: 717-533-8661
E-Mail: cust@igi-global.com • www.igi-global.com

List of Contributors

Table of Contents

Detailed Table of Contents

Internet has revolutionized the technological era, which has a significant impact on us by making communication much better not only with the living beings but also with non-living things through the medium of internet of things (IoT). Thus, this topic highlights how internet of things can minimize user intervention in controlling home appliances and monitoring its setting. Integrating IoT with cloud computing and web service helps us in providing feasibility in accessing home appliances (i.e., monitoring appliances and measuring home condition). The whole process of integration aims to create an intelligent system. Thus, smart home is one of the application of IoT aimed at improving comfort, safety, and wellbeing within our homes.

The emerging tunnel FET is analysed in terms of ON-state current, OFF-state current, subthreshold slope, switching capacitance to explore its applications for smaller size low-power high-speed digital and memory applications that are an integral part of portable intelligent devices for IoT applications. A large portion of IoT systems are associated with these embedded SRAM/DRAM memories that contribute to a major portion of power dissipation in systems-on-chip (SoCs) or digital design. Several SRAM cell-based memory designs with TFET structures are compared to focus their applications. The ambilpolar nature of TFET structures are investigated for highly random, unclonable secured hardware systems. New circuit designs with TFET were explored for turn-on voltage reduction, ON-state resistance reduction, and reverse leakage reduction techniques that plays an important role in designing efficient energy-harvesting systems.

It is a common saying that science is just a philosophy without precise engineering. Available monitoring systems for real-time monitoring of space, goods, or life stocks are either too complex or too costly for the majority of the population to operate so users have to revert back to visual inspection, which results in increased human effort. The proposed model works on internet of things-based technology for more accurate and secure real-time-based monitoring made accessible to the general public. Each node primarily contains an entire network of electronic sensors to measure the environmental parameter. Due to its compact size, this device is capable of deployment over remote areas. Older modules were mostly manual and could only be engaged for a singular task, but the proposed network of nodes is fully customizable and with minor changes can be used for a task that is entirely different from the other.

This chapter reveals the crucial challenges being faced by the society and the developers in making IoT reach to every nook and corner of the society. But as every coin has two sides, besides the boon of IoT, there are some banes too which can be now considered as the challenges open for the developers to make IoT a bigger success. Every application has its own issues in adopting IoT. Among them, the common and major ones are security and privacy concerns, ubiquitous network connectivity, societal reluctance to new technology, lack of standard protocols, gigantic data size, high-cost investment, robustness, and self-sustainability of IoT sensors connected in wireless sensor networks. Besides common challenges, challenges specific to different sectors of IoT applications like retail sector, healthcare sector, smart grid, smart city, agricultural sector, smart homes, transportation sector, smart metering, and power sector have been pondered in this chapter. These challenges need to be first combated so that IoT can bring the expected transformation at global level.

In China and Japan robots are used to serve in restaurants to serve or to just greet the customer. In restaurants robots need to understand the environment and make decisions accordingly, like changing the route if it finds any obstacle in between. To have this capability of making decisions and behaving like humans, robots need information from different sources from which they can understand the change in environment, position of other robots, and also the path to the destination. This can be done by integrating robots with IoT technology that allow the enhancement in the capabilities in robotic waiter. IoT also helps in monitoring the working parts of robots and movements. This chapter aims to discuss the real-time challenges and impact of IoT on robotics services in restaurants and hotels.

The mobile robotics industry is related to creating mobile robots that can move around in physical environments. Different types of mobile robot designs for obstacle avoidance have been experimented in the past based on different sensors, trajectory algorithms, etc. The chapter presents implementation details of different obstacle avoiding robots (OARs) using sensors, Bluetooth module, and IoT modules. The sensor-based obstacle-avoiding robots are designed using ultrasonic sensors and Arduino microcontrollers. Bluetooth-based obstacle-avoiding robots have been designed using Arduino mega and Bluetooth module and an Android application. IoT-based obstacle-avoiding robots can be designed in three different ways, using ethernet shield, node MCU, or Raspberry Pi. The IoT-based obstacle-avoiding robot using Raspberry Pi is the most popular mobile robot model that uses maximum on-chip modules in comparison to other designs, and also, the design can be extended by using cameras to use images for sensing the objects in order to avoid collisions.

Internet of things (IoT) is a technology that is constantly progressing and finding new applications. It is making its way into enormous sectors. One such sector where IoT has entered is healthcare. This sector is one of the most vital sectors of world. Health of each and every individual should be given utmost priority. Any person who is unwell requires vigilance. The patients in hospital are always kept under vigilance. The healthcare sector constantly strives to keep on finding new ways and methods to improve and upgrade the current methods used for vigilance of patients. The use of IoT in healthcare sector can drastically change the way this sector works now. In this chapter, the concept of IoT shall be explained in detailed. Along with it, the current vigilance system and methods used in healthcare will be stated and viewed. In addition, a detailed demonstration through case studies explaining the use of IoT in healthcare sector will also be presented in this chapter.

Around the world, city, state, governments, and public sector groups are leading the way in taking the internet of thing (IoT) to existence and improving the lives of citizens everywhere. Currently 21 billion devices are connected. For 2025, the installed base of IoT devices is forecast to grow to almost 76 billion worldwide. At the same time, the global smart agriculture market size is expected to triple by 2025, reaching $15.3 billion (compared to being slightly over $5 billion back in 2016). Because the market is still developing, there is still ample opportunity for businesses willing to join in. Building IoT applications, devices, and products for future agriculture set you apart as an early adopter, and as such, help you pave the way to success. Wireless sensor network has an ability to solve many problems of real world (i.e., agriculture, harvest, farm monitoring, post-harvest, remote irrigation control, warehouse, livestock monitoring, cold chain monitoring) and improve the decision-making capability of government in response to natural disasters.

IoT devices and their applications are supporting humankind in almost all domains. This chapter explores LoRaWAN and proposes to integrate the commodity cameras installed at the security points of most of the gates to residential areas. LoRaWAN is a media access control (MAC) protocol for wide area networks and it has been opted for its functional and architectural scalability. The intelligent inputs are transferred from the commodity cameras to the data concentrators (DC), on edge-based computing, the DC can transfer this input to fog, cloud, remote servers for machine learning integrations. This chapter demonstrates the basic architectural framework of the said implementation. However, the detailed implementation and prototype is beyond the scope of this chapter. The chapter has however demonstrated the architecture. The features of the commodity cameras have been listed that can serve as the feed to the concentrators that shall enable alarm generations at the local and remote policing sites.

The progressive research in the field of internet of things provides a platform to develop high performance and robust automated systems to control external devices via internet data transfer and cloud computing. The present emerging IoT research including user-friendly and easily-wearable sensors and signal acquisition techniques have made it possible to expand the IoT application areas towards healthcare sector. This chapter aims at providing a rationale behind development of IoT applications in healthcare, architecture details of internet of healthcare things (IoHT), and highlights a step-by-step development of IoT-based heart rate measurement and monitoring system using Arduino. The developed module has

Water is a vibrant source for life, and its management and treatment are a crucial issue at the present time. A smart solution for city-level water treatment plant is gaining importance with the development in communication technology. We can avoid the issues pertaining to water management and treatment, such as distribution, quality, monitoring of overflows, pH (pouvoir hydrogène) value of water, etc., with help of information communication and technology. Based on this standard, the authors proposed a smart water treatment plant for the entire city using IoT with business processes direction and decision support systems. The suggested IoT architecture not only provides real-time monitoring for supply of pure water being transferred to the citizens but maintains the quality and purity of water supplied along with automation of motor operation, automatic valve controlling, and health monitoring of equipment. The system also delivers an alert to a remote user when there is a deviation of water quality parameters from the pre-defined set of standard values.

Internet of things (IoT) is an emerging technology that is looked upon as a technology that can bring positive changes in development of the society. Using IoT, humans and things embedded with sensors, actuators, and other hardware can communicate in an intelligent manner, thus creating a sea of innovative ideas which can make life easy, simple, and smart. IoT has touched upon different areas such as healthcare, home security, and agriculture. In this chapter, the authors give an insight about the technology and its applications in various fields. The major focus is on emerging trends in education and potential applications of IoT in the field of education. IoT can bring a major shift in traditional educational practices and make the teaching-learning process more effective by actively engaging learners. Use of IoT devices can also help teachers in deciding their future plan of action and giving more attention to the slow learners. This chapter also highlights the challenges that this technology faces in the education sector.

Plants play a vital role in maintaining the ecological cycle, and thus, to maintain the plant's proper growth and health, adequate monitoring is required. Hence, the aim of the chapter is to create a smart plant monitoring system using automation and internet of things (IOT) technology. This topic highlights various features such as smart decision making based on soil moisture real-time data. For this purpose, sensors like soil moisture sensor, DHT11 sensor, level sensor, etc. are used. The soil moisture sensor measures the level of moisture (i.e., water content of different plants). The signal will be sent to Arduino

Preface

The main intent of this book is to explore the various applications and uses of "Internet of things". As "Internet of Things" is an emerging technology. The major emphasis of this book is on the impacts and challenges of internet of things especially in the field of automation and robotics applications.

Seventeen chapters are included in the book. Chapters include review on role of IoT in agriculture, measuring and analyzing power consumption. A very important role for industrialization of smart grids is discussed. A smart parking system for vehicles is also included. A chapter is on the design and implementation of an automatic street lighting system using a light resistance (LDR) sensor to turn on or off based on the intensity of sunlight.

A chapter covers the opportunities and challenges in virtualizing the on-premises Session Border Controller (SBC) implementation to Network Function Virtualization Cloud for better Cloud Communication scalability and automation in communication.

A greenhouse automation system has been designed by considering the environmental factors. A smart alert system is discussed for garbage clearance by sending an alert signal to web server present in the municipal for collecting the garbage from the dustbins by verifying the level of garbage filled.

Editors are thankful to the authors for their valuable contribution and publisher for the support.

Rajesh Singh
Lovely Professional University, India

Anita Gehlot
Lovely Professional University, India

Vishal Jain
Bharati Vidyapeeth's Institute of Computer Applications and Management (BVICAM), New Delhi, India

Praveen Kumar Malik
Lovely Professional University, India

Chapter 1
Internet of Things-Based Architecture of Web and Smart Home Interface

Jasjit Singh
https://orcid.org/0000-0001-9574-0324
University of Petroleum and Energy Studies, India

Ankur Kohli
University of Petroleum and Energy Studies, India

Bhupendra Singh
Schematic Microelectronics, India

Simranjeet Kaur
Reynolds Community Collage, Midlothian, USA

ABSTRACT

Internet has revolutionized the technological era, which has a significant impact on us by making communication much better not only with the living beings but also with non-living things through the medium of internet of things (IoT). Thus, this topic highlights how internet of things can minimize user intervention in controlling home appliances and monitoring its setting. Integrating IoT with cloud computing and web service helps us in providing feasibility in accessing home appliances (i.e., monitoring appliances and measuring home condition). The whole process of integration aims to create an intelligent system. Thus, smart home is one of the application of IoT aimed at improving comfort, safety, and wellbeing within our homes.

OVERVIEW

This chapter outlines IOT based Smart Home Interface. The wireless network is created by using ESP8266 NUTTYFI board, IOT Blynk platform. This network will allow the communication between the user and home devices wirelessly through Internet of Things. The system is designed to reduce the manual framework in operating home appliances and to ensure more safety by displaying the device status (ON/OFF) on Graphical widgets of Blynk. The data from the sensors (Temperature, Humidity) are collected

DOI: 10.4018/978-1-5225-9574-8.ch001

by NUTTYFI IOT board and uploaded to cloud, this uploaded data can be visualize on the blynk widget whereas the command to operate the Home devices can be send by using Blynk app which will furher process by ESP 8266 NUTTYFI board & Control the home devices.

The other operation of IOT based architecture of web are as follow:-

- Monitoring of indoor home temperature.
- Controlling Home devices wirelessly.
- Alert indictors on in abnormal conditions.
- Tracing the performance of device over the period of time.
- End-To-End IOT Security

IOT Based Architecture of Web & Smart Home Interface: Internet of Things technology provides reliability on controlling and monitoring home devices and suitable for energy saving without any loss of quality and convenience. It may support to control energy usage & wastage.

INTRODUCTION

Conservation of energy holds importance in today's era as we are exploiting energy more than its usage in large amount in day to day work routine for better comfort & convenience thus to equilibrate this energy usage smart system is required. Hence the aim behind making smart home interface using Internet of Things is to reduce energy wastage in hustle and bustle routine of life and provide smart decision making in operating and monitoring various home devices.

The embedded home interface system with IOT technology is reliable and economically sufficient in managing devices to turn on using sensor technology when requires. Adopting this smart home interface framework, we can exhibit the utilization of the energy can be reduced on various devices usage. The system framework reduces manual operation whereas increases automatic process which increases feasibility in handling the home appliances.

The framework of the chapter highlights the implementation of IOT in the field of Home automation. This will give us basic idea about how the IOT technology can be implemented to make smart home interface more efficient and reliable. The chapter covers the basic implementation of IOT technology henceforth allow beginners and keen learners to explore IOT domains.

Review of Literature

Anita,G discuss et al Hands-on Application of Internet of Things in Various Domain of Engineering explained various application of Internet of Things (IOT) in the domain of engineering such as health care, environmental monitoring, Fire & Safety Engineering, Home Automation etc (Piyush et al, 2018). Bhupendra,S discuss et al Wireless Methods & Devices for Home Automation explained various methods of interfacing home devices under local as well as global networks and various sensor technology for different embedded home devices (Lucio et al, 2016). Francseco,F discuss et al IOT Based Indoor Personal Comfort Levels Monitoring explained development of monitoring system to analyse the sensor data and to control devices using wireless IOT technology (Todica et al, 2016). Todica,M discuss et al Controlling Arduino Board with smartphone and Blynk via internet explained interface between Arduino board – ESP8266 and Blynk app by working on Firmware and Arduino SPI communication with ESP8266 via cloud (Todica et al, 2017). Todica,M discuss et al Home Security Alarm Notification with Smart Phone and Blynk explained working and interface of various alarm indicators with Blynk IOT platform via email notification on user smart phone with alert parameters (2017). Hung-yu,W discuss et al Two Paradigm in Cellular IOT Access for Energy-harvesting M2M Devices: push-based versus pull-based explained energy harvesting cellular communication, considering the properties of energy harvesting and M2M devices (Mei-Ju et al, 2016). Gaikwad,S discuss et al Smart Hospitals using Internet of Things highlights the open challenges that hinder the utility IOT in smart home (2016).Pundir,Y (2016) discuss Internet of Things: Challenges & Future Direction explained the unique addressing scheme of the Internet of Things which allow interaction with other things or objects to create new application, services to achieve common goals. Nazeem Basha,S (2016) discuss An Intelligent door system using raspberry pi and amazon web service IOT demonstrate smart system which will notify the intrusion by sending out email notification to the owner by logging all the intrusion data into the google spreadsheet of owner's drive account. Singh,R (2016) discuss A smart building automation system explained important role played by these systems in regular energy monitoring and management, therefore saves maximum energy and cost. Mohamed,M (2016) discuss Internet of Things for home automation explained home automation using smart phones and computers. Isa,E (2017) discuss Smart home automation: GSM Security Design & Implementation explained security system for the smart home automation supported by GSM embedded mobile module, which enables the alert messages transmission to both mobile devices of end user and central security office. Bikarama,Y (20180 discuss IOT Enable Fire Safety and Security devices for building explained the use to sensor technology in integration with the IOT platform & alert message are sent to user and fire station in case of hazardous scenario. Patel.J (2017) discuss Automatic home appliances and security of smart home with RFID,SMS,Email and real time algorithm based on IOT explained home computerize frame work using SMS and mail id based Home Appliances Control system and security with radio frequency identification (RFID). S.Yadav (2016) discuss Wireless Personal Area Network and PSO-Based Home Security System explain the monitoring and controlling unwanted events like burglars/ thieves at home,locality,city using security system proposed with Zigbee and GSM modem.

Related Work

The related research work was done by Yousef Ismael on Smart home using IOT based sensing and monitoring system in which the system maintains thermal comfort related to temperature and humidity, visual comfort related to colour and light and hygienic comfort associated with the air quality. The proposed system uses EmonCMS platform for collecting and visualizing monitoring data.

Yousef completed the research on Smart home using IOT based sensing and monitoring system. He did exploration in maintaining the comfort level in the smart home by automatically controlling various parameters such as temperature, intensity of light, air quality controller etc.

Problem Identified

While surveying, the problem identified that all the parameters of the Home automation are still not fulfilled and in most of the cases the risk factor related to Hackers attack, Data ownership risk, single point of Failure are most common. For that we must require a secure and feasible platform which is cheap and can be accessed by everyone in their smart phones. To eliminate these problems IOT based

Architecture of Web & Smart Home Interface is developed in this technological era which is economical and beneficial for the growth IOT based technologies.

Solution Proposed

The proposed solution of this chapter discusses the wireless way of monitoring indoor thermal comforts, device current status and operating its state(ON/Off) with the user friendly BLYNK IOT based platform. Thus, system is capable of sending alert message to the user smart phone via e-mail with the current status of all the devices when parameters value goes beyond safe range. The system consists of home appliances, embedded relay circuit, NUTTYFI IOT board (ESP8266), Power Supply (Bridge rectifier circuit), Filter (12V-5V) converter circuit. Figure 1 shows the simplified framework the system.

Keynotes: Number of devices attach to the relay circuit can be more than shown in the above simplified framework of the system.

Block Diagram and System Description

Figure 2 shows various home devices connected to the ESP8266 NUTTYFI microcontroller which connect and allow communicate of home devices with Blynk app via cloud.

Table1 shows the embedded components list require to establish the Smart Home interface. In table 1 different abbreviation used for jumper wires for internal circuit connection such as

M-M = Male to male
M-F = Male to Female
F – F = Female to Female

Figure 1.

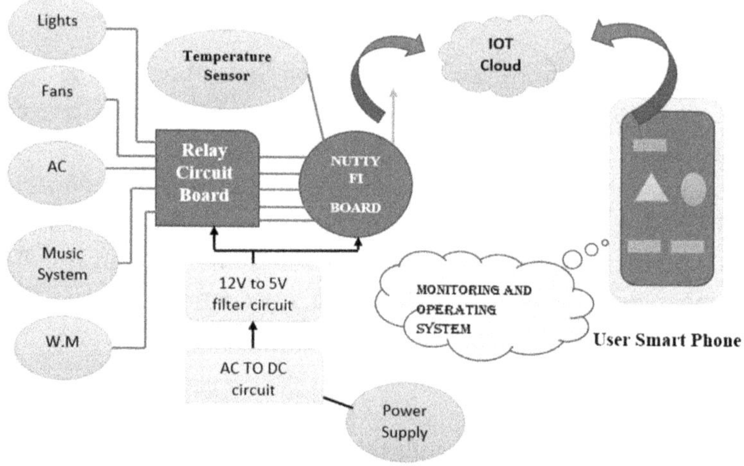

Figure 2. Block diagram of devices connected and system overview

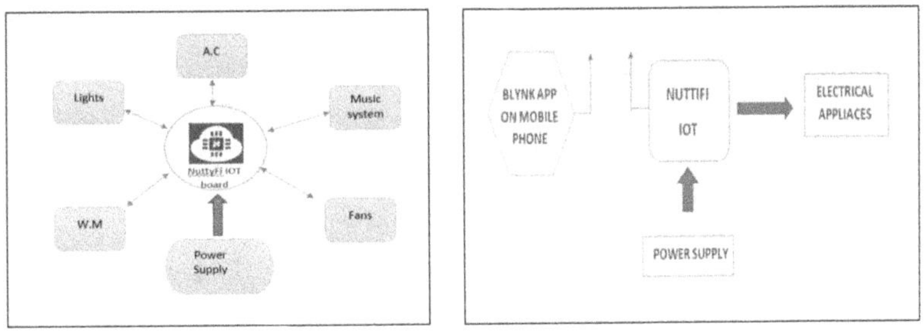

Table 1. Embedded components list

SNO.	Component Name	Quantity
1.	225J 400V capacitor	1
2.	1M .25W resistor	1
3.	100 ohm 2W	1
4.	1N4007 diode	4
5.	12V Ziner diode	1
6.	470 uF 25V capacitor	1
7.	Relay circuit board	5
8.	Filter Circuit	1
9.	NuttyFi IOT board	1
10.	Temperature sensor LM35	1
11.	Smart Phone with Blynk app	1
12.	Power Supply 230V	1
13.	Jumper Wire M-M	20
14.	Jumper Wire M-F	20
15.	Jumper Wire F-F	20

Methodology of the System

In the smart home interface system, all the home devices are connected and communicate with each other via cloud (IOT platform), this is accomplished by the means of NUTTYFI IOT micro-controller to which all the devices are connected and control by Blynk IOT application in the user smart phones. Blynk is one of the secure ways to communicate as it has specific authentication token for specific project, which prevent Data ownership risk.

The systematic of the system is as follow -:

Circuit Schematics and Detail Study of Software and Hardware Components

This section will discuss detail study of Power circuit (230V AC to 12V DC) & Filter Circuit Schematics.

Figure 3. Schematic of smart home interface

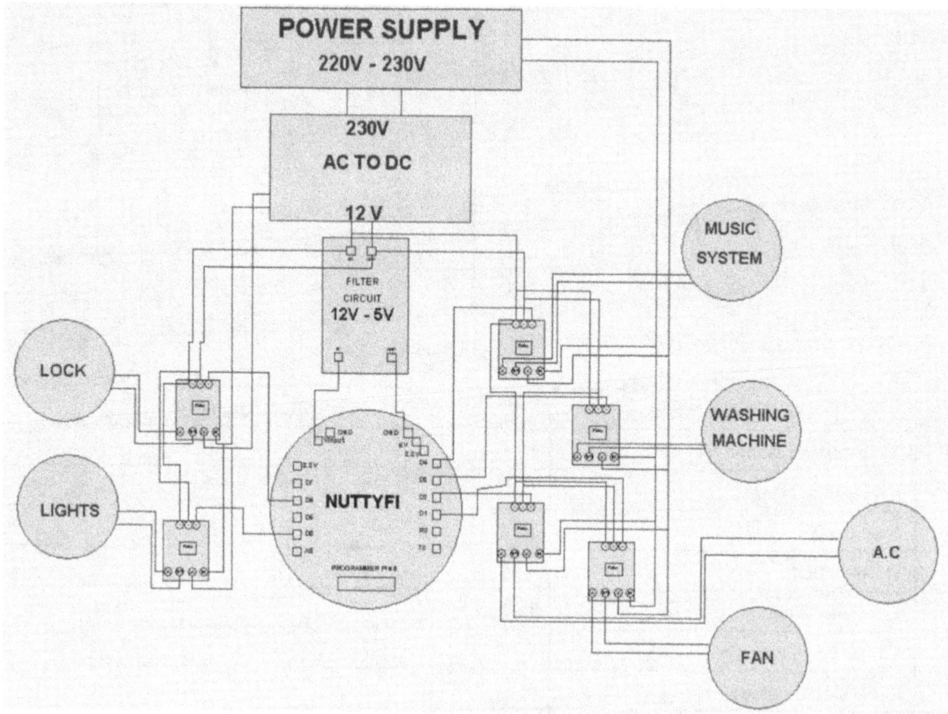

230V AC to 12V DC Convertor Circuit

The circuit is used to convert 230 V to 12V DC power supply which will further used to power all the components as shown in the Figure 3

Table 2shows the components list require to develop the above circuit.

Filter Circuit 12V DC to 5VDC

The circuit shown in the Figure 5 converts 12V DC to 5V DC which act as the input voltage to NUTTYFI microcontroller.

Table 3. shows the component list require to develop the above circuit.

Software Components of System

BLYNK is an IOT (Internet Of Things) platform which allow to control the devices and monitor senor data from any part of the world, it too allow graphical interface and app development by simply drag and drop widgets from Widget box.

The BYNK app can be install from play store in Android devices and App store in IOS devices.

Download Latest BLYNK Libraries:- https://github.com/blynkkk/blynk-libraries/releases/latest
Download Docs, Tutorials: http://www.blynk.cc

Figure 4. Schematic of 230V AC to 12V DC convertor circuit

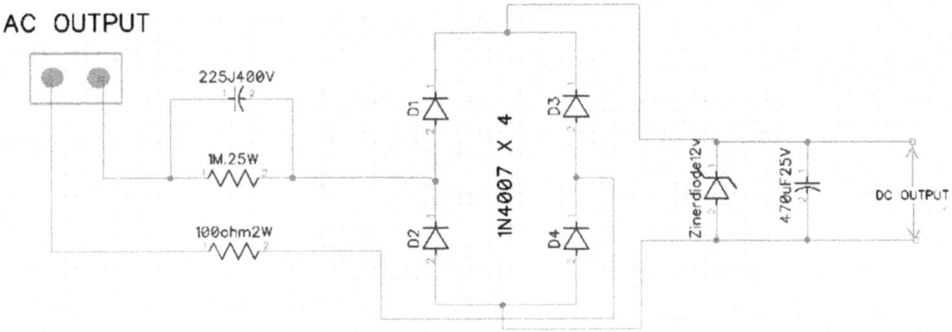

Table 2. Components list of AC to DC circuit

SNo.	Components Name	Quantity
1	225 J 400V Capacitor	1
2.	1M .25W Resistor	1
3.	100 ohm 2W Resistor	1
4	1N4007 Diode	4
5.	Zener Diode	1
6.	470uF 25V Capacitor	1
7.	Power Supply 220V-230V	1

Table 3. Component list of the Filter circuit

SNo.	Component Name	Quantity
1.	Male – Male Header pins	2
2.	1N4007 Diode	1
3.	50V 1000uF Capacitor	2
4.	7805	1
5.	330K ohm Resistance	1
6.	Led	1

Figure 5. Schematic OF Filter circuit (12v DC to 5V DC)

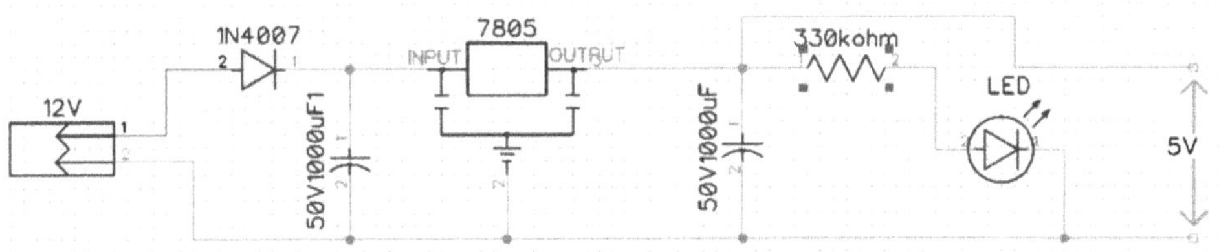

Installation of Blynk library in Arduino IDE

- **Method 1**
  - Open Arduino IDE > Sketch > Include libraries > Manage Library > Filter you search. (Type *Blynk*) > Install it.
- **Methods 2**
  - Web browser > Blynk.cc > Getting started > Download blink libraries (follow the instruction).

Figure 6.

Installation of Blynk app in Mobile Phones

After installing Blynk app from App store / play store, follow the instruction to operate it.

NUTTYFI IOT (ESP-12x Series) Board

NUTYFI IOT provides powerful IOT (Internet of Things) hardware platform which is based on ESP8266 12x series which allow user to build IOT products . NUTTYFI allow feasibility to user in managing, monitoring and controlling devices from any part of the world, it can interface with any IOT wen iCloud, Local or IOT mobile platform.

Pin Configuration

It consists of –

- 8 digital pins (D0 – D8)
- 1 Analog pin (A0)
- Ground
- 3.3V pin
- 5V pin

Figure 7. NUTTYFI IOT (ESP-12x Series) board

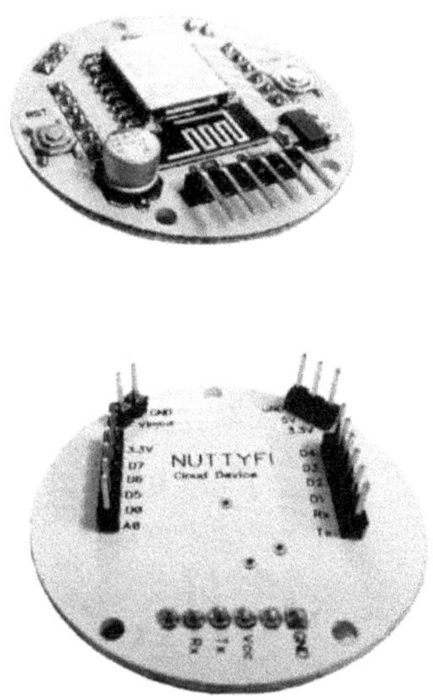

Input Power

Vin (Input Voltage)= 5 V

The NUTTYFI IOT board operate on 5 V and 3.3 V.

Programmer Used (FTDI – Basic USB to TTL Programmer)

Steps to Install NUTTYFI IOT Board in Arduino IDE

NUTTYFI is the IOT platform which allow the sensor and actuators too connect to the internet/cloud. This allow to store the data on the cloud as well the devices can be operated from any corner of the world. It consist if WiFi Chip Esp8266 which too allow other microcontrollers to connect to the internet.

Figure 8. Steps to install and start new project in Blynk
The Above Hardware Component can be Purchase From Official site of NUTTY Engineers: *https://www.nuttyengineer.com/product/nuttyfi/*

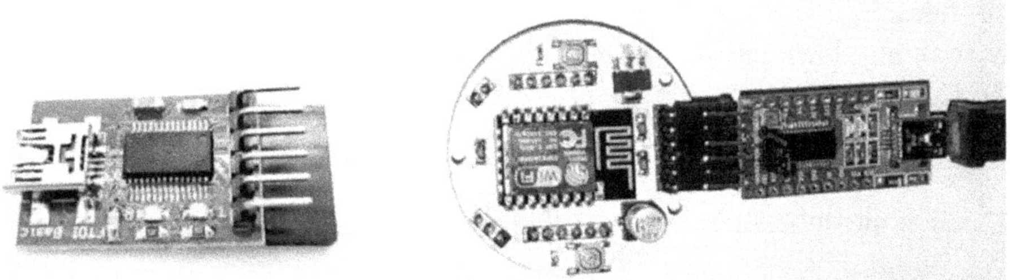

Figure 9. FTDI – Basic USB to TTL programmer

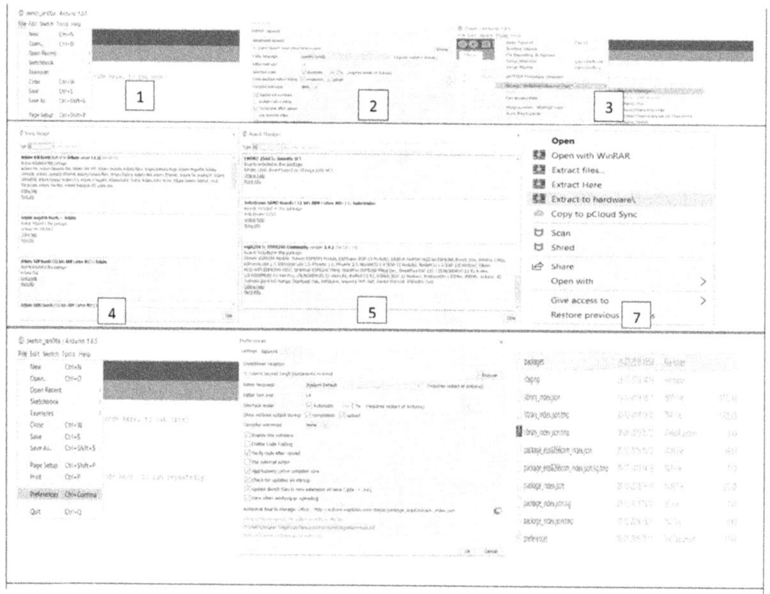

The following are the steps to install the NUTTYFI IOT board-

Essential Requirements

- Arduino version 1.6.5 or higher can be downloaded from the Arduino website - https://www.arduino.cc/en/Main/Software
    - Open Arduino, go to File > Preferences.
    - Paste the link given below into Additional Board Manager URL$_s$ as shown below -:http://arduino.esp8266.com/stable/package_esp8266com_index.jason

and press OK.

- Go to Tools > Boards > Board Manager., note that your computer/ laptop should be connected to the internet before executing this step .
- Click on Board Manager, a window as shown in the image below will appear, download all the packages index .
- Now by scrolling down cursor you will find NUTTYFI board package named as esps8266 by ESP8266 Community 2.4.2, click to install it, it will take 10 – 15 minutes depending upon the speed of your internet .

Note that version might be different depending upon latest updates.

- After that, click on the link https://www.nuttyengineer.com/download/hardware.rar

and download the file after downloading it you will receive the file "hardware.rar" , you require winrar ow extract this file.

- See in the Figure 9.
- After extracting to hardware, a hardware folder will appear, copy this folder .(paste this folder as shown in the step 13) .
- Now close Arduino software, reopen it, go to file > preferences.
- After clicking on preferences, a preference window will appear as shown in the image below, click on the highlighted link .
- A new window will appear, click on the package folder.
- Inside package folder esp8266 folder is present.
- Click on esp8266, two new folder will appear, delete the hardware folder as shown in the image & copy it as shown in the step 8.
- Restart Arduino software go to Tools > Boards > NUTTYFI IoT (ESP-12x Series).
- Board is successfully loaded in Arduino ide .

Fig. 2.0 Steps to install NUTTIFY IOT Board

Filter

Filter is a circuit which allow certain frequency to pass thorough it while attenuating unwanted frequency which produces noise in the system .

Depending upon the components used in designing filter they are categorize in one of two types i.e.

1. Active Filter
2. Passive Filter

Active Filter

1. Active filter consists of active components such as op – amps in addition to capacitors and resistors but not inductors .
2. Active filters deal with very low frequency (approaching 0 Hz)
3. Their limitation for the higher frequency is because of amplifier bandwidth limitation .

Use in communication system , biomedical instruments etc .

Figure 10. Filter used in designing architecture of web & smart home interface (passive filter, category – high pass filter

Passive Filter

Passive filter includes only passive components – inductors, resistors, capacitors .

They deal with high frequency range from 100 Hz to 300 MHz.

Their limitation to lower frequency is because of large capacitance and inductance at low frequency range.

Deal with radio frequency.

Depending Upon the Output Signals

The filters are classified into four types –

1. Low Pass Filter
2. High Pass Filter
3. Band Pass filter
4. Notch Filter

However, the term Low and high does not indicate the absolute value of frequency but relative value with respect to the cut off frequency.

Figure 11. a)Schematic of Low pas filter on Multisim b) Response Curve of Low pass filter

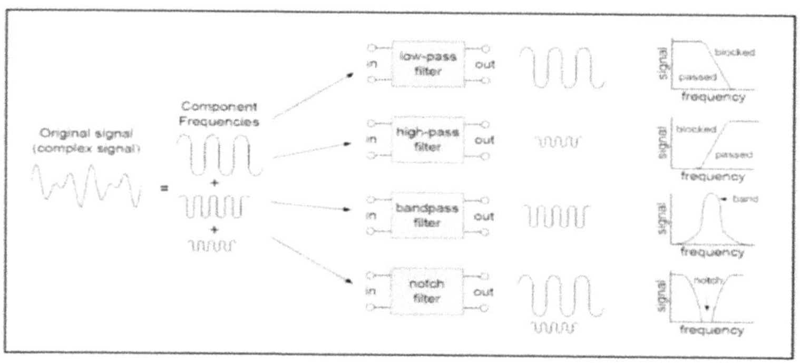

Figure 12. Four major types of filter

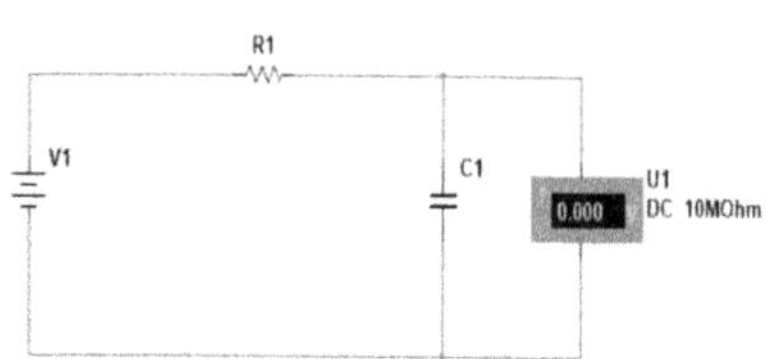

Fig A.8 a) Circuit Diagram of low pass filter

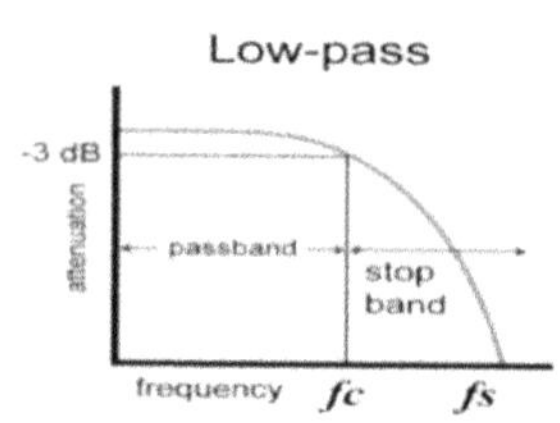

Fig A.8 b) Response curve of low pass filter

Low Pass Filter

A low pass filter passes low frequency and stops high frequency.

$$\mathbf{V_0} = v_S \frac{1}{1 + j\omega c_1 R_1} \Rightarrow H(\omega) = \frac{v_0}{v_S} = \frac{1}{1 + j\omega C_1 R_1}$$

where,

$\frac{1}{\omega c}$ V0

ω=0 → Vs

ω=∞ 0 → 0

V_0 = output voltage

V_s = source voltage

$H(\omega)$ = Gain of the system

High Pass Filter

High pass filter passes high frequency and reject low frequency.

$$v_0 = v_s \left(\frac{j\omega CR}{1 + j\omega CR} \right) \quad H(\omega) = \frac{v_0}{v_s} = \frac{j\omega CR}{1 + j\omega CR}$$

where

V_0 = output voltage

$\frac{1}{\omega c}$ V0

ω=0 ∞ → 0

ω=∞ 0 → Vs

V_s = source voltage

C = capacitance

Figure 13. a)Schematic of High pas filter on Multisim b) Response Curve of High pass filter

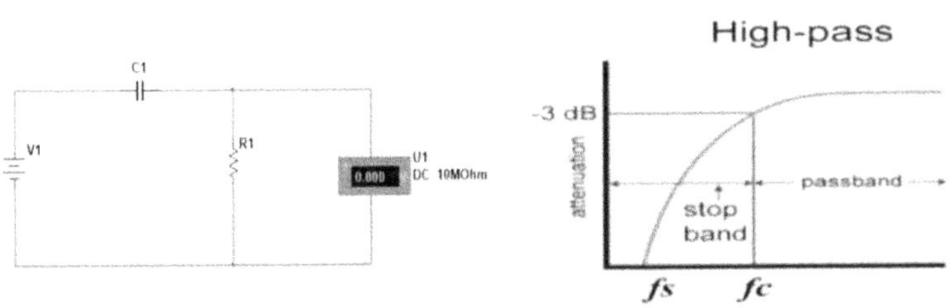

R = Resistance
$H(\omega)$ = Gain of the system

Different types of Band pass Filter are -

Bandpass Filter

This filter passes the frequency outside a frequency band and block the frequency inside the band.

$$\mathrm{H}(\text{É}) = \frac{\mathrm{V}_0}{\mathrm{V}_s} = \frac{\mathrm{j}\left(\text{É}\mathrm{L} - \frac{1}{\text{É}\mathrm{c}}\right)}{\mathrm{R} + \mathrm{j}\left[\text{É}\mathrm{L} - \frac{1}{\text{É}\mathrm{c}}\right]} \quad \frac{x_L}{R}$$

$\frac{1}{\omega C}$ ωL V0
ω=0 ∞ 0 Vs
$\omega=\omega_o$ 0 0 0
ω=∞ 0 ∞ Vs

$$\text{É}_0 = \frac{1}{\sqrt{\mathrm{LC}}} \quad \mathrm{Q}_s = \frac{\mathrm{X}_\mathrm{L}}{\mathrm{R}}$$

Here,

$H(\omega_0)=0$

$$H(\omega) = \frac{R_l}{R_l + R} \frac{1 + Q\Omega}{1 + jQ_s\Omega}$$

Figure 14. a) Circuit diagram of notch filter, b) response curve of the notch filter

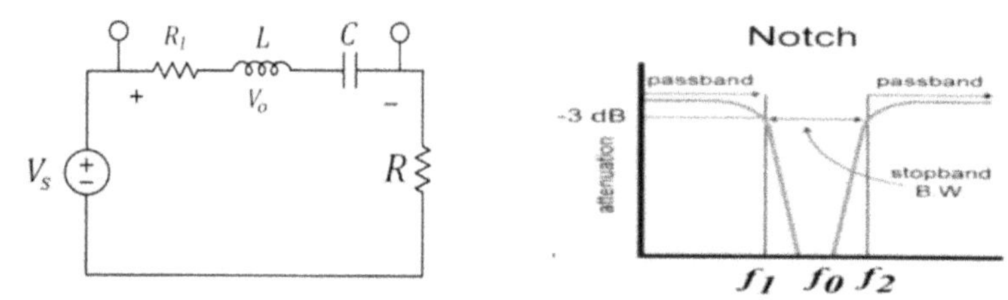

where

L = inductance
C = capacitance
R = resistance
$H(\omega)$ = Gain of the system
ω_0= Angular frequency

The filter which is used in making smart home interface using IOT based architecture of web falls under category of

Passive Filter, Type – High Pass Filter

The above circuit show that the filter converts 12 V to constant 5 V dc with no noise thus prevent the power loss in the further circuit .

Figure 15. Circuit & 3D model of DC jack

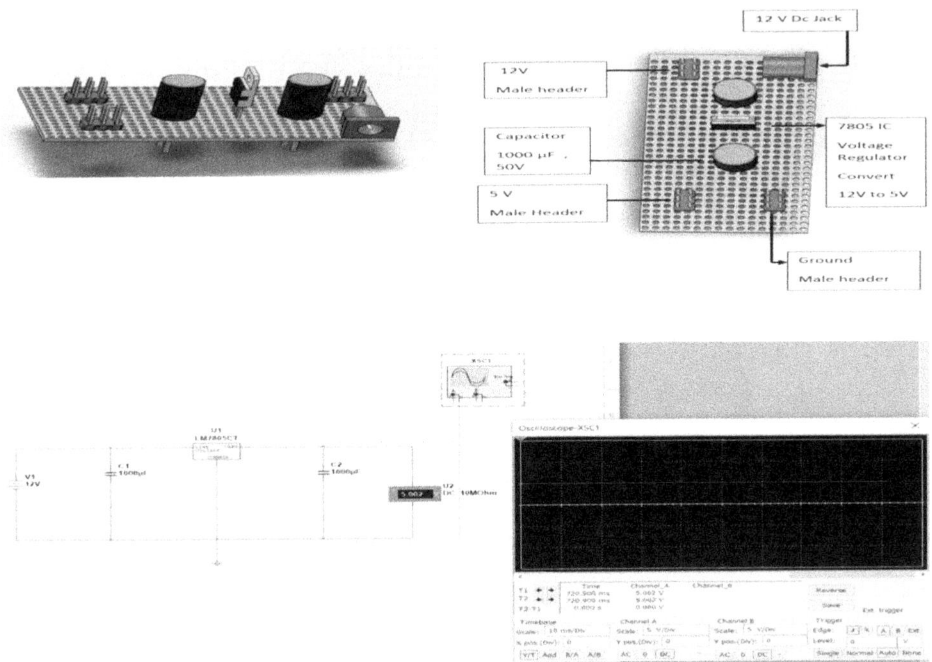

HARDWARE COMPONENT USED IN DESIGNING FILTER

Dc Jack

It is used to supply 12V to filter by the means of DC adapter(input: 100 – 240 V AC, 50/60Hz).DC adapter converts AC to DC .

7805 IC (Voltage Regulator)

It is one of the linear voltage regulator & convert 12 V dc to regulated 5 V DC as an output .

CAPACITOR

The function of capacitor in the in filter as shown in the Figure 19. a) .is to smooth the output by removing sudden spikes which are responsible for the noise in the system .

Temperature Sensor (LM35)

LM35 is precision centigrade Temperature sensor which directly calibrate the temperature in Centigrade with linear scale factor of +10-mV/°C Scale factor and its accuracy is 0.5°C (at 25°C).

- Its range varies from -55°C to 150°C
- Its operating voltage is 4V to 30V.

Figure 16. Circuit diagram of filter with constant 5 V output voltage which is display on an oscilloscope

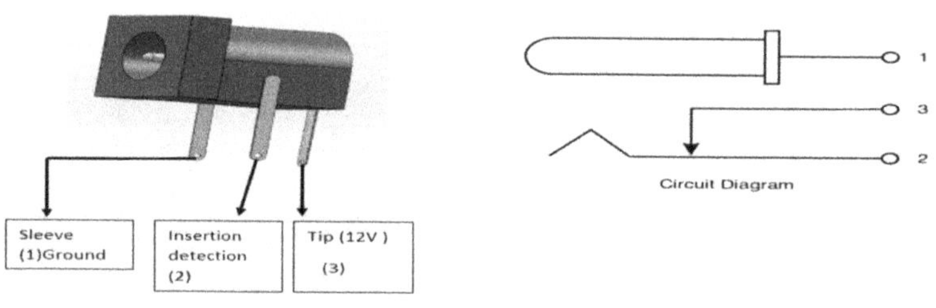

Figure 17.

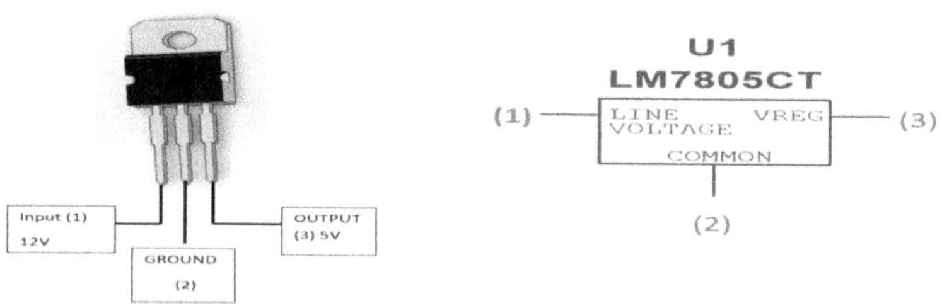

Figure 18. 7805 IC with simulator circuit jack

Figure 19. LM35 sensor (NUTTYFI product) with pin configuration and stimulator(proteus) view

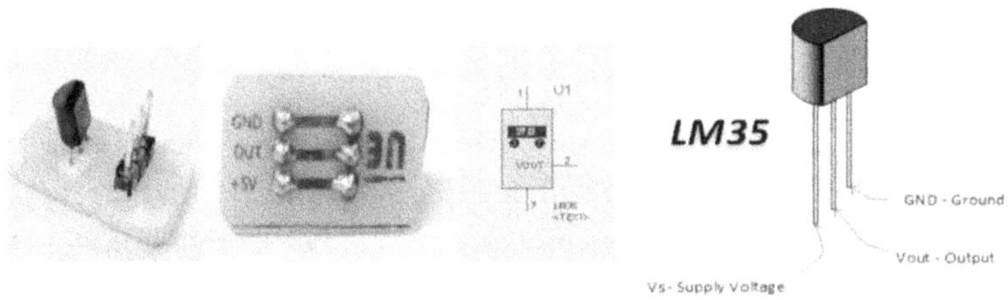

Pin Configuration (Fig A.16)
OUT – This pin will attach to the A0(Analog pin) of IOT NUTTYFI board .
GND – This pin will be attached to the GND of filter as shown in the Figure 6.
+5V - This pin will be attached to the +5V pin of the filter as shown in the Figure 6.
Datasheet of LM35 - http://www.ti.com/lit/ds/symlink/lm35.pd

```
# Programme to read Temperature sensor (LM35) and displaying its value in
Blynk widget gauge .
#define BLYNK_PRINT Serial
#include <ESP8266WiFi.h>
#include<BlynkSimpleEsp8266.h>
BlynkTimer timer ;                                                  // initiate
timer object from  blynk library
char auth[]="481a6f17df88429d92352fac7d663c95";                 // authentication
token
char ssid [] = "Singh 499";
// Hotspot/WiFi username
char pass[] = "singh124";                                  //// Hotspot/WiFi password
void setup() {
Serial.begin(9600);                                             // send data seri-
ally at the baud rate of 9600  Blynk.begin(auth,ssid,pass);
timer.setInterval(1000L, send_sensor);                }
void send_sensor()
{ int TEMP_SENSOR = ((analogRead(A0))/10);                         Blynk.
virtualWrite(V7,TEMP_SENSOR);              //   command to send sensor data from
hardware to Blynk app
```

```
}
void loop() {
 Blynk.run();
 timer.run();}
```

 # BJT Transistor as Switch (Used in Designing Relay Circuit which Will be Later Discuss in this Chapter)

The transistor (BC 109 BP) is used as switch to turn ON and OFF relay, by providing the control signal to the base of transistor by the NUTTYFI IOT (ESP 12x) series board.

Relays

Relays are the electromagnetic switch which can easily be control by small current given by microcontroller in the form of signal or by other means & can turn on and off much larger current . Thus, in this project all the electrical appliances are connected to relays which are further control by the IOT NUTTYFI (ESP – 12x series) board thus enable the smart interface with the devices around us .

Working of Relay

When the current flows through the primary coil which is winded around electromagnet, it activates electromagnet and generate magnetic field which attract the movable armature towards it as shown in the figure FigA.20 a) and thus activate the secondary circuit . When no current flows through the coil or electromagnet, the magnetic field is removed, and spring pulls back the movable armature thus make secondary circuit off .

Figure 20. Widget Gauge reading of temperature in Blynk app

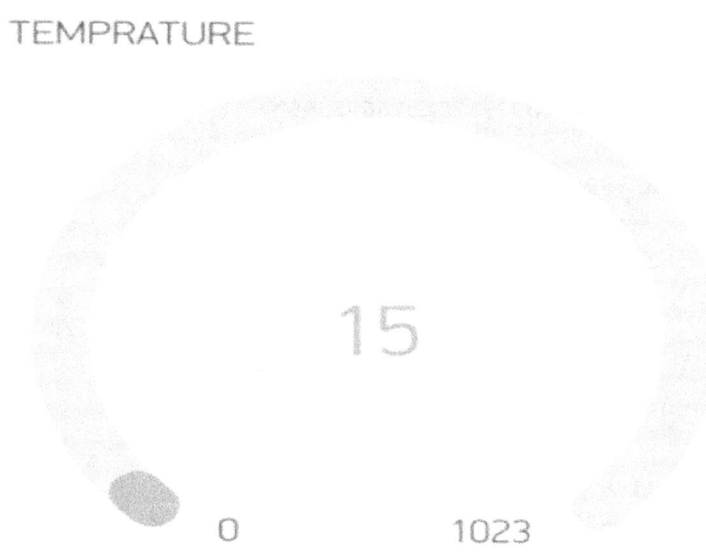

Figure 21. Schematic of BC109BP

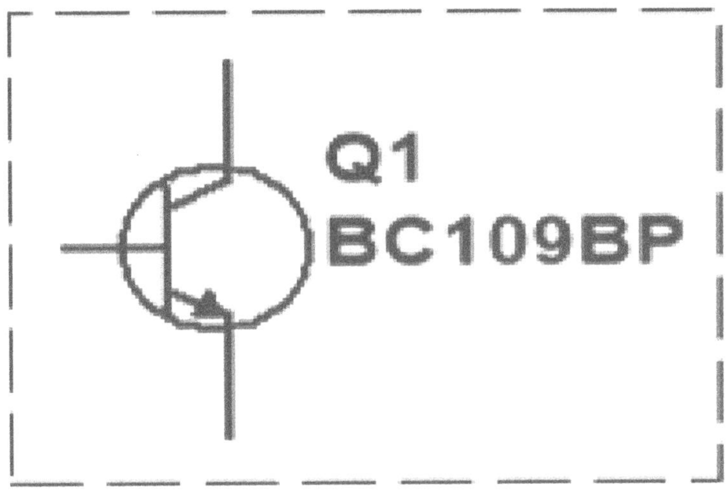

Thus, based upon above working principle the relay is classify into two types

1. **Normally Open (NO)Relay:** The relay in which load is connected to the secondary circuit which initially remain in off state.
2. **Normally Closed (NC) Relay:** The relay in which load is connected to the primary circuit which initially remain in ON state

 # 5V, 10 A Relay Module

The relay which we have used in this project falls under the category of **General-purpose relay** and controls the switching action of all the appliances of home.

It receives control signal form NUTTYFI IOT (ESP-12x series) board thus allow smart home interface through your smart phones through BLYNK app platform from any part of the world.

Above relay consist of NPN relay circuit and its switching action is shown in the figure below.

Working of NPN Relay as a Switch

1. Coil 1 of the relay is connected to the 12V supply V2 and coil 2 is connected to the collector of BC-109 transistor as shown in Fig A.21 a) & b) .
2. When the input signal Vin is given by the controller (NUTTYFI board) at the base of transistor (BC-109), the transistor act as close switch (saturation region) due to which the coil gets magnetize and secondary circuit to which the load is connected get ON and load directly get connected to the source.
3. When no signal Vin is applied by the controller at base of the transistor, it remains in cut off state (open switch) thus the coil gets demagnetize and the secondary circuit turn to OFF state (Load get disconnected from the source) .

Figure 22.

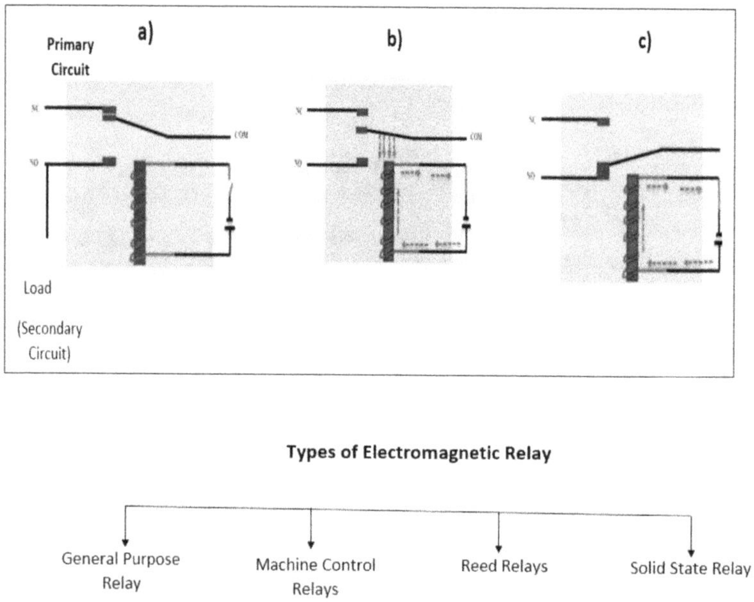

Figure 23. a) JQC_3F(T73) 10A 220VAC 2 channel relay module with PCB layout and pin configuration

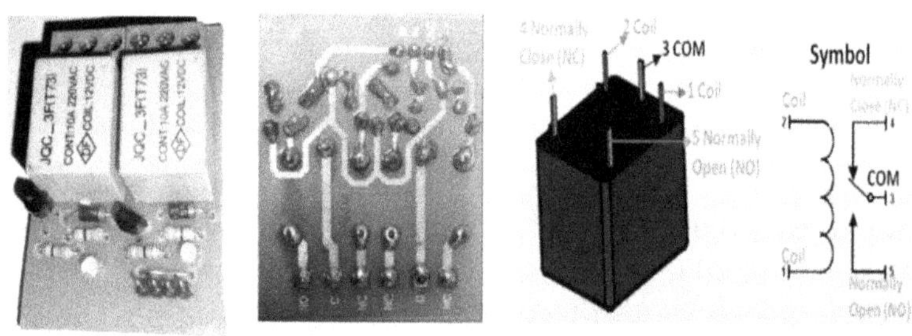

Figure 24. b) NPN relay circuit diagram

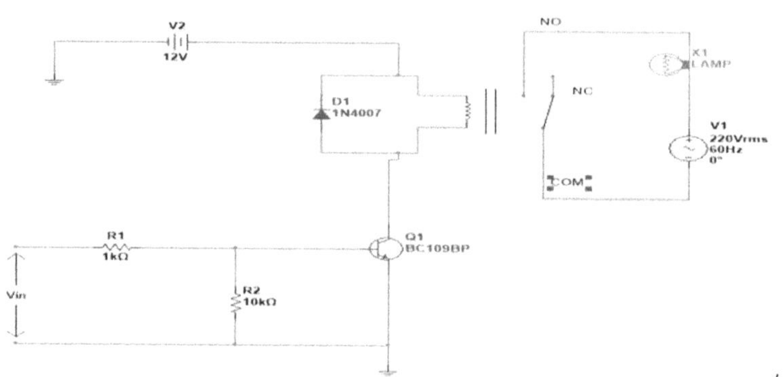

This how relay can be use as switch by applying small current at the base of the transistor and can connect and disconnect load from source .

Simple programme illustrating the working of relay using BLYNK App.

```
#define BLYNK_PRINT Serial
#include <ESP8266WiFi.h>                    // Header files
#include<BlynkSimpleEsp8266.h>
char auth[]="481a6f17df88429d92352fac7d663c95";     // authentication token of
Blynk project
char ssid [ ] = "Singh 499";                                          // WiFi
username
char pass[ ] = "singh124";
int LIGHT = D0;
int FAN   = D1;
WidgetLCD LCD_BLYNK(V0);        //  LCD widget connected to the V0(virtual pin)
of BLYNK app
BLYNK_WRITE(V1)                  // Writting data from virtual pin V1 of Blynk
app to D0 pin of NUTTYFI IOT
{int Lights = param.asInt();  //  varable reading state of (HIGH/LOW)virtul
pin V1
 if(Lights == HIGH)
 {LCD_BLYNK.clear();
 digitalWrite(LIGHT,HIGH);
 LCD_BLYNK.print(0,0,"LIGHTS ON");      // print commnd of LCD
}else
{ LCD_BLYNK.clear();
digitalWrite(LIGHT,LOW);
LCD_BLYNK.print(0,0,"LIGHTS OFF");
}}
BLYNK_WRITE (V2)             {
int Fans = param.asInt();         // reading status virtual pin V2 form blynk
app.
if(Fans == HIGH)
  {LCD_BLYNK.clear();
digitalWrite(FAN , HIGH);
LCD_BLYNK.print(0,0,"FANS ON");
  }
  else{
LCD_BLYNK.clear();
digitalWrite(FAN,LOW);
LCD_BLYNK.print(0,0,"FANS OFF");
  } }
void setup()
 {Serial.begin(9600);                         // send data serially at
```

```
the baud rate of 9600.
Blynk.begin(auth,ssid,pass);
pinMode(LIGHT,OUTPUT);
pinMode(FAN , OUTPUT);}
void loop(){
 Blynk.run();
}
```

How BLYNK Allow the smart Home Interface

Whenever we press any virtual key in BLYNK the message travel through BLYNK Cloud to the Hardware and vice-versa . Hardware components through IOT Nuttyfi board get connected to internet via BLYNK app .

IDENTITY OF SPECIFIC HARDWARE COMPONENT

Auth Token

Auth Token is the series of characters, digits generated by BLYNK for every new project and act as a unique identifier which connect your hardware to BLYNK app .

Figure 25. Connection of relay module with filter and NUTTYFI board and controlling appliance through BLYNK app

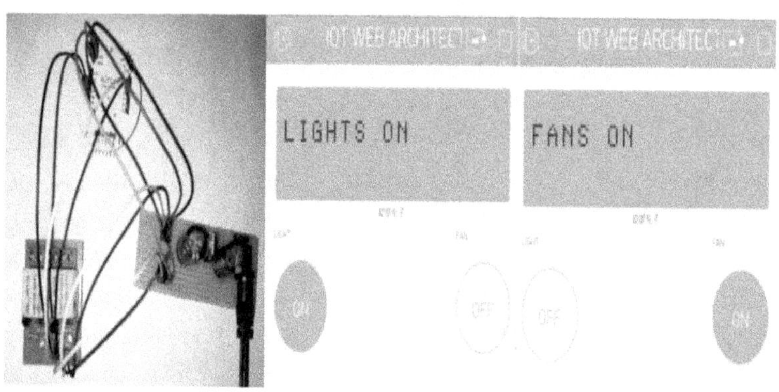

Figure 26. Transfer ofinformation from Blynk app to Hardware Components

The Auth Token generated by BLYNK for the (IOT WEB ARCHITECTURE) project is send through email and then copy to the programme, as shown in the image below .

char auth[] = "Your Auth Token "

BLYNK Cloud / Internet

The second step is to link your hardware components with the internet by copying ssid(username) and pass (password) into the programme as shown in the Figure 27 and 28.

- Upload the programme by clicking on small magnifying on the top right corner of the image by means of FTDI programmer as shown in the Figure 8.

Figure 28. Upload the programme to hardware by clicking on the magnifying glass on the top right corner

Now, all the home appliances connect to the NUTTYFI IOT (12 x series) board can interface with BLYNK app via BLYNK cloud only when internet connection is available .

BLYNK main operation and role in IOT Web Architecture

The RX (receiving) & TX (transferring) data from BLYNK to Hardware and Hardware to BLYNK is execute by following commands .

Send Data from app to Hardware

Data can be sent from widgets in Blynk app to the hardware by means of virtual pin to which devices are connected .

For e.g. in this project we have connected LIGHTS to D0 pin of Nuttyfi board which is control by blynk through V1 (virtual) pin .

Figure 27. (a) Interfacing of Blynk app with Home devices

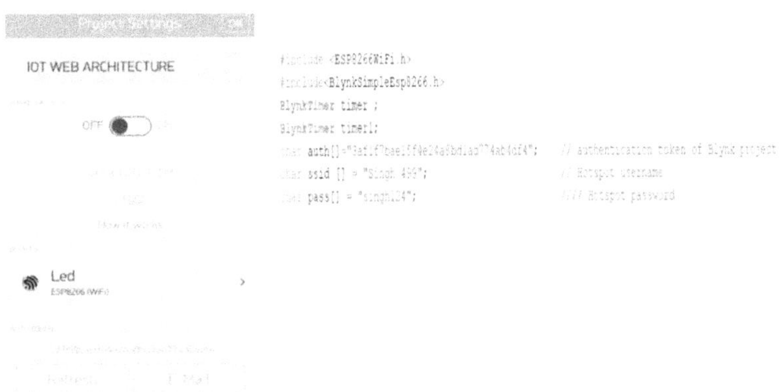

Figure 28. Auth Token generated by BLYNK app which is send through mail and copied in the programme

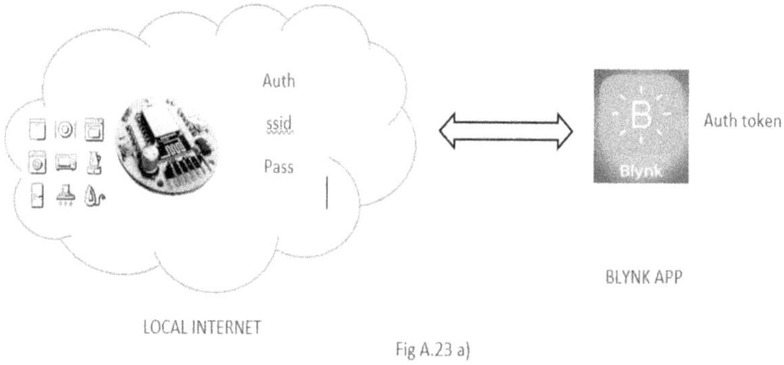

Figure 29.

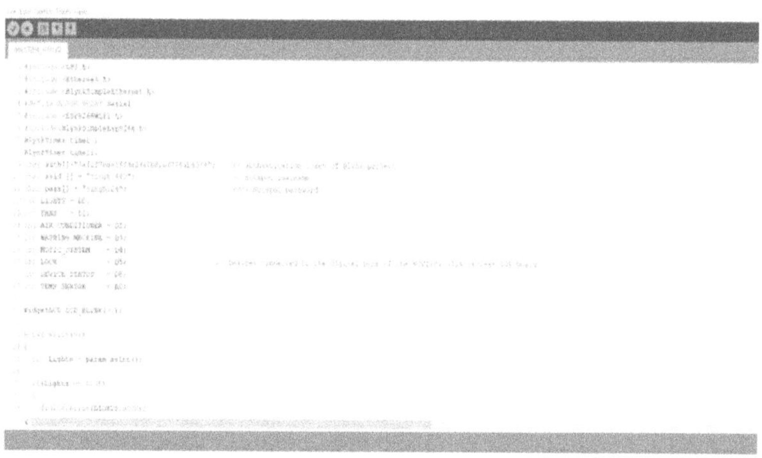

```
BLINK_WRITE (V1)          // Writing data from virtual pin V1 to pin D0(LIGHTS)
of NUTTYFI IOT BOARD
{
int Light = param.asInt();  // Reading status of virtual pin V1.

if (Light == HIGH)
{
digitalWrite(LIGHTS , HIGH);
}
else
{
digitalWrite(LIGHTS,LOW);
}}
```

Reading Data from Hardware to BLYNK App

Data from the hardware can by be read by BLYNK app through virtual pins . For e.g. in the smart home interface we are reading data from the temperature sensor connected to the analog pin A0 of NUTTYFI IOT board and sending to BLYNK app at virtual pin V7 (widget gauge) as shown in the fig 3.0.

Blynk.virtualWrite(V7,TEMP_SENSOR); // send data from hardware to app .

Thus, with the above mention two operation the IOT based Web Architecture allow the smart home interface from any corner of the world .

Notification About Device Turn on Period

BLYNK app provides special feature i.e. notification widget which allow user to receive notification status of device periodically. In designing smart home interface authors make use of this widget in notifying device turn on period with periodic time of 1 minute and when the email is sent .

Programme to receive notification after periodic time of 1 minutes .

```
#define BLYNK_PRINT Serial
#include <ESP8266WiFi.h>
#include<BlynkSimpleEsp8266.h>
char auth[]="481a6f17df88429d92352fac7d663c95";     // authentication token of
Blynk project
char ssid [ ] = "Singh 499";                                  // Hotspot/ WiFi
username
char pass[ ] = "singh124";
BlynkTimer timer ;
void notifyUptime()                        // function to notify the Device run-
ning satus per minute
                {
long uptime = millis() /60000L ;          Blynk.email("jasjitsingh4999@gmail.
com","Subject:Notify" , String("Device Running for ") + uptime + "minutes.");
// Blynk command to send email Blynk.notify(String("Running for" ) + uptime +
" minutes. ");                 // only notification till 15 seconds is allowed
// Blynk command to notify
   }
void setup()
              {Serial.begin(9600);
               Blynk.begin(auth,ssid,pass);
          timer.setInterval(60000L,notifyUptime);          // call notifyUptime
function with periodic time of 1 minute
}
void loop() {
```

Figure 30. Blynk notification status when device goes offline

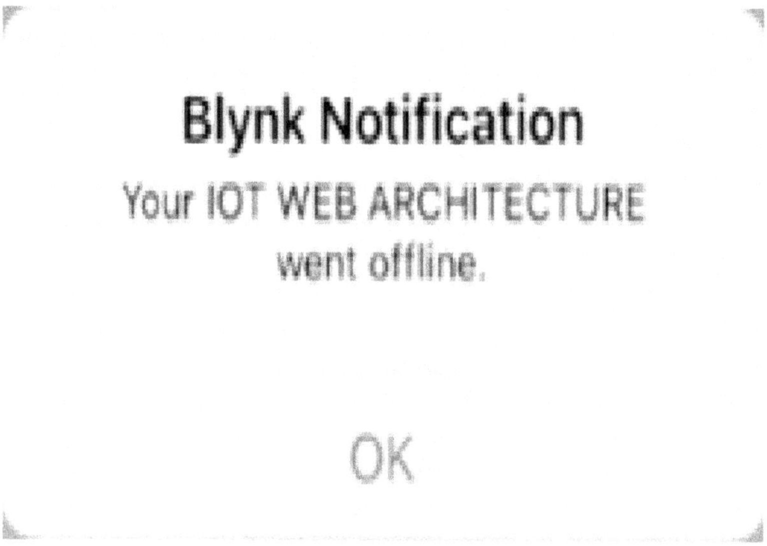


```
 Blynk.run();
 timer.run();}
```

Status of Devices on the Email

BLYNK app contains email widget which allow authors to design smart home interface by sending the status(ON / OFF) to the email of user (email entered in the BLYNK email widget) thus providing feasibility to monitor devices outside the BLYNK app .

```
Programme to receive device status in the email when the button is pressed
#define BLYNK_PRINT Serial
#include <ESP8266WiFi.h>
#include<BlynkSimpleEsp8266.h>
char auth[ ]="481a6f17df88429d92352fac7d663c95";     // auth token of Blynk 
project
char ssid [ ] = "Singh 499";                         // WiFi username & pass-
word
char pass[ ] = "singh124";
WidgetLCD LCD_BLYNK(V0);   // Blynk Lcd Connected to the V0 pin of Blynk app
BLYNK_WRITE(V8)            // Writting data from  virtual pin V8 of Blynk app 
to D6 pin of NUTTYFI IOT
{
 int statusE = param.asInt();
  if (statusE == HIGH)
{ LCD_BLYNK.clear();
  int TEMP_SENSOR = ((analogRead(A0))/10);              // Reading temrature from
```

```
the analog pin of NUTTYFI IOT
 LCD_BLYNK.print(0,0,"Status send");
 Serial.print(TEMP_SENSOR);
Blynk.email("jasjitsingh4999@gmail.com", "Subject: Device Status " , String("
Tempratrure ")+ TEMP_SENSOR  + '\n' + String(".") ) ;        // Blynk command
to send email
 Blynk.notify(String("Status Receive")) ;                      // Blynk command
to notify
   }else
{ LCD_BLYNK.clear();
}}
void setup() {
  Serial.begin(9600);
  Blynk.begin(auth,ssid,pass);  }
void loop()
{Blynk.run();    //  To run blynk commands
 }
```

Conceptual Overview of Programme

The IOT based architecture of web and smart home interface is used to control the devices of daily use such as Air conditioner, fans, lights, Dish washer, Washing machine, Music system, door locks more efficiently from distant place through our smart phones. User will too receive a status of all the devices through the medium of email .

Figure 31. Temp notification on the email

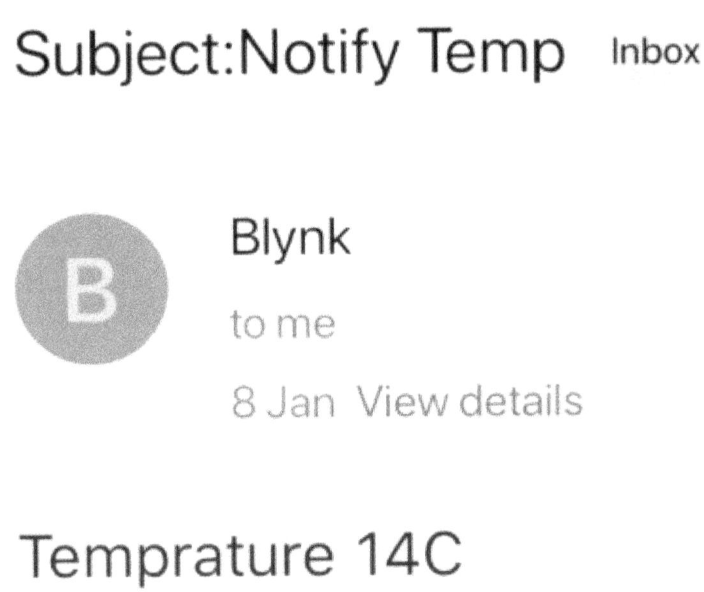

Figure 32. Devices connected to different pins of NUTTIFY IOT board

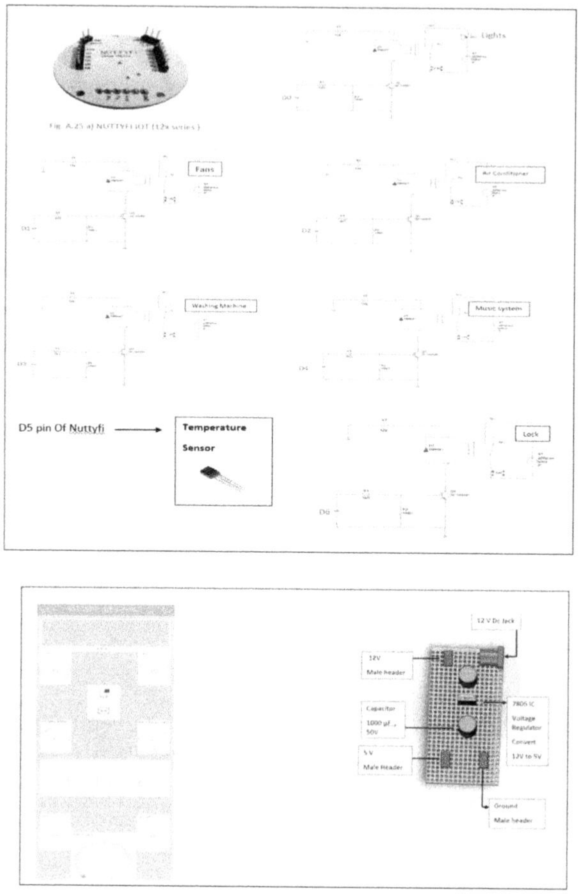

Circuit Diagram

The connection of external devices with IOT NUTTYFI (12 x) series board is given by -:

1. +5V pin of power supply from the filter as shown in Figure 25 c) to the V_{input} pin of NUTTYFI.
2. GND pin of filter is connected to the GND pin of the NUTTYFI.
3. LCD is connected to the virtual pin V0 of the BLYNK .
4. The D0 pin of NUTTYFI is connected to the Lights relay circuit .
5. The D1 pin of NUTTYFI is connected to the Fans relay circuit .
6. The D2 pin of NUTTYFI is connected to the Air Conditioner relay circuit .
7. The D3 pin of NUTTYFI is connected to the Washing Machine relay circuit .
8. The D4 pin of NUTTYFI is connected to the Music System relay circuit.
9. The D5 pin of NUTTYFI is connected to the Temperature Sensor .
10. The D6 pin of NUTTYFI is connected to the Lock relay circuit .

On merging the concepts of above mention sub programmes authors design have designed smart home interface and final programme is given by -:

```
# Programme
#define BLYNK_PRINT Serial                                    //Blynk serial
print header file
#include <ESP8266WiFi.h>                                               //ESP8266
header file
#include<BlynkSimpleEsp8266.h>                                   //Blynk header
file
BlynkTimer timer ;                                                  // Initial-
izing timer for the notification
BlynkTimer timer1;                                                // Initializ-
ing timer 1 for sending email

char auth[]="481a6f17df88429d92352fac7d663c95";     // authentication token of
Blynk project
char ssid [] = "Singh 499";                        // Hotspot username
char pass[] = "singh124";                          //// Hotspot password
int LIGHTS = D0;                   // Lights relay circuit connected to D0 pin
of NUTTYFI IOT board control by   virtual pin V1
int FANS    = D1;                    // FANS relay circuit connected to D1 pin
of NUTTYFI IOT board control by virtual pin V2
int AIR_CONDITIONER = D2;                         // AC relay circuit connected to
D2 pin of NUTTYFI IOT board control by virtual pin V3
int WASHING_MACHINE = D3;                   // WM  relay circuit connected to D3
pin of NUTTYFI IOT board control by virtual pin V4
int MUSIC_SYSTEM    = D4;


int LOCK              = D5;
int DEVICE_STATUS     = D6;                                  // Device status
connected to the D6 pin of IOT NUTTYFI board control by virtual pin V8
int TEMP_SENSOR       = A0;

int Lights1 ;
int Fans1  ;
 int Air_Conditioner1;
 int Washing_Machine1
 int Lock1 ;                                                       //variable
for lock
 int TEMP_SENSOR1

WidgetLCD LCD_BLYNK(V0);
BLYNK_WRITE(V1)
{
  int Lights = param.asInt();                              //  variable reading
```

```
state (HIGH/LOW) of virtual pin V1

  if(Lights == HIGH)
  {
LCD_BLYNK.clear();
digitalWrite(LIGHTS,HIGH);
LCD_BLYNK.print(0,0,"LIGHTS ON");  // print command of LCD
  }
  else
{
LCD_BLYNK.clear();
digitalWrite(LIGHTS,LOW);
LCD_BLYNK.print(0,0,"LIGHTS OFF");              //print command of LCD
}                                                        //end of else condi-
tion
}                                   //end of V1()
 BLYNK_WRITE (V2)
 {
  int Fans = param.asInt();
  if(Fans == HIGH)
  {
  LCD_BLYNK.clear();
    digitalWrite(FANS , HIGH);                          // Turning on Fan
 LCD_BLYNK.print(0,0,"FANS ON");
  }
 else  {
LCD_BLYNK.clear();
digitalWrite(FANS,LOW);
 LCD_BLYNK.print(0,0,"FANS OFF");
  }}
 BLYNK_WRITE(V3)             // Writing data from virtual pin V3 of Blynk app
to D2 pin of NUTTYFI IOT
 {
  int Air_Conditioner = param.asInt();

if(Air_Conditioner == HIGH)
  {
LCD_BLYNK.clear();                                         // Clearing lcd
screen to prevent overwrite condition for next statement on the LCD
  digitalWrite(AIR_CONDITIONER , HIGH );    // Turning on air conditioner at-
tach to relay
  LCD_BLYNK.print(0,0,"AC ON ");

  }
```

```
else
 {
LCD_BLYNK.clear();
digitalWrite(AIR_CONDITIONER , LOW);
LCD_BLYNK.print(0,0, "AC OFF");
  }
 }
 BLYNK_WRITE(V4)
{
  int Washing_Machine= param.asInt();
  if( Washing_Machine == HIGH)
  {
   LCD_BLYNK.clear();
   digitalWrite(WASHING_MACHINE , HIGH);     // turning on washing machine
   LCD_BLYNK.print(0,0,"WM ON" );
  }                                                              // end of if
statement
   else
   {
    LCD_BLYNK.clear();
    digitalWrite(WASHING_MACHINE , LOW);
    LCD_BLYNK.print(0,0,"WM OFF ");
 }
 }
 BLYNK_WRITE(V5)
 {
  String Music_System = param.asStr();

  if (Music_System == "play")
  {
  LCD_BLYNK.clear();
  LCD_BLYNK.print(0,0, "PLAY");
  }

  if (Music_System == "stop")
  {
  LCD_BLYNK.clear();
  LCD_BLYNK.print(0,0, "STOP" );
  }
  if (Music_System == "next")
  {                                           //  starting of if loop
LCD_BLYNK.clear();
LCD_BLYNK.print(0,0, "NEXT" );
  }
```

```
  if (Music_System == "prev")          //  if user press previous button on the
music player this condition will be satisfied
  {
   LCD_BLYNK.clear();
   LCD_BLYNK.print(0,0, "PREV" );
  }
  Blynk.setProperty(V5, "label", Music_System

   Serial.print(Music_System);

   Serial.println();
 }
   BLYNK_WRITE(V6)              // Writing data from  virtual pin V6 of Blynk app
to D5 pin of NUTTYFI IOT
{
   int Lock = param.asInt();       // variable read0ing state (HIGH/LOW) of
virtual pin V6
if (Lock == HIGH)                // If user press lock widget switch in the
blynk app ( this condition will be satisfied )
{
 LCD_BLYNK.clear();

  digitalWrite(Lock , HIGH);
  LCD_BLYNK.print(0,0,"LOCK ACTIVATED " );
}
else
{
 LCD_BLYNK.clear();
 digitalWrite(LOCK , LOW) ;  // Deactivating lock
 LCD_BLYNK.print(0,0   , "LOCK DACTIVATED");

}         // end of the else statement
}       // End of Blynk Write V6 loop
BLYNK_WRITE(V9)

{
 int ALL_OFF = param.asInt();


 if (ALL_OFF == HIGH)
{
  LCD_BLYNK.clear();                                          // initial ally
clearing LCD Screen
  digitalWrite(LIGHTS,LOW);
```

```
  digitalWrite(LOCK , LOW);
  digitalWrite(FANS,LOW);
  digitalWrite(AIR_CONDITIONER , LOW);  // Switching of AC
  digitalWrite(WASHING_MACHINE , LOW);   // Switching of Washing Machine
  LCD_BLYNK.print(0,0,"All_OFF " );         // Displaying on the LCD
  }                                                             //
end of I f condition
else
  {
 LCD_BLYNK.clear();                         //    If switch not press the LCD
will be clear
  }
  }
 void send_sensor()         // function to read sensor data on the widget gauge
of Blynk
 {
  int TEMP_SENSOR = ((analogRead(A0))/10);

Blynk.virtualWrite(V7,TEMP_SENSOR);

}
void notifyUptime()

{
long uptime = millis() /60000L ;
  Blynk.email("jasjitsingh4999@gmail.com","Subject:Notify" , String("Device
Running for ") + uptime + "minutes.");

 Blynk.notify(String("Running for" ) + uptime + " minutes. ");     // Notify
the running time of the device              // Blynk command to notify

}

BLYNK_WRITE(V8)

{
 int statusE = param.asInt();
   if (statusE == HIGH)
{                                          // beginning of if statement
LCD_BLYNK.clear();
       int Lights1=digitalRead(D0);
  int Fans1 = digitalRead(D1);                                     // Reading
state(HIGH/LOW) of the Digital pin D1  to which Fans  are attach
  int Air_Conditioner1=digitalRead(D2);                  // Reading state(HIGH/
```

```
LOW) of the Digital pin D2  to which AC  are attach
  int Washing_Machine1= digitalRead(D3);           // Reading state(HIGH/LOW)
of the Digital pin D3  to which Washing Machine   are attach
      int Lock1=digitalRead(D5);
      int TEMP_SENSOR = ((analogRead(A0))/10);          .

      LCD_BLYNK.print(0,0,"Status send");

  Blynk.email("jasjitsingh4999@gmail.com", "Subject: Device Status " , String("
Lights ") +  Lights1 + '\n' + String("  Fans ") + Fans1 + '\n' + String("  AC
") + Air_Conditioner1 + '\n' + String("  Washing Machine ") + Washing_Machine1
+ '\n' + String("  Lock " ) + Lock1 +  '\n' + String("  Tempratrure ")+ TEMP_
SENSOR  + '\n' + String(".") ) ;                          //      Sending state
(HIGH/LOW) to the email , here 0 → device is off
Here 1 → device is ON

   Blynk.notify(String("Status Receive")) ;                        // this command
will activate notify when the V8  is pressed
}
else
{                            // start of else loop
  LCD_BLYNK.clear();
}                            // end of else loop
}
void setup() {
  Serial.begin(9600);
  Blynk.begin(auth,ssid,pass);
  pinMode(LIGHTS,OUTPUT);
  pinMode(FANS , OUTPUT);
  pinMode(AIR_CONDITIONER, OUTPUT);
  pinMode(WASHING_MACHINE, OUTPUT);
  pinMode(MUSIC_SYSTEM , OUTPUT);
  pinMode(LOCK , OUTPUT);
  pinMode(D7, INPUT) ;
  pinMode(D6, INPUT_PULLUP) ;                     // Setting D7 (digital)pin of
NUTTYFI as INPUT_PULLUP i.e. initially LOW
  timer1.setInterval(60000L,notifyUptime);   // setting timer of 1 minute
  timer.setInterval(1000L, send_sensor);       // setting timer of 1 second

  /*int Lights1 = digitalRead(D0);
   int Fans1 = digitalRead(D1);
   int Air_Conditioner1= digitalRead(D2);
   int Washing_Machine1= digitalRead(D3);
   int Lock1 = digitalRead(D5);
```

```
    int TEMP_SENSOR1 = ((analogRead(A0))/10);*/
    } */
void loop()
{
 Blynk.run();      // initiate working of all blynk commands
 timer.run();      // initiate working of blynk timer
 timer1.run();   // initiate the working of blynk timer 2
 }
```

The Figure 33 show the interfacing of Blynk app with Devices .

- When the status button is pressed, current status of all the devices will be send to the email address as shown in the above Fig a.24 A) . Here 0 → Device is off

1→ Device is ON

- The notification will be sent after every 1 minutes (displaying the time on period turn on period of the device .
- The Gauge display the current temperature of the surrounding .
- All the devices can be control using virtual keys on the BLYNK app .

IOT too provide feasibility in monitoring and analysing various sensors, home equipment on different fields of Things Speak Server . Thus, collected data in the different field of Things Speak Server are store in its database and user can visualize it in the form of graphs.

Features of Things Speak Server

- All the data are collected in the private channels of the Things Speak Server.
- It too allows user to share their data with public channel so that everyone can visualize and get periodic updates.
- User can analyse their data on the MATLAB in different forms of graphs .
- It too sets alert to avoid uncertain circumstances.
- It allows users to schedule events on different dates, time etc.
- Things Speak Server allows user to interface with all their devices through the medium of android and IOS app. .

It is too compatible with –

- Arduino
- Raspberry Pi
- Mobile app
- Twitter
- Twilio
- MATLAB

- ESP8266 WiFi Module
- NUTTYFI IOT 12-x Series Board
- Particle Photon

Smart Home Monitoring with Things Speak Server

Authors have designed an alternative approach to collect, analyse sensor data and initiate trigger response according to the home environmental parameters by using Things Speak Server which is based upon IOT platform and too allow the transfer of its data to any website.

All the devices connected to NUTTYFI IOT 12-x series board are connected in same manner with IOT board (Arduino- Esp826 Wi-Fi board).

This board can be bought from the site of NUTTYFI engineer. It consists of Interfacing of Arduino Board with Node MCU Wi-Fi module.

1. Arduino – Act as the master. (sends data to salve(Wi-Fi module))
2. NodeMcu esp 8266 WiFi module - Act as slave (sends data to the server) .

Arduino Uno consist of

Digital pins – 14 pins (All the digital sensor are connected to these pins)

Analog pins –5 pin(All the analog sensor with analog outputs are connected to them)

User can create new account in https://thingspeak.com/ going through the tutorial given in the official website of Things Speak .

User can activate private/public channels with various field in which data can be monitor in the form of graph.

Circuit Layout of Sensors Connected to the Arduino Board

1. Temperature Sensor (LM35) which will monitor outside environmental temperature is connected to the A0 pin of Arduino board.
2. Temperature Sensor (LM35) which will monitor inside home environmental temperature is connected to the A1 pin of Arduino board.
3. LDR sensor(1) will monitor the intensity of light outside the home connected to the analog pin A2 of Arduino .
4. LDR sensor(2) will monitor the intensity of light inside the home, connected to the Analog pin A3 of Arduino. .

Figure 33. (Arduino+ Nodemcu) board to send data to the things speak server

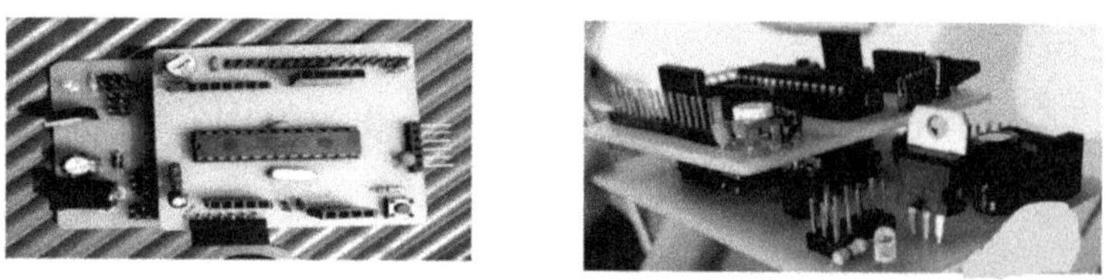

5. IR sensor (1) & IR sensor (2) is connected to the digital pin -2 of Arduino board .

 # Programme

1. Programme to send data from master(Arduino) to slave(Node MCU WiFi- Module)

```
#include <SoftwareSerial.h>                // Header file to send data serially
SoftwareSerial mySeries (6,7);     //  RX/TX through pin 6 & 7
int Temp1=A0;                              // Temperature sensor(1) connected to
A0 pin of Arduino
int Temp2=A1;                                  // Temperature sensor (2) connect-
ed to A0 pin of Arduino
int LDR1=A2, LDR2=A3;           // LDR1 & LDR2  connected to A2 pin & A3 pin of
Arduino
int IR1=2,  IR2=3;
int tstate1 =0, tstate2=0;      // tstate1 & tstate2 are the variables which
stores the input value of temperatures sensors
int Lstate1=0, Lstate2=0;
int Istate1=0, Istate2=0;
int X=0, Y=0;                               // Variable to set the predefine
value of X & Y .

void setup()
 {
  Serial.begin (9600);
  mySeries.begin(9600);

}
void loop() {
 tstatus1=(analogRead(Temp1))/2;

 tstatus2=(analogRead(Temp2))/2;
 Lstatus1= analogRead(LDR1);          // reading data of LDR sensor(1) & LDR
sensor(2)
 Lstatus2= analogRead(LDR2);
 Istatus1= digitalRead(IR1);                 // Reading  status of IR sensor(1)
 Istatus2= digitalRead(IR2);                       // Reading  status of IR
sensor(2)


 if (Istate1==HIGH)         // Checking state of IR sensor1.
 {
  X=10;                  // initializing vale to variable x
 }
 else
```

```
 {
  X=20;                      // initializing value 20 to variable x
 }

 if (Istate2==HIGH)      // Checking state of IR sensor 2
 {
  Y=10;                      // initializing value 10 to variable Y If IR sen-
sor state is high
 }      // end of if loop
 else
 {
  Y=20;                                  // initializing value  to variable Y If
IR sensor state is low
 }     // end of else statement

 Serial.print('\r');       // Serially displaying data in the form of package
 Serial.print(X);        // Send Value of IR sensor 1 to serial monitor  .
 Serial.print('|');
 Serial.print(Y);        .
 Serial.print('|');
 Serial.print(Lstatus1);

 Serial.print('|');
 Serial.print(Lstatus2);

 Serial.print('|');
 Serial.print(Temp1);              // Sending value of Temperature sensor 1 to
the serial monitor.

 Serial.print('|');
Serial.print(Temp2);   // Sending value of Temperature sensor2  to the serial
monitor.
 Serial.print('\n');       // End of the package

mySeries.print('\r');             //   starting of the package
mySeries.print(Istate1);           // sending status of ldr sensor1 to slave
board
mySeries.print('|');
mySeries.print(Istate2);
mySeries.print('|');
            mySeries.print(Lstate1);
            mySeries.print('|');
            mySeries.print(Lstate2);
mySeries.print('|');
```

```
mySeries.print(tstate1);
mySeries.print('|');
mySeries.print(tstate2);

mySeries.print('\n');                        // ending of the package.

 delay(1000);                                  // delay of 1 Second
 }
```

2. Programme to send data from the slave to Things Speak Server

```
#include <SoftwareSerial.h>      // Software serial header file
#include <ESP8266WiFi.h>         // Wifi module Esp 82866 header file
SoftwareSerial mySeries(D6,D7,false,256);
String apiKey1 = "5MWJ*************" ;
const char* ssid = "Schematic***********" ;     // username of the hotspot/
internet
const char* password= "hi@***************; // Password of hotspot/ Wifi
const char* server = "api.thingspeak.com";    // server web address
int data=0;          // initializing value of data variable to zero
int IR1=0,IR2=0,Temp1=0,Temp2=0,LDR1=0,LDR2=0;
WiFiClient client ;
void setup()
  {         // start of setup loop

  Serial.begin(9600);          // sending data to serial monitor at the baud
rate of 9600
  mySeries.begin(9600);    // send the data to the slave board (Wifi) at the
baud rate of 9600

  delay(10);                               // providing delay of 10 milli
second
   WiFi.begin(ssid,password);   // setting up hotspot password and username
   Serial.println();                         // this command jumps the cursor
to the new line
   Serial.println();
   Serial.print("Connecting to");   // printing status on the serial monitor
   Serial.println(ssid);                    // printing ssid on the serial
monitor
   WiFi.begin(ssid,password);    // setting up WiFi if not connected earlier
   while(WiFi.status() !=WL_CONNECTED)
 {
   delay(500);
   Serial.print(".");
```

```
  }
   Serial.println("");
   Serial.print("WIFI CONNECTED");
}
void loop() {
serialEvent_NODEMCU();   // calling of function inside the loop
display_function();         // calling function inside the loop
 data++;
 if (client.connect(server,80))            // api server port where database
is created
  {
 serialEvent_NODEMCU();              // calling function node mcu as declared
below

  if  (data>=10)                      // if value of counter increases above
10 then if loop will be executed
  {
  send1_TX_GROWTH_PARA();          // calling function
  data=0;                                 // resetting value of data
variable
  }
  }
  client.stop();
  Serial.println("Waiting");
 delay(1000);
}
void send1_TX_GROWTH_PARA()
{
          String postStr = apiKey1;              // assigning api key to post-
Str
          postStr +="&field1=";
          postStr += String(IR1);
          postStr +="&field2=";        // monitoring data of IR2 sensor in
field 2 of Things speak server
          postStr += String(IR2);
          postStr +="&field3=";                // monitoring data of LDR1
sensor in field 3 of   .
Things speak server
          postStr += String(LDR1);
          postStr += "&field4=" ;             // monitoring data of LDR2 sen-
sor in field 4 of .            .
Things speak server
          postStr += String(LDR2);
          postStr += "&field5=";
```

```
        postStr += String(Temp1);
        postStr += "&field6=";
        postStr += String(Temp2);
        postStr +="\r\n\r\n";       // ending of if statement
        client.print("POST /update HTTP/1.1\n");
        client.print("Host: api.thingspeak.com\n");
        client.print("Connection: close\n");
        client.print("X-THINGSPEAKAPIKEY: "+apiKey1+"\n");
        client.print("Content-Type: application/x-www-form-urlencoded\n");
        client.print("Content-Length: ");
        client.print(postStr.length());
        client.print("\n\n");
        client.print(postStr);
        Serial.print("Send data to channel-1 ");
        Serial.print("Content-Length: ");
        Serial.print(postStr.length());
        Serial.print("Field-1: ");  // Displaying data of IR1 to serial
monitor
        Serial.print(IR1);
        Serial.print("Field-2: ");   // monitoring value on Field 2
        Serial.print(IR2);
        Serial.print("Feild-3: ");
        Serial.print(LDR1);
        Serial.print("Feild-4
        Serial.print(LDR2);
        Serial.print("Feild-5");
        Serial.print(Temp1);
        Serial.print("Feild-6");
        Serial.print(Temp2);
        Serial.println(" data send");
    }           // ending of growth_Para function()
void serialEvent_NODEMCU()
{
  while (mySeries.available()>0)                 // when data is transferring
from master to slave
  {
  if (mySeries.available()<1)             // when no data is available if loop
activates
  return;                                       // compiler jump back to the
function
  char g=mySeries.read();             // reading character in the g variable
  if (g!='\r')                            // if value store in the g variable
do not match with '\r'
  return;
```

```
  IR1=mySeries.parseInt();                 // returning integer value to vari-
able IR1 of data transfer from the master
  IR2=mySeries.parseInt();                 // returning integer value to vari-
able IR2 of data transfer from the master
 LDR1=mySeries.parseInt();
 LDR2=mySeries.parseInt();
 Temp1=mySeries.parseInt();

 Temp2=mySeries.parseInt();                                     //  returning inte-
ger value of home
 delay(500);                          // Giving delay for half second
  }       // end of the while loop
  }      // end of loop of serialEvent_NODEMCU function
 void display_function()
  {
Serial.print('\r');          // start of the data package
Serial.print(IR1);          // Serially sending value of IR to the serial moni-
tor
Serial.print('|');          //  To separate two different  values of the sensors
Serial.print(IR2);      // Serially sending value of IR to the serial monitor
Serial.print('|');          // displaying value to the serial monitor
Serial.print(LDR1);  // Serially sending value of LDR1 to the serial monitor
Serial.print('|');
Serial.print(LDR2);      // Serially sending value of LDR2 to the serial moni-
tor
Serial.print('|');
Serial.print(Temp1);  // Serially sending value of Temp1 to the serial monitor
Serial.print('|');
Serial.print(Temp2);    // Serially sending value of Temp2  to the serial
monitor
delay(500);                          // giving value of delay of half second
  }                          // end of the programme
```

Uploading Programme to (Arduino – WiFi Board)

User need to separately upload programme to two different board .

1. Upload programme (master to slave) by choosing Arduino board from the tool in Arduino ide.
2. Upload the programme (slave to Wi-Fi) by choosing NUTTYFi IOT board from the tools in the Arduino Ide.

After uploading programme supply power to the board, IOT boards will be automatically connected to the Wi-Fi/ hotspot as soon as user log into the Blynk app and Things Speak Server .

Pros and Cons

Pros: There are various advantage for flexible smart home interface such as –

- **Reduces Manual Framework:** Smart home interface system allow easy access to all the devices of the home though smart phones from any part of the world thus reduces manual monitoring.
- **High Level of Safety Under Abnormal Conditions:** The system insure safety as it consist of automated self-trigger alert which will go on as soon as parameters goes beyond safe limit & alert message will be send to user email.
- **Real Time Monitoring:** System allow real time monitoring of all the devices of the house thus makes sure that maximum energy is conserve when devices are not in use.
- **Conservation of Energy:** System plays important role in preventing wastage of energy by tracking device working time in cloud server.
- **Flexibility:** System makes use of Blynk platform which provides safety to user against data ownership risk and is easy to operate through Blynk graphical interface.

Cons: There are few disadvantages of Smart home interface.

- **Skilled Workers**: The system need to be set up by skilled workers whose availability is difficult.
- **Installation cost:** The installation cost of the system is quite high as compare to other systems.
- **Availability of Internet** In most of the places where internet fails these devices fails to operate.

Figure 34. Analysing and controlling of devices through BLYNK app

RESULTS AND DISCUSSION

The Figure 35 shown the graphical interface of the Blynk app in user smart phone. The graphical display shows various Blynk widgets used control different home devices . The gauge bar displays the temperature reading of the home thus allow real time indoor monitoring of the thermal parameters.

User can check the status of all the devices by pressing the status button. This will send the status of all the home devices to the email as shown in the figure above.

CONCLUSION

- Thus, the chapter highlights the flexibility which user will experience in monitoring and controlling home devices using BLYNK app and Things speak Server based on IOT(Internet Of Things) platform .
- Blynk is the universal IOT platform, which is cheap, secure and easy to operate.
- Thus, system eliminates Hackers attack, Data ownership risk, single point of Failure of the system.

FUTURE SCOPE

The system can be modifies using more advance sensors and artificial intelligence to increases comfort level as well as high safety and security of home. By monitoring devices performances, we can develop more efficient devices for home so as to conserve maximum energy. Most of the industries are adopt-

Figure 35.

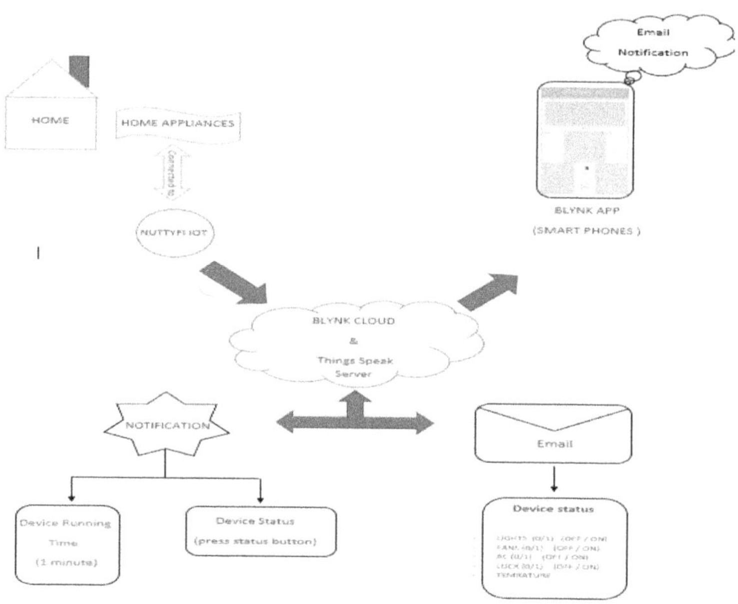

ing IOT as it combines automation with wireless network thus insure high reliability in monitoring and operating devices. Thus, by conserving energy through IOT technology may help us create better future for next generation and preserve nature.

REFERENCES

Anita, G., Rajesh, S., Mamta, M., Bhupendra, S., Ravindra, S., & Vikasgarg, G. (2018). *Hands on approach on Application of Internet of Things in Various Domain of Engineering*. Delhi: International Research Publication House.

Anita, G., Rajesh, S., Piyush, K., & Yadav, S. (2016). Wireless Personal Area Network and PSO-Based Home Security System. Academic Press. doi:10.1007/978-81-322-2523-2_24

Chasta, R., Singh, R., Gehlot, A., Mishra, R., & Choudhury, S. (2016). A smart building automation system. *International Journal of Smart Home, 10*(9), 217-224. . doi:10.14257/ijsh.2016.10.9.20

Isa, E., & Sklavos, N. (2017). Smart home automation: GSM Security Design & Implementation. *Journal of Engineering Science and Technology Review.*, *3*(10), 170–174. doi:10.25103/jestr.103.22

Kanase, P., & Gaikwad, S. (2016). Smart hospitals using internet of things (IOT). *International Research Journal of Engineering and Technology, 3,* 1-3.

Lucio, C., Francesco, F., Gianluca, I., Longhi, S., & Giacomo, T. (2016). *IOT Based Indoor Personal Comfort Levels Monitoring*. Academic Press. . doi:10.1109/ICCE.2016.7430548

Mei-Ju, S., Hung-yu, W., & Guan-Yu, L. (2016). Two Paradigm in Cellular IOT Access for Energy-Harvesting M2M Devices: Push-Based Versus Pull-Based. *IET Wireless Sensor System.*, *6*. doi:10.1049/iet-wss.2015.0040

Mohamed, M. (2016). *Internet Of Things for Home Automation*. Academic Press. doi:10.13140/RG.2.2.15237.35046

Nazeem Basha, S., Jilani, S.A.K., & Arun, S. (2016). An intelligent door system using raspberry pi and amazon web services IOT. *International Journal of Engineering Trends and Technology, 33*(2), 1-6.

Piyush, K., Bhupendra, S., Rajesh, S., & Anita, G. (2018). *Wireless Methods & Devices For Home Automation*. Delhi: International Research Publication House.

Pundir, Y., Sharma, N., & Singh, Y. (2016). Internet of things (IOT): Challenges & future directions. *International Journal of Advanced Research in Computer and Communication Engineering, 5*(3), 1-5.

Rajesh, S., Anita, G., Bikarama, Y., & Bhupendera, S. (2018). *IOT Enabled Fire Safety and Security devices for building. In Pen2Print*. EduPedia Publication.

Shingala, K., & Patel, J. (2017). Automatic home appliances and security of smart home with RFID, SMS, Email and real time algorithm based on IOT. *International Research Journal of Engineering and Technology, 4,* 1-7.

Todica, M. (2016). *Controlling Arduino board with smartphone and Blynk via internet.* Academic Press. . doi:10.13140/RG.2.2.23956.30080

Todica, M. (2017). *Home Security alarm notification with smartphone and Blynk.* Academic Press. . doi:10.13140/RG.2.2.32529.43369

Chapter 2
Low-Power High-Performance Tunnel FET With Analysis for IoT Applications

Suman Lata Tripathi
Lovely Professional University, India

ABSTRACT

The emerging tunnel FET is analysed in terms of ON-state current, OFF-state current, subthreshold slope, switching capacitance to explore its applications for smaller size low-power high-speed digital and memory applications that are an integral part of portable intelligent devices for IoT applications. A large portion of IoT systems are associated with these embedded SRAM/DRAM memories that contribute to a major portion of power dissipation in systems-on-chip (SoCs) or digital design. Several SRAM cell-based memory designs with TFET structures are compared to focus their applications. The ambilpolar nature of TFET structures are investigated for highly random, unclonable secured hardware systems. New circuit designs with TFET were explored for turn-on voltage reduction, ON-state resistance reduction, and reverse leakage reduction techniques that plays an important role in designing efficient energy-harvesting systems.

TRANSISTOR DESIGN FOR IOT SYSTEMS

The significant growth in the amount of data to be processed, transmitted, and stored by IOT systems focuses the concern toward low power secured hardware structures. A low energy secured system requires new researches at different level of micro-architectures, gate level/register transfer level and transistor level design (Taheri et al, 2017). The increasing demand of Internet of Things (IoTs) as a future perspective of electronics industry requires highly energy efficient design of sensor nodes. Several IOT based applications such as sensing hazardous environmental conditions, wearable and portable health monitoring system, bio-medical sensors, remote surveillance, traffic monitoring and other sensing network require ultra-low high speed, low power and energy-efficient devices to maintain long battery life as well as embedded self energy harvesting technique (Ahmad et al, 2018). A large portion of IOT systems

DOI: 10.4018/978-1-5225-9574-8.ch002

are associated with these embedded SRAM/DRAM memories that contribute to major portion of power dissipation in Systems-on-Chip (SoCs) or digital design (Lee et al, 2013). The flip-flops are critical components in any SoC design, so Some of literature shows flip-flop designs with FETs especially for microprocessor based systems and digital logic where flip-flops are used as pipeline registers, register files and buffers (Rasouli et al, 2006; Gupta et al, 2016). These flip-flop supports voltage scaling and works for supply voltages from 0.3V to 0.6V with improved leakage by 4 to 7 decades with advance transistor designs. The designer task is achieving secured low power micro-architectures by taking consideration for functionality, performance, power, area and overall cost. The new transistors are coming into picture with advance features to replace CMOS technology removing the scaling barriers. The technology must have some distinguished feature and performance that can be implemented for secured IOT systems with low energy. The transistors based on new technology must design by taking care for, fault models, manufacturing defects and reliability issues.

ADVANCE TUNNEL FET STRUCTURES

Increasing need for smart devices based on internet-of-thing (IOT), focuses the designer concern for new emerging solid state transistors other conventional MOSFET. This work deals with the analysis of Tunnel FET to optimise device performance for low power, high switching speed circuit enabling its use in smart applications. Low voltage application of Tunnel FET with steep subthershold slope, has potential to replace its MOSFET counterpart for future scaling due to thermal limits imposed on nano-level transistors. Longer channel region increases the tunneling area results in increasing tunneling current and decreasing miller capacitance to improve device switching performance for digital application. The Tunnel FET has significant potential to overcome limitations imposed due the scaling in low voltage region because of its steep subthreshold slope that leads to abrupt transition from ON-state to OFF-state. Use of $Si_{0.7}Ge_{0.3}$ material as pocket region (0.5nm) enhances band-to-band tunneling by decreasing tunneling distance. The $Si_{0.7}Ge_{0.3}$ pocket region also provide additional barrier to OFF-state leakage leading to the reduction in leakage current drastically resulting in to low static power consumption. Si-SiGe interface is utilised because of its better matching and easier fabrication steps as compared to other available hetero-junction structures. The proposed pocket ultra-small pocket $Si_{0.7}Ge_{0.3}$ Junction-less TFET (JLTFET) exploits advantage junction-less regions and p-type pocket region to improve device performance in subthreshold region showing improvement in subthreshold slope and better ON and OFF-state drain current ratio as compared to other similar TFET structures. Proposed ultra-small pocket JLTFET shows high value I_{ON}/I_{OFF} ratio and good subthreshold behaviour even with 2nm gate length and body thickness 0.5nm that can be part of efficient digital and memory applications with lesser occupancy of silicon area compatible to portable and wearable smart devices. All the analysis and verification for proposed JLTFET carried out on Visual TCAD 2D/3D device simulator.

Technology scaling imposes conditions on subthreshold performance of transistors. The subthreshold slope (SS) limitation of 60mV/decade is the major challenge for CMOS at room temperature and, that imposes restrictions on further reduction of static power consumption in CMOS based memory designs. The MOSFET switching process based on the temperature-dependent model of electrons injection over the barrier that restricts the transition slope steepness to be scaled further in terms of ON/OFF currents (Taur et al, 2013). A novel inter-band tunnel FET was introduced to provide steeper sub-threshold slope (<60 mV/decade) (Fan et al, 2016) and low static power consumptions. Hetero-junction vertical p and n TFET structures were proposed with very small subthreshold slope, low threshold voltages and OFF-state current up to the value of 10^{-15} A/um that is an important design consideration while down scaling of device (Nirschl et al, 2004). Inclusion of narrow band gap material such as SiGe, Ge and InAs with silicon were utilized to enhance band to band tunneling in TFET with better ON and OFF current ratio (Khatami et al, 2009). A thin Pocket regions between source and channel region, reduces tunneling distance and increases band to band tunnel effect resulting in more steep subthreshold characteristics (Avci et al, 2012). InAs interface with GaSb was explored to design 10nm pTFET to get 60mV/decade subthreshold slope for several orders of current in comparison to its performance with FinFET structure (Uygar et al, 2014). Longer channel region like L, T or U-shaped Tunnel FET implemented with narrow band gap material Si-Ge pocket region to support band bending resulting into more BTBT tunneling (Sharma et al, 2015; Li et al, 2017; Wang et al, 2014). ZS-TFET further increases ON-state current and low subthreshold slope as compared to other vertical TFET provided by enlarged tunneling area as well as suppressed lateral parasitic and improved analog/RF performance. But fabrication of above mentioned shapes of tunnel FET is very complex and difficult to implement. Usually the tunnel FET shows considerable ON-state current with longer channel length that indicates the compromise in terms of dimensions that is again unsuitable for future smart portable devices. Another way to improve ION/IOFF ratio is to decrease off state current. A dual material gate TFET designed with metal gate of high work function, overcome the problem of drain induced barrier lowering and improves ION/IOFF ratio with suitable high-K dielectric material under gate contact below 20nm channel length and low Vdd (0.5V) (RouzbehMolaeiImenabadi et al, 2017). Inclusion of junction-less source-channel and channel-drain region controls the parasitic resistance and improves ON-state current (SnehSaurabh et al, 2012; Sahay et al, 2016) and also reduces the effect of random dopant fluctuations. Junction-less tunnel FET shows increase in both ON and OFF-state current with lesser improvement in ION/IOFF ratio that is a major drawback of junction-less transistors (Kumar et al, 2016). Buried oxide region (SOI) under the channel region substantially improves subthershold performance (Sahay et al, 2016) by reducing leakage current and further ION/IOFF ratio. TDJLT (AvinashLahgere et al, 2017) was proposed with thin insulating layer at the centre channel region, showed improvement in ON and OFF current ratio by decreasing OFF-state leakage current drastically.

IMPORTANT DESIGN PARAMETERS

Some of important design parameters that are responsible for efficient and reliable device performance are need to be discussed. These parameters are:

ON-state current (I_{ON}): ON-state transistor drain current with gate-source voltage (V_{gs}) and drain-source voltage (V_{ds}) equal to the supply voltage. A higher I_{ON} is preferred achieve higher speed with less turn-on resistance.

OFF-state current (I_{OFF}): OFF-state transistor drain current with V_{gs} equal to zero and V_{ds} equal to the supply voltage. A lower I_{OFF} indicates lower leakage current. It has more adverse affect on short channel transistors performance in comparison to longer channel transistors. The OFF-state leakage current depends on mainly two physical phenomena described as:

- The limitations imposed on electron drift characteristics in the channel,
- The threshold voltage variations due to the shortening channel length.

The major short channel effects are:

1. Channel Punchthrough
2. DIBL
3. V_{th} Roll off
4. Surface scattering
5. Velocity saturation
6. Impact ionization
7. Hot electrons

Channel Punchthrough

The expressions for the drain and source junction widths are:

$$x_{dD} = \sqrt{\frac{2\epsilon_{si}}{qN_a}\left(V_{DS} + \varnothing_{Si} + V_{SB}\right)}$$

$$x_{dS} = \sqrt{\frac{2\epsilon_{si}}{qN_a}\left(\varnothing_{Si} + V_{DB}\right)}$$

where V_{SB} and V_{DB} are source to body and drain to body voltages. When the depletion regions surrounding the drain extends to the source, so that the two depletion layer merge (i.e. $x_{dS} + x_{dD} = L$), punchtrough occurs. Punchthrough can be minimized with thinner oxides, larger substrate doping, shallower junctions, and obviously with longer channels.

Drain-Induced Barrier Lowering (DIBL)

The increased drain voltage decreases potential barrier in the channel region, leading to *drain-induced barrier lowering* (DIBL). The DIBL effect increases with the decrease of channel length because of increase in effect of drain voltage in channel region.

V_{th} Roll Off

In smaller channel length, the source and drain field affect deeply into the middle of the channel, which lowers the potential barrier between the source and drain. This causes the increase in subthreshold current that result into decrease of V_{th} or V_{th} roll off. The effect of source and drain field on depletion region is also known charge sharing effect.

Surface Scattering

As the channel length becomes smaller due to the lateral extension of the depletion layer into the channel region, the longitudinal electric field component ε_y increases, and the surface mobility becomes field-dependent. The electrons accelerated toward the interface by ε_x suffered collisions due to the increasing effect of longitudinal field ε_y. This phenomenon is known as *surface scattering* which causes reduction of electron mobility.

Velocity Saturation

The performance of short channel devices is also affected by velocity saturation, which reduces the transconductance in the saturation mode. At low ε_y, the electrons drift velocity v_{de} in the channel varies linearly with the electric field intensity. However, as ε_y increases above 10^4V/cm, the drift velocity tends to increase more slowly, and approaches a saturation value of $v_{de}(sat)=10^7$ cm/s around $\varepsilon_y = 10^5$ V/cm at 300 K. Hence, the drain current is limited by velocity saturation instead of pinch off.

Impact Ionization

Another undesirable short channel effect, especially in NMOS, occurs due to the high velocity of electrons in presence of high longitudinal fields. When high velocity electron impact on silicon atom, generate electron-hole (e-h) pairs by impact ionization.

Hot Electrons

Another problem, related to high electric fields, is caused by so called hot electrons. These high energy electrons can enter the oxide, where they can be trapped, and make oxide negatively charged. It can accumulate with time and degrade the device performance by increasing V_{th} and lead to poor gate's control on the drain current.

JUNCTION-LESS TUNNEL FET (JLTFET) MODEL

In this work, an ultra-small pocket $Si_{0.7}Ge_{0.3}$ junction-less TFET (JLTFET), is proposed and performance has been optimized with different gate contact material/oxide region in terms of better subthreshold performance. The use of p-type $Si_{0.7}Ge_{0.3}$ pocket region offers additional barrier to OFF-state leakage current with considerable amount of ON-state current showing improvement in ON/OFF current ratio. The narrow band gap material $Si_{0.7}Ge_{0.3}$ pocket region shows sharp changes in drain current depending upon different bias conditions. A comparison made between proposed pocket JLTFET and Junctions based Pocket Tunnel FET similar in structure and dimensions. The proposed design of ultra-small Pocket Si0.7Ge0.3 nJLTFET and pJLTFET can be further implemented in future with superior memory design of DRAM (Sahay et al, 2016) cell and SRAM (Makosiej et al, 2016; Amir et al, 2016) cell.

DEVICE DESIGN AND STRUCTURES

2-D visual TCAD (Cogenda) has been used to implement the novalnJLTFET. The source/drain/channel (n-type) doping is kept at 1e20cm^{-3} for source/drain and channel is kept at 1e16 cm^{-3}. The channel doping is kept low to improve carrier mobility in this region. Figure 1 indicates the noval 2D nJLTFET

Figure 1. 2D Structure of nJLTFET with $Si_{1-x}Ge_x$ pocket region

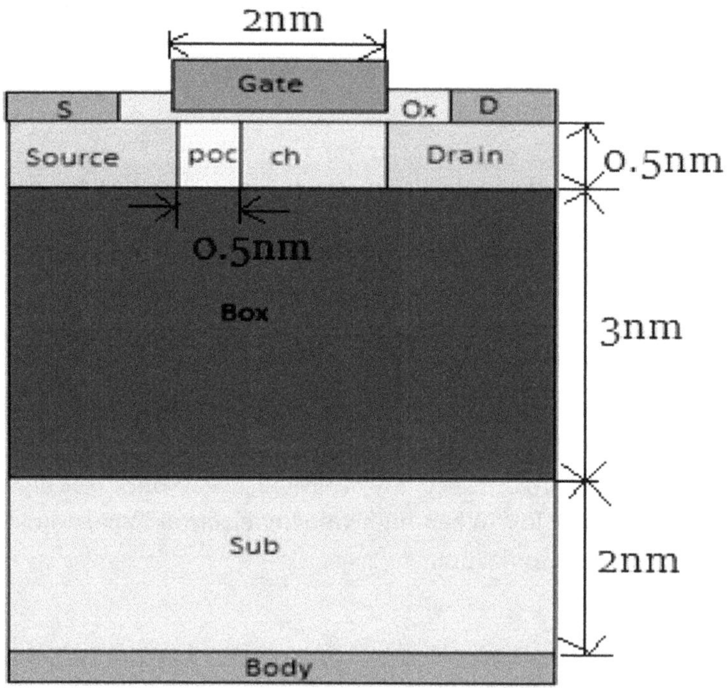

structure with body thickness 0.5nm, buried oxide region with 3nm and p-type substrate region of 2nm. Thickness of oxide region under gate is kept 0.1nm. A metal gate with high value of work function Pt (5.7eV) used as gate contact material with gate length 2nm. Oxide region made with High dielectric constant material HfO_2(25) used replacing SiO_2(3.9) under gate and performance compared. To increase band to band tunneling property a pocket region of $Si_{0.7}Ge_{0.3}$ of 0.5nm thickness is included near the source region under influence of gate. Figure 2 shows 3D structure of pocket JLTFET indicating device region view and material region view.

Figure 3 represent the energy band diagram of nJLTFET obtained with Visual TCAD simulation. Addition of p-type pocket region reduces off-state current drastically due to the presence of an extra tunneling barrier to off-state leakage current. A sharp peak appeared in the energy band diagram representing the raised barrier through $Si_{0.7}Ge_{0.3}$ pocket region. Figure 4 shows the channel potential under gate indicating sudden dip in source–pocket interface.

Electric Field variation depicted in Figure 5 indicates sharp variations in source-pocket interface. An additional electric field peak under pocket region is observed in comparison to the middle of channel in TDJLT (Kumar et al, 2016). Hole density offers high value in off-sate conditions that means the channel region suppresses the parasitic BJT action. But a low value of hole density in on-state exploits better switching speed as shown in Figure 6. Figure 7 represents electron density increases in channel region with increase in gate voltage. Therefore, the proposed nHTFET supports parasitic BJT triggering with a raised barrier height between source and channel region.

Figure 2. 3D simulation view of nJLTFET Device material view

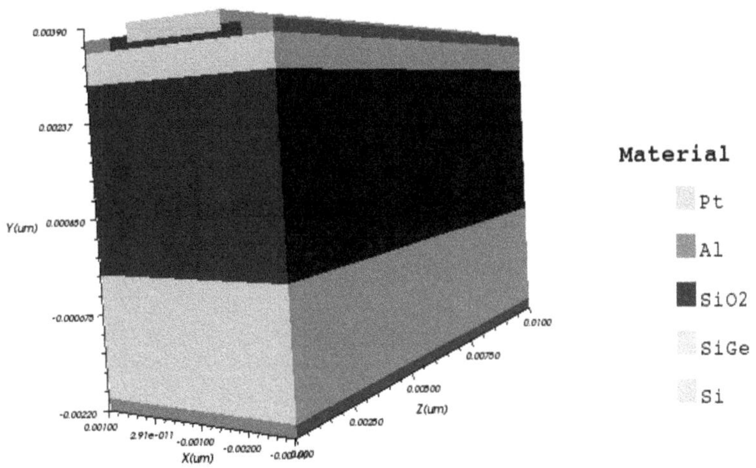

PERFORMANCE PARAMETERS

To compare the performance of different transistors following are the main parameters of interest.

1. ON State Current (I_{ON})
2. OFF State Current (I_{OFF})
3. Subthreshold Slope (SS)
4. Drain Induced Barrier Lowering (DIBL)
5. Propagation Delay (T_d)

Figure 3. Energy band diagram of nJLTFET along channel region

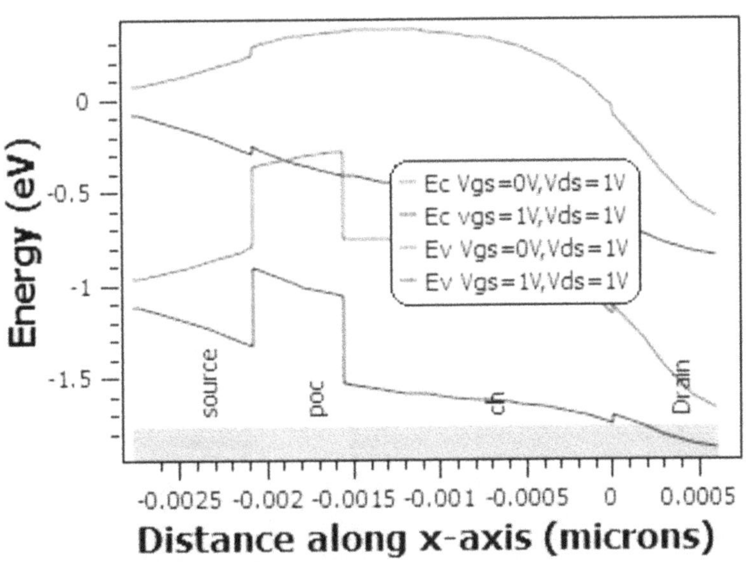

Figure 4. Channel potential of nJLTFET

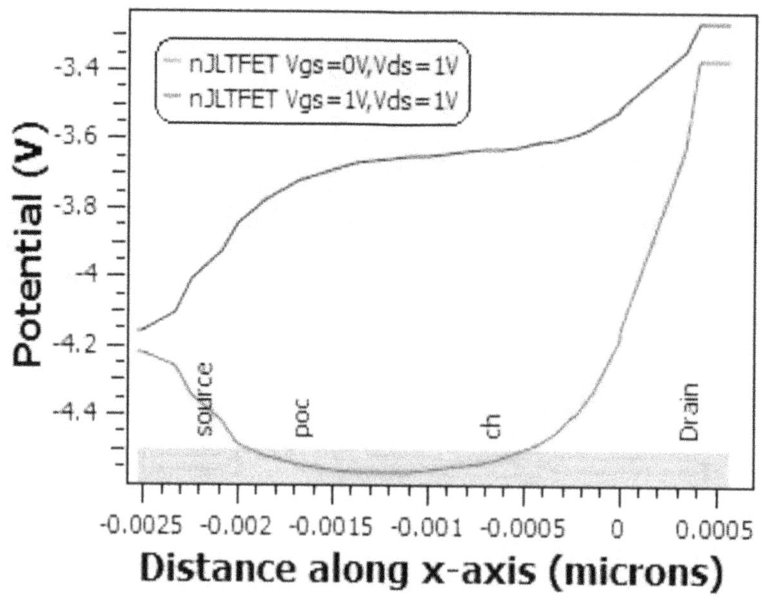

1. 5. 1. **ON State Current I_{ON}:** On state current is a important parameter of MOSFET. It directly affects speed of circuits made by device. Value of I_{ON} should be high. I_{ON} is drain current at

$V_{GS}=V_{DS}=V_{DD}$

Figure 5. Electric field of nJLTFET

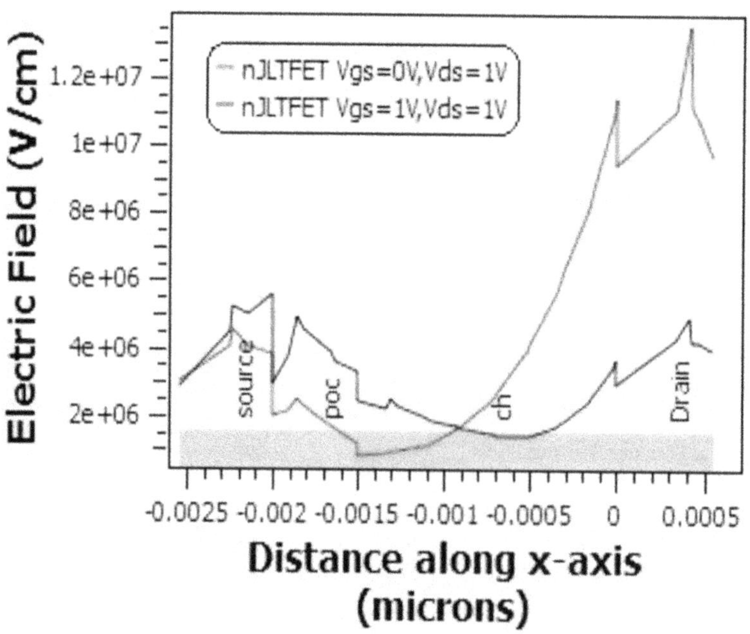

Figure 6. Hole density along channel region

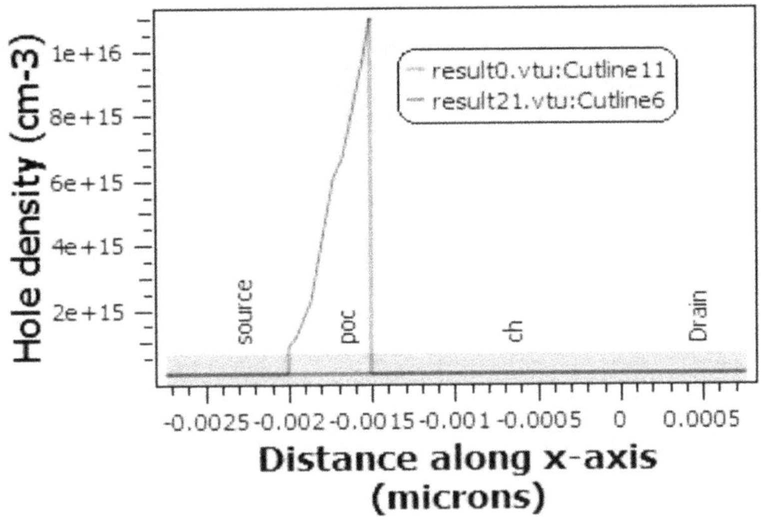

2. **OFF State Current I_{OFF}:** It is the current when transistor is at cut off state. OFF-state current is measured at V_{GS}=0V when V_{DS}=$V_{DD.}$ High value of I_{OFF} results in high leakage power consumption. Value of I_{OFF} should be as low as possible.
3. 5. 3. **Subthreshold Slope:** The subthreshold slope (SS) gives the percent variation of the drain current as a function of the gate voltage in weak inversion.

$$SS = \frac{dVgs}{d\left(\log I_D\right)} \text{(mV/decade)}$$

Figure 7. Electron density along channel region

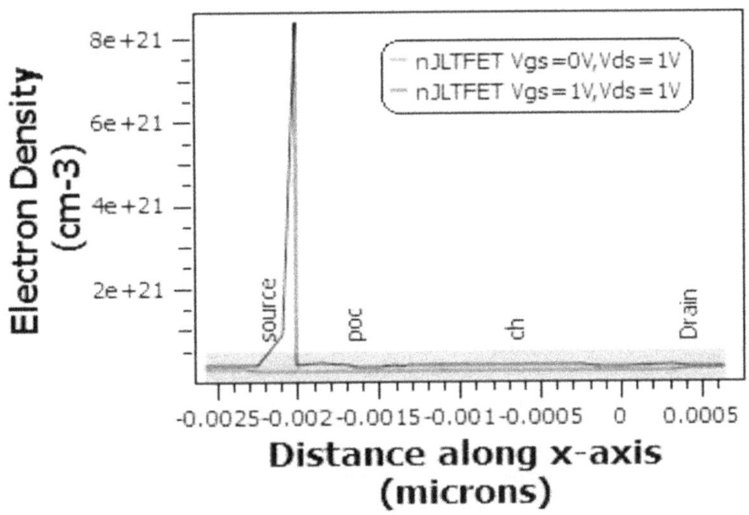

4. 5. 4. **Drain Induced Barrier Lowering (DIBL):** DIBL is measured as change in threshold voltage (V_{th}) with respect to change in V_{DS}. So for measurement of DIBL, threshold voltage of MOSFET for two different V_{DS} is measured and DIBL is calculated by following expression;

$$DIBL = \frac{\Delta V_{th}}{\Delta V_{DS}} = \frac{V_{th1} - V_{th2}}{V_{DS1} - V_{DS2}}$$

5. 5. 5. **Propagation Delay:** In this work propagation delay is measured for inverter circuits. Average of rise time delay and fall time delay is known as propagation delay.

$$T_{PD} = \frac{T_R + T_F}{2}$$

Where T_{PD} is propagation delay, T_R is rise time delay and T_F is fall time.

DEVICE SIMULATIONS AND RESULTS

DC Analysis

Firstly simulation performed for conventional junction based Tunnel FET then performance compared with pocket tunnel FET. Then proposed pocket JLTFET is simulated in linear (Vds=0.1V) and saturation region (Vds=1V) with varying gate voltage. Figure 8 shows sharp I_d versus V_{gs} characteristics of Junction-less TFET with or without pocket region. $Si_{0.7}Ge_{0.3}$ Pocket region introduced reduces OFF state leakage current by maintaining ON current and shows sharp subthreshold characteristic. I_d versus V_{gs} characteristics of JLTFET (with or without pocket region) shows almost similar behaviour in linear (V_{ds}=0.1V) and saturation region (V_{ds}=1V). JLTFET with or without pocket region shows good performance in linear and saturation region but comparatively JLTFET with SiGe pocket region has better I_{ON}/I_{OFF} current ratio of the order of 2.64×10^8. The subthreshold performance of JLTFET are under limit with subthreshold slope 65.7 mV/decade and DIBL 19.5mV/V even with ultra-small dimensions (L_g=2nm, Body thickness=0.5nm). Similar characteristics of pocket JLTTFET obtained in saturation region (V_{ds}=1V) comparing other TFET structures as shown in Figure 9.

Comparison of JLTFET with different gate and oxide material have been done in Figure 10 which shows advantage of Pt/HfO_2 interface as metal-gate/oxide region over other metal gate/oxide interface. Therefore, Pocket JLTFET has higher I_{ON}/I_{OFF} current ratio and sharp subthreshold slope with Vgs variations in range of 0-0.5V (V_{ds}=0.1V).

Proposed ultra-small pocket $Si_{0.7}Ge_{0.3}$ JLTFET exhibit high value of I_{ON}/I_{OFF} current ratio (Table1) above than 10^8 in comparison junction based Tunnel FET with similar dimensions with or without pocket regions. Although the dimensions are very small even then the device subthreshold performance is under limit with tremendously sharp subthreshold slope of 56mV/Decade and DIBL have a range of 23mV/V. This support the further the scaling trend with advancement of technology incorporating future applications.

Figure 8. I_d versus V_{gs} of Pocket $Si_{0.7}Ge_{0.3}$nJLTFET in comparison to junction based TFET (V_{ds}=0.1V)

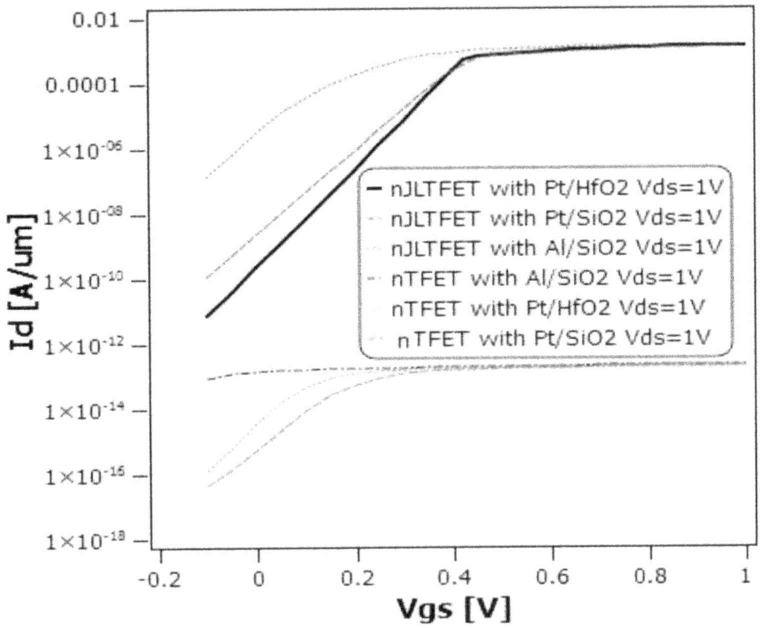

AC Analysis

The analysis of JLTFET structure is carried out for 10MHz frequency range on Visulal TCAD device simulator.

Figure 9. I_d versus V_{gs} of Pocket $Si_{0.7}Ge_{0.3}$nJLTFET in comparison to junction based TFET (Vds=1V)

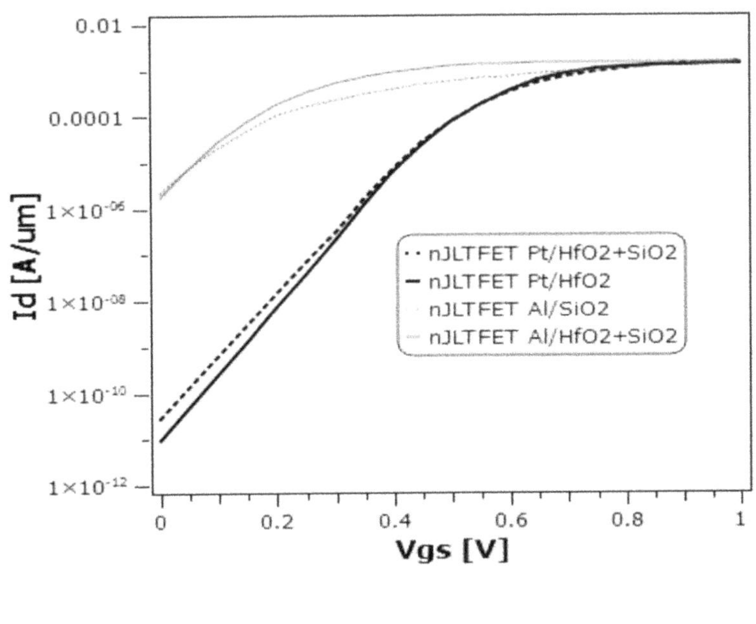

Figure 10. I_d versus V_{gs} of Pocket $Si_{0.7}Ge_{0.3}$ nJLTFET with varying gate contact and oxide material (V_{ds}=0.1V)

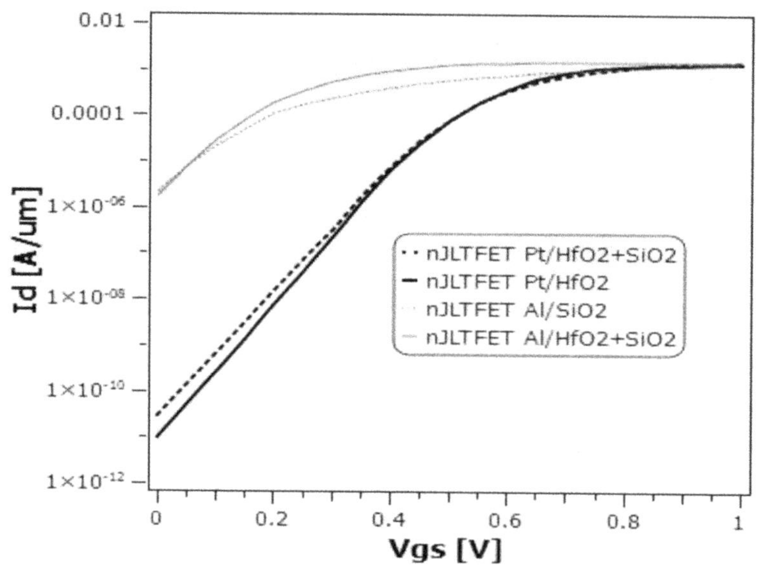

SWITCHING CAPACITANCE

The proposed JLSGTFET shows a very high gate to drain capacitance(C_{gd}) due to increased barrier arised with thin SiGe p-type pocket region at source side. Incresed total capacitance (C_T=C_{gs}+C_{gd}) increases the overall Miller capacitance of proposed TFET in comparison to MOSFET structures due to the dominating feature of C_{gd}. Figure 11 describe gate to drain capacitances as a function of gate voltage. The value of gate to drain capaciatnce (C_{gd}>2E-15 F/um) is higher than gate to source capacitance (C_{gs}>3E-16 F/um) as depicted in existing litrature (Mookerjea et al, 2009; Yang et al, 2010; Mallik et al, 2012). Therefore the switching characteristics of transistor is more dependent on C_{gd} that can be reduced with smaller

Table 1. Performance comparison between TFET and JLTFET with or without Pocket region

Device type	I_{OFF}	I_{ON}	$I_{ON/IOFF}$
TFET with pocket$Si_{0.7}Ge_{0.3}$(HfO2/Pt)	8.75E-17	4.06E-14	4.64E2
TFET with Pocket$Si_{0.7}Ge_{0.3}$(SiO2/Pt)	4.48E-17	4.96E-14	1.11E3
JLTFET with Pocket$Si_{0.7}Ge_{0.3}$(HfO2/Pt)	2.65E-12	7.00E-04	2.64E8
JLTFET without Pocket	3.51E-13	7.65E-05	2.17E8

Table 2. Subthreshold performance comparison of JLTFET

Device type	SS(mV/ddecade)	DIBL(mV/V)
TFET with pocket$Si_{0.7}Ge_{0.3}$(HfO2/Pt)	112	2
TFET with Pocket$Si_{0.7}Ge_{0.3}$(SiO2/Pt)	65.7	23
JLTFET with Pocket$Si_{0.7}Ge_{0.3}$(HfO2/Pt)	56	19.6

Figure 11. Gate to drain capacitance of $Si_{0.7}Ge_{0.3}$ JLTFET

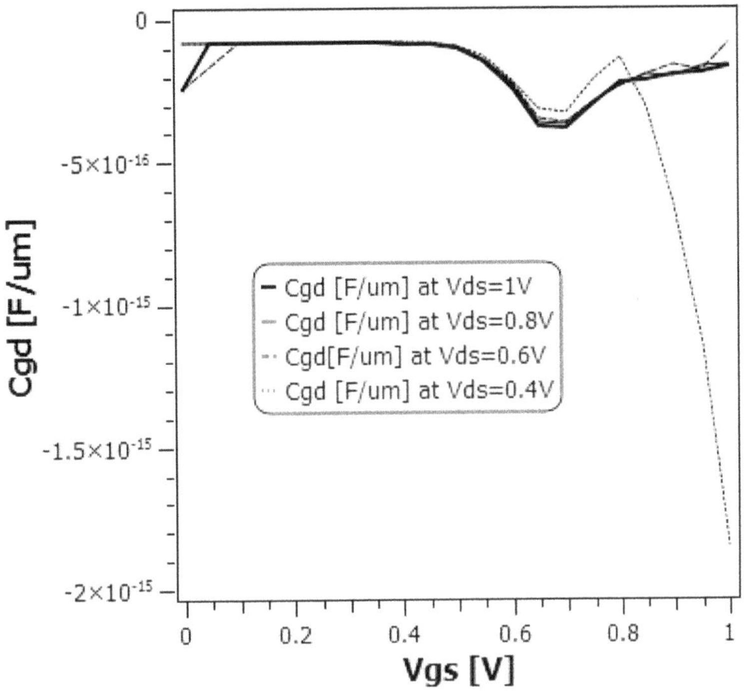

band gap $Si_{0.7}Ge_{0.3}$ region that results in higher ON state current. It also improves the circuit speed by reducing delay due to capacitive effect of transistors.

USE OF TUNNEL FET IN ADVANCE MEMORY DESIGN

A larger embedded system memories specially SRAM, are required to support a huge range of applications and also to store a large sensor data along instructions (Patel et al, 2016). High percentage of energy is consumed in these embedded system memories in that SRAM will be important consumer to the power dissipation for Systems-on-Chip (SoCs) (Lee et al, 2013). So, designing of SRAM cell focused for such system become key to the success of such IOT systems. Since the tunnel FET are already explored for its low static power consumption and high I_{ON}/I_{OFF} ratio, make this suitable in design of advance SRAM cell. Voltage scaling plays important role in energy consumption. So, a proper optimization of supply voltage is required. The total energy consumption follows the expression:

$$E_{total} = E_{switching} + E_{leakage}$$

$$E_{switching} \alpha C_{load} V_{dd}^{2}$$

$$E_{leakage} = \Gamma V_{dd} I_{leaakge}$$

Here, Γ is execution time and $I_{leakage}$ is leakage current. Eqn. shows that optimization in energy consumption can be done with following parameters:

1. V_{dd} scaling
2. Leakage current reduction
3. Switching time.

Initial part of this chapter focuses about the design of advance tunnel FET structures with optimization in terms of I_{ON}/I_{OFF} for better switching and subthreshold parameters under limit to overcome the limitation imposed to power consumption per unit chip. The Tunnel Field Effect Transistor (TFET) has an tremendous potential to reduce the static and switching power losses. Tunnel FET is mainly based on band-to-band tunnelling. In TFET, the SS is not limited to 60 mV/decade compared to the case of CMOS. TFETs with SS less than 60 mV/dec, have been already designed and analysed in available literature (Datta et al, 2014; Saripalli et al, 2010) proposed mixed TFET-MOSFET

An extra pairs of transistors were used in addition with conventional 6T and 8T cells to design 8T TFET and 9T TFET SRAM cells (Makosie Et al, 2016; Amir et al, 2016). The pair of TFETs facing inward and outward make read and write operations easier in these designs. The stable read operation is still an important concern in 8T TFET cell (shown in Figure 12a) as in conventional 6T cell due to its conflicting design requirements. However, 9T TFET (Amir et al, 2016) cell handled this issue by decoupling read operation using a separate read bit line along with extra transistor, as shown in Figure 12b. However, the poor drive current in TFETs is still a concern in these cells. The write-ability of TFET SRAM cells (Chen et al, 2013; Strangio et al, 2015) has been improved with write-assist employed to these cells. A 7T driverless TFET SRAM cell connected with virtual-ground write-assist was proposed in (Chen et al,2 013) which improves the write-ability. Authors in (Strangio et al, 2015) proposed TFET bit cell with low write contention (shown in Figure 13) using V_{DD}-collapse assist.

The unidirectional characteristic of TFET were utilised in 8T SRAM cell to overcome the half selected cell (HS cell) stability issues at the expense of degradation in write-ability (Morris et la, 2015). 9T TFET SRAM (Chen et al, 2014) cell consists of a cell core using cross-coupled inverter pair along with power cut off switches in pull-up path of each inverter. Additional write access transistors AL and AR and read buffer with a single transistor, ND. The power cut-off switches, PL1 and PR1 are used here to drive BL and BLB and exploited to open the pull-up network during write operation that provide '0' going node without any error. A RBL bit line is used to do read operation. The read and write operations are carried out using one WL word line opposing separate word lines for write and read operation. The 7-T TFET cell (Figure 14) uses read buffer similar to 9-T TFET cells but it uses same BLB bit line for read/write operation. This results in to increase in the bit line capacitance and therefore, read delay. Ahmad etal (Ahmad et al, 20180 proposed (Figure 15) a 9T TFET based SRAM bit cell with improvement of 1.15 times higher write margin (WM), 25% lower write delay, 7% smaller standby leakage power measured consumes 73% lower write (average=57%) energy at V_{DD}=0.3 V. The proposed cell shows a tremendous improvement in the read/write operations as compared with existing 7T and 8T TFET cells. Reducing leakage in SRAM is an important element in the design of low power systems. The 9T SRAM cell TFET cell reduces the leakage in the cell core using stacked transistors in pull-up path.

Figure 12. a) 6T SRAM b) 8T TFET SRAM (Saripali Et al, 2011)

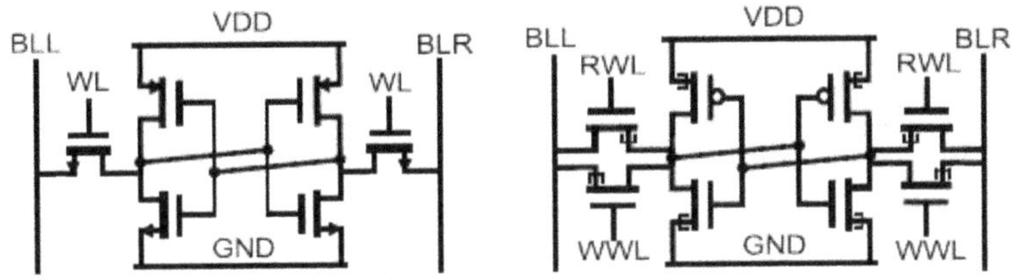

Figure 13. 9T SRAM (Amir et al, 2016)

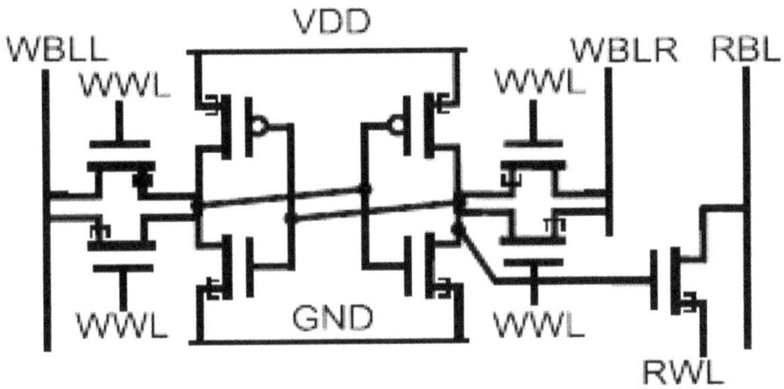

TFET FOR ENERGY EFFICIENT SECURED IOT APPLICATION

Radio-frequency (RF) powered energy harvesting systems have supplemented new perspectives in several applications such as health monitoring, bio-signal acquisition, and battery-less data-transceivers (Ahmad et al, 2018; Li et al, 2014). Figure 16 shows energy harvesting and power management block that consists impedance matching networks, dc-dc converter for voltage boosting, RF rectifier, voltage regulator, and the energy storage devices (Li et al, 2014). Since the overall system is powered by the dc

Figure 14. 7T SRAM (Morris et al, 2015)

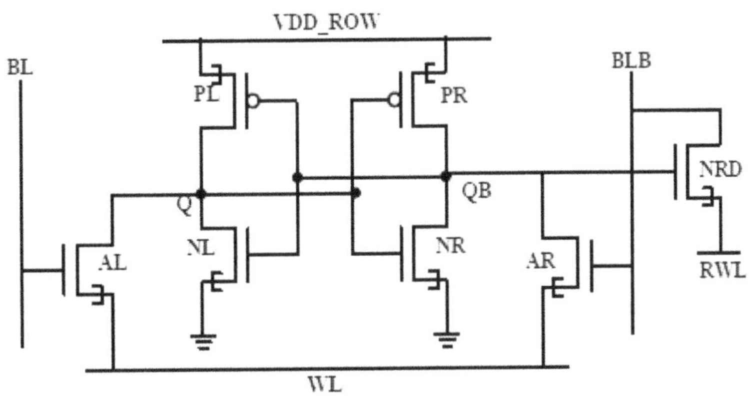

Figure 15. 9T SRAM (Ahmad et al, 2018)

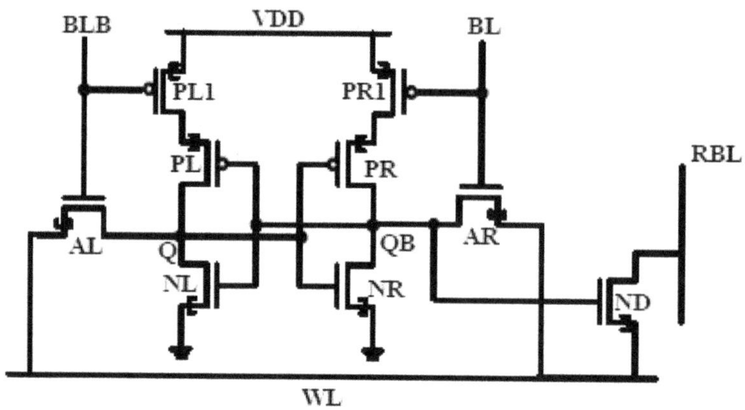

output from the rectifier circuits, the design of a high-sensitivity, high-PCE RF rectifier is critical for these applications that increase the available dc power in entire system operation (Francesc et al, 2017).

TFET Advantages for RF Rectifier Design

In an RF rectifier design, the dc output power (P_{dc}), dc output voltage (V_{dc}) and power conversion efficiency (PCE) are the key performance metrics (At a given input RF power (P_{RF}) with an rms voltage amplitude (V_{ac}) of and a load resistance (R_L), and PCE of a rectifier are expressed as

$$V_{dc}=2V_{ac} - V_{drop}$$

Here, V_{drop} is the lumped voltage loss across the rectifier circuit.

Power conversion Efficiency, $PCE = \frac{P_{dc}}{P_{dc} + P_{loss}}$

Figure 16. RF powered energy harvesting system (Li et al, 2014)

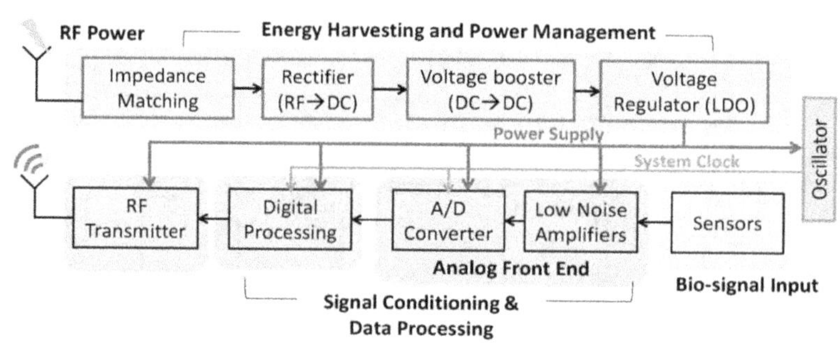

Figure 17. HTFET RF rectifier designs: (a) H2T, (b) H4T, and (c) H4N

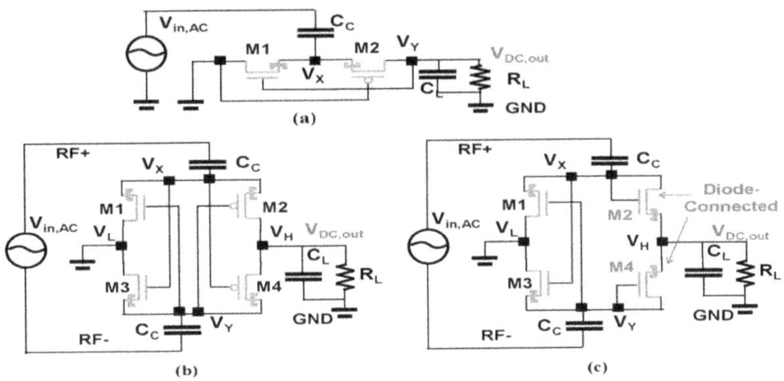

Average Power Loss, $P_{loss} = P_{leakage} + P_{reverse} + P_{switchin} + P_{redistribute} + P_{resistive}$

where, $P_{leakage}$, $P_{switching}$, $P_{reverse}$, $P_{redistribute}$ and $P_{resistive}$ are transistor off-state leakage loss, average parasitic capacitive switching loss, reverse conduction loss due to the current from the output node to the input, power loss in charge redistribution between the coupling capacitors and the transistor network, and the resistive power loss due to nonzero transistor on-state channel resistance respectively. The HTFET (Liu et al, 2014) device designed have characteristics such as steep slope, high on-state current at low supply voltages and uni-directional conduction can be employed to reduce the power losses in the rectifier design. New circuit designs with HTFET by Turn-On voltage reduction, On-State resistance reduction and reverse leakage reduction techniques plays important role in designing efficient energy harvesting system.

The HTFET rectifier topologies using the following three designs: n-type and p-type HTFETs based 2-T SVC (H2T) and 4-T cross-coupled (H4T) designs and the 4-T N-HTFET-only design (H4N) are shown in Figure17. H2T is similar to the conventional diode connected rectifier, except that it uses the dc output voltage to statically “cancel” the transistors threshold voltages. The H4T rectifier is similar to the standard cross-coupled CMOS rectifier with “cancelation” technique of dynamic threshold voltage. H4N is RF rectifier with only N-HTFET by replacing the diode-connected N-HTFETs in place of P-HTFETs to avoid degraded pHTFET performance.

TFET shows ambipolarity nature that indicates the flow of current in both highly positive and negative bias voltages. Therefore, ambipolarity characteristics of TFET make it suitable for highly random and unclonable circuit and thus improve the hardware security (Aditya et al, 2018). The ambipolar delay line was introduced to reduce the probability of hardware attacks and increase the unpredictability nature of circuit (Figure 18). A significant ambipolarity nature present in TFET device makes the TFET digital circuits functionality unconventional. Taking the unique device characteristics of TFETs into consideration, TFET based transceiver design exhibits enhanced performance in terms of overall energy efficiency that is realizable with CMOS 3D IC designs at low V_{DD}.

Figure 18. TFET Transmission gate based ambipolar delay line

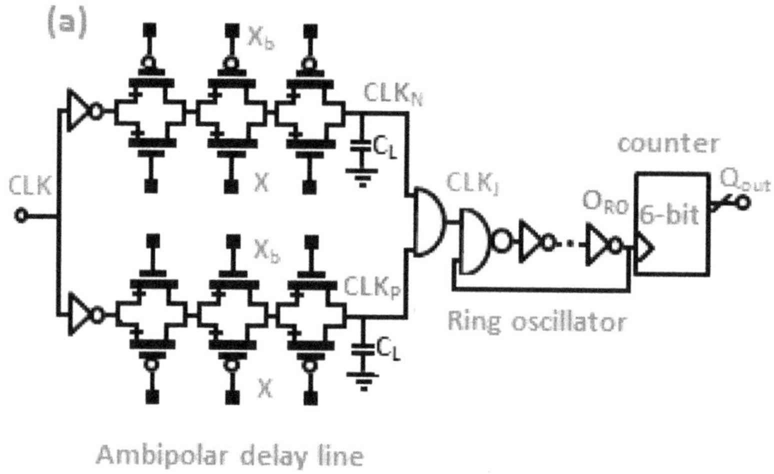

CONCLUSION AND FUTURE CHALLENGES

The present work focuses about transistor design considerations for energy efficient high speed digital and memory applications that is a major challenge for IOT systems. The work compares improvement in ON/OFF performance of JLTFET with other existing transistors showing advantage over conventional CMOS technology. The proposed JLTFET has high value of I_{ON}/I_{OFF} ($>10^8$), ideal subthreshold slope (<60mV/decade) and DIBL (<24mV/V) compared to available literature. A suitable value transistor capacitance with different bias conditions results into high speed operations with lesser delay. JLTFET structures are also showing reliable performance with varying operating conditions with <10% variations for important design parameters. The work also gives emphasis on transistor size reduction (<5nm) keeping subthreshold parameters under limits with low static power consumption and high component density per unit area of integrated circuit. Analysis of different SRAM cell architecture using TFET has been explored. The presented transistor can be utilised for similar 7-T, 8-T and 9-T memory architectures with further enhancement in optimised circuit technology that support huge demand of memory with low power consumption in portable and smart IOT applications. The analog/RF circuit designs with ultra-low power TFET have recently gained a lot of interest, benefited from the steep subthreshold performance to drain current ratio improvement, turn-on voltage reduction, uni-directional operation, as well as its desired low-voltage and high-frequency characteristics. The ambipolarity nature of TFET structures, make this suitable for highly random, unclonable circuit and thus enhance the hardware security. Taking the unique device characteristics of TFETs into consideration, TFET based transceiver design exhibits enhanced performance in terms of overall energy efficiency at low V_{DD}. The TFET are now introduced in power devices used as rectifiers or other places in highly efficient energy harvesting systems. The overall chapter is a brief discussion of TFET based applications focused for future portable IOT based highly efficient, low power, low voltage, unclonable secured systems. Several IOT based applications such as sensing hazardous environmental conditions, wearable and portable health monitoring system, bio-medical sensors, remote surveillance, traffic monitoring and other sensing network require ultra-low high speed, low power and energy-efficient devices to maintain long battery life as well as embedded self energy harvesting technique

REFERENCES

Aditya, Nagateja, Vishvakarma, Yellappa, Cho, & Vaddi. (2018). Tunneling Field Effect Transistors for Enhancing Energy Efficiency and Hardware Security of IoT Platforms: Challenges and Opportunities. *IEEE International Symposium on Circuits and Systems (ISCAS).*

Ahmad, S., & Alam, N. (2018). Robust TFET SRAM cell for ultra-low power IoT applications. *AEU*, *89*, 70–76.

Ahmad, S., Alam, N., & Hasan, M. (2018). Robust TFET SRAM cell for ultra-low power IoT applications. *Int. J. Electron. Commun.*, *89*, 70–76. doi:10.1016/j.aeue.2018.03.029

Amir, M. F., Trivedi, A. R., & Mukhopadhyay, S. (2016). Exploration of Si/Ge tunnel FET bit cells for ultra-low power embedded memory. *IEEE J Emerg Sel Top Circuits Syst*, *6*(2), 185–197. doi:10.1109/JETCAS.2016.2547818

Amir, M. F., Trivedi, A. R., & Mukhopadhyay, S. (2016). Exploration of Si/Ge tunnel FET bit cells for ultra-low power embedded memory. *IEEE J Emerg Sel Top Circuits Syst*, *6*(2), 185–197. doi:10.1109/JETCAS.2016.2547818

Ankit Sharma, A. (2015). *P-channel Tunneling Field Effect Transistor (TFET): Sub-10nm Technology Enablement by GaSb-InAs with Doped Source Underlap. In Device research conference (DRC).* IEEE.

Avci, U. E. (2012). Understanding the feasibility of scaled III–V TFET for logic by bridging atomistic simulations and experimental results. *Proc. VLSI Technol. (VLSIT) Symp.*, 183–184.

Cavalheiro, D., Moll, F., & Valtchev, S. (2017). TFET-Based Power Management Circuit for RF Energy Harvesting. *IEEE Journal of the Electron Devices Society*, *5*(1), 7–17. doi:10.1109/JEDS.2016.2619908

Chen, Y. N. (2014). Ultra-low voltage mixed TFET-MOSFET 8T SRAM cell. *International symposium on low power electronics and design*, 255–8.

Chen, Y. N., Fan, M.-L., Hu, V. P.-H., Su, P., & Chuang, C.-T. (2013). Design and analysis of robust tunneling FET SRAM. *IEEE Transactions on Electron Devices*, *60*(3), 1092–1098. doi:10.1109/TED.2013.2239297

Datta, S., Liu, H., & Narayanan, V. (2014). Tunnel FET technology: A reliability perspective. *Microelectronics and Reliability*, *54*(5), 861–874. doi:10.1016/j.microrel.2014.02.002

Fan, M.-L., Chen, Y.-N., Su, P., & Chuang, C.-T. (2016). *Challenges and designs of tfet for digital applications. In Tunneling Field Effect Transistor Technology* (pp. 89–109). Springer.

Gupta, N. (2016). Ultra-Low-Power Compact TFET Flip-Flop Design for High Performance Low-Voltage Applications. *IEEE Symposium on Quality Electronic Design.*

Imenabadi, Saremi, & Vandenberghe. (2017). A Novel PNPN-Like Z-Shaped Tunnel Field-Effect Transistor With Improved Ambipolar Behavior and RF Performance. *IEEE Transactions on Electron Devices*, 64.

Khatami, Y., & Banerjee, K. (2009). Steep Subthreshold Slope n- and p-Type Tunnel-FET Devices for Low-Power and Energy-Efficient Digital Circuits. *IEEE Transactions on Electron Devices*, 56.

Kumar, M. J., & Sahay, S. (2016). Controlling BTBT-induced parasitic BJTaction in junctionless FETs using a hybrid channel. *IEEE Transactions on Electron Devices*, *63*(8), 3350–3353. doi:10.1109/TED.2016.2577050

Lahgere, A., & Kumar, M. J. (2017). A Tunnel Dielectric-Based Junctionless Transistor With Reduced Parasitic BJT Action. *IEEE Transactions on Electron Devices*, *64*(8), 3470–3475. doi:10.1109/TED.2017.2713423

Lee, Y., Yoon, D., Kim, Y., Blaauw, D., & Sylvester, D. (2013). Circuit and system design guidelines for ultra-low power sensor nodes. *IPSJ Trans Syst LSI Des Method*, *6*(2), 17–26. doi:10.2197/ipsjtsldm.6.17

Lee, Y., Yoon, D., Kim, Y., Blaauw, D., & Sylvester, D. (2013). Circuit and system design guidelines for ultra-low power sensor nodes. *IPSJ Transaction on System LSI Design Methodology*, *6*(2), 17–26. doi:10.2197/ipsjtsldm.6.17

Li, Liu, H., Wang, S., Chen, S., & Wang, Q. (2017). The Programming Optimization of Capacitorless 1T DRAM Based on the Dual-Gate TFET. *Nanoscale Research Letters*, *12*(1), 524. doi:10.118611671-017-2294-3 PMID:28875269

Li, W., Liu, H., Wang, S., & Chen, S. (2017, March). Reduced Miller Capacitance in U-Shaped Channel Tunneling FET by Introducing Heterogeneous Gate Dielectric. *IEEE Electron Device Letters*, 38.

Li, W., Liu, H., Wang, S., Chen, S., & Yang, Z. (2017). Design of High Performance Si/SiGe Heterojunction Tunneling FETs with a T-Shaped Gate. *Nanoscale Research Letters*.

Li, X. (2014). RF-powered systems using steep-slope devices. *Proc. IEEE Int. New Circuits Syst. Conf.*

Liu, Li, Vaddi, Ma, Datta, & Narayanan. (2014). Tunnel FET RF Rectifier Design for Energy Harvesting Applications. *IEEE Journal on Emerging and selected topics in Circuits and Systems, 4*, 400 – 411.

Makosiej, A., Gupta, N., Vakul, N., Vladimirescu, A., Cotofana, S., Mahapatra, S., ... Anghel, C. (2016). Ultra-low leakage SRAM design with sub-32 nm tunnel FETs for low standby power applications. *Micro & Nano Letters*, *11*(12), 828–831. doi:10.1049/mnl.2016.0442

Makosiej, A., Gupta, N., Vakul, N., Vladimirescu, A., Cotofana, S., Mahapatra, S., ... Anghel, C. (2016). Ultra-low leakage SRAM design with sub-32 nm tunnel FETs for low Stand by power applications. *Micro & Nano Letters*, *11*(12), 828–831. doi:10.1049/mnl.2016.0442

Mallik, A., & Chattopadhyay, A. (2012). Tunnel field-effect transistors for analog/mixed-signal system-on-chip applications. *IEEE Transactions on Electron Devices*, *59*(4), 888–894. doi:10.1109/TED.2011.2181178

Mookerjea, S., Krishnan, R., Datta, S., & Narayanan, V. (2009). Effective capacitance and drive current for tunnel FET (TFET) CV/I estimation. *IEEE Transactions on Electron Devices, 56*(9), 2092–2098. doi:10.1109/TED.2009.2026516

Morris, D. H., Avci, U. E., & Young, I. A. (2015). Variation-tolerant dense TFET memory with low VMIN matching low-voltage TFET logic. *Symposium on VLSI technology*, T24–T25.

Nirschl & Wang. (2004). *The tunneling field effect transistor (TFET): the temperature dependence, the simulation model, and its application.* IEEE.

Patel, H. N. (2016). Optimizing SRAM bitcell reliability and energy for IoT applications. *17th international symposium on quality electronic design (ISQED)*, 12–17. 10.1109/ISQED.2016.7479149

Rasouli, S. H. (2006). *Double Edge Triggered Feedback Flip-Flop in Sub-100nm Technology*. ASP-DAC.

Sahay & Kumar. (2016b). Realizing efficient volume depletion in SOI junctionless FETs. *IEEE J. Electron Devices Soc., 4.*

Sahay, S., & Kumar, M. J. (2016a). Controlling L-BTBT and volume depletionin nanowire JLFETs using core–shell architecture. *IEEE Transactions on Electron Devices, 63*(9), 3790–3794. doi:10.1109/TED.2016.2591588

Saripalli, V. (2011). Variation-tolerant ultra-low-power heterojunction tunnel FET SRAM design. *Proceedings IEEE/ACM international symposium on nanoscale architectures*, 45–52. 10.1109/NANOARCH.2011.5941482

Saurabh & Kumar. (2012). Novel Attributes of a Dual Material Gate Nanoscale Tunnel Field-Effect Transistor. *IEEE Transactions on Electron Devices, 58*(February), 1023–1029.

Strangio, S., Palestri, P., Esseni, D. A. V. I. D., Selmi, L., Crupi, F., Richter, S., ... Mantl, S. (2015). Impact of TFET unidirectionality and ambipolarity on the performance of 6T SRAM cells. *IEEE J Electron Devices Soc, 3*(3), 223–232. doi:10.1109/JEDS.2015.2392793

Taheri, S., & Yuan, J.-S. (2017). Security Analysis of Tunnel Field-Effect Transistor for Low Power Hardware. *International Journal of Computer Science and Information Technologies, 8*(2), 271–275.

Taur, Y., & Ning, T. H. (2013). *Fundamentals of modern VLSI devices*. Cambridge University Press.

Uygar, E. (2014). *Tunnel Field-Effect Transistors: Prospects and Challenges. Journal of Electron Device Society.*

Wang, W. (2014, January). Design of U-Shape Channel Tunnel FETs WithSiGe Source Regions. *IEEE Transactions on Electron Devices*, 61.

Yang, Y., Tong, X., Yang, L.-T., Guo, P.-F., Fan, L., & Yeo, Y.-C. (2010). Tunneling field-effect transistor: Capacitance components and modeling. *IEEE Electron Device Letters, 31*(7), 752–754. doi:10.1109/LED.2010.2047240

Yang, Z. (2016, July). Tunnel Field-Effect Transistor Withand L-Shaped Gate. *IEEE Electron Device Letters*, 37.

Chapter 3
Customizable Multipurpose Nodes Based on Internet of Things for Various Applications

Shubham Tiwana
Independent Researcher, Coventry, UK

Vaibhav Mathur
UPES, Dehradun, India

Sushabhan Choudhury
UPES, Dehradun, India

ABSTRACT

It is a common saying that science is just a philosophy without precise engineering. Available monitoring systems for real-time monitoring of space, goods, or life stocks are either too complex or too costly for the majority of the population to operate so users have to revert back to visual inspection, which results in increased human effort. The proposed model works on internet of things-based technology for more accurate and secure real-time-based monitoring made accessible to the general public. Each node primarily contains an entire network of electronic sensors to measure the environmental parameter. Due to its compact size, this device is capable of deployment over remote areas. Older modules were mostly manual and could only be engaged for a singular task, but the proposed network of nodes is fully customizable and with minor changes can be used for a task that is entirely different from the other.

INTRODUCTION

Wireless sensor network is the main part of any IoT based research. They are the most effective way to fetch and transmit data on real time basis, in accurate manner. Due to their ubiquity, the data obtained on the cloud can be very useful for research and census purposes. Network used here for transmission is secured. Various other ways of fetching data and transmission can be used. Zigbee and Bluetooth

DOI: 10.4018/978-1-5225-9574-8.ch003

connectivity is the most common for such purposes. With the help of this technology, we can monitor and visualize the conditions of any of the remote areas which could prove to a great boon for human society. The secured data transmission also reduces the chances of stealing and misuse of data to the lowest possible level (Chakrabarty, 2016). This concept of data collection through the sensor based embedded system, can also be used for other useful purposes, such as water treatment. In USA, in the last few decades, with the help of such sensor and SCADA based system, the authorities have proper analysis of data. The sensors are placed at a multiple locations and form a network which is useful for determining the condition of water in real time basis at multiple locations. Without this system, this task would require manual sample collection and manual comparison with nominal values of water from time to time. Due to such systems, water related diseases and problems in the USA have been decreased and complete analysis of data is possible (Koo, 2015). Role of IoT based systems is also very crucial in other applications such as home automation systems. The system could reduce the human effort and is ideal for a coffee making world. From basic light, fan automation to more complex appliances like heating or cooling can all be controlled using an IoT based system. They can be efficiently used for multiple applications, which require real time attention. For home related uses, these systems can be effectively use normal Wi-Fi connectivity. The result of the process can be readily accessed through online servers or Android based applications. Customizable feature of the system makes them to be used at any place like the Gardening, intruder detection, fire alarm, Gas detection etc (Pirbhulal, 2017). These kind of reliable systems are also very useful and efficient regarding the safety angle. Such IoT systems are capable to recognize the unknown intruders, with the help of their infrared sensors, and unique Radio Frequency identity distribution. The presence of unwanted element would be recorded by such elements with respective methods. This information will then be transferred accurately and instantly to servers and notifications to the user and local authorities can also be generated by the system. The size and cost of the system would be so small that it can be used by anyone, even without the prior knowledge of IoT or technical knowledge. This can be used as an anti-theft system, in the house. Due to its customizable characteristics, it can be installed anywhere, and can be put in any shape and size according to the requirements of the user (Kodali, 2016). Due to the secured network and protocols, we can also connect these systems through our Smartphone. If Wi-Fi is not providing good network, then we can also provide the ZigBee connectivity. The ZigBee network can be in star or delta formation and one end of the network could access the Internet. Every other nodes would then record their individual parameters and transfer the data accordingly to this last node where it could be uploaded on the net. This advantage of system makes it ideal for a conditions like forests, mountainous regions or villages (Pavithra, 2015). To connect the appliances and the automation system to the main Web, many options are useful and feasible. In case of network failure of Wi-Fi or Zigbee, LTE network can be used. With the help of the GSM Module, effective networking can be provided to the system. 5G connectivity nowadays can play good role to handle multiple IoT nodes and other electronic appliances. Due to presence of one of these readily available methods the overall implementation of the system is much easier so that it can be accessed and implemented from any part of the world. The overall system would require only a little bandwidth for data uploading which would not hinder other uses of the internet connection. The data uploaded on the servers can be accessed from anywhere from time to time, this could also be replaced by other methods like instant notification when the system detects some unusual data parameter (Mandula, 2015). Strong security is an important factor to operate a network. In case of some Malware or virus in the network, whole working and dependency on the home appliances and home security can fail. There are many kind of approaches to infiltrate a Wireless Sensor Network. Therefore avoiding such situations and handling the network securely is an important task. In today's world, security is an important while transferring data because of personal data presence, unwanted elements can poach the data which could then be used for non-ideal usage of data. This problem can be resolved by proper authentication of the data by sender and receiver. Asymmetric Elliptic curve cryptography is one of the solutions which could eliminate the possibility of poaching of data (Santoso, 2015). For the usage and requirements of industries also, the IoT based nodes will be proven very effective. With the help of infrared sensors and other modern electronic sensors, they can observe and monitor the movements in the restricted areas in real time. The security of the area will be maintained at all times and the results could be made available on online servers. The user can access the remote surveillance from any part of the world and perform any necessary actions. Regarding the industrial accidents, they can play a vital role, by monitoring the levels of oxygen, sulphide, hydrogen, LPG and many other gases. Effective monitoring of such parameters would help in saving lives and maintain the overall productivity of the Industry. At the control room, zigbee transceiver module could be used for the receiving purpose and Bluetooth module could be used for the transmission purposes (Choudhury, 2015). For the safety regarding issues in the house and especially for the old people, these nodes can be deployed at the entry and exit points of the house. Safety of senior citizens nowadays is a big cause of worry, and especially for those, who live alone in the house. These nodes can work for their safety of them. Due to their customizable properties, these systems can be fitted with the sensors, concerned directly to the anti-intruder purposes. They can be deployed at the points of the house, where the probability of intrusion is maximum. The sensor network will detect parameters and transfer the data to microcontroller which in turn will transfer the data to an online server which can be readily accessed by the family of the senior citizens. This can transmit signals directly to nearest police station, and also put on the warning buzzer. The system implementation and maintenance require no prior technical knowledge therefore can be utilized by senior citizens with ease. The system makes the task of real time monitoring of an area much easier than the available solutions (Agarwal, 2016). The implementation of such research will also be crucial for the domain of civil engineering and the construction site safety management. It can provide computational intelligence to the entire safety network, active over the site. The system require very small space for deploying and can monitor much larger distances through a single node only. The network ofsuch nodes can also be connected to Zigbee if the internet connectivity is not available at that location, one of the nodes can act as gateway to the entire network and upload the data in real time basis on online servers. This will provide real time based activity recognition facility, by smart network, equipped with sensors, controlled by a single processor unit. The device can implement both tasks. It can scan for human presence in the high risk areas, and also check for the overloading in the machines been used over there. The system thus helps in preventing industrial accidents and save precious lives. The cost of implementation of system is also negligible compared to the task performed by the model. The system also does not hinder the productivity of the industry and can be implemented and maintained without prior technical knowledge. The result of the surveillance would be readily available on the net and can be easily accessed from any part of the world (Rossi, 2019). Nowadays, in the era of hi-tech buildings, and skyscrapers, safety is also a very major concern. Threat of fire, due to overloading and several other issues always show its presence. In such accidents, timings are very important. With the help of the proposed device, real time transmission can be performed to the local rescue crew. The network of sensors will detect any unusual levels of smoke or temperature variations and send the data to the microcontroller. The microcontroller will process the individual data from all the sensors and perform athematic operations to get the final parameters in standard units. This data will then be transferred over to web and can be seen in form of graphical forms which are user friendly and any abnormalities can be easily intercepted. An alarm system can also be controlled with the device with notification sending facility. The system would be cheap and reliable which will provide safety to the building (Okokpujie, 2019). Apart from industrial accidents, one of the major cause of worry is the accidents that happens during the extraction of natural resources. Workplaces like the coal mines are amongst the high risk zones. In the past years, due to frequent accidents, requirement of pre warning systems was felt. The sensor network, with reliable and secured network has proven an effective solution. Wireless sensor network has helped the authorities multiple times, to predict such accidents before and always issued a pre warning, with help of the electronic sensor, and always active network. China is the most mines –accident prone country of the world, facing huge life losses, in the past years. WSN technology not only made the lives of common people easier, but also made its image as a life saving technology. Due to usage of very elementary electronic sensors and components, the issue of power consumption never came as hurdle. The sensors would be measure the level of oxygen and harmful gases in the environment and transfer the data over to a microcontroller. The microcontroller would then process this data and after mathematical operations convert the parameters into individual standard units. The microprocessors will convert this data into packets and transfer to the master node present at the ground level through Wi-Fi modules. The master node at ground will combine the data from all nodes and upload to online servers. The monitoring of this data can result in timely evacuation and saving of lives of workers (Yinghua, 2012). In the same mining industry, there was another requirement, of wireless sensor networks or the WSN came up. This time, the concern was not only from the safety measure, but it was also for the environment related issue. Tailing disposal is very crucial for the mining industry, anywhere in the world. Tailing dam failure is the most dangerous disaster in the mining industry. Huge loss of lives and money have taken place did by the tailing dam failure incidences all over the world. In this accident, weather also plays a key role. Deformation of tailing dam is been done in maximum cases by the suddenly changed weather. Therefore, there is a requirement of the IoT and cloud computing based system, which can monitor not only the water levels of the tailing dam, but can measure the weather of the area, in real time basis. The system is needed to observe swiftly the minute change in the weather. The cloud server has an important function here. The officials or the authorities of the mining firm, or the government authorities, and even the safety crew can monitor the conditions of water level and dynamically changing weather in an ubiquitous fashion. The access of this data, retrieved on the cloud, which is been fetched on real time basis, is globally, and can be seen from anyplace in the world, and at any time. This system has been installed at many tailing dam areas and has proven itself successful. The network of nodes combined with electronic sensors will acquire the individual parameters in real time basis and after conversion in standard units upload the data onto online servers from where the overall monitoring and productivity of the system can be done. The system can also be connected to a control room where the change in parameters can be easily seen and intercepted. This monitoring of buildings and dams would reduce the human efforts in the field and help in conservation of natural vegetation as well as human lives (Sun, 2012). In the field of agriculture also, the usage of these systems is been recommended. Crop monitoring and soil monitoring is very crucial for the farmers to get the exact information about the condition of their crop. Agriculture of our country contributes a huge amount in the GDP. Due to lack of proper monitoring and other scientific methods, farmers are losing hope regarding the crop production. In this digital era, e-farming is been heavily recommended and required. Wireless sensor networks, could help the farmers and the agriculture industry, to manage the farms and fields more effectively and easily. The requirement is there to deploy the network of sensors, with an easily implementable design, which can measure the various parameters, which affect the crop growth and crop security. The proposed model fulfils these requirements by measuring the parameters like soil moisture, water level, sunlight, oxygen and carbon dioxide levels on real time basis. The system can be controlled by anyone even without the technical knowledge. The network of sensors and microcontrollers will convert the measured digital voltages into appropriate standard units. Only requirement here will be the presence of an active Wi-Fi network or a Bluetooth network to connect the sensor system. The final processed data will be uploaded onto a cloud server and will be made available to farmers through android based applications or websites directly and easily. The data can also be converted into graphical form for better understanding by the farmers. The system is compact and can be deployed at any desired locations (Mohanraj, 2016). Requirement for the Wireless sensor technology in field of global food security is heavily required. Usage of Radio Frequency identification distribution is also recommended along with the electronic sensors, and network connectivity. The platform of Internet of things here is very accessible, affordable, and interactive for the users to get easy to use. The cloud computing technologies used along with the data collection methods also provide feasibility to the agriculture industry. The live condition of crops can be viewed directly to the cloud server. We have live examples of many countries all over the world, where the combination of multiple technologies, Wireless sensor network, Internet of things, and cloud computing have changed the lives of the people, associated to the agriculture industry. Even many Asian countries like the China, Japan and even India, people started adopting the latest technologies, to make the vegetation and crop farming more effective, smart and secured. And also, the network used here for the data transmission for the sensor network is much secured and there is almost no possibility to temper the data, which shows the condition of the crop. The system will require handful of sensors and microcontrollers and would provide the identification of batches of crops. The sensors can monitor if the environmental parameters are not suitable for storage of the crops and notify the users when the levels exceed of drops the nominal ranges which would save the products from becoming spoiled goods (Patil, 2012). Cloud computing and big data analytics are one of the most helpful technologies regarding the agricultural sector. With the help of the electronic sensors, which is an integral part of the internet of things technology, provide the real time data to be fetched and been collected to the cloud. The big data analytics can also be used for the analysis regarding the crop growth, agricultural growth, and various other analysis and census purposes. The sensors used here in the motes are also able to provide the accurate information regarding the soil fertility, soil moisture, and other parameters, concerned to growth of the crop. Professionals of the agricultural sector and the agricultural scientists and even various environmentalists can analyze the data available on the cloud, to identify the various impacts, which are affecting the farming of the particular area. Various other challenges are affecting the fertility of the soil and the farming. One of those factors is global warming. Our environmentalists can easily analyze the data of the agriculture sent by the wireless transmission, to the cloud server, and consult the farmers regarding the necessary steps required to be taken, so that the efficiency of crop production by the farmers can be improved effectively. A series of sensors work together to measure the surrounding parameters and transfer the respective voltages level to connected microprocessor. The microprocessor after performing athematic operations convert these voltage levels into respective standard units. The converted data provides the real time parameters of the environment and provides better results than the traditional methods like visual inspections. The final data is uploaded on the cloud servers from where it can be easily accessed by the farmers without the prior technical knowledge of microcontrollers or electronic sensors (Gill, 2017). The sensor based system is also helpful to avoid and pre warn the natural disasters. An important fact to be observed here is that some of the disasters are quite predictable. Timely measure of parameters like soil moisture, rainfall level, sunlight, temperature and humidity can help predict and prevent such disasters. Problem of landslides during the monsoon and heavy rainfall is a life ailing accidents, especially in the northern parts of India, and state of Kerala. Although we currently have various systems and methods to determine the amount of rainfall, and local administration still have to close some roads after heavy rain. But sometimes, an unexpected landslide can occur in a civilian area. The IoT enabled system successfully can be deployed here also, for timely detection and prevention of these accidents before their occurrence. These systems are composed with multiple electronic sensors and cloud computing technology. These systems are capable of measuring various parameters that can indicate the possibility of a landslide before the accident. Moisture content in the soil, slope of the hill, total rainfall, temperature etc. are the factors that can help the professionals to receive the early warning before the accident. All the data can be monitored live in the control room by the cloud server. Here role of the network provided to the IoT system plays a very important role. The connectivity can be either Wi-Fi or Bluetooth so that the data can be received constantly without any disturbance so that the officials of the local administration can monitor the data on a real time basis, and the necessary steps could be taken to close the road and avoid accidents. The sensor network can be easily deployable in such areas with much ease and can provide real time data monitoring of the area by measuring the environmental parameters with respective electronic sensors and transferring the data to microprocessor for further mathematical operations. The microprocessor acts as hub for the data and convert individual data in their respective standard values. This data is then relayed to master node using Wi-Fi modules. The master node compiles all of the individual data from every node and timely uploads the data onto the online servers. The data on the cloud server can be readily accessible to local residents and authorities for taking timely action which would prevent the loss of precious human lives (Ramesh, 2009). Current techniques are also very effective for monitoring the conditions of the steep slope and to predict the occurrence of landslides or other natural disasters. But each of them have their own limitations. One of the methods is the remote sensing technique. This technique includes space borne, aerial and terrestrial survey of the slope and soil. The technique provides timely monitoring of environmental parameters but this technique is a costly affair, and discontinuity in data acquisition, due to resolution issues make it less effective for implementation at much larger levels. Another such method is the ground based techniques. In this technique, surveying method is been mainly used. Some of the sensors like accelerometers, inclinometers, strain gauges, wire extensometers, GPS and pressure transducers are used. The main issue with this technology is the inadequate method of data transmission, like the GPRS modem, wired connections, data loggers etc. This technology of data transmission is not helpful to deliver the data in real time in a precise way. But the embedded system, fused with the IoT technology and cloud computing described above, is completely suitable for this purpose. The sensors measure the individual parameters in form of digital levels and send this data to a central microprocessor which applies mathematical operations and convert the data from voltage levels into preferred standard units. The microprocessor then send this data to online servers in form of packets. The security of data is maintained by proper encryption and authentication of data. The data send by the electronic sensors here are very precise. The data sent here is on completely real time basis. Many lives can saved from this system, especially in the remote areas of the country like the Uttarakhand, JK, Himachal Kerala etc. The system helps in predicting the occurrence of natural disaster and is better than the other methods in terms of cost, reliability, precision and simplicity (Rosi, 2011). In the recent years, dependency towards the electronic systems have be observed in the field of forest department and anti-poaching solutions. An animal behavior is always a point of focus of the scientists, doctors, and the forest guards. It is impossible to be present with the animal at every point of time. This challenge is been resolved by the IoT technology. With the help of small sized electronic system, the doctors, forest guards and other authorized people can track the concerned animal on real time. The data fetched by the device, wearing by the animal, is been sent to the cloud at every point of time. But there is another problem felt that connection of the Wi-Fi can be lost anytime, as it is a forest region. But in the customizable motes, we can connect the zigbee module, which use the radio frequency technology that ensures the secured and strong connectivity, between the embedded system and the cloud server. Similarly, we can also monitor the health and behavior of the animals by this system. Already these systems are in use in the forests of Kenya with the help of Cisco. Receiving the real time data at the cloud server to the experts makes it helpful to take swift decisions regarding animal health at correct time. This system also acts as security system for animals against the intrusion of illegal poaching. The motes are been deployed at various places of the forests to detect the illegal and unauthorized presence. The sensors in each motes records the parameters like the location of the animal or intruder. This information is later transferred over to its counterpart microprocessor which converts the digital voltage levels into physical parameters after applying mathematical operations. The converted data is later converted into packets and send to cloud servers. The authorities can then access the data and monitor the area more easily and effectively. The security of data is also maintained by appropriate encryption and authentication of the data at transmitter and receiver. The network can be easily implemented to large areas with low maintenance requirements (Liu, 2015). This system can play an important role to preserve the endangered species. And not only the animals, but this system can also be used for the protection of forests, against the illegal cutting of trees. These systems can be deployed in the area, where the forest guards need to monitor the endangered animal specifically. The main advantage of this device is that due to its customizable properties, we can easily change its design, according to the requirement of the user. Same here in this case, the device can be designed specifically according to the need of the forest guards. These systems can play important role to preserve the trees and other natural resources of the forests, which are been continuously harmed by the external factors, like those of poachers, and the tree cutters for individual profits. The zigbee module, present in the system will ensure the connectivity of the device with the cloud server, so that proper and regular transmission of data can be ensured. Tempering with data is not possible due to proper encryption and authentication at every stage of transmission. With the help of this, not only forest guards, but the animal doctors will also be able to determine and analyze the behavior of the animals. Changing behavior of animals, with changing conditions like weather, external disturbance, health issues etc. can be viewed and accessed easily. The sensors will sense individual parameters in terms of quantized voltage levels and regularly send the data to the connected microprocessor. The microprocessor on receiving the data converts it into required physical parameters standard unit. This processed data is in turn transferred over to the online servers from where it can be plotted in form of graphs. The system is reliable and cheap and can be deployed easily to form a network of nodes that can cover an entire forest (Guo, 2015). In the sector of the poultry farms, there is a requirement of such innovations which can monitor environmental parameters precisely and effectively. Many people lose their lives due to lack of hygiene and other unidentifiable diseases due to absence of a reliable monitoring system which can measure and notify when the conditions of an area is dropping the nominal value. There is a requirement of monitoring of the health and behavior of the chickens and pigs, which are been stored in the poultry farm . The IoT is amongst the most trending and emerging technologies, which created a successful impact towards the resolution of these issues. The health regarding parameters were been successfully fetched by the electronic sensors and then these parameters, like the moisture contents, temperature, smog etc. are been transmitted to the cloud server. The health regarding parameters and several other factors can be viewed at the cloud server. Farm owner, and even the doctors can monitor the condition of pigs and chickens been kept in the farm. If required, government authorities can also check the condition of the farm form anyplace in the world, and at any time. This can ensure the safety of both the animals, and the customers, which are going to buy the pork and chicken, so that no disease can be transmitted. The sensors sense the level of harmful gases in the environment in terms of digital values and transfer this data to the connected microcontroller. The microcontroller combines all the individual data from the sensors and convert them into respective standard units. This data is then converted into packets and transferred over to cloud servers using Wi-Fi modules. The data at the online servers is readily available and accessible to user from any part of the world. The timely monitoring of such parameters can save the lives of chickens and pigs and also human lives if required. With the help of such system, transmission of diseases can be controlled effectively. The system is cheap and effective solution for real time monitoring of the poultry farms (Mahale, 2016). The research proposed here can be proved very effective in case of weather monitoring and the analysis regarding the global warming. Using the temperature, humidity, carbon monoxide, and other environment concerned sensors, a complete database can be collected over the cloud server, so that research and census purposes can be performed easily. The device is also capable to measure the parameters, which are able to indicate the pollution content of the area. Smoke sensors can be able to express the amount of smoke in the atmosphere. All the data retrieved by the device is in real time. This data and the network used for the data transmission purpose, is very secured, and can be connected easily to the Wi-Fi network, or the Bluetooth network, depending upon the location of deployment. If this system is deployed on a busy street, the pollution and levels of other harmful gases can be easily identified. The sensors measure the amount of harmful gases in environment in terms of voltage levels. This data is then transferred over to microprocessor which compiles all the individual data from every sensors and convert them into respective standard units after applying algebraic operations. This final data is sent to cloud servers in through data packets. The information at the cloud servers is easily accessed and available to administrator users which can analyze the data through numerical or graphical form (Ram, 2016). Internet of things will be the most effective solution to measure the pollution per day and in terms of cost also. Due to usage of very elementary sensors and basic electronic circuits this device will be the best in terms of cost . The network used here is a simple Wi-Fi network, with the help of which, this system can be easily connected to the Wi-Fi commonly used at the home. But the access to the device is restricted to administrator users only . The sensor network will measure the pollution level in the environment in terms of digital voltage levels . This data is then transferred to connected microprocessor which in turn compiles the individual data from every sensors and convert into respective standard values . These values are obtained using mathematical operations wrt to a measured value . The standardized values are then transferred over to cloud servers through Wi-Fi modules . The data on the cloud servers are secured and can be accessed through any part of the world . The transfer of data is secured by encryption and authentication thus eliminating the possibility of poaching of data by undesired elements for unfair uses (Xiaojun, 2015). In the era of online shopping, IoT is the most effective and reliable way to track the logistics. The unique Radio frequency identification distribution tag of each carton helps to relocate its position, even during its travelling. The position of the logistic is uploaded to the cloud server immediately with the help of the network connectivity. Not only for logistic tracking, but when food and vegetation are been travelling, their conditions are also viewed graphically on the cloud server. The sensor network will be able to sense any unusual and unwanted change in the parameters by sensing the parameters in terms of voltage. This data of voltages are then send to respective microprocessors. The microprocessors will then compile and convert the voltages into respective parameters. This conversion is made through mathematical or algebraic operations taking a standard value in account. This final data is then sent to remote servers and later can be accessed through android based applications or web browsers from any parts of the world . Thus the motes discussed above are very helpful to sense the conditions of the vegetables and fruits travelling, and then send to the appropriate authorities, so that in case of any unwanted presence of any gas, or increasing pressure which could damage or spoil the goods, the authorities can respond to the situation and take appropriate decision (Qu, 2016). In ancient times, food related diseases have a strong bond with humans. Only Europe based food illness affects 1% of the world population. In 2011 16.7% population of USA got ill due to food related issues. Present day, long distance animal and vegetable transportation has been increased. This not only increased the risk of infections, but also gave antisocial elements an opportunity to conduct biomedical attacks against common people. Food related contamination and diseases also hits the economy of the nation. The proposed network of nodes will be able to detect any unusual value of parameters which could either be dropping below or rising above the nominal range. The sensor network detects the parameters such as temperature, humidity, oxygen and carbon dioxide levels. The sensing in done in form of quantized levels of voltages. These information of voltage levels is then transferred over to microprocessor. Microprocessor in turn compiles all the individual data from every sensors and convert the data into standard physical values through mathematical operations, further athematic operations can also be applied for extra conversions. This processed data is then transferred over to cloud servers. The security of data is also kept in mind by encrypting and authentication the packets of data transferred and received. The data can be accessed using android based applications or web browser in numerical or graphical form in real time basis (Bosona, 2013). Tremendous changes are also observed during improper storage and Transportation of potatoes like changes in specific gravity, starch content, and sloughing. The improper storage or monitoring of products results in damaged or soiled goods which reduces the overall profitability and can also result in loss in some cases. Thus there is a need for system which can monitor these environmental parameters in real time basis and is affordable for household implementation. Proposed IoT based motes fulfil these needs and thus are ideal for such situations. The data is easily available for authorised users which can take necessary steps when container parameters are not in required range (Whittenberger, 1951). The old and conventional methods and techniques used were insufficient to save the vegetation during storage time and the transportation period. Dehydration & fermentation were ancient techniques, but thermal processing and freezing technologies were developed in 20th century. But these methods also had certain demerits. In case of Freezing, due to low thermal exposure, damage to tissues of vegetation begins, causing the nutritional loss to the eatables. The Internet of things based solution overcomes these demerits due to absence of requirement of dehydration or freezing due to the presence of real time data monitoring system by electronic sensors. The sensors sense the environmental levels and send them to microprocessor. The microprocessor convert the voltage levels into physical parameters and combines the result to transfer it on cloud servers (Barrett, 2012). Due to High costs, loss of aroma and texture, method of drying was also not good option for conservation of fruits and vegetable goods. The method of drying is not useful in today's world for conservation of goods. Therefore a cheap and affordable system is required for monitoring of fruits and vegetables. The system should be affordable to be implemented at household level but could be upgraded also for industrial purposes. An IoT based system overcomes these difficulties and have proven to be a great alternative for real time based monitoring of goods (Sagar, 2010). In absence of modern transportation methods, almost half of the vegetation is received rotten during standard arrival time.

Thus there is a requirement for using modern and scientific technology for not only storage but also for the transportation methods to store and deliver the vegetation in eatable form. The sensor network monitors the parameters in real time basis without using extreme methods like freezing or dehydrating so the vegetables and fruits can be transported in eatable form. The microcontroller controls and compiles all the data received by individual sensors which was in form of voltages. The microcontroller then converts the voltage levels in standard physical units by mathematical operations. The sensors can range from temperature sensors to humidity sensors to moisture sensors to oxygen level sensors to carbon dioxide level sensors to air quality sensors. The sensing of environmental parameters does not require to change the properties of goods thus is the ideal solution for such methods (Zhang, 2011). One more factor is required to deliver and preserve the fruits and vegetables. This factor is the monitoring of these eatables and to control the atmosphere, associated to the cold storage parameters, like the temperature, humidity, thermal exposure, water requirement, presence or isolation from other gases etc. To execute this solution, there is a requirement of a real time monitoring system, which can fetch and upload the current parameters to a centralized place (which can be a server/cloud/mobile app) which can then in turn be used for proper monitoring of fruits and vegetables. The sensor will work in parallel to each other to observe the change in the environmental parameters and send the results to the microprocessor. The microprocessor handles the conversion of data send by sensors to physical standard units through mathematical algorithms. The IoT based model is cheaper than the traditional methods and provides better resolution and monitoring of goods than other available methods. The security of the data packets is also a priority and a number of algorithms can be applied to encrypt and decrypt the data by authorised personals only (Tsang, 2017). The research proposed here is based on the Wireless sensor network. A network of sensors are connected to a central microprocessor. The sensors are tasked to sense respective parameters and transmit the observations to microprocessor. The factors such as Data privacy (to prevent receiving of data in wrong hands), Data integrity (to avoid losses or errors in parameters), and Data freshness (to ensure real time data) must also be taken into account when designing such a system. An IoT based sensor network with layers of encryption and decryption capabilities at sender and receiver respectively overcomes these problems. The appropriate authentication algorithm can also be included in the implementation to increase the overall security of the model (Gope, 2016). In today's Android based era, some basic observations and activities associated with humans can be easily sensed by very fundamental sensors. These electronic sensors are low cost, small size and are easy to operate. The environmental sensors like temperature and humidity and rainfall level and air quality levels to distance measuring sensors like ultrasonic and infrared sensors to fire or metal detection sensors to human interaction sensors like fingerprint sensors and voice detection sensors are all examples of wide range of sensors readily available in the market. The sensors helps in reducing the overall human efforts and are basic building blocks in an automated system. The sensors sense the individual parameters in terms of quantized voltage levels . They are very effective and successful medium to observe and record human activities as well as requirements. Even some huge corporate such as Google has also invested an amount of $3.2 billion on Intelligent Temperature Controllers (Fan, 2015). The WSN based system proposed here must consume less energy. Several parameters involved here such a temperature, humidity, GPS coordinates etc. are parameters which could change in very short time. Thus power source used here must be reliable and can withstand by the load of high energy consumption. The IoT based network require less power consumption and can be operated using a power bank for remote deployment or a direct AC power line through a dc adaptor with necessary current ratings . The overall size and power consumption of the model is small and therefore can be easily deployed at any location (Jedermann,

2006). The cloud server used here for the application, is a very reliable storage component. The final result is transferred by microprocessor to the cloud server through Wi-Fi modules . The information stored in cloud server can be accessed by authorised users from any parts of the world in default numerical form or a graph can also be plotted with the respective values for easier understanding of the results . This provides ubiquity to the data and parameters, recorded by the sensors inside the containers, storing the food. The cloud to store the parameters is much secured and the access to it must be the authorized one (Tongke, 2013). This experiment of fusion of various sensors with microcontroller, to connect it with a Wi-Fi network, was done in the early stages of 2000s. The results shown that the system was successfully connected with the external world, even after putting inside the container of vegetables and fruits, whose monitoring was been done. The sensor network sense the individual parameters in form of digital voltages and transfer this data to connected microcontroller. The microcontroller compiles all these data from each sensors and convert the data from digital voltages into respective physical values. The microcontroller then sends this finalized data to an online server through a Wi-Fi module . The data transfers take into account the security of the data by encrypting and decrypting the data at transmitter and receiver side. Only authorised personals would be able to access the data thus eliminating the possibility of poaching of the data (Ruiz-Garcia, 2008). The ESP 8266 or Node MCU module used here, which communicates with the available Wi-Fi using various communication protocols. A complete layer of protocols are been used in the Wireless Sensor Network. It uses Application layer, Transport layer, Network Layer, and Link Layer. Once Wi-Fi is enabled, ESP 8266 acts as part of LAN and can connect with any other component of LAN (Kaur, 2012). IoT technology was been built to reduce the human efforts. This technology aimed to carry out the state of the object upon the Internet network. Due to this, computers will be able to interact directly with the objects, without the help of human. There were many remote locations, not only on earth, but also in our cities, where human access is very dangerous, but their monitoring is mandatory. One such field is the underground drainage and manhole. Several cleaning workers lose their lives to keep these drainage systems clean. For this challenge, IoT is the most suitable solution. The motes, having multiple sensors, interfaced with the Arduino microcontroller, and the network connecting module can be deployed in the underground drainage system. Network connectivity, whether it can be Bluetooth, or Wi-Fi or LTE needed to be provided. The main purpose of these motes is to sense the various parameters, like the mainstream flow, and the water level, which are the main reasons, to affect the drainage. The motes can also measure the amount of gases, and the chocks, so that, before the cleaning workers enter the manhole. In several cases, the drainage workers died, due to heavy presence of gases, inside the manhole. All the data, like the mainstream flow, water level, presence of chocks, and the amount of gases, would likely to be fetched by the connectivity module (Node MCU or XBEE or GSM Module), and will be sent to the cloud server . The cloud server is accessible to the control room, and the higher officials of the local administration . Thus, first of all, with the help of IoT, the administration can get the visualization of what and where is the problem in the underground drainage system. (Muragesh, 2014).

Block Diagram

Figure 1. represents the block diagram of configuration 1. It consists of two LEDs, microcontroller (Arduino Uno), Temperature sensor (LM35), Light sensor (LDR), gas sensor, air quality sensor, touch sensor and infrared sensor with power supply.

Figure 1.

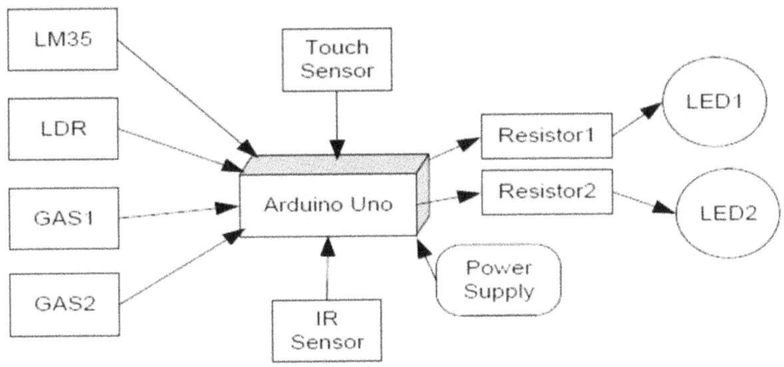

Figure 2. Represents the block diagram of node to sense the data using multiple sensors. It consists of LED, controller(Arduino Uno), Temperature sensor, Soil sensor, Water level sensor, Humidity sensor, Light sensor, smoke sensor, RF modem(XBee) with power supply.

Methodology

The proposed system consists of Arduino Mega microcontroller, 2200mAH power banks, to fulfill the requirement of power supply, ESP 8266 module and Xbee modules, to provide connectivity to the network of Wi-Fi. Sensors and actuators can be used according to the requirement of the user, as the system proposed here is a customizable system. But various sensors can be used here for the sensing purposes, like the light intensity sensor, temperature sensor, humidity sensor, carbon monoxide sensor, Liquefied petroleum gas sensor, smoke sensor, rainfall sensor, soil moisture sensor, Infrared sensor for motion detection, fire detection sensor etc. The selection of sensors for this IoT enabled device can be decided by the requirement of the user. The requirement of active Wi-Fi or the active Bluetooth is always there, so that data fetched by the system can be sent to the cloud server, or even on the android app, without any disruption. The data is been stored on the cloud server in a ubiquitous manner. The data on the cloud can be accessed from anywhere in the world, and at any time. The data obtained on the cloud is been shown in the graphical manner. To control the entire embedded system, which includes the network of various sensors used and the connectivity modules used is been controlled by the Arduino Mega microcontroller. This is a programmable microcontroller, which is been programmed by the C/C++ language

Figure 2.

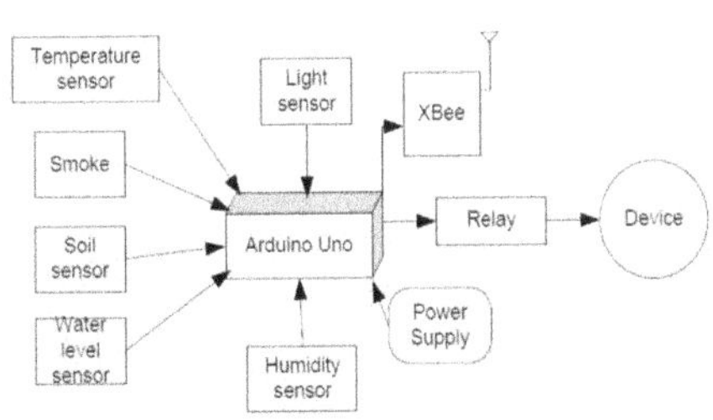

Figure 3. UML Activity Diagram / Flowchart

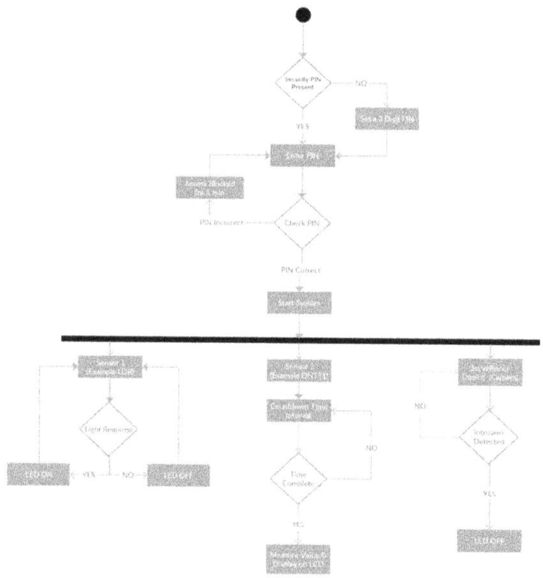

in the Arduino Integrated Development Environment or the Arduino IDE. In the embedded system, various circuits are also been used, using the resistors, pull up resistors, op amps, inductors, capacitors, diodes, for voltage conversion purposes, AC rectification purposes, circuit safety purposes, and output enhancement purposes. The power supply, which is been used here is very reliable. It is a power bank, which have a standard output of 2200 mAH. But as it is been discussed, the power supply can be varied according to the requirement of the user. The user can use it in many ways. They can fix the fire sensor, on the area, which has highest possibility of flammability. They can be used for receiving the early warning regarding the industrial safety. Fire presence can be confirmed by the sensor and warning can be transmitted to the rescue team stations. Wide range of gas sensors can be interfaced easily with the Arduino microcontroller, to monitor the level of gases in the industry, so that the industrial accidents of gas leakage can be detected earlier, and rescue teams can get warning call earlier by the data transmission technology of IoT. Not only for industrial users, but for the farmers also, these motes will make farming easier. Farmers can monitor their crops more easily and accurately, through their mobile app. This time the sensors, which are concerned with the parameters with the agriculture could be interfaced with the Arduino microcontroller . With the help of soil moisture sensor and temperature sensor and humidity sensor and rainfall sensor and infrared proximity sensor to confirm the condition of soil and temperature and humidity and rainfall and presence of a human or animal respectively. Similarly, the local authorities of a country can deploy the landslide detection system, at the slopes, where the probability of landslides is maximum. With the help of those motes, early warning of a landslide can be received, during the unfavorable weather condition. In this case, soil moisture sensor and rainfall sensor and temperature and humidity sensor could be deployed, so that the landslide can be detected and possibly prevented in time. In the control centers of the administration, they have the access of the cloud server, and the officials can identify it beforehand. The graphs obtained on the cloud server, which shows various different parameters, will help the officials to study and visualize the condition of the slope . For the restricted areas, where every human presence is not authorized, at those areas also, these motes can be applicable, and effective. They can be interfaced with the infrared sensor, proximity sensor, and ultrasonic sound sensor, along with a unique Radio frequency identity distribution scanner. Before a person enters the area, these scanners will scan the unique RFID tags, given to only authorize personnel . If scanner does not scan any RFID card, and the other three sensors discussed above confirmed some presence, then the buzzer attached to the system would get triggered. And also, in the security control room, the cloud server will show the presence of some unauthorized access. For the study and census purposes of the pollution and global warming, these motes can be deployed. With the help of the gas sensors and dust sensors and air quality sensors and temperature sensors, the pollution and other gas amounts can be measured and recorded in the cloud server on regular basis. Usage of power banks is been preferred more over here, because of the easier mobility. In many cases, the deployment on these motes can be at a remote location, or at locations, where less space is been provided. In this case, if power banks were used as power supply, then there will be no hindrance in the deployment of these motes. There is a wide range of application, where these motes are most suitable, like forest fire detection, intrusion detection, poultry and farming concerned applications. In these areas, space for deployment is a cause of worry.

- **Simulation**: Before fabricating the actual hardware, simulation on the computer is mandatory. By performing this step, the possibility of damage to the hardware, by the current, due to improper connection in the hardware can be finished. For the simulation purposes, proteus simulation software is been used. Each and every connection is been checked very minutely. The sensors, used here were been successfully working, and were transmitting the data properly.
- **Hardware Design:** All the sensors, which are been required by the user, are connected to the Arduino microcontroller. The ESP 8266 Node MCU, and the xbee module is also been connected to the Arduino. In the embedded system described here, other circuits like the voltage converter, alternating current rectifier circuit, and the other circuits, for safety of embedded system are also been used. The system was transmitting the data to the cloud successfully.
- **Software Design:** The Arduino is the microcontroller, which is been programmed in the C/C++ language, in the Arduino integrated development environment. All the coding and algorithm related tasks were performed here. Although this system is customizable, changes are required in the code, corresponding to each version of the embedded system used here. All the codes, corresponding to each version are been successfully checked, and compiled by the Arduino IDE.
- **Data Analysis:** As per the task of this IoT based system, the data is transmitting successfully and continuously upon the cloud server, with the help of the Node MCU or the xbee module, depending upon the connectivity. The data received upon the cloud is in the graphical form. And it is been received on real time basis. Due to the customizable characteristics, and multi-parameter sensing ability, many factors, and readings can be received on the cloud. Having the ubiquitous characteristics of IoT, the data can be accessed, from any place and anytime. The data received here is analyzed successfully.
- **Troubleshooting and Debugging:** During various stages of manufacturing this system, i.e. hardware designing, coding for Arduino in IDE, software simulation, hardware and circuit connections, and the fabrication process, various errors and technical and logical hindrances came up. All the errors were been taken care of.

Libraries

The codes written in the Arduino microprocessor are one of the most important element of proposed research. Libraries provide extra functionality for use in the code written for the microprocessor in the Arduino environment. Below are some Libraries, which are been used in thr proposed research

- **Liquid Crystal Library:** This library function is used for predefining and providing extra functionality to the LCD Screen of the system. While operating an LCD, user needs to put data into the data registers, corresponding to, what they want to see on the display. This work is been resolved by the LCD library. The LCD contain Enable pin, R/W pins, and 8 data pins. All the working and corresponding timing diagrams are been managed and controlled by this library.
- **DHT11 Library:** To operate DHT11 temperature and humidity sensor, this library is been used. With help of this library, extra features can be provided to sensor. For the DHT11 temperature, it has its own single wired protocol for transferring data. These protocols require that there must be very accurate timing and timing diagram for the transfer of data from sensor, to the microprocessor. These timings and timing diagrams were been managed by the DHT library.
- **MQ 135 Sensor Library:** MQ 135 Sensor, is mainly used for detecting the air quality, and other smoke related parameters. It convert ppm of particle into corresponding voltage. But while transferring that current, it require very precise handling of timing diagram, and conversion techniques, so that proper voltage can be obtained for corresponding ppm. This task is been easily assisted by the MQ 135 library.
- **Fire sensor Library:** Fire sensor is mainly used to detect the flames. It works by detecting the UV flame of corresponding wavelength. Gives logic 1 if detected wavelength above threshold level, otherwise gives logic 0. To control these wavelength detection task, and precise serial communication of the sensor with microprocessor, the fire sensor library is been used by the user.
- **SPI Library:** SPI is a protocol, used for the serial communication between two devices. SPI library control the data traffic, frequency, and clock pulses, to check and update system whether data shifted in and out with rising or falling edge of clock. Also, slave line, master line and serial clock is managed by this library.
- **Wire Library:** This library is majorly used in the I2C communication. With the help of this library, user can synchronize the data transfer between the devices on the I2C bus and it's generated by the master device.

Observations

See Figure 4.

CONCLUSION

The research done here is been done with the vision to implement the internet of things successfully, in a secured manner. Although network security is a point of concern, but overall, the transmission pathway, and the networking gateway is very safe. Due to the customizable properties of the system, its execution in maximum field is possible, and it proves the monopoly, and has its unique place, in many challenges.

Figure 4.

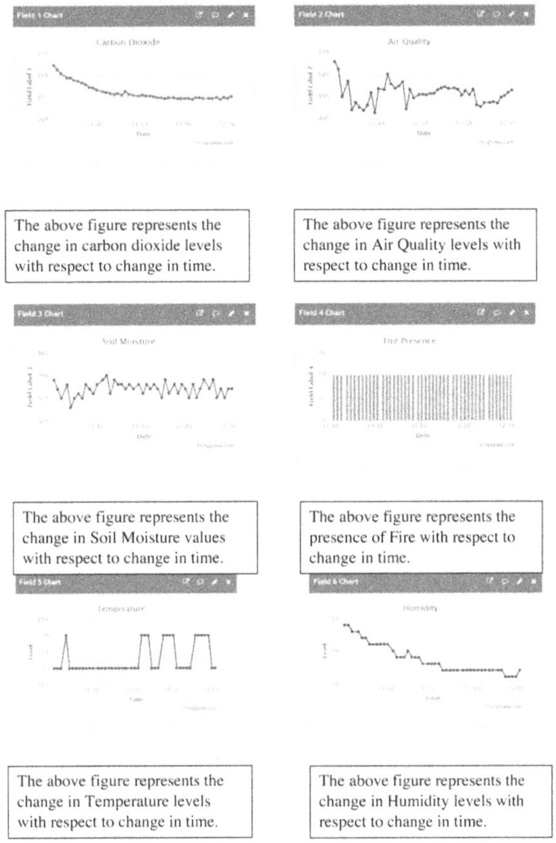

The main target was to take the concerned data, whether it can be temperature, humidity, moisture content of soil, amount of certain gas, motion detection, to a central place, which has the ubiquitous characteristics. The data is being successfully fetched by the embedded system, and provide a lossless transmission to the cloud server, with the help of a continuous network, which can be LTE, or WiFi or even Bluetooth. The device been discussed above helped in successful monitoring of various challenges, like the farming, agriculture, forests, home security, logistic tracking, healthcare, landslides etc.

As we saw, some existing technologies were not sufficient to deal with the challenges effectively. The system of motes has the biggest advantage of its flexibility in many key factors, like the no. of sensors, used to measure required characteristics, the type of network connectivity, required for data transmission, and the most crucial factor, size of the mote. With the help of circuit designing techniques, the size of the system can be reduced to the minimum level.

REFERENCES

Agarwal, A., Mehandiratta, E., Sanket, R., Samkaria, R., Gupta, T., Singh, R., & Gehlot, A. (2016). Smart Door Lock System for Elderly, Handicapped People Living Alone. *International Journal of Smart Home*, *10*(6), 155–162. doi:10.14257/ijsh.2016.10.6.16

Barrett, D. M., & Lloyd, B. (2012). Advanced preservation methods and nutrient retention in fruits and vegetables. *Journal of the Science of Food and Agriculture*, *92*(1), 7–22. doi:10.1002/jsfa.4718 PMID:22102258

Bosona, T., & Gebresenbet, G. (2013). Food traceability as an integral part of logistics management in food and agricultural supply chain. *Food Control*, *33*(1), 32–48. doi:10.1016/j.foodcont.2013.02.004

Chakrabarty, S., & Engels, D. W. (2016, January). A secure IoT architecture for Smart Cities. In 2016 13th IEEE annual consumer communications & networking conference (CCNC) (pp. 812-813). IEEE. doi:10.1109/CCNC.2016.7444889

Choudhury, S., Kuchhal, P., Singh, R., & Anita. (2015). ZigBee and Bluetooth network based sensory data acquisition system. *Procedia Computer Science*, *48*, 367–372. doi:10.1016/j.procs.2015.04.195

Fan, X., Huang, H., Qi, S., Luo, X., Zeng, J., Xie, Q., & Xie, C. (2015). Sensing home: A cost-effective design for smart home via heterogeneous wireless networks. *Sensors (Basel)*, *15*(12), 30270–30292. doi:10.3390151229797 PMID:26633424

Gill, S. S., Chana, I., & Buyya, R. (2017). IoT based agriculture as a cloud and big data service: The beginning of digital India. *Journal of Organizational and End User Computing*, *29*(4), 1–23. doi:10.4018/JOEUC.2017100101

Gope, P., & Hwang, T. (2016). BSN-Care: A secure IoT-based modern healthcare system using body sensor network. *IEEE Sensors Journal*, *16*(5), 1368–1376. doi:10.1109/JSEN.2015.2502401

Guo, S., Qiang, M., Luan, X., Xu, P., He, G., Yin, X., ... Fang, D. (2015). The application of the Internet of Things to animal ecology. *Integrative Zoology*, *10*(6), 572–578. doi:10.1111/1749-4877.12162 PMID:26338071

Jedermann, R., Behrens, C., Westphal, D., & Lang, W. (2006). Applying autonomous sensor systems in logistics—Combining sensor networks, RFIDs and software agents. *Sensors and Actuators. A, Physical*, *132*(1), 370–375. doi:10.1016/j.sna.2006.02.008

Kaur, P., & Katiyar, M. (2012). The energy-efficient hierarchical routing protocols for WSN: A review. *International Journal of Advanced Research in Computer Science and Software Engineering*, *2*(11).

Kodali, R. K., Jain, V., Bose, S., & Boppana, L. (2016, April). IoT based smart security and home automation system. In *2016 international conference on computing, communication and automation (ICCCA)* (pp. 1286-1289). IEEE.

Koo, D., Piratla, K., & Matthews, C. J. (2015). Towards sustainable water supply: Schematic development of big data collection using internet of things (IoT). *Procedia Engineering*, *118*, 489–497. doi:10.1016/j.proeng.2015.08.465

Liu, X., Yang, T., & Yan, B. (2015, November). Internet of Things for wildlife monitoring. In *2015 IEEE/CIC International Conference on Communications in China-Workshops (CIC/ICCC)* (pp. 62-66). IEEE.

Mahale, R. B., & Sonavane, S. S. (2016). Smart Poultry Farm Monitoring Using IOT and Wireless Sensor Networks. *International Journal of Advanced Research in Computer Science*, *7*(3).

Mandula, K., Parupalli, R., Murty, C. A., Magesh, E., & Lunagariya, R. (2015, December). Mobile based home automation using Internet of Things (IoT). In *2015 International Conference on Control, Instrumentation, Communication and Computational Technologies (ICCICCT)* (pp. 340-343). IEEE.

Mohanraj, I., Ashokumar, K., & Naren, J. (2016). Field monitoring and automation using IOT in agriculture domain. *Procedia Computer Science*, *93*, 931–939. doi:10.1016/j.procs.2016.07.275

Muragesh, S. K., & Rao, S. (2014). Automated internet of things for underground drainage and manhole monitoring system for metropolitan cities. *International Journal of Information & Computation Technology*.

Okokpujie, K. O., John, S. N., Noma-Osaghae, E., Okokpujie, I. P., & Okonigene, R. E. (2019). A wireless sensor network based fire protection system with SMS alerts. *International Journal of Mechanical Engineering and Technology*, *10*(2), 44–52.

Patil, V. C., Al-Gaadi, K. A., Biradar, D. P., & Rangaswamy, M. (2012). Internet of things (Iot) and cloud computing for agriculture: An overview. Proceedings of Agro-Informatics and Precision Agriculture (AIPA 2012), 292-296.

Pavithra, D., & Balakrishnan, R. (2015, April). IoT based monitoring and control system for home automation. In *2015 global conference on communication technologies (GCCT)* (pp. 169-173). IEEE.

Pirbhulal, S., Zhang, H. E., Alahi, M., Ghayvat, H., Mukhopadhyay, S., Zhang, Y. T., & Wu, W. (2017). A novel secure IoT-based smart home automation system using a wireless sensor network. *Sensors (Basel)*, *17*(1), 69. PMID:28042831

Qu, T., Lei, S. P., Wang, Z. Z., Nie, D. X., Chen, X., & Huang, G. Q. (2016). IoT-based real-time production logistics synchronization system under smart cloud manufacturing. *International Journal of Advanced Manufacturing Technology*, *84*(1-4), 147–164. doi:10.100700170-015-7220-1

Ram, K. S. S., & Gupta, A. N. P. S. (2016). IoT based Data Logger System for weather monitoring using Wireless sensor networks. *International Journal of Engineering Trends and Technology*, *32*(2), 71–75. doi:10.14445/22315381/IJETT-V32P213

Ramesh, M. V. (2009, June). Real-time wireless sensor network for landslide detection. In *2009 Third International Conference on Sensor Technologies and Applications* (pp. 405-409). IEEE. 10.1109/SENSORCOMM.2009.67

Rosi, A., Berti, M., Bicocchi, N., Castelli, G., Corsini, A., Mamei, M., & Zambonelli, F. (2011). Landslide monitoring with sensor networks: Experiences and lessons learnt from a real-world deployment. *International Journal of Sensor Networks*, *10*(3), 111–122. doi:10.1504/IJSNET.2011.042195

Rossi, A., Vila, Y., Lusiani, F., Barsotti, L., Sani, L., Ceccarelli, P., & Lanzetta, M. (2019). Embedded smart sensor device in construction site machinery. *Computers in Industry*, *108*, 12–20. doi:10.1016/j.compind.2019.02.008

Ruiz-Garcia, L., Barreiro, P., & Robla, J. I. (2008). Performance of ZigBee-based wireless sensor nodes for real-time monitoring of fruit logistics. *Journal of Food Engineering*, *87*(3), 405–415. doi:10.1016/j.jfoodeng.2007.12.033

Sagar, V. R., & Kumar, P. S. (2010). Recent advances in drying and dehydration of fruits and vegetables: A review. *Journal of Food Science and Technology*, *47*(1), 15–26. doi:10.100713197-010-0010-8 PMID:23572596

Santoso, F. K., & Vun, N. C. (2015, June). Securing IoT for smart home system. In *2015 International Symposium on Consumer Electronics (ISCE)* (pp. 1-2). IEEE.

Sun, E., Zhang, X., & Li, Z. (2012). The internet of things (IOT) and cloud computing (CC) based tailings dam monitoring and pre-alarm system in mines. *Safety Science*, *50*(4), 811–815. doi:10.1016/j.ssci.2011.08.028

Tongke, F. (2013). *Smart Agriculture Based on Cloud Computing and IOT*. Academic Press.

Tsang, Y. P., Choy, K. L., Wu, C. H., Ho, G. T. S., Lam, H. Y., & Koo, P. S. (2017). An IoT-based cargo monitoring system for enhancing operational effectiveness under a cold chain environment. *International Journal of Engineering Business Management*, *9*, 1847979017749063. doi:10.1177/1847979017749063

Whittenberger, R. T. (1951). Changes in specific gravity, starch content, and sloughing of potatoes during storage. *American Journal of Potato Research*, *28*(10), 738–747. doi:10.1007/BF02851330

Xiaojun, C., Xianpeng, L., & Peng, X. (2015, January). IOT-based air pollution monitoring and forecasting system. In *2015 International Conference on Computer and Computational Sciences (ICCCS)* (pp. 257-260). IEEE. 10.1109/ICCACS.2015.7361361

Yinghua, Z., Guanghua, F., Zhigang, Z., Zhian, H., Hongchen, L., & Jixing, Y. (2012). Discussion on application of IOT technology in coal mine safety supervision. *Procedia Engineering*, *43*, 233–237. doi:10.1016/j.proeng.2012.08.040

Zhang, Y., Chen, B., & Lu, X. (2011, August). Intelligent monitoring system on refrigerator trucks based on the internet of things. In *International Conference on Wireless Communications and Applications* (pp. 201-206). Springer.

Chapter 4
Challenges in the Area of IoT

Manish Kumar Saini
Department of Electrical Engineering, Deenbandhu Chhotu Ram University of Science and Technology, Murthal, India

Akanksha Aggarwal
Department of Electrical Engineering, Deenbandhu Chhotu Ram University of Science and Technology, Murthal, India

Sunita Saini
Department of Management Studies, Deenbandhu Chhotu Ram University of Science and Technology, Murthal, India

ABSTRACT

This chapter reveals the crucial challenges being faced by the society and the developers in making IoT reach to every nook and corner of the society. But as every coin has two sides, besides the boon of IoT, there are some banes too which can be now considered as the challenges open for the developers to make IoT a bigger success. Every application has its own issues in adopting IoT. Among them, the common and major ones are security and privacy concerns, ubiquitous network connectivity, societal reluctance to new technology, lack of standard protocols, gigantic data size, high-cost investment, robustness, and self-sustainability of IoT sensors connected in wireless sensor networks. Besides common challenges, challenges specific to different sectors of IoT applications like retail sector, healthcare sector, smart grid, smart city, agricultural sector, smart homes, transportation sector, smart metering, and power sector have been pondered in this chapter. These challenges need to be first combated so that IoT can bring the expected transformation at global level.

INTRODUCTION

Internet of Things (IoT) has taken a giant leap in the world of things. It has seen a paradigm shift since two decades when it was named by Kevin Ashton, executive director of Auto-ID Center in the year 1999. In earlier times, internet used to connect people; after that, it connected so-called things with people and now it connects things with each other in a huge world of things interconnected for transmission of

DOI: 10.4018/978-1-5225-9574-8.ch004

any sort of information targeted to transform everything into a smarter version. It has brought a revolutionary change in each and every industry leading to inception of smart grid, smart home, smart cars, smart city, smart meters, smart TVs, smart vehicles, smart appliances and smart order and delivery systems to name a few. That day is not very far when your refrigerator will place an order or send you the notification for the vegetable or other items got out-of-stock; your oven will get pre-heated before you enter your kitchen for dinner preparation; control of alarm placed in your child's room will be in your phone; your phone will warn you about fuel level and possible delay in the route before you start your journey; your phone will send you notification when your favorite items comes again in the store shelf; your door will get unlocked if your relatives come; car GPS screen will show you possible place of parking once you enter in parking zone; your watch will show your heart's performance and many more unimaginable things. This imagination is corroborated by the statistics given by CISCO IBSG that by 2020, the Internet connected devices will reach 50 billion and unquestionably pervade every aspect of our life (Tripathi & Anuradha, 2018).

IoT does all of these works by enabling smart devices to observe the events, collecting and merging the information collected from different sensors, applying the learnt knowledge to decide the most appropriate action and then taking action to affect or control the things or objects in the real world connected through wireless sensor networks (WSNs). In this way, IoT facilitates machine-to-machine interfacing supported by highly sophisticated networks depending on the distance between the machines or sensors, e.g., wireless personal and local area network (WPAN/LAN) for short to medium range distance and wireless wide area network standards (WWAN) for long range. WPAN/WLAN offers many different standards for communications, such as Wi-Fi, ZigBee and 6LoWPAN which supports IEEE 802.15.4 protocol for low-power consumption applications (Barnaghi & Sheth, 2016). The type of network and sensors is chosen depending on the need of particular applications.

At present, there is plethora of applications of IoT aimed at making the things smarter. IoT has become the integral part of concept of smart city in which every function is performed in a smarter way with more efficiency and less labour cost for making the life of human being more comfortable.

ISSUES COMMON TO ALL IOT APPLICATIONS

Though, IoT is becoming the new brightening star of every industry, but large plexus of IoT is creating many hurdles in successful adoption of IoT in different paradigms. A long deployment cycle, innate complexity of IoT applications and high cost of a large scale deployment can easily doom a nascent deployment. So, many challenges need to be solved before designing and deploying any IoT application to help IoT applications survive and grow (Jeyanthi, Thandeeswaran, 2017). Though, every application of IoT has some different issues particular to that application only, but in this plethora of problems being faced by IoT, there are some prominent issues common to each and every application as discussed below.

1. **Ubiquitous Presence of Internet:** The first and foremost requirement of IoT is internet facility for seamlessly connecting millions of objects present in the network. Internet connectivity is as much important for IoT as oxygen for us. So, any IoT based application can be fully used at any place if and only if there is easy availability of internet to the users at that location and that too at no cost or easily affordable cost for everyone.

2. **Data Size:** Among millions of connected objects, there are millions of sensors deployed for sensing their environment which varies in different applications. In some applications, sensors sense and communicate the readings to the network after some regular or irregular intervals. But in some real-time applications in industries, sensing and communication process incessantly goes on which results into production of voluminous data on centre servers and thereby, causes overloading of the internet servers.
3. **Data Security:** Smart applications work smartly if and only if they are provided with all the related data and information. But when the confidential information goes to the network, it becomes vulnerable to different security hacks like malware attack, data tampering, spoofing, denial of service etc.; although all applications use some security mechanisms while transmitting any type of authenticated information to the network. Despite of these mechanisms, there are high chances of security hacks causing leakage of information to wrong hands and many bad consequences thereafter. There are many countermeasures for these illicit activities but techniques which detect the malicious entry into the network are more common rather than techniques which prevent the malicious entry (Misra, Maheswaran & Hashmi, 2017). So, it is still an open opportunity for developers to protect illegal sensor network deployment.
4. **High Cost:** Though, market of smart devices has seen many advancements in the development of IoT based smart sensing and actuation systems. Still, the cost of smart IoT based systems is very high which becomes hurdle for IoT in reaching every corner of the society. The costly deployment, costly maintenance of smart systems and unpredictability about return on investments are viewed as some of the biggest obstacles in accepting IoT based smart systems.
5. **Energy Requirements:** Different sensors and actuators connected across the network should be low power consuming devices. As all things present in WSNs are portable, they cannot be supplied power by wires continuously and therefore, they should be battery powered. But, batteries also need frequent charging at regular intervals. Another problem occurs when energy sources connected in the sensor nodes malfunction or get damaged completely (Mahmoud & Mohamad, 2016). In both scenarios, recharging and replacement of the batteries in billions of devices connected across the world is obviously not at all an easy task. Many sensors are present at such remote locations or in such a harsh environment where frequent access is not practical. Therefore, such sensor devices should be designed and deployed which consume minimal power or accommodate some sort of energy harvesting feature from its surrounding environment. In other words, self-sustaining sensors are the best solutions to be used in WSNs.
6. **Data Analytics:** Enormous amount of data are captured from large number of sensors. Analysis of huge data with high accuracy and speed is required to initiate actuation devices in accordance with analyzed data. Correct and fast data analysis needs expert knowledge and information of the particular field embedded in the computational devices so that devices can take prompt and accurate actions.
7. **Interoperability:** Various types of sensors and devices operating on different mechanisms and following different protocols are present in the network. For efficient operation of different interconnected applications and transferring various types of information among the smart devices, all devices should follow a common protocol and standard so that there is no issue in use of sensors if deployed in different applications.
8. **Data Integration:** Multiple sensors are used for collecting the information about surrounding environment. Sensors should follow the common methodology or standards for transferring data over

WSNs so that integration of data received from multiple sources can be actualized. Semantics of data should be part of data itself and it should not be based on the application logic so that sensor data are not application dependent.

9. **Fault Tolerance:** IoT devices may be deployed in harsh conditions and hostile environment. Therefore, there are high chances of sensor error or malfunctioning, causing problems to whole application. Debugging of the connected system may also be a tedious job in such cases. So, the IoT devices should be tolerant and capable enough to handle and rectify any exceptions that occur. They should be able to adapt to the changes occurred during the communication and to act appropriately.
10. **Device Management Scalability:** All smart applications have wireless sensors deployed at local level. In case of interconnection or upgradation of the applications, these sensors could not be extended to large-scale environments. This generates the need of development of scalable IoT devices which can be easily accommodated in bigger environment.
11. **Acceptability Among the Society:** IoT devices are new for the society. Their complex operations and costly technical requirements prohibit them from being the common man's choice. On top of that, sharing of their personal information, continuous watch on their activity, high costs and consumer unawareness are the biggest obstacles that makes common man reluctant of accepting this new technology. People will surely not like that someone is watching them, controlling them and passing over their information to a common network throughout the day. The society will have to be made aware of boons of IoT based smart technology so that they unhesitatingly adopt IoT based smart systems.
12. **Lack of IoT Administration and Government Initiatives:** Many IoT based smart systems have already pervaded the industrial market, but, still there is no Government administration regulating the rules for IoT devices and applications. Hitherto, IoT has been the toy of private players and designers in the market who are more interested in their self-profit. Widespread adoption of IoT needs Government initiatives in this direction.
13. **E-Waste:** Technology of IoT has deployed millions of sensors in our surrounding environment. But, unfortunately, the electronic devices do not have long lifespan. They demand replacement at regular interval so that complete system could continuously work without any fault. Firstly, replacement of millions of devices is an issue. Secondarily, let us suppose that they are replaced but then, disposal of large number of faulty and unrepairable sensors and components becomes the main concern which bothers the system developers and stakeholders. The place of dumping also needs thoughtful selection as WSN components consist of hazardous and toxic materials that pose environmental risks in future if they are land filled or incinerated. Therefore, only biodegradable and non-hazardous sensor nodes should be designed and deployed.
14. **Transmission and Communication:** The mode of communication in all IoT devices involved in the wireless network needs to be chosen carefully depending on the distance from which they transmit or retrieve information. Various examples for communication within short distance are Radio Frequency Identification (RFID) devices, Near Field Communication (NFC) units etc.
15. **Deployment of IPv6:** The network of IoT consists of many tiny devices, all of which require unique IP address for interconnection and communication with connected devices. Generation of such a large number of unique IP addresses hampers the IoT's progress and moreover, more security features and protocols like IPv6 are required for securing and maintaining large number of IP addresses.

16. **Flexibility and Evolution of Applications:** Every day, sensors and devices are evolving with new capabilities and new designs. This leads to development of new techniques and algorithms to analyze the data. But making old sensors compatible with the latest versions is a difficult job. So, IoT devices should be made compatible to easy upgradation with the technological advancement.

The full potential of IoT will be unlocked only if all these issues are addressed. Many issues have been solved to some extent but still there are many open challenges in the path of full acceptance of IoT based technology like self-sustaining and biodegradable sensors, internet connectivity to each and every nook of the world, security and privacy concerns and consumer awareness for social acceptance of IoT. Besides these common issues, every application of IoT has some unique problems to be dealt with before hoping for handsome return on investments from IoT systems.

ISSUES IN DIFFERENT IOT APPLICATIONS

As IoT has not left any corner of our life untouched, it has sneaked into every facet of our daily life whether it is healthcare, car parking, remote surgery, shopping, agriculture, electricity generation, weather monitoring and a long list to continue with. In recent times, IoT continues to amaze the masses but IoT applications also have some challenges and issues in the dazzling world of IoT. These challenges leave the door wide open for engineers to make IoT fully successful without any blots. Some issues unique to the IoT applications are discussed in brief below.

Retail Sector

IoT revolutionized the entire world of retail business when it stealthily entered into multiple domains of business like supply chain management, customization of shopping applications, asset tracking and management, financial decision making, inventory management, smart fulfillment of stores and warehouses, logistics, foot traffic monitoring, customer behavior analysis, predictive equipment maintenance, automated check-outs etc. (Singh & Singh, 2015). With the inception of IoT in the business sector, every business process transformed into a more efficient process. The data collection from warehouses and factory floor became fast and more accurate; time from warehouse to market reduced; tracking of all the assets improved whether it is physical asset, product or human resource. All these IoT based business solutions which digitized brick and mortar retail shops and malls, facilitate improved customer experiences with high return on investment too. But, the implementation of IoT in retail business is not getting accelerated due to many obstacles as briefed below:

1. **Staffing Needs:** The present business employees are not having the knowledge required for operation of IoT based processes. Successful implementation of IoT in retail sector first requires massive training of employees. But, in present scenario, the bigger criticality is that employees are not even willing to learn the new technology and get conversant with IoT embedded business processes. The possible but not completely successful solution for this issue is recruitment of technically skilled labour in place of old employees. But that option is not suitable as per business ethics. It will also lead to massive unemployment. Therefore, training of the existing staff is more feasible and potential solution to this issue although it may take time and face employees' reluctance. The same issue is

more problematic at higher level of management where senior managers having good knowledge of IoT technology and commitment towards implementation of IoT in their business processes, are required the most so that entire organization can smoothly adopt IoT under their vision.

2. **Large Investment:** Implementation of IoT requires building of IoT enabled infrastructure which demands a huge financial investment. The required sensors, RFID tags, IoT embedded kiosks, etc. at every place would cost the company a huge amount. The financial investment on building up and maintenance of IoT infrastructure will certainly have high returns in future (if implemented successfully) but it generates the sudden demand of huge money, possibly leading to cost-cutting in other domains of the sector.
3. **Incompatible Products/Services:** In present context, the products or services are not having compatibility to accommodate IoT elements. As a result, IoT technology could not be adopted in retail sector with the existing infrastructure. This generates the need of IoT enabled products/services which may be fulfilled by manufacturing of new IoT enabled products or by introducing IoT elements like sensor, RFID tags, etc. in the existing products only. Both solutions, i.e., production of such new elements or services and upgradation in existing products will consume the time and other resources that too in large amount.
4. **Undeveloped Consumer Awareness:** Although, IoT applications have been started in retail sector, but for now, that is at minor scale only. Consumers are, therefore, less aware about how to utilize the IoT enabled products and services. Less awareness prevents the customers to fully utilize the facilities provided to them by IoT services. Low utilization or mis-utilization may lead to reduction in return of investment that is expected after successful launch of IoT enabled products or services.
5. **Lack of Security Regulations:** The biggest question over adoption of IoT technology in retail sector is of data security and privacy. To facilitate customers during shopping, IoT enabled services know the customers' data including their personal details, preferences, choices and previous transactional details. But this knowledge may go to wrong hands in case of data breach and cause fraudulent activities. The same risk occurs with the company's product data. Information about where, how much and how the products are being transported from warehouse to the market etc. may also cause trouble in case of data breach. There are still no standards being properly framed and followed for maintaining data security.
6. **Lack of Technological Infrastructure:** Adoption of IoT in the retail sector has few primary technical network pre-requisites among which the main thing is availability of internet services at every place and at no cost. Everything whether product or services needs the internet connection to be connected with the network of other things. But, there is deficiency of fast internet connection facility either in form of LAN/WAN or wi-fi at every place at minimum cost. To make IoT reach to every corner of business, the internet facility will have to first reach there to enable the IoT technology.
7. **Lack of Industrial IoT Standards:** Although more than two decades have passed since the beginning of word IoT, but not a large span has passed after it became the buzzword. Still, there is lack of common standards and regulations to be followed for IoT technology. As there would be different products and different types of sensors working on different platforms, a common standard is required to connect all different types of sensors on a common platform.

All these issues don't let the IoT solutions thrive in the business industry. If IoT has to create its unbeatable position in the retail market, significant attention will have to be devoted to address these challenges.

Healthcare Sector

The domain of healthcare has also been benefitted from IoT based development. Many IoT devices ranging from highly sophisticated IoT based pacemaker implanted into our body to simple health monitoring wearables have arrived in the market. Today, IoT has brought such a revolution that a surgeon need not be near to the patient for a major surgery; he may even remotely operate upon the patient. Besides this, many healthcare applications employing IoT based interconnected medical devices have been developed recently, e.g., pharmacy management systems (PMS), remote patient monitoring, health informatics, telemedicine, clinical decision support, electronic health/medical records (EHRs/EMRs), maintenance of personal health records (PHR), fall detection and many ambient assisted living applications (AAL). But, in reality, there are many barriers which are hindering the adoption of e-health based on the IoT. Those barriers are listed as below.

1. **User's Concern About Security and Privacy:** Most of the non-users of smart IoT based healthcare system believe that their confidential EHR and PHI are not as much safe with electronics systems as they are in paper records. Large volume of sensitive data traversing through WSNs is susceptible to interception and manipulation while in transit causing serious concerns or fatalities.
2. **Concern About Patient Acceptance:** In IoT based healthcare, during visit to physician, the patient may not feel taken good care of and well treated as physicians would spend more time interacting with the computer than the patient. But, this issue will not be so always as it can be overcome with experience of smart systems which will increase physician's speed on the smart systems.
3. **High Cost:** Cost of different products and services ranging from small to medium-sized practices is very high. So, even for big hospitals, cost is one of the biggest barrier to usage of IoT based systems without large budgets sanctioned for IT development. Small-size gadgets and wearables are also having high initial cost which makes common public hesitate to use them.
4. **Time and Complexity:** IoT based medical systems are little complex but the physicians having no experience with smart systems, consider them very complex and time-consuming. Physicians don't have time to learn and get conversant with the smart systems.
5. **Absence of Computer Skills:** Medical data entry staff, sometimes, don't have advanced computer knowledge and they don't have concern with patients too. This may cause incoordination between the actual patient data and data entered by them into the system. Inadequate training of medical staff members in working with smart tools and technology may also cause exposure of vulnerable surfaces.
6. **Unpleasant Experience of Physicians:** Some physicians don't like working with smart systems and EHRs due to irritating search for different menus and buttons while checking the patient records. According to them, this hunting wastes their time which they should spend talking with the patient for better understanding and analyzing their medical issues. Sometimes, physicians wonder if the software will suit their style of practice.
7. **Lack of Usability by Patients:** Due to device complexity, patients may be unable to properly manage the monitoring devices. Patient's inability to use the devices correctly can interrupt the monitoring system and lowers their reliability on these devices.
8. **Concern About Data Entry:** Patients go to physicians seeking for their counseling abilities. But at many places, their medical information is first asked and entered into the system by data entry staff. Time consumed in making data entries makes the patient feel dejected and neglected.

9. **Low Awareness:** One of the biggest obstacles is lack of communication among users about the benefits of smart IoT based healthcare systems. People's positive reviews and the advantages of smart systems should be shared with everyone over social networks so that people could be motivated to use and rely upon these systems.
10. **Physical Space Barriers:** EHRs need many computer devices and concerned accessories at the clinic or hospitals which may cause space shortage and put financial burden on the clinics' or hospitals' administration.
11. **Legal Requirements:** As this is a new field, legal requirements are not fully known for hospitals setting. However, pilot studies are going on in tertiary care centers in Chennai and some other cities.
12. **Low power Operation:** Due to size constraint and miniaturization, monitoring devices have limited memory and processing power which can deteriorate its performance and cause failure in getting accurate results. These small devices have battery problem too. Though, they are less power consuming devices, but, still, they need frequent replacement or recharging.
13. **Right Access Control Permissions:** Access control permissions give any person the authority to check the patient's personal data and also the device's control code. The access control permissions should be granted in such a manner that no malicious user can access the devices but there should be no problem for the medical staff in accessing the data.
14. **Absence of Encryption on Device:** Many healthcare IoT devices could not accommodate the standard encryption protocols due to limited on-board memory and if encryption is accommodated, it also consumes more power for doing complex computations. This makes the device vulnerable to early recharging or replacement of the batteries.
15. **Security:** There are different security risks in smart healthcare systems, e.g., dropped data between our smartphone and hospital's network, data counterfeit, patient's data disclosure to health insurance companies by gaining unauthorized access, frequency jamming, data flooding, data eaves dropping, spoofing of patient's data on publicly available communication channels etc. Patient monitoring devices are generally used in a public environment which can also pose security risks. Generally, clear text data is used in EHRs which can give the chance to hacker to misuse patient's information.

As networked medical devices have become common in the healthcare sector, but with its growing popularity, security threats are also increasing which may undermine the IoT based advanced technology if not treated and mitigated well and may also cause large financial instability in the existing system.

Agriculture Sector

The need of smart agriculture is felt by the growing needs of increasing population as per the prediction of the Food and Agricultural Organization of the United Nation (FAO) which states that by 2025 and 2050, 8 billion and 9.6 billion mark will be crossed by global population respectively (FAO, 2009). Therefore, food production will also have to be increased by 2050 worldwide; as without food, there is no use of smart city or smart systems. With present resources, the rising demand of food production can only be met through intensification of agricultural practices by embedding smart technology in it and embracing IoT. There are many challenges which further intensifies the problem of growing food demand, e.g., extreme weather conditions, gradual climate change and environmental impact. To combat these issues and meet the growing demand, we will have to reduce waste generation and increase the productivity of each agricultural activity. For this, we need smart farming along with IoT technologies. In today's world, IoT is being envisioned for multiple applications, such as smart farming, weather monitoring and forecast, sensor based precision farming, asset monitoring, crop monitoring, soil monitoring, livestock monitoring, smart warehousing, logistics and distribution, sensing of temperature, soil moisture content and humidity measurement, field soil preparation and fertilization requirement calculation, agricultural drones and many more to count. Despite the possible utilities of IoT in smart agriculture, some of the issues are being faced in using IoT.

1. **High Initial Hardware Cost:** The smart IoT embedded farming devices are of high initial cost as of now. Unfortunately, most of our farmers cannot afford the high initial cost of smart devices. As a result, farmers continue relying on less advanced farming technologies which limit their productivity.
2. **Maintenance Cost:** Beside high initial cost, smart devices frequently demand maintenance also either in form of battery replacement or sensor replacement. For operation of IoT devices, network connectivity is also a compulsory requirement which may not be affordable for every farmer.
3. **Training:** Farmers need to be trained for operating smart IoT devices in their agricultural practices. Learning the complex operations of IoT devices is a tedious job especially for less educated farmers. Farmers should be made aware of benefits of smart practices over their normal practice so that they can willingly adopt new technology of IoT.
4. **Poor Internet Connectivity in Farms:** Network connectivity in field areas is still not very sufficient. Most of the agricultural areas and livestock farming areas have slow or no internet connections as compared to main urban areas. So, there are two problems in usage of IoT devices: slow internet connectivity and lack of connectivity.
5. **Lack of Awareness:** There is lack of awareness in farmers as far as technology-based farming solutions and their applicability are concerned. This stems from lack of knowledge and fear of upgrading to a technology of higher level. In order to develop commercially viable solutions, it is critical to keep these factors in mind. The solution must function in local languages and have interfaces that are easy to understand for laymen.
6. **Fear From new Technology:** The important thing to be considered when developing IoT solutions for the farming industry is to take away the fear of new technology. Farmers may not be adopting IoT technology because they feel apprehensive of trying something new in their fields as they have emotional attachment with their resources. They believe that new devices will not care and work as efficiently as they do in their fields.
7. **Robustness to Variability:** IoT devices should be designed keeping in mind the fact that in any farming environment, no two cases are identical. This dynamic scenario makes the development, testing and validation of digital technologies in agriculture a lot more difficult as compared to other industries. In other industries, the data is more static, hence, new technologies can be tested easily.
8. **Robustness to Harsh Environmental Conditions:** The sensors and other connected devices employed in smart farming are deployed in external environment where they are directly exposed to harsh conditions like scorching heat, high rainfall, high solar radiations, high humidity, strong winds and other physical dangers capable to damage the smart devices. Therefore, electronic devices should be strong enough so that they can withstand harsh environmental conditions and function incessantly.

9. **Low Power Consumption:** Reliable operation of smart end-devices demands that they should consume low power for long periods because it is not an easy task to frequently replace the batteries deployed in open fields at large scale. This problem can be solved to some extent by power harvesting but then also, power consumption should be low because small harvesting modules like solar panels and wind turbines will generate small amount of power.
10. **Large Amount of Data:** Smart farming includes large number of interconnected IoT devices which generates unbelievable amount of data which is difficult for small-scale server infrastructure to handle. Such a large amount of heterogeneous data coming from enormous amount of small sensors also cause high cost and data latency while transmitting the data over cloud.
11. **Interruption in Internet Connectivity:** In agricultural practices, WSN nodes are deployed in open space where wireless link quality gets highly disturbed by environmental effects like multi-path propagation effects and resultant background noise. Smart transceivers deployed for farming get also affected by temperature, human presence, humidity and other obstacles. So, strong and reliable technologies should be developed for transmission especially for rural areas.
12. **Clashes with Vendors:** Vendors of smart IoT solutions put very restrictive demands for the farmers. In these types of agreements, cash-strapped farmers are forced to get their equipment repaired from vendors only which they can themselves fix at less cost. Farmers also don't want that vendors share or resell their fields data without their permission or compensation. The ownership issues also exist with the farmers regarding the ownership of expensive equipment deployed in their fields.

Every new IoT solution targeted for smart farming should keep these issues in mind and new technologies should be designed and developed according to the compatibility with the rural environment and farmer's needs.

Smart Metering

In today's smart world, metering operation of energy resources or any other resources has started using IoT devices for facilitating metering in remote areas and reducing the human error and labour cost involved in this activity. In smart meters, information about consumption gets transmitted to the utility over the network and corresponding bills are generated and sent wirelessly to the consumers. Bill payment is also done using the internet connectivity. The smart metering solutions are more efficient and more tamper-proof compared to conventional metering alternatives. But, there are certain issues in smart IoT based metering solutions which need to be first considered before designing and deploying smart solutions.

1. **Large Investment Cost:** Deployment and maintenance of smart meters in the energy distribution system require several million dollars which are not guaranteed to be compensated with return-on-investments. In smart meters, integration of several devices required for modulation, demodulation and additional memory for storing the data logs could increase the overall deployment costs.
2. **Replacement and Synchronization with Old Techniques:** For adopting new IoT based smart meter solutions, all the existing meters will have to be replaced which is incredibly tough job for utility companies. If meters are replaced then the synchronization of smart meters with other existing infrastructure will also be a greater challenge. Lack of infrastructure required for synchronization between old and new devices may make the installment of smart meters unproductive exercise and investment in vain.

3. **Requirement of Communication Network:** It is sometimes difficult to provide and deploy the robust communication network at some locations due to difficult terrains and hostile climatic conditions. In those areas, IoT devices may fail in smooth and efficient operations.
4. **Difficult Maintenance:** After installation and energization of smart solutions, maintenance of all connected devices in case of any fault will be more serious challenge. Any fault correction in the networked devices needs serious consideration towards each connected device.
5. **Data Management:** IoT based smart meter solutions require transfer of massive amount of data between a smart meter and the server located at the base station in both unidirectional and bidirectional way. Management, storage and analytics of this much data is really a tedious job. Something like what data, how much data can be and should be transmitted, time of transmission, type of modulation and demodulation of data, also need to be judged for maximum efficient operation of smart metering solutions. The data transmitted should be sufficient so that customer and utility companies can access all the intended information like energy consumption.
6. **Devices Interconnectivity:** There are many interconnected devices deployed for different purpose in the smart meter network. These devices can be synchronized in the best possible manner if and only if every connected device and appliance deployed in the distribution and metering network is compatible for interconnection. This problem becomes more critical with growing number of devices to be interconnected in the network. As the number of devices increases, compatibility issue becomes more pronounced due to presence of multiple standards and protocols which are followed by those devices.
7. **Privacy Issue:** Data transfer in smart meters and smart appliances deployed in residential houses poses risk to privacy of end-user by misuse of their personal information. The data transmitted over the network can give more confidential information about the consumers besides just the energy consumption. Keen analysis of data can divulge some information about end-user's habits, for instance, their wake up time, sleeping time, departure time, arrival time, when they are on vacation, number of members at home etc. An attacker can easily plan the mischievous activity if he has eavesdropped on the energy consumption of households.
8. **Impersonation/Identity Spoofing:** Communication network is prone to malware attack which aims at communicating on behalf of a legitimate thing or person or consumer in an illegal manner, by illicitly using his or her identity. Wrong-doer can con the identity of some consumer's smart meter and can tamper his or her energy consumption readings.
9. **Low Bandwidth:** In most of the IoT solutions, low bandwidth is used for smart meter communication networks leading to generation of large traffic and putting limits on the quantity of data which can be feasibly transmitted.
10. **Other Issues:** Besides above discussed issues, there are some challenges encountered by the utility companies, e.g., propagation problems, limitations in network coverage and data capacity. In addition, data concentrators which are opted as solution of some of the network coverage problems may lead to accommodation, safety issues and high cost investment.

Though, IoT is being successfully used in smart metering solutions but they can't sustain for long duration on account of these challenges. These challenges will hamper the utilization of IoT in metering applications irrespective of the field whether it is electricity, water, gas or oil etc.

Smart Homes

IoT has revolutionized the human life by fusing smartness and high efficiency into the systems. Many chores of our daily life varying from lighting system, control of all connected equipments, doors and lock systems, refrigerators, TV and other devices etc. have been made smart and automatic by using IoT devices. Enormous applications are emerging day by day that make our home a smart home but still it has not reached to the expected scale of employability due to few challenges which interfere with widespread usage of IoT applications of smart homes.

1. **Customers' Unwillingness to use:** The biggest challenge in ubiquitous adoption of smart homes is the customers' unwillingness to use new technology. Customers do not want to opt for smart solutions due to their complex usability and also the high cost. The interface of smart solutions should be made as simple as classic solutions which consumers can easily understand and use. They have become habitual of old systems, so, they are little reluctant in trying smart IoT based solutions.
2. **Responsiveness:** Smart solutions should execute the user commands with immediate effects and also immediately give the corresponding feedback on the user interface. Consumers feel irritated when they have to wait for response after giving the input command to the smart system. This is a real technical challenge for the smart systems whose architecture involves multiple layers of marshalling, event conversion and dispatching events in a distributed network.
3. **High Up-Front Investment Cost:** Smart home solutions cost very high to the consumers which deject them from installing IoT solutions in their new homes. Same problem occurs for both old and new buildings and both homeowners and tenants. Existing homeowners don't find it comfortable and beneficial to get smart devices installed in their current buildings as it will require retrofitting wiring in the home. And new homeowners have already spent so much in buying or constructing the building, they don't wish to invest more on their homes due to money shortage. In case of tenants also, they are not willing to invest on buildings which they do not own. Finally, most of the consumers groups find the smart solutions unfavorable to them. Therefore, the smart solutions will have to be designed and developed in minimum cost and easy usability like easy-to-install wireless devices if IoT solution providers want to attract every consumer group.
4. **Architectural Readiness:** Consumers accept only those smart solutions which facilitate them in upgrading their existing infrastructure by augmenting them with smart devices rather than to make it smart from the scratch.
5. **Security & Privacy of Data:** IoT technology connects all the devices through a network which makes data transmitted through network vulnerable to spoofing and misuse. Some hacker can hack the network of our home and control the actions of IoT devices which can lead to many fatalities. Data disclosure and data eaves dropping can make home prone to burglary as thief can extract the confidential information about the residents and their daily routine from the intercepted data from the network.
6. **Expectability, Acceptance and Reliability:** People especially elder people hesitate for any change in their daily routine and become slave of older machines. Due to this nature, they think that IoT based new home automation systems are useless and money wastage. So, smart home solutions should be designed in consumer-friendly manner so that elder people can find smart solutions as their friend not a supervisor or intruder keeping keen watch on them and controlling their actions

and decisions regarding lighting, heating, entertainment etc. Friendly smart home solutions will be acceptable to the residents without any fear.

7. **Context Awareness and Adaptiveness:** Many smart home solutions are not context-aware and adaptive in their actions. They become the reason of frustration for some people who are not of do-it-yourself type. IoT devices should be smart enough, easy to operate and keep running in the background to make adaptive and smart decisions. For instance, lowering of temperature due to switching off of lights should not unintentionally switch on the air conditioner.
8. **Extendibility:** Most of the IoT devices are being designed for specific applications; they could not be extended to large scale or different application even if same operations are required in other applications. So, all IoT devices should be configured and deployed in such a way that it can be extended for other solutions also.

In present time, smart home solutions are in reach of only rich and affluent people. If these challenges are addressed successfully, then smart home solution can be the global choice leading to overall development of the society.

Smart Grid

IoT has efficaciously set the wheels of smart grid rolling with the advent of WSNs and IoT devices as sensor nodes. Smart grid enables us to optimize our energy consumption whereas IoT embedded in smart grid allows us to control and monitor energy usage in our home in real-time. IoT devices facilitate us to customize our energy demands according to peak load on grid and make it possible for the consumers to interact with grid more efficiently using an user interface (Bekara, 2014). But despite various benefits of IoT in smart grid, there are few major challenges also which hinder the easy adoption of IoT by the users.

1. **Training of Users:** Smart grid operations are very complex even for technical staff. For full utilization of all IoT devices deployed in smart grid, exhaustive technical training is required for everyone because there are many intricacies in operation of all connected devices in smart grid which can be understood only after gaining skilled knowledge of complex operations and with experience in working on different IoT networks.
2. **Legacy Devices:** The electric grid comprises of many devices which have very long lifespan. These already deployed systems and devices may not have any special security features more than communication facility. Integration of these legacy devices into the smart grid is an important challenge that should be taken care of while designing new IoT devices to be embedded in smart grid because their integration to the IoT based smart grid is the only option as they can neither be replaced with new devices nor updated to newer secured version.
3. **Field Locations of Devices:** There are millions of field devices in the power grid, e.g., switches, meters and transformers. Although their physical security is a critical design consideration, as they are potentially vulnerable due to their field location. On account of their vulnerability, it should be taken care of that design of network security does not depend on these devices for grid integrity.
4. **High Complexity:** Each operation of IoT based smart grid is very complex and its complexity is one of the major factors which undermines its successful and widespread applicability. Many people refrain themselves from using IoT based small solutions for distributed smart grid only due to its complex design and operations.

5. **Theft of Assets:** IoT based smart grid comprises of huge number of devices which are deployed in open space which makes them prone to theft. So, security of IoT devices should also be kept as primary consideration while designing the security mechanism of the complete smart grid.
6. **Scale:** Complete communication infrastructure designed to support entire global smart grid is of very large scale like the Internet. As it is observed in the case of Internet, security design and management of such a huge network has its own challenges like identity management for a large number of entities, management of keys for data integrity and confidentiality, segmentation and integration of multiple wired and wireless communication networks.
7. **Data Tampering:** Another important hurdle in smart grid is data tampering where an illicit person can change the transmitted data like dynamic electricity prices sent prior to peak periods and change them to the lowest prices. Consequently, consumers can increase their power consumption in form of different loads (like charging e-cars, etc.) rather than reduce it. This situation will further overload the network of smart grid.
8. **Authorization and Control Access Issues:** In IoT supported smart grid, several devices like smart meters and field deployed sensors and actuators in distribution substations, are monitored and configured remotely. Taking advantage of this phenomenon, an attacker or even an angry employee, can attempt to get the unauthorized access rights, to manipulate them or damage physical assets which may cause power outages.
9. **Heterogeneity:** There is large discrepancy in the characteristics of different devices/objects present in the smart grid, for instance, their bandwidth, time-sensitivity, memory, energy autonomy, computation ability etc. and their employed protocols and communication stacks. This makes getting end-to-end secured communication a challenging work which may even need the adaptation of existing solutions or even using gateways.
10. **Availability and DoS Issues:** The facility of internet connectivity in different vital assets of smart grid like electricity meters, transformers, substations, circuit breakers, cables etc. makes them vulnerable to security threats and also creates the possibility of DoS attack which makes these assets partially or totally unavailable to the complete network. Though, this fact is incredible but this is a reality. Because, in conventional grid system, no one could easily locate and target these types of assets, especially at large scale. Thus, these devices were safe in the classical power grid systems.

Smart grid can be made more acceptable if the interface is customized according to the customers' needs, comfort and understandability. All these challenges which are faced while integrating IoT with the smart grid, need serious attention for seamless utilization of IoT based solutions for operations of smart grid. Any failure occurred due to any of the above discussed challenges can halt the entire power system network. On top of that, power failure will disrupt whole IoT network connecting huge number of different devices and applications.

Power Sector

IoT has become the integral part of all three domains of power sector which are generation, transmission and distribution. IoT based smart solutions have been extensively used in power generation sector for different purposes like power scheduling, fuel handling, process optimization, generation control, detection of abnormalities, synchronization, alert & alarm, plant control etc. IoT can also upgrade the efficiency of power generation using distributed energy resources, thus, should be used in different power generation phases like operations, energization, maintenance and optimization. On the other hand, IoT plays a key role in smooth power transmission as it is being employed for transmission line state monitoring, reliable transmission, intelligent processing, real time monitoring, maintenance assist and fault location detection. IoT facilitates intelligent monitoring of existing electric power transmission and distribution networks and also helps plant operators to proactively take action in cases of blackout, customer queries' redressal and supervise integration of different distributed energy resources (Bedi, Venayagamoorthy & Singh, 2016). Moreover, IoT can also lower down the losses occurring due to power theft by putting a strict check on electrical parameters (voltage, current, power and phase) in real-time and by identifying the accurate location of theft. Like other applications, usage of IoT in power sector also, faces some issues which are still open for some development.

1. **Power Management:** In electric power and energy systems, IoT devices may be deployed at locations which are hazardous, toxic, or hard-to-reach areas, thus, causing difficulty in grid connection or battery replacement to power these devices. So, only those IoT devices should be installed at such locations which are self-sustaining for long period of time.
2. **Network Bandwidth Constraints:** Huge amount of data gets transmitted through the network connecting complex grid environment making it difficult to adjust with low bandwidth.
3. **Reliable Sensors:** Many IoT sensors available in the market don't have few important features like strong cyber security, efficient power management, situational intelligence, context awareness and enhanced cyber security. IoT sensors deployed in power sector must have these features incorporated into them for high efficiency and productivity.
4. **Coexistence Challenge:** Different connectivity protocols like Wi-Fi, Bluetooth, ZigBee etc. are used by IoT applications in electric power and energy systems for data transfer. These different connectivity protocols face the coexistence challenge, i.e., interferences caused due to interaction between different wireless connectivity protocols which share the same frequency band. The quality of network gets significantly degraded due to these interferences.
5. **Big Data Challenges:** Large amount of data is generated from thousands of iot devices connected across electrical power systems. Enormous amount of data makes it challenging to store, track, analyze, capture, search, share, transfer, secure, visualize and interpret the generated data.
6. **Energy-Efficient Communication Networks:** With rapid increase in number of IoT devices having networking features and growth of information and communication technology, energy requirement is also increasing at the alarming rate. This situation demands effective power management solutions for the communication networks.

Among all these challenges, reliable and self-sustaining sensors are the primary requirement as they may be deployed in hostile environment. There may be enormous applications of IoT devices in electrical power and energy systems if all these challenges are solved in entirety.

Smart City

Evolution of IoT has led to a spurt in the number of applications which transform city to the smart city having automatic street lighting system, automatic smart parking system, smart weather monitoring systems, automatic biometric based security systems, automatic traffic monitoring, smart waste management, pollution controlling etc. (Ahlgren, Hidell & Ngai, 2016). Though, IoT has not become successful till now in converting every city of the world into a smart city due to some constraints which are listed as follows.

1. **Network Connectivity:** Due to the ever-increasing volume of sensors and their data, robust connectivity technology is the necessary requirement for success. It is also often limited by a city's budget.
2. **Being Socially Inclusive:** Many smart city solutions are present in the market but these are not readily accepted by masses. For instance, smart transit programs that give riders real-time updates are useful only when more than half of the population uses mass transit or Uber. IoT based smart city solutions are of no use for group of elder people who doesn't use mobile devices or apps. So, smart solutions should be designed such that it could reach and benefit all groups of people not just tech-savvy users.
3. **Digital Security:** IoT based solutions upload the personal data into the network which is further shared among multiple users. This feature makes it very critical to safeguard the data from malicious attackers. So, suitable security measures should be adopted to save the private and proprietary information of citizens, Government, research partners, universities and digital infrastructure.
4. **Privacy Concern:** The working style of IoT dejects the citizens from happily accepting the IoT based smart city solutions. Nobody wants to share his or her personal data with others. Also, nobody want to be watched by some other person like big brother, their Government or other mischievous element. Cameras installed on every street corner may help deter crime, but they can also install fear and paranoia in law-abiding citizens. Therefore, IoT solutions must be designed considering the allowable invasion of privacy.
5. **Huge Investment:** There are many financial constraints in widespread usage of smart city solutions like limited funds released from Government, large up-front investment, long term delays before maturing and lack of resources to sustain the project over time. These financial problems remain the most difficult challenges for smart cities. One of the potential and popular solutions to combat these financial challenges is public-private partnerships.
6. **Sensor Compatibility:** IoT sensors have no standards to be followed. It all depends on the vendors that which sensor is to be used for particular application and how the data will be acquired from these sensors.
7. **Training of Common Users:** Smart city solutions are comparatively having more complex infrastructure due to the interconnection of millions of devices. This complexity generates the need of effective and practical training of users for unlocking the full potential of IoT based solutions and making IoT solutions reach to every group of the society.
8. **Legislative Issues:** Besides technical and social challenges, there are some legislative issues being faced by IoT based solutions, for instance, searching out workable policies to regulate stakeholders, unleashing economic development, maintaining benefits for the city's population, permitting growth in research-and-development investment and formulating the right governance structure given organizational barriers to supporting smart city initiatives.

IoT based solutions can be successfully deployed in different applications of smart city only after addressing these challenges. These challenges may be application specific and needs to be considered application-wise. To make IoT based smart city solutions win the common man's heart and be adopted by everyone irrespective of their stature, social issues first need to be addressed by the designers and developers.

Transportation Sector

One of the vision of smart transportation is undoubtedly, the autonomous vehicles, in which IoT does wonders in various functions like adaptive cruise control, blind spot detection, parallel and rear parking assist, driver monitoring system, traffic sign recognition, night view assist system, collision avoidance system, antilock braking system, airbags control, navigation system, adaptive headlights, lane departure warning and lane keep assist. Other applications of IoT in smart transportation sector are Vehicle-to-Vehicle solutions which help in avoiding accidents and warning the drivers of any impending crashes, e.g., left turn assist and intersection movement assist (Sheehan, Murphy, Mullins & Ryan, 2018). It can also be ensured that drivers adhere to the safety compliance regulations using global positioning system tracking. Few other areas where IoT is being applied are smart parking solutions, traffic management solutions, passenger entertainment, fleet management and telematics solutions, security solutions, traffic control and electronic toll collection. In spite of so many applications, there are many challenges caused by IoT devices and connected network in smart transportation.

1. **Fast Processing of Huge Data:** All smart transportation applications render an enormous and unimaginable set of data which need not be just stored but analyzed critically to take the quick decisions. Such a technology which can perform on such huge set of data and that too in milliseconds is still a challenge for the developers of smart IoT solutions.
2. **Sensor Accuracy:** Among technical challenges, sensor accuracy is the foremost challenge as the sensors output which acts as input to the actuators must be a fully accurate and valid to ensure the safe working of vehicle. But, unfortunately, perfect accuracy is rarely guaranteed in cases of sensors and any misdirection may lead to great mishaps.
3. **Environmental Challenges:** Sensors installed in the open environment are in direct exposure to the environmental conditions which may be sometimes harsh to the sensor's health like extreme snowfall, rain, heat or windy situations. Weather around each sensor cannot be always predicted with surety. In these type of situations, sensors' stability cannot be ascertained putting the travelers' life in danger.
4. **Security Challenge:** The smart vehicles can be forced to perform what it is not supposed to do under the commands of malicious user who can cause life risk to the travelers. As the network is unsecure, there may be unintentional attacks also which occur due to malfunctioning of system components.

All these issues prevent the full utilization of IoT technology in developing smart transportation sector irrespective of the application. It may be the case of autonomous vehicle, real-time tracking of public transit, smart parking management and guidance, traffic management, vehicles health monitoring, fleet management, support to ride-sharing system etc.

CONCLUSION

Internet of Things, a buzzword in today's ubiquitous smart world of technology has begun benefitting all the market segments but developers building the IoT based smart devices need to ponder several issues before harnessing the potential benefits of IoT. This chapter focused on all the preliminary concerns and vulnerabilities to be analyzed by the users and developers so that more effective and successful IoT device may be built and smartly connect the whole world. All these issues have been briefed with their corresponding potential solutions to unlock the whole gamut of IoT. This discussion is necessary prior to design and deploy any IoT device in any industrial and customer relationship applications and to understand the applicability of designed sensors and devices for that particular domain.

Few instances of common hindrances are cost, interoperability, adaptability, data latency, acceptability, security, legal rights and ethical issues and sensor energy. Data problem is becoming more aggressive day by day with tremendous rise in the volume of data and consequent rise in computational complexity with ever-increasing sensor and terminal nodes. Lack of internet connectivity at many remote areas is another one of the biggest obstacles. Internet is blood of IoT connected world without which none of its system can function. On the other hand, sensors and actuators work as body organs, so, to help body work properly, these organs also need proper attention. Also, the complete system or network depends on information sharing which can only be acquired, collected and dispersed through these sensors (or actuators). So, these sensors should be self-sustaining over a long period of time to make entire IoT network robust. Many severe ethical issues may also arise due to absence of any IoT administration, common protocols and standards, thus, hampering the seamless usage of IoT in different fields of automation. Some of these issues like security, if not taken care of with caution, may also prove IoT lethal for society. These issues have not arisen overnight; but the work is going on for getting more security to the IoT products. But the main problem is high cost which customers are forced to pay in lieu of securing their data. If secured personal computers can be developed then why secured IoT products cannot be developed that too at minimal price so that customers would not need to compromise with security due to cost factor. Oversight of these few prominent issues can easily make IoT turn out to be a mere wastage of financial and human resources.

Besides these common issues, every sector whether it is transportation, healthcare, agriculture, smart homes, smart city, smart grid, retail business management, or smart metering, has few domain-specific challenges which need to be dealt prior to deploy the new IoT solutions in particular domains. Some of these issues are being given full consideration now-a-days but some are still being overlooked which can drastically impair the usefulness of IoT from common man's perspective. With efficient management of all these discussed issues, IoT seems to envisage a world which is fully connected with such excellent devices which can do anything and everything better than human and all the present technological advancements will appear as dwarfed. Moreover, it is the only thing which anyone can blindly rely upon to beat the competition and be future-proof.

REFERENCES

Ahlgren, B., Hidell, M., & Ngai, E. C. (2016). Internet of Things for Smart Cities: Interoperability and Open Data. *IEEE Internet Computing*, *20*(6), 52–56. doi:10.1109/MIC.2016.124

Barnaghi, P., & Sheth, A. (2016). On Searching the Internet of Things: Requirements and Challenges. *IEEE Intelligent Systems*, *31*(6), 71–75. doi:10.1109/MIS.2016.102

Bedi, G., Venayagamoorthy, G. K., & Singh, R. (2016). Navigating the challenges of Internet of Things (IoT) for power and energy systems. *Proceedings of 2016 Clemson University Power Systems Conference (PSC)*, 1-5. 10.1109/PSC.2016.7462853

Bekara, C. (2014). Security Issues and Challenges for the IoT-based Smart Grid. *Procedia Computer Science*, *34*, 532–537. doi:10.1016/j.procs.2014.07.064

Jeyanthi, N., & Thandeeswaran, R. (2017). *Security Breaches and Threat Prevention in the Internet of Things – common issues*. IGI Global. doi:10.4018/978-1-5225-2296-6

Mahmoud, M. S., & Mohamad, A. A. H. (2016). A Study of Efficient Power Consumption Wireless Communication Techniques / Modules for Internet of Things (IoT) Applications. In Advances in Internet of Things (pp. 19–29). Academic Press.

Misra, S., Maheswaran, M., & Hashmi, S. (2017). Case Studies of Selected IoT Deployments. In *Security Challenges and Approaches in Internet of Things*. Cham: Springer. doi:10.1007/978-3-319-44230-3_6

Sheehan, B., Murphy, F., Mullins, M., & Ryan, C. (2018). (in press). Connected and autonomous vehicles: A cyber-risk classification framework. *Transportation Research Part A, Policy and Practice*.

Singh, S., & Singh, N. (2015). Internet of Things (IoT): Security challenges, business opportunities & reference architecture for E-commerce. *Proceedings of 2015 International Conference on Green Computing and Internet of Things (ICGCIoT)*, 1577-1581. 10.1109/ICGCIoT.2015.7380718

Tripathi, B. K., & Anuradha, J. (2018). *Internet of Things (IoT) Technologies, Applications, Challenges, and Solutions*. CRC Press.

Chapter 5
IoT Impact and Challenges on Robotic Waiters in Automation of Restaurants and Hotels

Neelima Mishra
Suresh Gyan Vihar University, Jaipur, India

Dinesh Goyal
Poornima Institute of Engineering and Technology, Jaipur, India

Ahmed A Elngar
https://orcid.org/0000-0001-6124-7152
Faculty of Computers & Artificial Intelligence Beni-Suef University, Beni Suef City, Egypt

ABSTRACT

In China and Japan robots are used to serve in restaurants to serve or to just greet the customer. In restaurants robots need to understand the environment and make decisions accordingly, like changing the route if it finds any obstacle in between. To have this capability of making decisions and behaving like humans, robots need information from different sources from which they can understand the change in environment, position of other robots, and also the path to the destination. This can be done by integrating robots with IoT technology that allow the enhancement in the capabilities in robotic waiter. IoT also helps in monitoring the working parts of robots and movements. This chapter aims to discuss the real-time challenges and impact of IoT on robotics services in restaurants and hotels.

CONCEPT OF INTERNET OF THINGS (IOT)

This is known in industries that IoT is improving flexibility (Lennvall et al., 2017) in automated industries. Not only in industries but in homes also Internet of Things (IoT) had become an essential part. According to some reports, IoT (How Internet, 2016) (Turcu et al., 2012) refers to all devices that are connected to each other smartly and have the feature to communicate with each other via existing internet

DOI: 10.4018/978-1-5225-9574-8.ch005

technology or any dedicated communication protocol. These devices in IoT infrastructure can be any object/thing like human, smart machine, animal, any physical or virtual thing etc. having unique address (Ray, 2016). In simple language Internet of things (IOT) is basically a network where physical devices or things are connected to each other (here physical devices are not just computer systems) for sharing information smartly via Internet (Patel et al., 2016) (Ali et al., 2015).

Vision of Internet of Things (IoT)

- IoT is a standard that spreading widely among the things by either connection, wired or wireless. And for communication unique addresses of things are in used. This address helps in communication among the things to make new application and services that develop a smart scenario or environment where all uniquely identified living, non-living, real or virtual are transforming information smartly to achieve a goal.
- IoT brings all the things under one roof, it mention the entire thing that can communicate via Internet with any means of communication. It not only include computer system, mobile or any electronic device as a thing but also non-electronic thing for example books, human, plants etc.. This is how IoT brings new paradigms in Internet, as this is allowing things to communicate intelligently because they can provide information about themselves. They can gain the information and meanwhile they can be the source of information.

Application of Internet of Things (IoT)

- **Home Automation:** In home automation or smart home IoT providing the facility to owner to control the doors, switches and electronic gadgets at home anytime and anywhere via Internet.
- **Personal Healthcare:** Connecting personal health care gadget with IoT helps a sports person to monitor his/her pulse rate, heart beats etc.
- **Smart City:** IoT makes cities smarter by integrating sensors at different places. For example the time when any road side dustbin is full completely then IoT enable sensor gives signal to municipal. Also person can check the available free parking using IoT.
- **Smart Watch (Wearable Gadgets):** Smart Watch gather all the information of user by asking to log in and then user can make calls, see notifications and mails on watch itself.
- **Learning at anytime and anywhere:** IoT brings boom in education sector. Using IoT communication is possible between teacher and students and they can also check the other information of school without present in the class.
- **Digital and connected Factory:** In factories some time worker has to go for the field jobs also, at that time the IoT enabled machines can transmit the information via any mean like mail or message to the worker about the status of job or condition of machine.
- **Connected Cars:** This technology is the integration of different great technologies like communication system for navigation, software, and different sensor. This combination allows the driver to let the car drive autonomous in a consistent speed (Huang et al., 2010).

Figure 1. Internet of things (IoT, 2015)

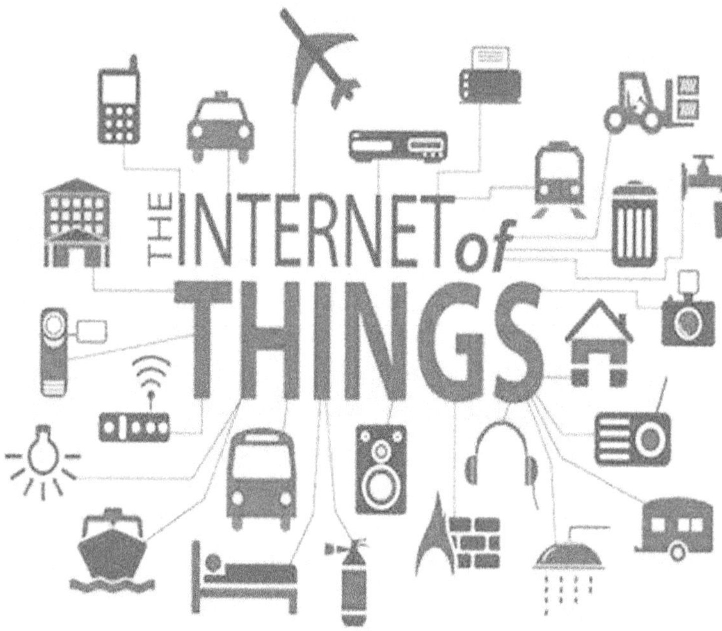

INCORPORATING ROBOTS WITH INTERNET OF THINGS (INTERNET OF ROBOTIC THINGS)

While talking and watching the market of IoRT, the first and most important question that comes in our mind is "what is IoRT and from where it comes from?" After understanding IoT and learning about intrinsic and indispensable quality of IoT, it comes that objects with unique IDs are connected with sensors, tags and actuators (iScoop, retrieved 2019). So the process of communication and transmission of data among devices of IoT carries:

- Data collection.
- Analyzation and communication among devices or to hub.
- Cloud platform for analyzing data.
- IoT platform (as middleware).
- And providing somewhat advantage to sensors (connected passively) to take decision.

The above simple explanation of IoT brings us to notice about two important things, which are:

1. Machine to Machine (M2M) communication.
2. Information Gathering.

One of the main and intelligent machines among many machines in Industries is Robots, not only in manufacturing industries but also in different sectors of consumer industry to assist and giving service to customer. So, connecting this physical intelligent machine called Robot with digital world of Internet of Things open the ways to achieve new services and brings the ability of taking decision by machine itself for given scope of job. Because the robots are not only connected with Internet of Things but also take action according to the data analyzed and received.

Internet of Things (IoT) and Robotics

It is already stated in many papers that robots are the most intelligent machine. They took the maximum market of automation industries and improved many of the issues they are experiencing with human in repetitive, tedious and life taking jobs. Same as industries there are some public sectors also where robots are handling jobs very perfectly, like airports, hospitals, homes and restaurants etc..

Earlier there are some limitations with robots like they are designed and programmed for dedicated purpose only and at the time of change in any condition or environment they do not response. They are design and assign for a specific task only that create a thick wall of limitation with robots because environment and applications are changing very fast. But now by using Artificial Intelligence (AI), robots are becoming more intelligent and can respond to the change in environment. But still there are some issues situated with robots like remote monitoring and many more. For making the robots more responsive and be the decision maker as per situation they should be kept updated by information and instructions, this can only happen by internet. This need of internet with robots then introduce the term "Cloud Robotics", Google self-driving car (Guizzo, 2011) is the best example for this. Then ABI Research (ABI Research, retrieved 2019) has introduced Internet of Robotics Things (IoRT) that was further used in many researches (Turcu et al., 2012) (Ray, 2016) (Simoens, 2018) (Chowdhary, 2017), this is a merging technology of robotics and IoT and an advanced version of Cloud Robotics. In industries the jobs are fixed but still companies are adopting IoT with robotics services to make their jobs flexible, remote monitoring of jobs, maintain the components (here it means predictive maintenance) and the most important is for the safety of human workers. Likewise consumer industries are totally dynamic and the need of IoT is very high here. Airport and restaurants are the best areas of consumer industry to understand the robotics services as they do navigate around to serve customers and communicate with them.

The general reason behind assimilating IoT with Robots (Simoens, 2018):

- Firstly, information services from IoT were helping in sensing and remote monitoring of robots.
- Later, it enables the robots to behave autonomously and can interact with other things also over Internet.
- Here, robots are connected to the IoT as a thing and can interact with other things over Internet and behave in two ways; either the source of information or the consumer of information.

Robot is connecting to the IoT in either two modes:

- Active Mode: Robot is connected to the internet, allowing sending real time information to the Internet.
- Passive Mode: Robot is not connected to the Internet, but can be uniquely identified through a unique address.

Figure 2. Overview of IoRT

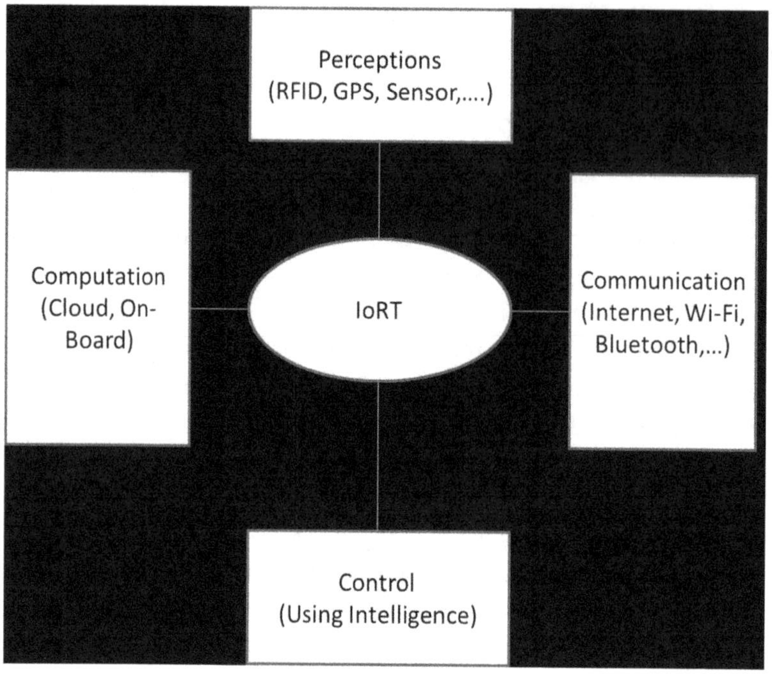

Two Standard technology in Internet of Things and Robotics: Sensors and Actuators

For making robots more intelligent and capable to interact with human it is important to understand the two standard technologies other than Internet of Things, which are, Sensors and Actuators. Every human interaction functions like voice or face identification or control of mobility etc. that helps in making robots intelligent enough to understand the environment is implemented using the combination of Sensors and Actuators. LIDAR – Light Detection and Ranging, Radar, Motors, Cameras, Microphones are some of the example of Sensors and Actuators.

Lidar, Radar and Cameras are the most common sensors (Henrik et al., 2016) used in robots, these sensors helps robot in understanding the surroundings. In spite of this microphone sensor is also used for voice recognition. Sensor helps mobile robots in collection of data for different position and at change of time (Luo et al., 2002).

The feature to adopt the change in outer surrounding or environment and perform accordingly is comes under actuator part. In malls and offices, the automatic process of opening and closing of door by sensing human presence or any movable object, is the best example to understand the actuation (LaValle, 2006).

Figure 3. Navigation of Robot to Serve on Table in Restaurant

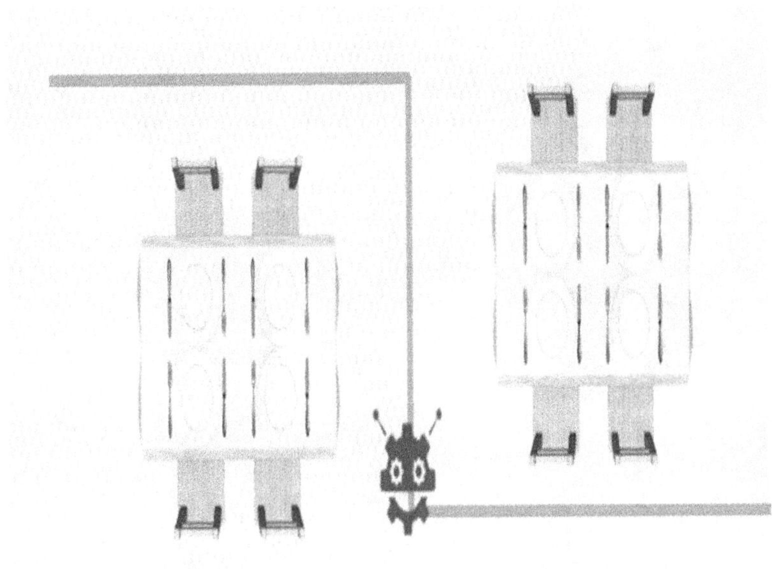

General Capabilities Enhance in Robots after Incorporate with Internet of Things

Observation Capability

The information analyzing feature of Internet of Things (IoT) broadens the perspective of Robot. The main element of observation capability is to getting information of its own location (i.e. self-localization). This is done by interact and communicate with other things connected via Internet.

For example: In restaurant after serving to customer robot need to go back to its folding station to get charged. In this situation the cameras in restaurant (that are connected with Internet of Things) can helps robots to locate its folding station (Thrun et al., 2008).

Navigation Capability

The movement is the most important part of Robot and IoT assist the robots by providing information about shortest path to be taken and the obstacles in between.

For example: In restaurant, robotic waiter needs to navigate to customer table to serve food. For this it needs two information, firstly is the desire table and secondly the shortest path up to that table to serve food. Here Internet of Things help robots by fetching the information about environment of restaurant that include path and location of tables and can easily navigate to the desired table. If an obstacle comes in between then robot can change the path to another shortest one using IoT itself.

There is one best example to understand the navigation using Internet of Things i.e. Google Car or driverless car. The Google car (Sharon et al., 2014) is a system that use detail information of map to navigate from source to destination. It is using the combination of software and hardware technology that consist of radar sensors, laser range finder to visualize live traffic and radar sensor.

Figure 4. Robot is taking decision to do not enter the kitchen till presence of human inside (Cirillo et al., 2010)

Performance Capability

The objective of Internet of Things is to understand the environment and provide relative information to robot and act to modify their environment. This is helping robots in knowing the object, pick and place the object and can also hold the object.

For Example: (Rusu et al., 2008) describe an intelligent serving robot that helps in kitchen to locate the kitchen ware. It can grasp, move and then release the object in kitchen. The knowledge about location and details of object is obtaining using Internet of Things.

Decision Making Capability

The decision making capability refers to decide the best action according the situation. This can be happen with the pre known model of environment and some predefined action. Both information (about environment and action) should be perfect because the quality of decision is completely depend on these two factors. Like (Cirillo et al., 2010) propose an intelligent robot that having the ability to plan the day cleaning activity of house. It is having the intelligence of taking decision to not to clean the bed room till any human is inside. This sensing capability and knowledge of action is provided by connecting it with Internet of Things.

For example: Smart Floor Cleaning Robot (CLEAR) (iRobot, 2015), a robotic vacuum cleaner is an autonomous electronic device that is intelligently programed to clean a specific area through a vacuum cleaning assembly. The best product of robotic vacuum cleaner is iRobot.

Interactive Capability (IoT, 2015)

Here interaction refers to interact with human or any other system along with understanding the mood by reading expression. Robot can also interact with doing action in response towards human, here robot do read the gesture or voice command (Cambria, 2016) (De Ruyter, 2005) (McColl et al., 2016).

For example: (McColl et al., 2016) discussed about some survey of robots that can interact with human and also socially acceptable. To interact with human, robot should be known with the current state of human by reading his/her behavior. There are number of autonomous robot the researcher discussed in his paper that develops to identify the behavior of human by any of the below or combination of some of the below factors: Facial expressions, gesture, body language, voice command etc.

LITERATURE REVIEW

Integrating robots with Internet of Things makes the robot more intelligent to serve public without bounded in any constraint of environment. In public sector or consumer industry they can understand the gesture and do react as well as they are serving human in carrying their goods also. They can read the outer surrounding and provide the information to its server via Internet, as well as robots are behaving as a consumer of information also that it receives from the server somewhere via Internet again. Like consumer industry, the manufacturing industries are also adopting robots along with Internet of Things to get flexibility in their jobs with low cost and for the sake of human worker safety. In manufacturing industries the robots are helping human in transporting heavy materials and tools from one place to another with avoiding the obstacle and understanding the surrounding environment.

There are some research papers that providing information about robots which are serving in many sectors leveraging with Internet of Things.

The K5 by Knightscope is the best example describe in research paper (Bendel, 2017). This robot is used for the security purpose in shopping malls. Basically it surveillance the customer and take care of any suspicious activity around it. In case of any unwanted activity happen K5 quickly gives the information in the control room of shopping mall.

There is one more service robot that Care-O-Bot (Graf et al., 2004) that is used in homes and trade fairs. As name is describing it is used for nursing purpose at home and having the ability of communication also.

Then another robot name Enon (Frontech, 2007) was introduced in 2007. The purpose of introducing Enon is to escort the visitors in Kyotara Nishimura Museum.

Like Enon there are some other robots also that deploy in museum to guide public. The need of these robots to integrate with Internet of Things is to navigate safely without any collision in an indoor crowded environment. Wolfram Burgard (Burgard et al., 2000) proposed a robot in his paper that guide customer in the “Deutsches Museum Bonn”. And Sebastian Thrun (Thrun et al., 1999) proposed a Museum tour guide robot called MINERVA. This robot was deployed in Smithsonian museum for few weeks only.

Not only in Shopping Malls and Museum but at homes where elder people do live alone and need someone to interact with, there a robotic solution called ElliQ (Ackerman, 2019) which eliminate the loneliness in elder generation. This robot is using AI (Artificial Intelligence) and Internet to assist elder generation by first learning the behavior of elder person with his/her gesture. It can quickly connect to the internet or digital data to play music and search things for them and can also monitor the health too.

Along with elder persons there is a robot for children too called Nao (Boboc et al., 2014). This is used for education purpose, here robot do interact with children and help them in learning of new things. In spite of learning it also works as tour guide to the laboratories.

Internet of things is in use from many years to make industries automated even the restaurants also. S. Dharani (Dharani et al., 2018) has proposed the Internet of Things (IoT) based restaurant that is fully autonomous and his vision is to no make the customer wait for long to order the food. He uses Arduino, IR (Infrared) sensor and motor to change the traditional system of restaurants. There is one more paper that proposed the low cost touch screen based Restaurant Management System using an android Smartphone or tablet as a solution (Jakhete et al., 2015). This system is for smart ordering from E-Menu. Here customer does not facilitate with any hard copy of menu, he/she can easily order food using their smart phones or tablets by connecting it to Internet.

In restaurants of Japan and China, robotic waiters are already giving service to public like Hajime (Steven, 2011) is the restaurant in Bangkok, Thailand that converts the theme into robotic theme. Here, public can order food by the touchscreen on table there and then a line follower robot come to serve their desire meal. Here the robot not only serve but do dance also to entertain the customer.

In restaurants and hotels robots are not only helps in serving the food but helps in room service and in kitchen too. Flippy (PYMNTS, 2017) is the robot that helps chefs in kitchen side. It cannot cook food but it does help in flip the burger, this is how its name comes from. David Zito is the inventor of Flippy gives a statement that it relies on Artificial Intelligence (AI), Internet of Things (IoT), sensors and cameras to perform the task.

Two robots name Jeno and Jena (Soo-Jin, 2017) is doing room servicing in one of the hotel in Japan. Both robots are equipped with sensor that help them in avoiding the obstacles comes in between and they also connect with the Wi-Fi of Hotel to communicate with the lift and do phone call to customer to ask him/her to open the door.

Abhishek (Patil et al., 2016) has proposed a model to control the robot using different hand gesture. The movement of hand gesture is provided using MEMS Sensor to robot. The Sensor is situated on hand and the movement is sensed by MEMS Sensor. This MEMS Sensor convert the movement into electrical signal and transmit it by Internet of Things (IoT) to the receiver side (there is a controller at receiver side) and the receiver provides these signals as a direction to the robot, this communication (between receiver and robot) is also done through Internet of Things (IoT).

This research paper (Robotics, 2017) describing the best example to understand Internet of Robotic Things (IoRT) real time application. In amazon warehouse thousands of robots are in use for picking and packing the merchandise order by customer. These robots used to carry boxes and move to the destination in warehouse without any collision with other robot, for this they are using combination of different technologies like machine learning, depth sensing and recognition of object. At the time of order placed by customer at amazon store, the software send the order to the nearest mobile robot and then the robot recognize the merchandize and brings the box for packing and all.

ROBOTIC WAITERS IN RESTAURANT

Numbers of robots are there providing services in different areas, especially in public areas. Recently a robot Troika had been introduced in South Korea airport which help passenger find their gate. (Lee, 2017) The guiding bot responds to its name. Travelers can insert their tickets into its scanner to get flight information, and Troika will then ask if they want to be escorted to their gates, warning laggards to "Please stay closer so I can see you." Like Troika there is one more robot that is used to clean the airport floor. Basically it is an autonomous vacuum cleaner by LG (Read, 2017), the Airport Cleaning Robot is a large vacuum cleaner fitted with cameras, light sensors, and sensor-laden bumpers which moves around the airport vacuuming up rubbish. Like airport, there is another most popular public place i.e. restaurants.

Background

Serving robots are recently used in China and Japan. These countries are using robotic services in restaurants in many ways; they are serving sushi through conveyor belt and serving robots are used in place of waiters that are navigating through the fixed path to serve. To overcome from the line following technology to free indoor navigation and to complete the process from ordering food to billing, it is required to connect robot with Internet.

Objective

In most of the restaurants in India, traditional method exists where customers give their orders to waiter and waiter do write them on paper which is then prepared and served to customer, and give paper bill after completing order. The researchers had already proposed Robotic waiter in our previous papers (Mishra et al., 2018) (Mishra et al., 2018) that will navigate autonomous to serve. There customers do order their desire food via smart mobile phones using local Wi-Fi that will directly reflect on the screen at kitchen side and then served by robotic waiters using shortest path identification to the correct table and then bill will either mail or message to the customer. This complete process is facilitating by local Wi-Fi of restaurant. The main objective of this chapter is to discuss the impact and challenges of using Wi-Fi as the communication channel on serving robot.

CHALLENGES AND IMPACTS OF IOT ON ROBOTIC WAITER

There are various wireless technologies (Nayyar et al., 2018) used for connection between things and Internet, they are RFID (Radio Frequency Identification), Wi-Fi, Bluetooth and microcontroller board. The researchers are using Local Wi-Fi connectivity for the communication so that customer can only order food after occupying the seat in Restaurants and Hotels. The major challenge while installing the robots and serving is good connectivity of Local Wi-Fi. Restaurants are having multiple dongles for Wi-Fi and multiple floors. The biggest challenge with serving through robot is to connect the robots with Wi-Fi of different dongles and creating maps of different floors because it needs to upload these maps of all floors on server for further navigation. Internet of Things becomes important when multiple robots are navigating through single map. In this case Internet of Things has the responsibility of having information about the path of each and every robot so that robots cannot comes in between any other robot. Using Internet of Things robot must either wait while other robot is passing through the same path.

Issues Arise at Low Connectivity of Local Wi-Fi

Mapping Issue

The serving robot is using the visual imagine via camera for mapping. Robots then broadcast the indoor map of restaurant to the dedicated server for further navigation. Then for proper navigation of robots, server will send the data from map on the primary memory of robot. This process of mapping is done via local Wi-Fi connection, at the time of weak signal it can be possible of missing of some data.

Single robot can cover the range of Wi-Fi at particular range easily but after moving forward the range of Wi-Fi shifts to another router, in this case robot has to update the map of that next range. There are two parts in mapping;

- The predefined indoor map that is uploaded by the robot itself using LIDAR – Light Detection and Ranging Sensor does not have any obstacle. This LIDAR sensor is used to provide the information about surrounding as 360 degree view i.e. orthogonal view.
- Then the already uploaded map start updated by robots while real time movement with movable and non-movable obstacle like human or another robot comes in between or any static chair etc. This updated map with obstacle helps other robots to avoid obstacle and can navigate freely. This update is done by each and every robot on Global Map to keep the other robots updated with coming obstacle.

The weak signal of Wi-Fi may cause disturbance in either updating data on server or while transferring updated map to robot.

Order Placing and Billing Issue

At the time of entering of customer in restaurant they use the local Wi-Fi to access the menu of that restaurant and do order the desired that will reflect on screen at kitchen side and then after asking for bill, customer receives the bill via message or mail. At the time of weak signal there can be a chance of missing the process.

Here is the block diagram in Figure 5 of smart ordering system, where customer can access the web ordering system and that order will retrieve by the restaurant employee. This is the basic operation of smart ordering system, further the order will be serve by Robot to the desired table.

The smart ordering system can be divided into 3 main components:

- **Web Ordering System:** Customers can place their order via their smart phone.
- Menu Management System- The menu management will be authorized by the authorized person from restaurant. In this they can manage their menu and can hide/delete/add the required item.
- **Order Retrieval System:** In this system, restaurant employee will keep track on each and every order receive/serve.

The weak connectivity of Wi-Fi may delay the order or may be may be missed before displayed on KOT window screen and make the customer wait.

Figure 5. Block Diagram of Smart Ordering System (Mishra et al., 2018)

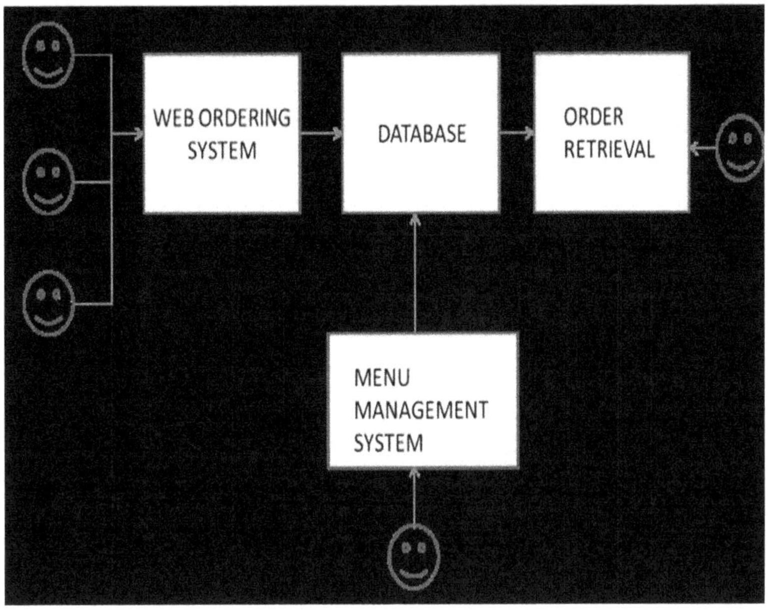

Position Allocation Issue

To allocate the real time position of robot, it acts as a thing and connected to the Internet for real time sharing. The camera, digital compass and sensor are used to localize the position of robot on map. This

Figure 6. Ordering Application via Smart Phone

Figure 7. Position of Robot on Map

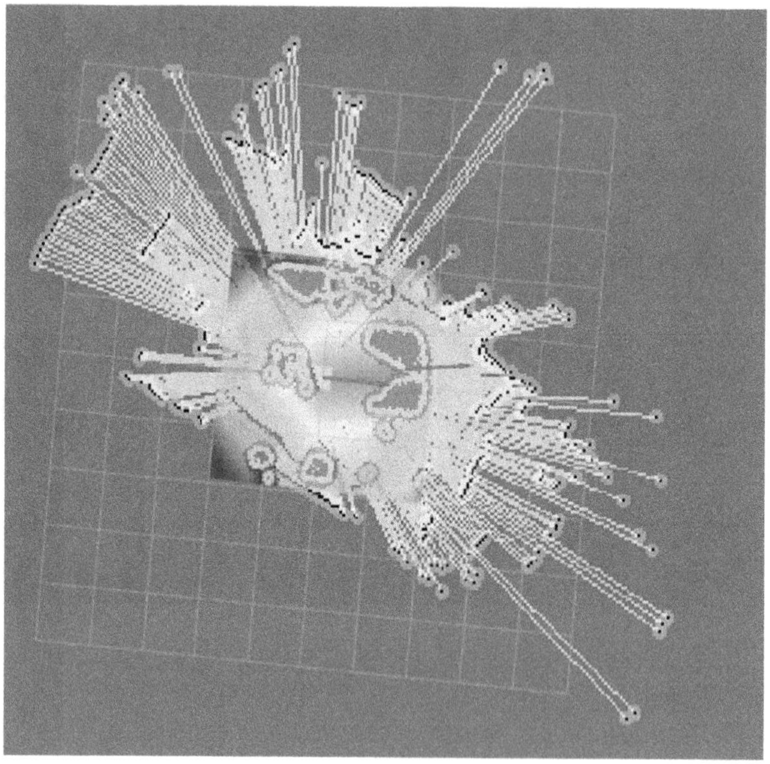

local position of robot need to be updates every time on server so that the robot can navigate properly and serve on correct table. Also at the time of any obstacle or intruder comes in way of serving robot and it gives a notification by alarm to clear the way and then in case of no clearance it change the path by accessing the alternate path by server from that current position. In below screenshot of live map by serving robot square box in grey color is the current position of robot.

Robot updating its position and current behavior according to time and if Wi-Fi connectivity become weak disconnect then after re-connecting with Internet robot either stay on the previous state (the last updated time) or the map will overlap between two time span. There are two types of maps one is the Global Map that is on Server and another is the Local Map that is in Robot memory. Global map shows the indoor map of restaurant and local map gives the position of robot. In this case time frame of both Global and Local Map should be common that gives combine information of at what time the robot is at which position at Restaurant. The disconnection with Wi-Fi may misguide the robot and that then start providing wrong positional information of that particular robot to other robots also.

Navigating to the Table Issue

Navigation process is connected with mapping. The robot capture the route of each and every table during the mapping process and send it to server, then at the time to select path for serving food to table robot receive the location of table by server using local Wi-Fi connectivity. The whole process of navigation

Figure 8. Path to serve (In green line)

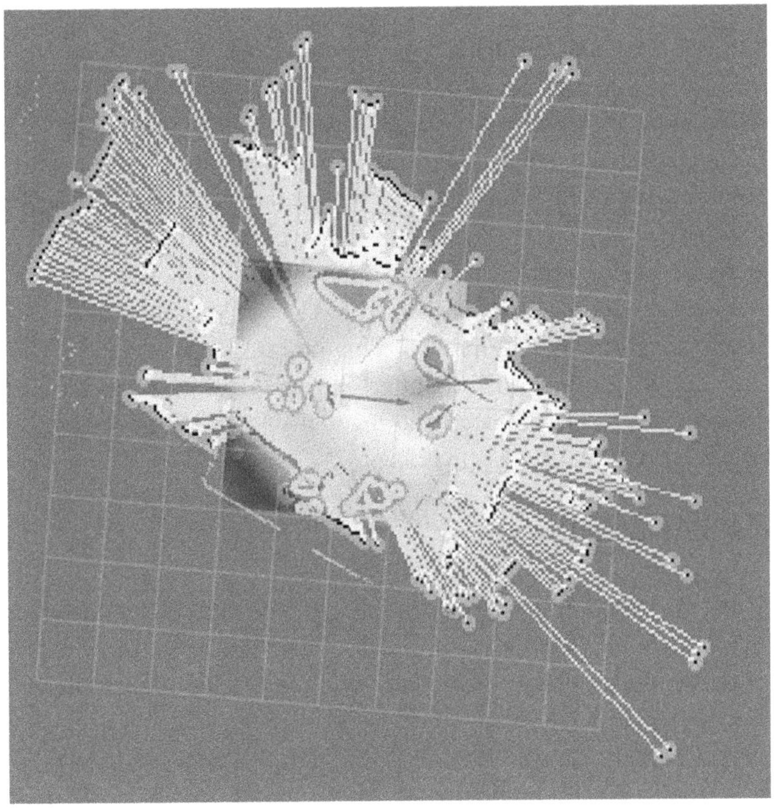

and finding the path from source to the table cannot be accomplish if Wi-Fi connectivity is not strong. In Below image, the green line from one red arrow to another is the path to go for serving.

While navigation of serving robot two factors are to be defining, one is the x-y coordinates of table and Odometry of Robot. This information is provided by Global map to robot. In this case the main function of Internet of Things is to update both factors on time so that robot can determine the position of table as well as position of itself. Otherwise at any point of disconnection with Wi-Fi robot get misguide and cannot able to identify and serve at correct table.

Proposed Solution for Described Issues

Major processing is performing on main server, that's the reason of robot dependence on local Wi-Fi for communication and updating. So it needs a good coverage of local Wi-Fi while mapping, navigation and ordering food. The weak Wi-Fi is the major issue comes in between while serving. To overcome from this issue researchers are proposing a standalone serving robot. In this the complete processor is integrated on robot body that allow robot to do the processing on its own without depending on local Wi-Fi.

Impact of Connecting Serving Robots with Internet of Things

Remote Monitoring of Serving Robots

Using local Wi-Fi, the complete data of indoor map, location of robot and path for navigation is stored on Global server. Each robot is having its own unique machine ID that distinguished them with each other. Using these IDs the researchers can remotely monitor the working of robots and can troubleshoot them before its breakdown.

In real time navigation every robot navigate while updating the obstacles comes in between and update it at Global Map, as all robots are gaining the path from Global Map then in this case other robots are also able to know about the obstacle that updated. This process is performed by each and every robot while navigating. This identification of position of robot helps while troubleshooting, as researcher can gain the current position through Global Map and manually instruct that particular robot to comes to folding station and then troubleshoot the issue arise.

Making Restaurants Autonomous

Not only automobile industries but consumer industries are also adopting smart autonomous machines. In restaurants hiring robots as a waiter is best way to make the restaurants and hotels autonomous. Adopting IoT with robots is the best way by connecting smart phones of customer for ordering food and serving robots to serve on table.

The complete process of automated restaurants takes place in following steps:

Step 1: Firstly, robots make a complete map of restaurant and different floors (if there) of hotel and update it on the server i.e. global map, here no obstacle is present on map and this map is access by every robot serving in restaurant to navigate.

Step 2: First customer engages a table in restaurant and use the local Wi-Fi of restaurant for the e-menu on his/her smart phone.

Step 3: Customer do order the food by smart phone that reflect on the screen at kitchen side where the chefs do prepare the order and keep the ordered food tray on available free robot (at folding station). Let that robot is "A".

Step 4: Now meanwhile some other robots are also serving on different tables or on different floors. At that moment the other robots do update the global map with moving and still obstacle comes in between.

Step 5: The robot "A" now navigate by accessing the path from global map to the desire table from where the order comes from. The updates of obstacle by other robots helps in deciding the free shortest path to the destine table.

Step 6: To avoid the collision between serving robots, they update their current location on server by local map that is in the memory of robot itself.

Step 7: By accessing all the updated information on Global map Robot "A" then deliver the food to the table and come back to its folding station again.

Step 8: After completing the meal, customer do ask for bill with his/her smart phone that they receive as a message or mail on the same and they can do digital payment.

Step 9: This complete process is depending on the good communication connectivity among the robotic waiters and the strong Wi-Fi in restaurant.

FUTURE SCOPES

The scope of improvement is still there. Room service or escorting the guests to the room or to the help desk is a part of hospitality in renowned hotels. Here robots from different floors do update their location on main server and communication among robots is also required to be updated with map of floors. Also while doing room service robots have to communicate with lift to select the floor and can serve to desire room using the updated floor map on server.

CONCLUSION

Internet of Things (IoT) is a network of many smart devices connected and communicates via Internet and robot is one of the most intelligent machines that already capture the large area of industries. This combination of IoT and Robots opens the door for more development in future in the field of every industry. Coming from the time of making industries automated we are in the era of automating the consumer industries like airports, restaurants etc. also. Here, researchers had discussed about Internet of Things, reason of connecting robots with IoT that then term as Internet of Robotic Things (IoRT). In previous papers of researchers they propose and compare the serving robot where robots do serve food to customers autonomous. Introducing serving robots in restaurants is not enough to call restaurant an autonomous. There is a need of approach to make the complete working of waiters in restaurants and hotels autonomous and paper free. This is done by adopting Internet of Things (IoT). Here researches connect the serving robots are with local Wi-Fi as a thing of restaurant for sending and receiving data online. This combination of Internet of Things (IoT) and Robots comes with many challenges and impacts. Also the researchers had discussed about free indoor navigation (without following any line) in one of their previous paper, so this also brings a major challenge and Impact on serving robot while navigating without any collision or breakdown to serve in restaurant and at different floors for food and room service. In this chapter researchers have discussed the real time challenges and impacts while connecting smart phones (to order food) and serving robots (to serve food) with local W-Fi.

REFERENCES

ABI Research. (n.d.). *Internet of robotic things*. Retrieved from https://www.abiresearch.com/market-research/ product/1019712-the-internet-of-robotic-things

Ackerman, E. (2019). *ElliQ, A Social Home Robot for Older Adults, Now Available for Pre-Order*. Retrieved from https://spectrum.ieee.org/automaton/robotics/home-robots/elliq-a-social-home-robot-for-older-adults-now-available-for-preorder

Ali, Z. H., Ali, H. A., & Badawy, M. M. (2015, October). Internet of Things (IoT): Definitions, Challenges and Recent Research Directions. *International Journal of Computer Applications*, *128*(1), 37–47. doi:10.5120/ijca2015906430

Bendel, O. (2017). *Service Robots in Public Spaces*. Retrieved from https://www.heise.de/tp/features/Service-Robots-in-Public-Spaces-3754173.html?seite=all

Boboc, R. G., Horaţiu, M., & Talabă, D. (2014). An Educational Humanoid Laboratory Tour Guide Robot. *Procedia: Social and Behavioral Sciences*, *141*, 424–430. doi:10.1016/j.sbspro.2014.05.074

Bugard, W. (2000). Experiences with an Interactive Museum Tour-Guide Robot. *Artificial Intelligence*, *114*(1/2).

Cambria, E. (2016). Affective Computing and Sentiment Analysis. *IEEE Intelligent Systems*, *31*(2), 102–107. doi:10.1109/MIS.2016.31

Chowdhury, A. R. (2017). *IoT and Robotics: A Synergy*. Peer J Preprints; doi:10.7287/peerj.preprints.2760v1

Cirillo, M., Karlsson, L., & Saffiotti, A. (2010). Human-aware task planning: an application to mobile robots. *ACM Trans Intell Syst Technol*, *1*(2), 15:1–15:26. Retrieved from https://dl.acm.org/citation.cfm?id=1869404

De Ruyter, B., Saini, P., Markopoulos, P., & van Breemen, A. (2005). Assessing the Effects of Building Social Intelligence in a Robotic Interface for The Home. *INT COMPUT*, *17*(5), 522–541. doi:10.1016/j.intcom.2005.03.003

Dharani, S. (2018). IOT Based Smart Restaurant. *National Conference on Emerging Trends in Information, Management and Engineering Sciences.*

Frontech, F. (2007). *Service Robot Enon from Fujitsu Acts as Visitors' Guide at Museum of Best-Selling Mystery Novelist.* Retrieved from http://www.frontech.fujitsu.com/en/release/20070920.html

Graf, B., Hans, M. & Schraft, R.D. (2004). *Care-O-bot II—Development of a Next Generation Robotic Home Assistant.* Academic Press.

Guizzo, E. (2011). How Google's Self-Driving Car Works. *IEEE Spectrum.* Retrieved from https://spectrum.ieee.org/automaton/robotics/artificial-intelligence/how-google-self-driving-car-works

Henrik, I., & Christensen, G.D. (2016). Sensing and Estimation. In Springer Handbook of Robotics (2nd ed.; pp. 91-111). Springer International Publishing.

How Internet of Things effects on our daily lifestyle. (2016). Retrieved from https://www.futurescope.co/internet-of-things-iot/

Huang, Y., & Li, G. (2010). *Descriptive Models for Internet of Thing. In IEEE International Conference on Intelligent Control and Information Processing* (pp. 483–486). ICICIP.

IoT. (2015). Retrieved from https://dzone.com/articles/the-internet-of-things-gateways-and-next-generation

iRobot.com. (2015). *iRobot Corporation: We Are The Robot Company.* Retrieved from http://www.irobot.com/

iScoop. (n.d.). *The Internet of Robotic Things (IoRT): Definition, Market and Examples.* Retrieved from https://www.i-scoop.eu/internet-of-things-guide/internet-robotic-things-iort/

Jakhete, M.D., & Mankar, P.C. (2015). Implementation of Smart Restaurant With The E-Menu Card. *International Journal of Computer Applications, 119*(21).

Kehoe, B., Patil, S., Abbeel, P. & Goldberg, K. (2015). A Survey of Research on Cloud Robotics and Automation. *IEEE Transactions on Automation Science and Engineering*, *12*(2), 398-409.

LaValle, S. M. (2006). Planning Algorithms. Cambridge University Press.

Lee, Y. (2017). *Robots at South Korea Airports to Help Passengers Finding Their Gates*. Retrieved from https://skift.com/2017/07/04/robots-at-south-korean-airports-to-help-passengers-find-their-gates/

Lennvall, T., Gidlund, M. & Akerberg, J. (2017). Challenges When Bringing IOT into Industrial Automation. *IEEE Africon,* 905–910.

Luo, R. C., Yih, C. C., & Su, K. L. (2002). Multisensor fusion and integration: Approaches, applications, and future research directions. *IEEE Sensors Journal*, *2*(2), 107–119. doi:10.1109/JSEN.2002.1000251

McColl, D., Hong, A., Hatakeyama, N., Nejat, G., & Benhabib, B. (2016). A Survey of Autonomous Human Affect Detection Methods for Social Robots Engaged in Natural HRI. *Journal of Intelligent & Robotic Systems*, *82*(1), 101–133. doi:10.100710846-015-0259-2

Mehta, Y. (2015). *What Is The Internet Of Things Definition – IoT Definition | Details*. Retrieved from http://iotworm.com/what-is-internet-of-things-definition/

Mishra, N. (2018). Automation in Restaurants: Ordering to Robots in Restaurant via Smart Ordering System. *International Journal of Technologies and Management.*

Mishra, N. (2018). *Issues in Existing Robotic Service in Restaurants and Hotels. 3rd International Conference on Internet of Things and Connected Technologies.*

Mishra, N. (2018). Spine in Robot: Designing of Lightweight Robot with the Spine to Provide Balancing and Load Bearing Capability. *15th International Conference on Recent Trends in Engineering, Applied Science and Management (ICRTESM-18).*

Nayyar, A. (2018). Smart Surveillance Robot for Real-Time Monitoring and Control System in Environment and Industrial Applications. Information Systems Design and Intelligent Applications, 229-243.

Patel, K.K. & Patel, S.M. (2016). Internet of Things-IOT: Definition, Characteristics, Architecture, Enabling Technologies, Application & Future Challenges. *International Journal of Engineering Science and Computing, 6*(5).

Patil, A. T. (2016). IOT Based Remote Access Human Control Robot Using MEMS Sensor. *International Journal of Computer Science and Mobile Computing*, *5*(3), 816–826.

Poczter & Jankovic. (2014). The Google Car: Driving Toward A Better Future? *Journal of Business Case Studies, 10*(1).

PYMNTS. (2017, August 3). *Robot v. Food.* Retrieved from https://www.pymnts.com/intelligence-of-things/2017/miso-robotics-flippy-the-iot-restaurant robot/

Ray, P. P. (2016). Internet of Robotic Things: Concept, Technologies, and challenges. *IEEE Access: Practical Innovations, Open Solutions*, *4*, 9489–9500. doi:10.1109/ACCESS.2017.2647747

Read, B. (2017). *Rise of the Airport Robots*. Retrieved from https://www.aerosociety.com/news/rise-of-the-airport-robots/

Robotics+IoT=The Internet of Robotic Things. (2017, November 20). Retrieved from https://blog.techdesign.com/robotics-internet-of-robotic-things/

Rusu, R. B., Gerkey, B., & Beetz, M. (2008). Robots In The Kitchen: Exploiting Ubiquitous Sensing and Actuation. *Robotics and Autonomous Systems*, *56*(10), 844–856. doi:10.1016/j.robot.2008.06.010

Simoens, P. (2018). The Internet of Robotic Things: A review of concept, added value and applications. *International Journal of Advanced Robotic Systems*, 1–11.

Soo-Jin, K. (2017). *Room Service is Now Delivered by Robots at These Hotels in Singapore*. Retrieved from http://www.scmp.com/magazines/style/travel-food/article/2119123/room-service-now-delivered-robots-singapores-hotel-jen

Steven. (2011). Restaurant Staffed By Dancing Robot Samurai Waiters. *Sora News 24*. Retrieved from https://soranews24.com/2011/06/06/restauraunt-staffed-by-dancing-robot-samurai-waiters/

Thrun, S. (1999). MINERVA: A second-generation museum tour-guide robot. *Proc. of the International Conference on Robotics and Automation (ICRA' 99).*

Thrun, S., & Leonard, J. J. (2008). Simultaneous localization and mapping. In B. Siciliano & O. Khatib (Eds.), *Springer handbook of robotics* (pp. 871–889). Cham: Springer. doi:10.1007/978-3-540-30301-5_38

Turcu, C., Turcu, C., & Gaitan, V. (2012). *Merging the Internet of Things and Robotics*. Recent Researches in Circuits and Systems.

Chapter 6
Experiments on Design of Obstacle Avoiding Robots Based on Sensors, Bluetooth, and IoT

Shirshak Kumar
CSIR-CSIO, Chandigarh, India

Suraj
CSIR-CSIO, Chandigarh, India

Sahil Sandhu
CSIR-CSIO, Chandigarh, India

Narinder Singh Jassal
CSIR-CSIO, Chandigarh, India

Jitendra Virmani
CSIR-CSIO, Chandigarh, India

Kriti
Thapar Institute of Engineering and Technology, India

ABSTRACT

The mobile robotics industry is related to creating mobile robots that can move around in physical environments. Different types of mobile robot designs for obstacle avoidance have been experimented in the past based on different sensors, trajectory algorithms, etc. The chapter presents implementation details of different obstacle avoiding robots (OARs) using sensors, Bluetooth module, and IoT modules. The sensor-based obstacle-avoiding robots are designed using ultrasonic sensors and Arduino micro-controllers. Bluetooth-based obstacle-avoiding robots have been designed using Arduino mega and Bluetooth module and an Android application. IoT-based obstacle-avoiding robots can be designed in three different ways, using ethernet shield, node MCU, or Raspberry Pi. The IoT-based obstacle-avoiding robot using Raspberry Pi is the most popular mobile robot model that uses maximum on-chip modules in comparison to other designs, and also, the design can be extended by using cameras to use images for sensing the objects in order to avoid collisions.

DOI: 10.4018/978-1-5225-9574-8.ch006

In this chapter, three different techniques for the design of an obstacle avoiding robot has been discussed.

INTRODUCTION

Robotics is considered to be an interdisciplinary branch of science and engineering dealing with the design and use of robots to replicate human actions. Nowadays robots are finding use in many situations e.g. automation of manufacturing process, space programs, environments that are dangerous to humans etc.

Robots are increasingly being put to use in various industrial environments like automotive, aerospace, healthcare, quality control, and inspection, warehousing to perform repetitive tasks that require consistent precision and accuracy or heavy payload capacity e.g. material handling, order fulfillment, lifting trucks, automatic guided vehicles, packaging, quality control etc.

One of the recent advancements in robotics is the design of mobile robots. Automated guided vehicles follow predetermined paths while autonomous mobile robots can move from point A to B using alternative routes in case of congestion or obstacles. The autonomous mobile robots can be used in security, on-shelf inventory, production inspection, automated facility cleaning among other tasks.

LITERATURE SURVEY

Many attempts have been made in the past for the design of autonomous mobile robots. All these differ basically in the choice of sensors used, mapping of the path to be followed, the operations used to set the operational parameters and the hardware platform utilized.

The designs of the obstacle avoiding robots (OARs) discussed in the chapter are shown in Figure 1.

Tabassum et al. (2017) presented the design of an obstacle avoiding robot using Arduino microcontroller and three ultrasonic sensors for obstacle detection.

Chen et al. (2009) designed two models of a wheeled mobile robot for obstacle avoidance. The two models were named: Model 1: short-distance obstacle avoidance model and Model 2: target-driven obstacle avoidance model. The first model makes use of the ultrasonic sensors for avoiding the obstacles while the target-driven model makes use of fuzzy logic along with the sensor signals for speed control.

Yang et al. (2010) presented an intelligent mobile robot design for avoiding obstacles using multiple ultrasonic sensors and fuzzy control rules. All the programming has been performed in the VC environment using C and Colbert language.

Xiong et al. (2011) analyzed the movement of the designed robot under complex unknown environments. The robot's obstacle avoiding algorithm is based on MIMO fuzzy rules on the basis of the shape analysis of the obstacle.

Boujelben et al. (2013) proposed an approach that is based on a hierarchical Fuzzy controller for designing an obstacle avoiding robot which can be placed in an unknown environment containing mobile obstacles.

Yamada et al. (2013) proposed a robot design that follows a designated path by avoiding mobile obstacles. In order to scan the region for obstacles, two laser sensors (one at the front and one at back) have been used.

Bhagat et al. (2016) designed an obstacle avoiding robot using ultrasonic sensors and ATMEGA-8 microcontroller.

Figure 1. The designs of the obstacle avoiding robots (OARs) discussed in the chapter

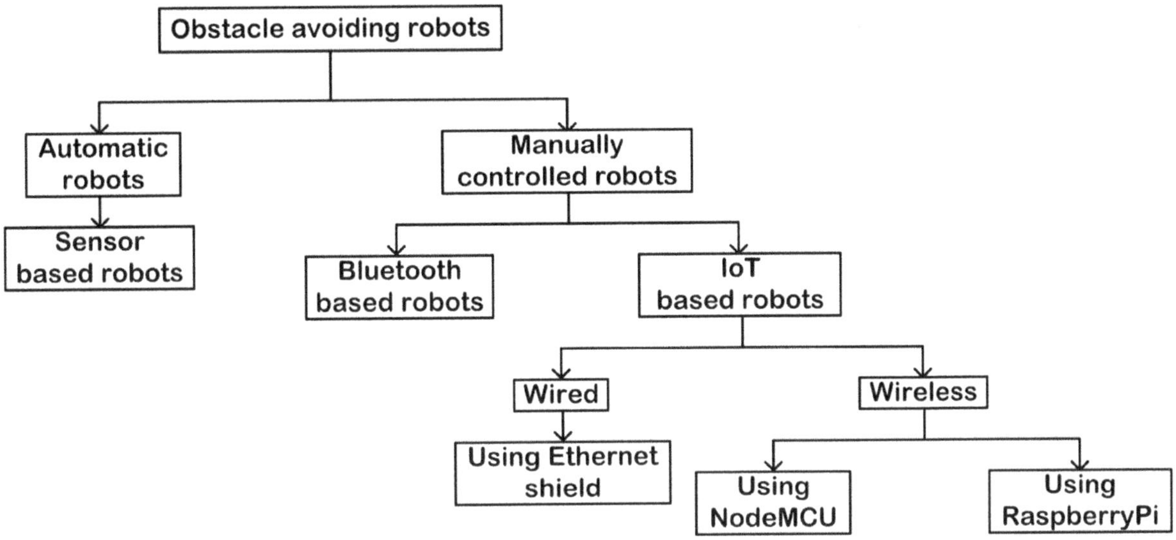

Wu et al. (2014) used 6 ultrasonic sensors and ATMEGA 162 microcontroller to design an obstacle avoiding system on the wheeled robot that continuously scans its surroundings to avoid the obstacles while moving towards the desired target. A PD controller was also integrated into the design for the wall-following method so that the robot could avoid large obstacles.

Jathavara et al. (2017) designed an obstacle avoiding robot based on ultrasonic sensors and a biologically inspired learning method to compute the location and speed of the obstacle such that robot could compute a time window within which it completes its movement to avoid the obstacle.

Zhao et al. (2018) designed a hexapod (6-legged) robot for obstacle avoidance. It uses many sensors like compass, IMU, RGB-D camera, GPS system, and force sensor. The force sensor is connected to the core computer through an Ethernet while the GPS system and compass help the hexapod in navigation.

Nasucha et al. (2015) designed an obstacle avoiding robot using proximity sensors and ATMEGA 328P microcontroller.

Raghav et al. (2017) proposed the design of an obstacle avoiding robot using Bluetooth module, ultrasonic sensors.

Vaghela et al. (2016) proposed the design of an Android app controlled obstacle avoiding robot using ultrasonic sensors and Bluetooth module.

Hanumante et al. (2013) proposed a low-cost obstacle avoidance robot with a dynamic steering algorithm to navigate in unknown environments using IR sensors and ATMEGA-8 microcontroller.

Shahdib et al. (2013) devised a new scheme to detect an obstacle, measure its distance and calculate its size using the camera and an ultrasonic sensor to design an autonomous mobile robot.

Ryther et al. (2009) designed an automatic mobile robot to navigate in an unfamiliar indoor environment. The data of the surroundings is collected using laser scanners and odometer mounted on the robot. The path to be followed for avoiding obstacles is determined using the wavefront path planning algorithm.

Heidarsson et al. (2011) proposed the design of an autonomous surface vehicle that uses profiling sonar for the scanning of surroundings to avoid obstacles in lakes and harbors.

Gray (2000) designed and implemented an obstacle detection and avoidance system that could be placed on an automatic farm tractor. GPS and fiber optic gyroscope (FOG) have been used as sensors for providing the location and direction information to the tractor. For deciding the path to be followed by the tractor, path planning has been done using the terrain map, possibility graph and an "A" search algorithm.

Ahasan et al. (2012) proposed the design of an autonomous robot for equipment management purposes in the household or industry using GPS navigation and IR sensors.

Guanshan (2009) proposed the design of an autonomous wheeled robot for obstacle avoidance using neural networks based on information fusion technique to solve the motion planning problem. The information regarding the obstacle is captured using eight ultrasonic sensors and infrared sensors.

PROBLEM FORMULATION

Based on the survey of the previous works, it can be noted that an efficient design of autonomous obstacle avoiding robot is required in the industry that is cost-effective and is able to find use in situations like automation of manufacturing process, space programs, environments that are dangerous to humans etc. Thus the main aim of this chapter is to evaluate the capabilities and performance of three different designs of obstacle avoiding robots (*a*) Sensor-based OARs, (*b*) Bluetooth based OARs and (*c*) IoT based OAR.

1. EXPERIMENT DESCRIPTION

In the present work, three different experiments have been conducted to design different obstacle avoiding robots. The description of the different experiments along with their components is given in Table 1.

Description of Components

The description of the common components used in the three experiments is given below

1. **Arduino Microcontroller:** Arduino is an open source prototype platform that is based on easy-to-use hardware and software. It has a circuit board, which can be programmed (called a microcontroller) and ready-made software called Arduino IDE that is used to write and upload the computer code to the physical board. The labeled Arduino board is shown in Figure 2.

2. **Motor Driver:** The motor driver used in the present work is L298N. It is a dual H-bridge driver used to control the speed and direction of two DC motors simultaneously. The operating voltage range of motors that can be controlled by L298N is between 5V-35V with a peak value of current being 2A (Boxall, 2014). The labeled L298N module is shown in Figure 3.

The H-bridge mode of the motor driver operation for controlling the motors is shown in Figure 4.

For controlling the direction of motion of the DC motors, pulses are sent to En1, En2 In1, In2, In3 and In4 pins of the motor driver.

Table 1. Experiment description and list of components

Experiment No.	Description	Component List
Experiment-1	Design of an automatic obstacle avoiding robot using ultrasonic sensors	(*i*) Power Supply Section, (*ii*) Arduino Microcontroller, (*iii*) Ultrasonic sensors (*iv*) Motor Driver (L298N) with a base, (*v*) DC Gear Motor, (*vi*) Single Core Connecting Wires, (*vii*) LEDs.
Experiment-2	Design of an obstacle avoiding robot using Bluetooth	(*i*) Power Supply Section, (*ii*) Arduino Microcontroller, (*iii*) Bluetooth Module (HC-05), (*iv*) Motor Driver (L298N) with base, (*v*) DC Gear Motor, (*vi*) Single Core Connecting Wires, (*vii*) LEDs.
Experiment-3	Design of an IoT based obstacle avoiding robot	(*i*) Power Supply Section, (*ii*) NodeMCU/ Ethernet shield/ Raspberry Pi, (*iii*) Motor Driver (L298N) with base, (*iv*) DC Gear Motor, (*v*) Single Core Connecting Wires, (*vi*) LEDs.
Experiment-3*a*	Design of an IoT based obstacle avoiding robot using Ethernet shield	
Experiment-3*b*	Design of an IoT based obstacle avoiding robot using NodeMCU	
Experiment-3*c*	Design of an IoT based obstacle avoiding robot using Raspberry Pi	

For controlling the speed of the motor, PWM signals can be applied to the enable pins and the speed varies according to the width of the pulses.

Figure 2. Labeled Arduino UNO board

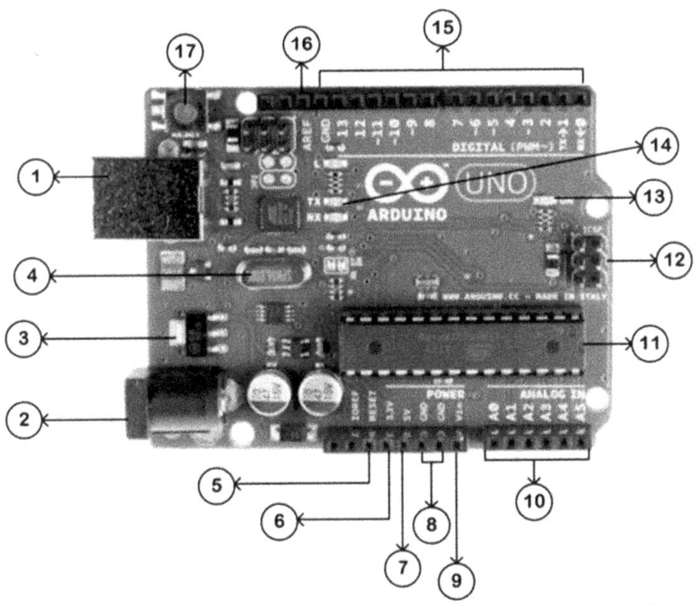

1: Power USB, 2: Power (Barrel Jack), 3: Voltage Regulator, 4: Crystal Oscillator, 5 & 17: Auduino Reset, 6: 3.3 V Supply, 7: 5 V Supply, 8 & 9: Ground, 10: Analog Pins, 11: Main Microcontroller, 12: ICSP Pin, 13: Power LED Indicator, 14: Tx and Rx LEDs, 15: Digital I/O, 16: AREF

Figure 3. Labeled L298N module

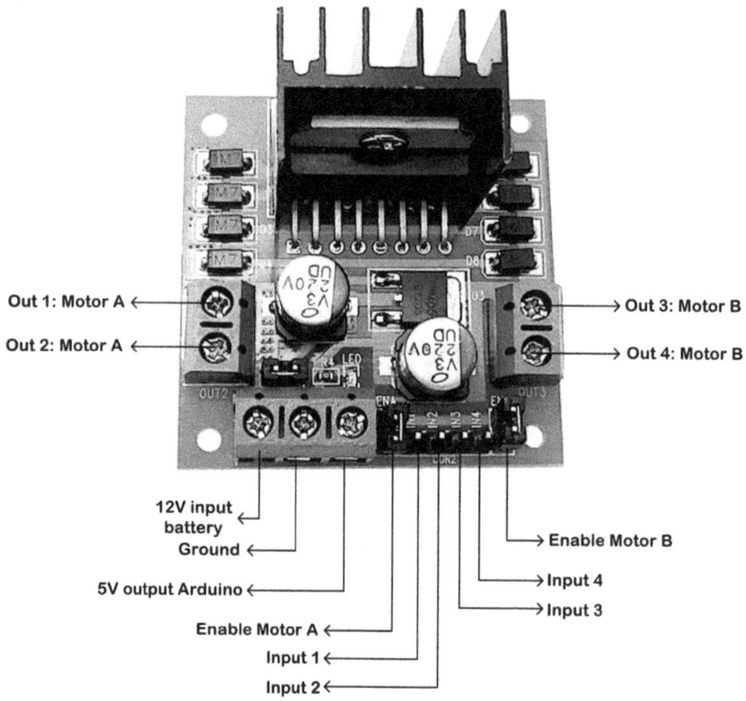

Figure 4. H-bridge

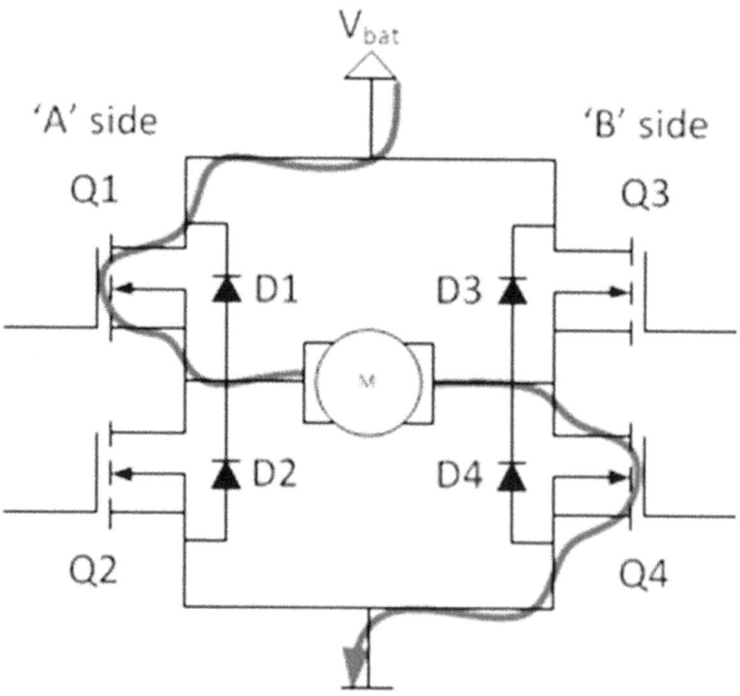

3. **DC Gear Motors:** The DC gear motor and BO (toy) motor are shown in Figure 5.

Motion Control of DC Gear Motors

For controlling the forward and backward movement of the motors, motor driver L298N has been used.

For controlling the motors on the left side, pins EnableA, Input1, and Input2 have been used while for controlling the motors on the right side, pins EnableB, Input3, and Input4 have been used. The corresponding truth table has been given in Table 2.

Based on the truth table presented in table 2, different motions of the robot can be represented in Table 3.

Experiment-1: Design of an automatic obstacle avoiding robot using ultrasonic sensors

The manually controlled obstacle avoiding robot can be converted to an autonomous obstacle avoiding robot by placing ultrasonic sensors that measure the distance of the obstacle and accordingly the robot turns left or right to avoid the collision.

The basic block diagram of a sensor-based obstacle avoiding robot is shown in Figure 6.

The basic connection diagram of the sensor based obstacle avoiding robot is shown in Figure 7.

The operation of the robot can be understood on the basis of the flowchart shown in Figure 8 and the Arduino code used for controlling the motion of the robot is shown in Figure 9.

Figure 5. DC motors (a) Labelled DC gear motor, (b) BO motor

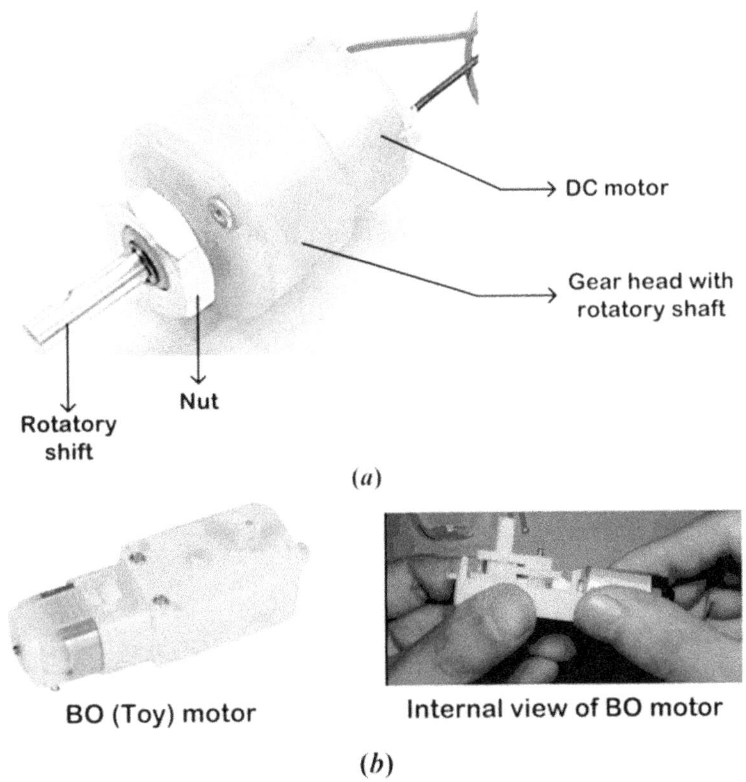

Table 2. Truth table indicating motor operations

Left Motor			
EnableA	**Input1**	**Input2**	**Motor Operation**
HIGH	0	1	Forward
HIGH	1	0	Backward
Right Motor			
EnableB	**Input3**	**Input4**	**Motor Operation**
HIGH	0	1	Forward
HIGH	1	0	Backward

Table 3. Different motor movements for robot motions.

EnableA	EnableB	Input1	Input2	Input3	Input4	Left Motor	Right Motor	Motion
HIGH	HIGH	0	1	0	1	Forward	Forward	Forward
HIGH	HIGH	1	0	1	0	Backward	Backward	Reverse
LOW	HIGH	0	0	1	0	OFF	Backward	Left turn
HIGH	LOW	0	1	0	0	Forward	OFF	Right turn
HIGH	HIGH	0	1	1	0	Forward	Backward	Clockwise rotation
HIGH	HIGH	1	0	0	1	Backward	Forward	Counter-clockwise rotation
LOW	LOW	0	0	0	0	Stop	Stop	Stop

Figure 6. The basic block diagram of sensor-based obstacle avoiding robot

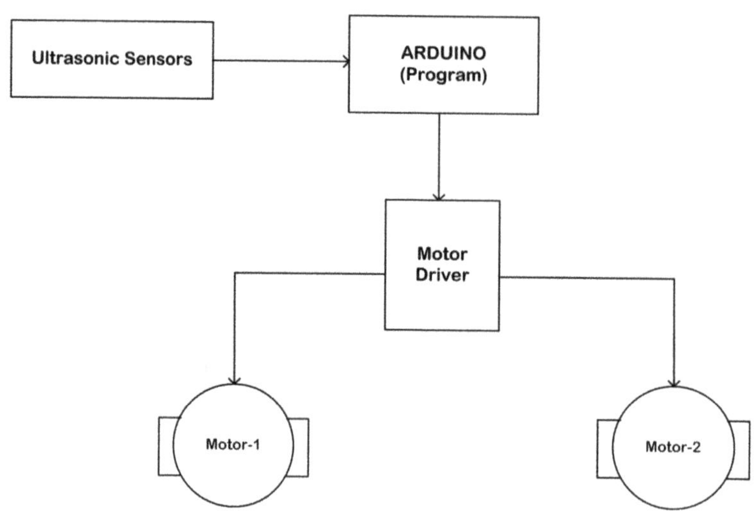

Figure 7. Basic connection diagram of sensor-based obstacle avoiding robot

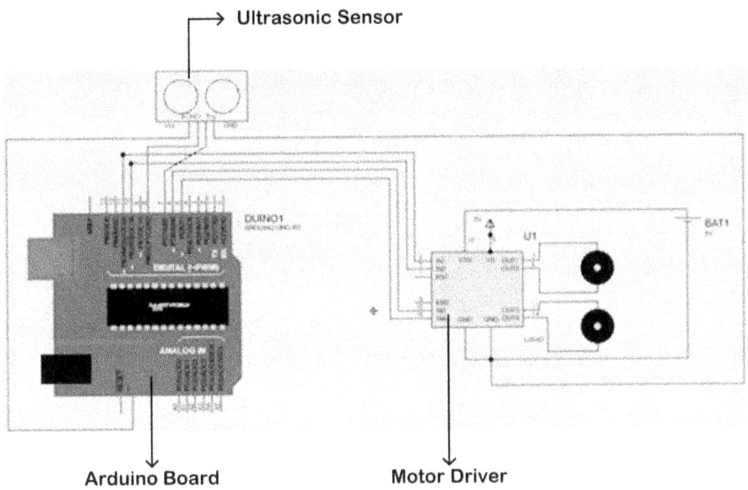

Experiment-2: Design of an obstacle avoiding robot using Bluetooth module

In the present work, a Bluetooth module based obstacle avoiding robot design has been implemented whose movements can be controlled by a user over a short range.

The Bluetooth module used in the present work is HC-05. This module has been designed for establishing a wireless serial connection. The operating voltage of HC-05 is 3.3 V. It is an IEEE 802.15.1 standardized protocol that can be used to build a wireless personal area network. The serial port of the module is Bluetooth V2.0+EDR (Enhanced Data Rate) having 3Mbps modulation with complete 2.4 GHz radio transceiver and baseband. The communication to the microcontroller is established using serial communication (USART). The module works in 2 modes: (i) Data mode- Exchange of data take place between devices and (ii) Command mode: Used to send AT commands to module serial port. The labeled Bluetooth module is shown in Figure 10.

The block diagram of the system is shown in Figure 11.

An android application named Bluetooth electronics from Keuwlsoft has been used for sending signals to Bluetooth module. The application comes with a large selection of controls like buttons, switches, pads, sliders, accelerometers, terminals, graphs etc. The sample of the interface showing the control pads and the accelerometer is given in Figure 12.

The connection diagram for the circuit of Bluetooth based obstacle avoiding robot is shown in Figure 13.

1. For controlling the motion of the robot, the Android application was downloaded into a Bluetooth enabled phone using PlayStore. In the app, different functions were configured on the panel to send corresponding signals to the Bluetooth module e.g. for a forward press, character '*f*' is transmitted via the Bluetooth of a mobile phone to the receiver of the Bluetooth module (HC-05).
2. Arduino software was used to write a switch case program in embedded C to perform the corresponding function with reference to the character received by the Bluetooth module. This program was compiled and uploaded to the Arduino board that was connected to the PC using a B-type USB cable.

Figure 8. The flowchart depicting the functionality of an automatic obstacle avoiding robot using ultrasonic sensors

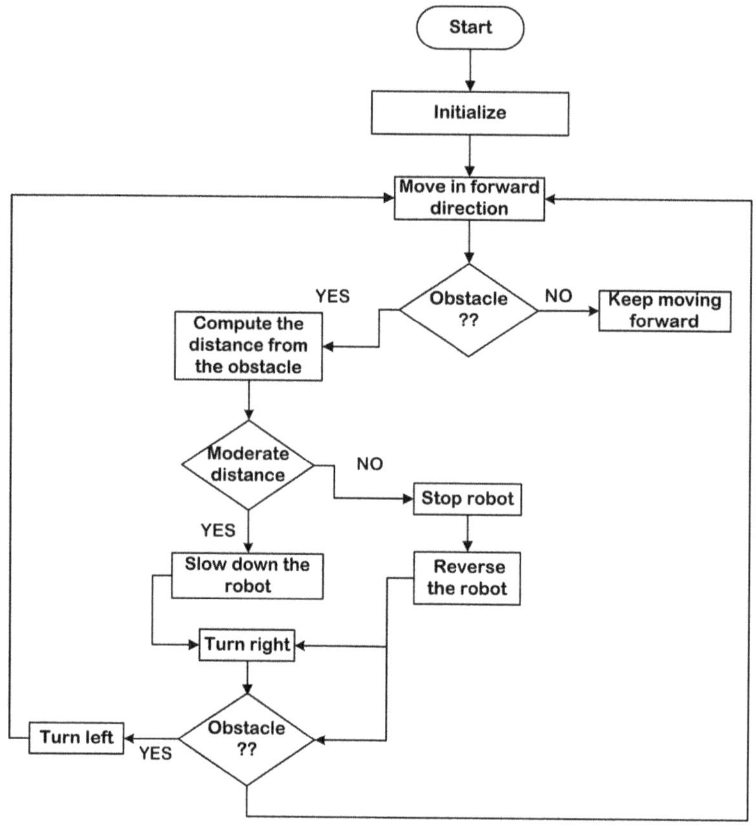

3. The output from the Arduino board was sent to the motor driver that in turn controls the motion of the DC gear motors connected to the wheels of the robot.

The same set of operations can also be performed by using the accelerometer instead of the control pads.

The motion of the phone controls the signal being transmitted to the Bluetooth module e.g. for a forward movement of the phone, character '*f*' is transmitted via the Bluetooth of a mobile phone to the receiver of the Bluetooth module (HC-05). The corresponding switch case statement is executed in the Arduino and the motion of the DC gear motors is controlled.

The flowchart depicting the functionality of a Bluetooth based obstacle avoiding robot is shown in Figure 14.

The Arduino code used for controlling the robot motion is given in Figure 15.

Experiment-3: Design of an IoT based obstacle avoiding robot

An IoT based robot can be designed in either of the three ways: (*i*) using Ethernet shield, (*ii*) using NodeMCU or (*iii*) using Raspberry Pi.

The basic flowchart for the working of an IoT based obstacle avoiding robot is shown in Figure 16.

Experiment-3 (*a*): Design of an IoT based obstacle avoiding robot using Ethernet shield

Figure 9. The flowchart depicting the Arduino code used for controlling motion of the sensor based robot

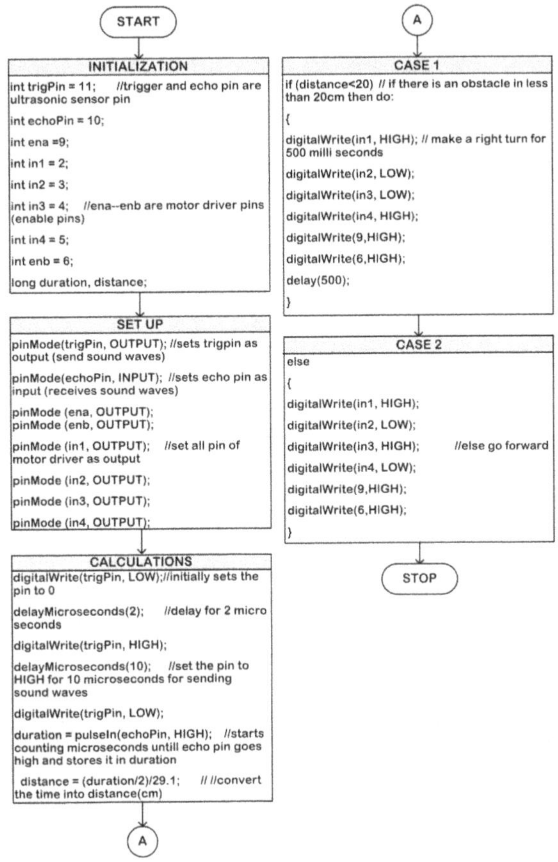

The Arduino Ethernet Shield 2 is used to establish connectivity between the Arduino board and the internet using the Ethernet library. Connector RJ45 is used for connection. The Ethernet shield is shown in Figure 17.

Figure 10. Labeled HC-05 module

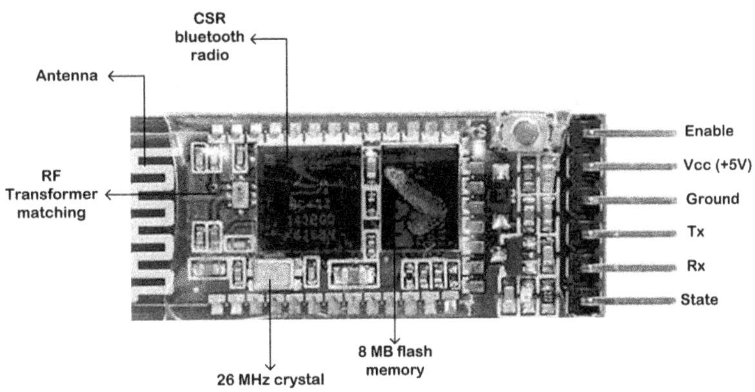

Figure 11. The basic block diagram of Bluetooth based obstacle avoiding robot.

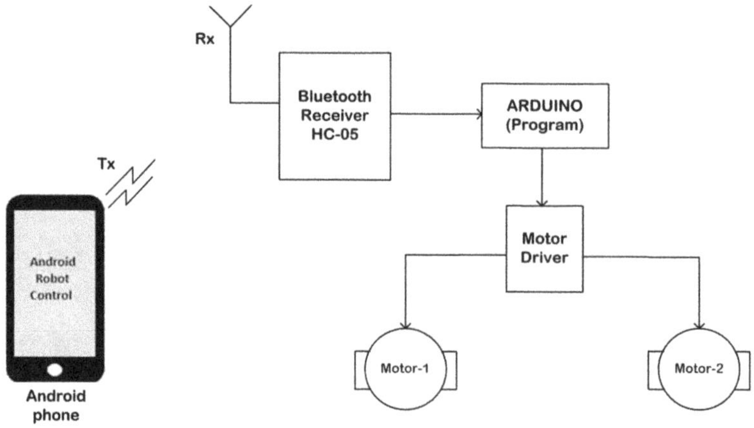

Figure 12. Bluetooth electronics application interface (a) accelerometer control OFF, (b) accelerometer control ON

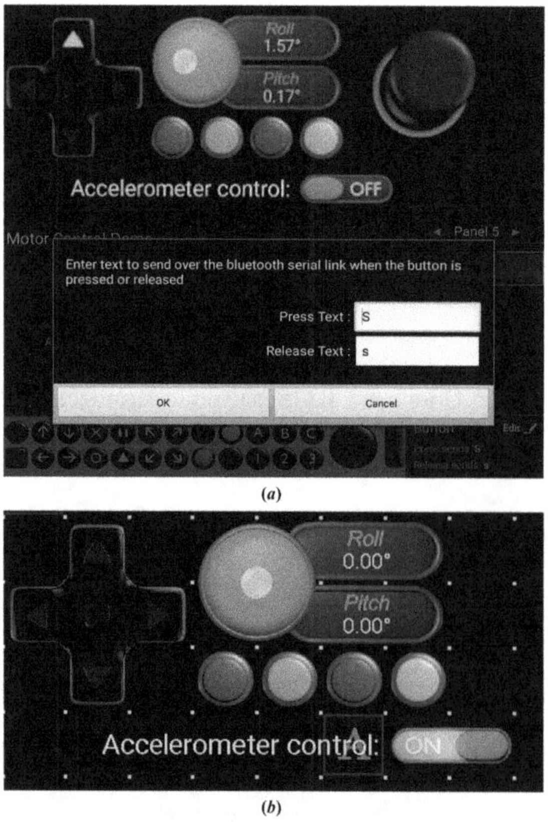

Figure 13. Connection diagram of the Bluetooth based obstacle avoiding robot

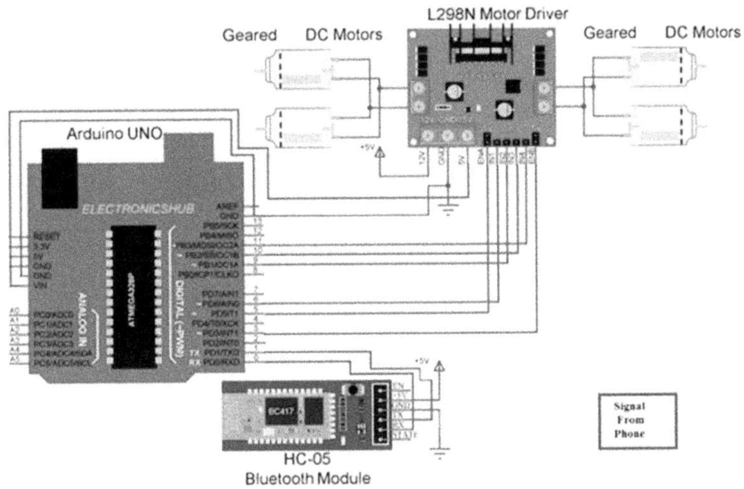

Figure 14. The flowchart depicting the functionality of a Bluetooth based obstacle avoiding robot

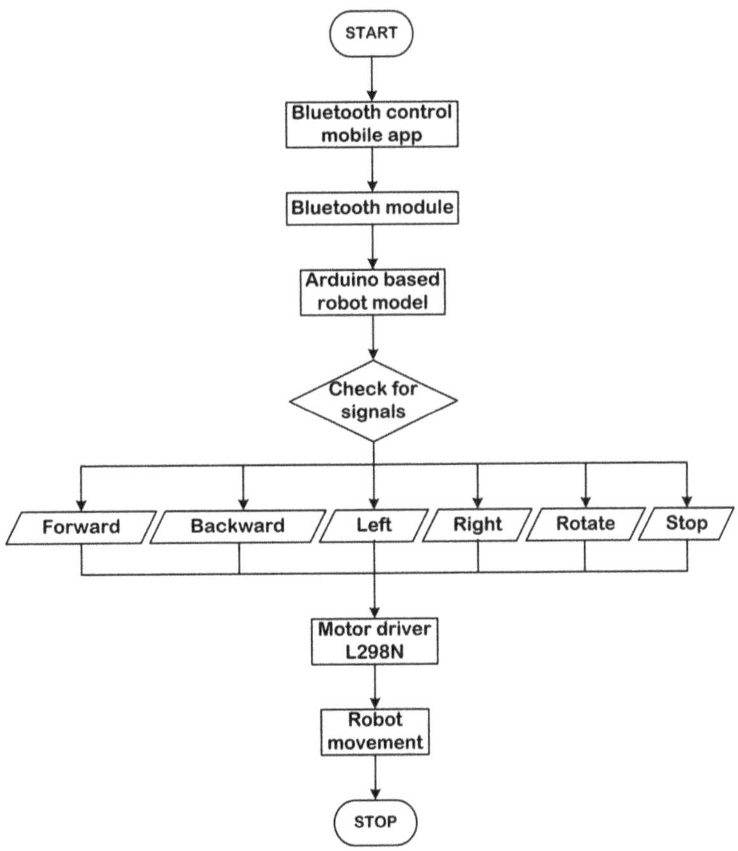

Figure 15. The flowchart depicting the Arduino code used for controlling the motion of the Bluetooth based robot.

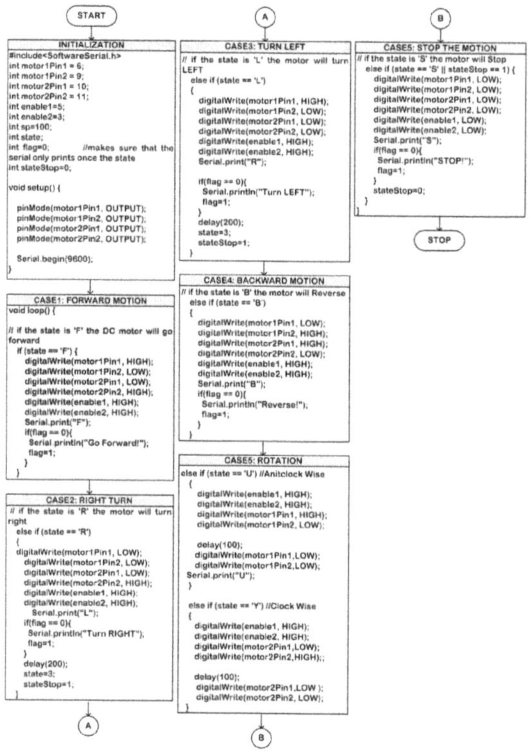

Figure 16. The flowchart of an IoT based obstacle avoiding robot

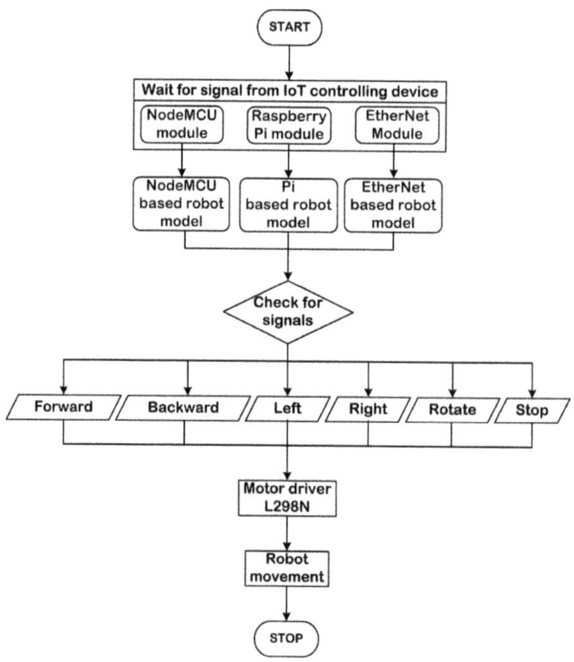

Figure 17. Labeled diagram of Ethernet shield

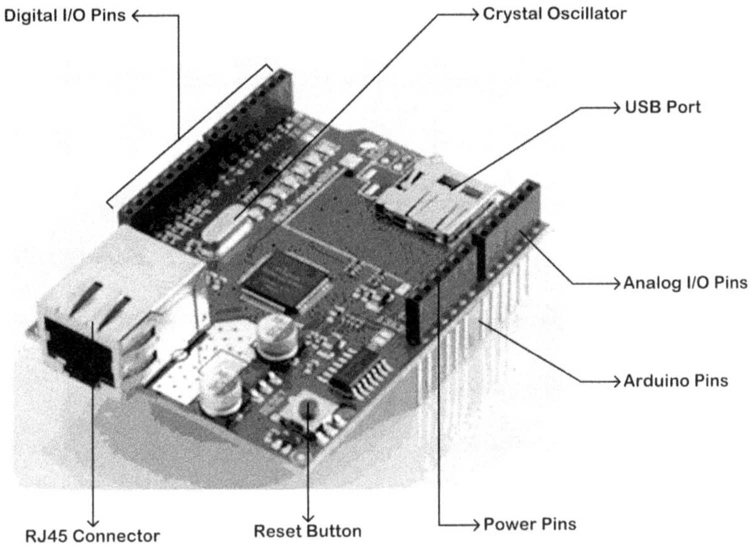

Figure 18. The basic block diagram representing an IOT based robotic system using an Ethernet shield

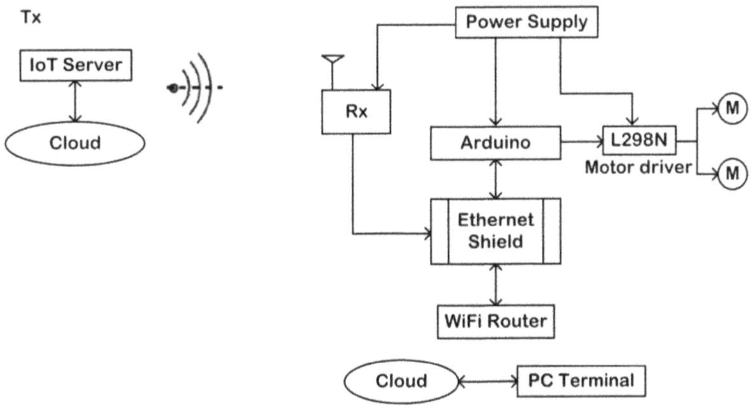

Figure 19. Connection diagram of an IoT based obstacle avoiding robot using Ethernet shield

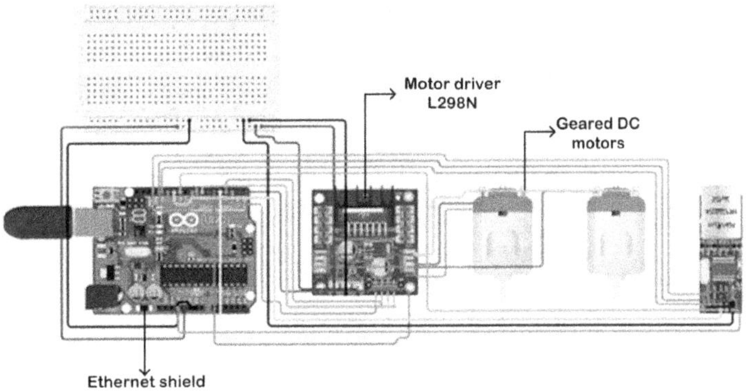

Figure 20. Arduino code for IoT based obstacle avoiding robotic using Ethernet shield

Figure 21. Pin diagram of NodeMCU

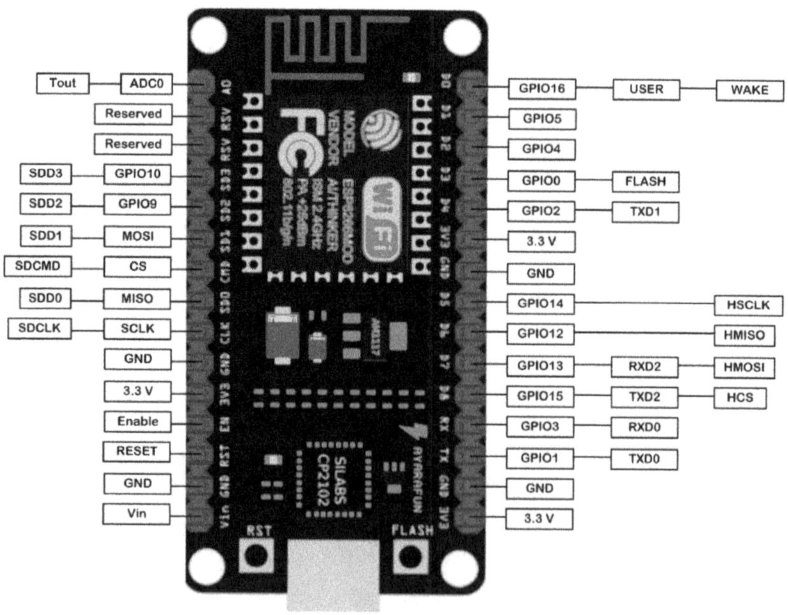

Figure 22. The basic block diagram representing an IoT based obstacle avoiding robot using NodeMCU

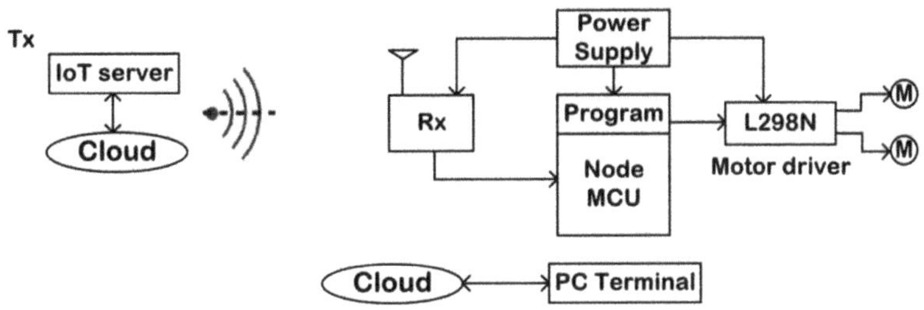

Figure 23. Connection diagram of an IoT based obstacle avoiding robot using NodeMCU

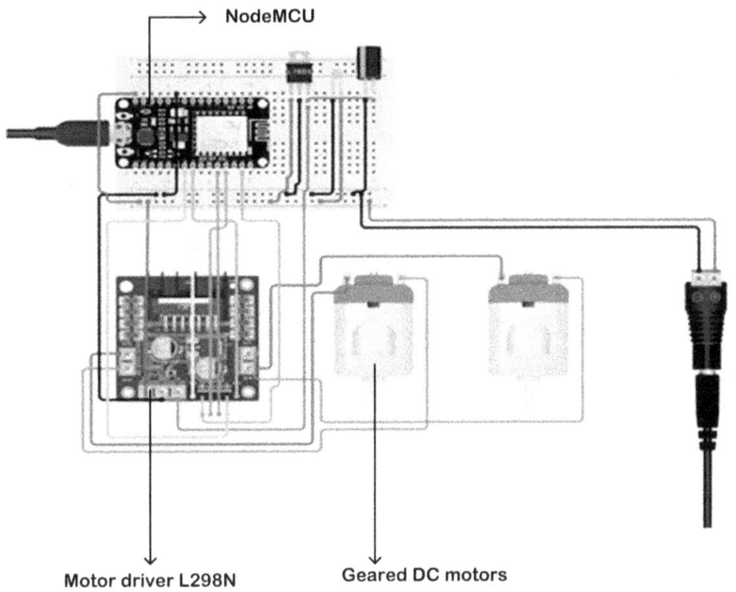

The basic block diagram representing an IoT based obstacle avoiding robot using an Ethernet shield is shown in Figure 18.

The connection diagram for the circuit of IoT based obstacle avoiding robot using Ethernet shield is shown in Figure 19.

The Arduino code for controlling the robot motion is given in Figure 20.

Experiment-3 (*b*): Design of an IoT based obstacle avoiding robot using NodeMCU

NodeMCU is an open source IoT platform for the ESP8266 Wi-Fi SoC. It is an integrated chip that provides internet connectivity. It can be programmed through the USB port and Arduino programming. It combines the features of a microcontroller with a WiFi access point and station.

Figure 24. The code for an IoT based obstacle avoiding robot using NodeMCU

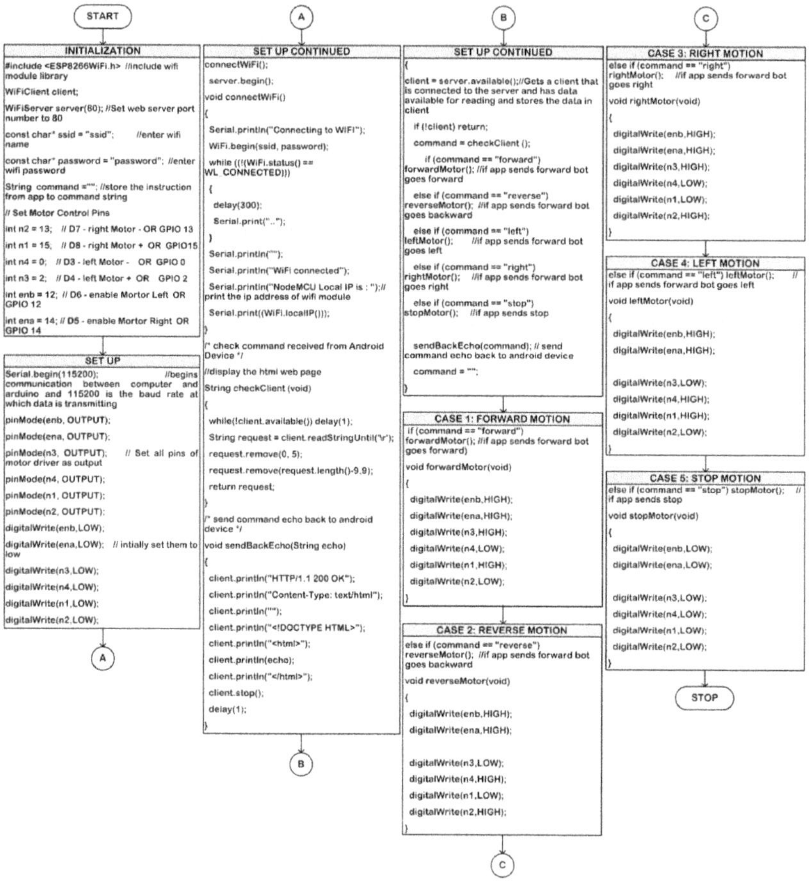

Figure 25. Labeled Raspberry Pi board

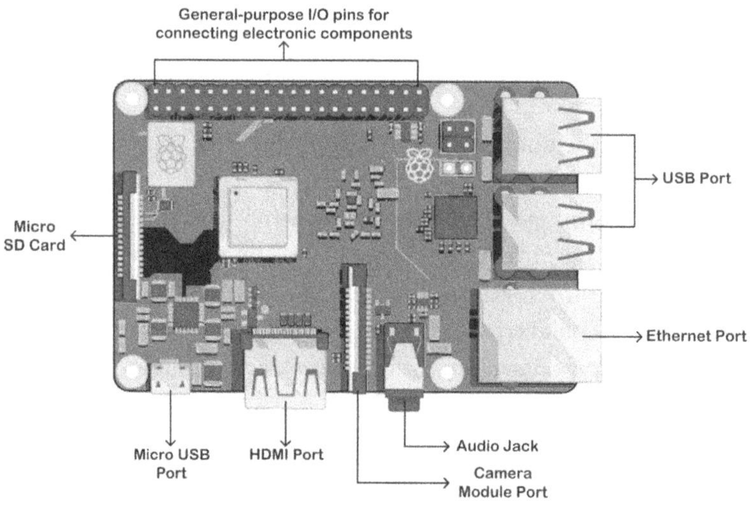

Figure 26. The basic block diagram representing an IoT based obstacle avoiding robot using Raspberry Pi

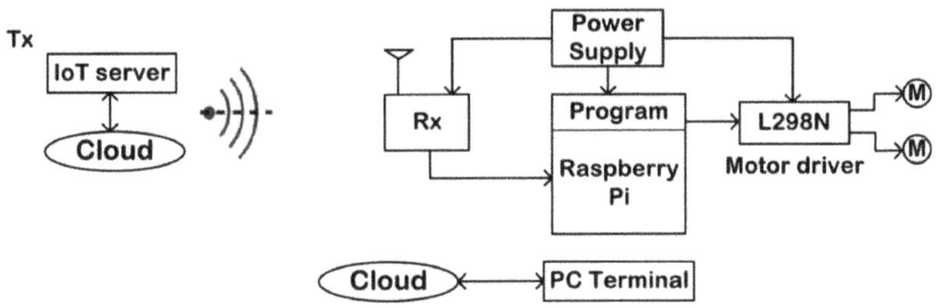

Figure 27. The basic requirements and steps followed for developing the Raspberry Pi based system

The kit has Analog pins (A0) and Digital (D0-D8) pins on its board along with PWM functionality, SPI communication, etc. The pin diagram of NodeMCU is shown in Figure 21.

The basic block diagram of an IoT based obstacle avoiding robot system using NodeMCU is shown in Figure 22.

Figure 28. Connection diagram of an IoT based obstacle avoiding robot using Raspberry Pi

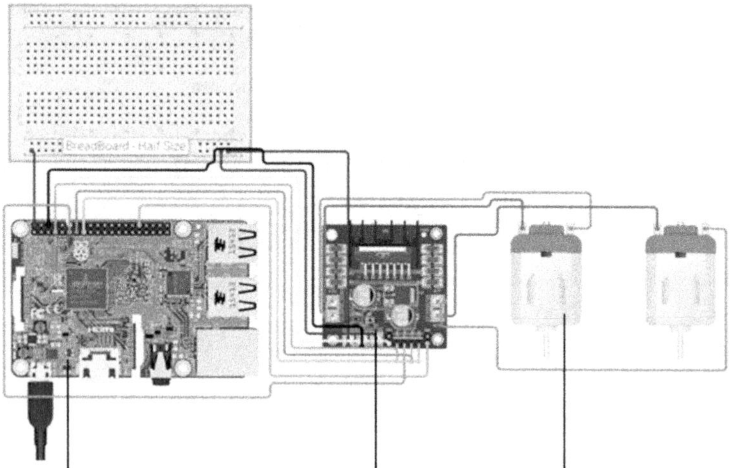

The connection diagram for the circuit of IoT based obstacle avoiding robot using NodeMCU is shown in Figure 23.

The code for controlling the motion of the robot is shown in Figure 24.

Experiment-3 (*c*): Design of an IoT based obstacle avoiding robot using Raspberry Pi

By using Raspberry Pi and corresponding RASPBIAN operating system, an obstacle avoiding robot can be designed to avoid the collision. The general labeled Raspberry Pi board is shown in Figure 25.

The basic block diagram representing an IoT based obstacle avoiding robot using Raspberry pi is shown in Figure 26.

The basic requirements and steps followed for developing the Raspberry Pi based system are shown in Figure 27.

The connection diagram for the circuit of IoT based obstacle avoiding robot using Raspberry Pi is shown in Figure 28.

The Python code used for controlling the robot motion using Raspberry Pi is given in Figure 29.

CONCLUSION AND FUTURE SCOPE

In the present work, different experiments have been performed for the design of an obstacle avoiding robot using sensors, Bluetooth module and IoT. These robot designs can be used largely in the automation industry for scenarios where human intervention can be reduced to a minimum or in harmful environments etc. The Raspberry Pi based obstacle avoiding robot model can be the most widely used design for the industry using onboard cameras and other sensors for the design of an autonomous obstacle avoiding robot.

Figure 29. The flowchart depicting the Python code used for controlling the motion of the robot using Raspberry Pi

REFERENCES

Al Ahasan, Hossain, Siddiquee, & Rahman. (2012). Obstacles invariant navigation of an autonomous robot based on GPS. *Proceedings of IEEE International Conference on Computer and Information Technology*, 524-532.

Bhagat, K., Deshmukh, S., Dhonde, S., & Ghag, S. (2016). Obstacle avoidance robot. *International Journal of Science, Engineering and Technology Research*, *5*(2), 439–442.

Bluetooth Module HC-05. (n.d.). Retrieved from https://www.electronicwings.com/sensors-modules/bluetooth-module-hc-05-

Boujelben, M., Rekik, C., & Derbel, N. (2013). Hierarchical fuzzy controller to avoid mobile obstacle for a mobile robot. Proceedings of 10th International Multi-Conference on Systems, Signals & Devices (SSD), 1-8. doi:10.1109/SSD.2013.6564094

Boxall, J. (2014). *Tutorial - L298N Dual Motor Controller Module 2A and Arduino*. Retrieved from https://tronixlabs.com.au/news/tutorial-l298n-dual-motor-controller-module-2a-and-arduino/

Chen, Y. S., & Juang, J. H. (2009). Intelligent obstacle avoidance control strategy for wheeled mobile robot. *Proceedings of ICROS-SICE International Joint Conference*, 3199-3204.

Gray, K. W. (2000). *Obstacle detection and avoidance for an autonomous farm tractor* (Master of Science thesis). Utah State University.

Guanshan, H. U. (2009). Neural network based on information fusion using for avoiding obstacle robot. *Proceedings of WASE International Conference on Information Engineering*, 565-568.

Hanumante, V., Roy, S., & Maity, S. (2013). Low cost obstacle avoidance robot. *International Journal of Computing and Engineering*, *3*(4), 52–55.

Heidarsson, H. K., & Sukhatme, G. S. (2011). Obstacle detection and avoidance for an autonomous surface vehicle using a profiling sonar. *Proceedings of IEEE International Conference on Robotics and Automation*, 731-736. 10.1109/ICRA.2011.5980509

Jathavara, V., & Ashwath, P. (2017). Obstacle avoiding robot: a Lego EV3 robot. *Proceedings of International Conference on Recent Advances in Electronics and Communication Technology*, 43-49. 10.1109/ICRAECT.2017.61

Nasucha, M. (2015). Development of an obstacle avoiding robot. *Jurnal Sistem Komputer*, *5*(2), 55–63.

Ragahv, A., Mishra, P., Verma, P., Randhawa, K. S., & Thakur, T. (2017). Obstacle avoiding robot. *International Journal of Advanced Research in Computer Science and Software Engineering*, *7*(4), 415–417. doi:10.23956/ijarcsse/V7I4/0250

Ryther, C. A., & Madsen, O. B. (2009). *Obstacle detection and avoidance for mobile robots* (Bachelor's thesis). Technical University of Denmark.

Shahdib, Ullah, Hasan, & Mahmud. (2013). Obstacle detection and object size measurement for autonomous mobile robot using sensor. *International Journal of Computers and Applications*, *66*(9), 28–33.

Tabassum, Lopa, Tarek, & Ferdosi. (2017). Obstacle avoiding robot. *Global Journal of Researches in Engineering: H Robotics & Nano-Tech*, *17*(1), 19-24.

Vaghela, A., Patel, J., & Vaghela, S. (2016). Obstacle avoidance robotic vehicle using ultrasonic sensor, android, and Bluetooth for obstacle detection. *International Research Journal of Engineering and Technology*, *3*(2), 339–348.

Wu, T. F., Tsai, P. S., Hu, N. T., & Chen, J. Y. (2014). Use of ultrasonic sensors to enable wheeled mobile robots to avoid obstacles. *Proceedings of 10th International Conference on Intelligent Information Hiding and Multimedia Signal Processing*, 958-961. 10.1109/IIH-MSP.2014.240

Xiong, Y., Xiao, S., & Zha, F. (2011). An avoiding obstacle algorithm of mobile robot under complex unknown environments. *Proceedings of 2nd International Conference on Mechanic Automation and Control Engineering*, 1019-1022. 10.1109/MACE.2011.5987106

Yamada, T., Sa, Y. L., & Ohya, A. (2013). Moving obstacle avoidance for mobile robot moving on designated path. *Proceedings of 10th International Conference on Ubiquitous Robots and Ambient Intelligence*, 509-514. 10.1109/URAI.2013.6677324

Yang, D., Xia, J., & Xiang, L. (2010). The research on intelligent mobile robot's avoiding obstacle by ultrasound. *Proceedings of International Conference on Artificial Intelligence and Computational Intelligence*, 77-79. 10.1109/AICI.2010.23

Zhao, Y., Chai, X., Gao, F., & Qi, C. (2018). Obstacle avoidance and motion planning scheme for a hexapod robot Octopus-III. *Robotics and Autonomous Systems*, *103*, 199–212. doi:10.1016/j.robot.2018.01.007

Chapter 7
Intelligent Health Vigilance System Using Internet of Things (IoT)

Bhushan Patil
https://orcid.org/0000-0003-2407-0022
Mumbai University, India

Manisha Vohra
https://orcid.org/0000-0001-8010-0076
Mumbai University, India

ABSTRACT

Internet of things (IoT) is a technology that is constantly progressing and finding new applications. It is making its way into enormous sectors. One such sector where IoT has entered is healthcare. This sector is one of the most vital sectors of world. Health of each and every individual should be given utmost priority. Any person who is unwell requires vigilance. The patients in hospital are always kept under vigilance. The healthcare sector constantly strives to keep on finding new ways and methods to improve and upgrade the current methods used for vigilance of patients. The use of IoT in healthcare sector can drastically change the way this sector works now. In this chapter, the concept of IoT shall be explained in detailed. Along with it, the current vigilance system and methods used in healthcare will be stated and viewed. In addition, a detailed demonstration through case studies explaining the use of IoT in healthcare sector will also be presented in this chapter.

WHAT IS IOT?

IoT is exchange or communication of data between different objects or devices that are connected to each other through internet. IoT helps in establishing connectivity between different devices and objects through the usage of internet.

DOI: 10.4018/978-1-5225-9574-8.ch007

IoT is a link which helps in attaining all the necessary data and information about the connected device or object that one would require to know. It makes the information and data of the connected device or object available at just one click. It actually bridges the gap between the things connected through it.

IoT removes barriers of distance by connecting devices and objects virtually and wirelessly through internet without the need for the devices to be physically present in front of each other. The users can control and access their IoT connected devices from any location in the world. This gives the users a great flexibility and proves as a major advantage.

BACKGROUND OF IOT

The background of this technology is dated back in the early 1980's when this technology was first used at Carnegie Mellon University of Pennsylvania in United States. At that time the term Internet of Things (IoT) was not even coined and people were not even aware that how helpful this technology can be in future. In Carnegie Mellon University a Coca Cola machine was present. This machine was internet connected and it would indicate regarding the stock of Coca Cola inside the machine and would also indicate whether the Coca Cola bottles placed inside the machine are cooled or not.

WORKING OF IOT

The IoT working is a simple process. It involves contributions from sensors and cloud. The working process is initiated by sensors. The sensors gather all the information and data from its surroundings. This data is safely passed on to the cloud for processing. After processing of data, it is available for the user to view. The user also has the facility to intervene and control the IoT process. The working of IoT can be considered as a fixed flow of events taking place one after the other. For better understanding, the working of IoT can be considered in parts. The first part is action of sensor.

1. **Action of Sensor:** The sensor's job is to sense all the possible data from its vicinity. The sensor acts even on a very negligible change happening in its vicinity and always actively keeps on sensing each and every change happening in its range.
2. **Preparation for Sending Data to Cloud:** The sensor has all the data it has received through sensing. The data is in raw form and this data collected from the sensor needs to be sent to the cloud. The sensor makes preparation for sending the data collected to cloud by ensuring all the data remains unaltered and in it exact raw form in which it is collected.
3. **Transition to Data Processing Stage:** The data undergoes a lot of changes after transition to data processing stage. The raw data is processed and after being processed it is ready to be sent to the user.
4. **Empowering the User:** The user can at any time and from any location view and access the data of the device or object connected through IoT. The user is also empowered with the ability to control the device or object connected through IoT.

CHARACTERISTICS OF IOT

The characteristics of IoT are as follows:

1. Heterogeneity
2. Interconnectivity
3. Connectivity
4. Compatibility
5. Safety
6. Authentication
7. Flexibility
    1. **Heterogeneity:** The IoT technology enables its devices to be heterogeneous. They are able to interact with various other devices belonging to different software platforms and networks.
    2. **Interconnectivity:** The IoT allows different objects and devices to remain interconnected.
    3. **Connectivity:** The IoT devices are such developed that it is highly hassle free to connect them with other devices.
    4. **Compatibility:** All IoT objects and devices are compatible with each other.
    5. **Safety:** The data being sent to the users through IoT should remain safe. It should not get attacked by hackers.
    6. **Authentication:** The IoT devices are able to allow authentication feature for the users ensuring the end users receiving the data are authenticated.
    7. **Flexibility:** The IoT technology provides the flexibility of connecting devices and providing access to the users from any part of the world.

IOT PROTOCOLS

There are different IoT protocols present. They are as follows:

1. Message Queuing Telemetry Transport (MQTT).
2. Constrained application protocol (CoAP).
3. Extensible messaging and presence protocol (XMPP).
4. Internet Protocol Version 6 (IPv6) over low power wireless personal area networks (6LoPWPAN).
    1. **Message Queuing Telemetry Transport (MQTT):** It is a quite simple protocol used for messaging. It is used for devices have low bandwidth. MQTT protocol has many advantages associated with it. For example, it enables its users to give commands to control output. This helps in providing the users total control. It also helps in reading and publishing the data that the sensor nodes are receiving. It is a highly efficient protocol.
    2. **Constrained Application Protocol (CoAP):** CoAP is a service layer protocol. This protocol is utilized for devices which are on the same network. Along with it, this protocol is also used for nodes and devices that are connected to internet. Along with it, CoAP is also used for devices that belong to different networks. These devices can be joined to each other through internet. CoAP is also intended for use in wireless sensor network_nodes.

3. **Extensible messaging and presence protocol (XMPP):** This protocol allows an almost real-time exchange of data which is structured but extensible. This data exchange can take place between two or more network entities. Since the protocol is extensible, it can be used for various applications such as for file transferring, gaming applications, smart grid, etc.
4. **IPv6 Over Low Power Wireless Personal Area Networks (6LoWPAN):** The 6LoWPAN protocol has emerged after gaining inspiration from internet protocol. The thought that internet protocol can be used with small low power devices having few processing capabilities made way for development of 6LoWPAN protocol and this is how this protocol came into existence.

The 6LoWPAN protocol has defined encapsulation mechanisms. These mechanisms are very useful. Along with it, header compression mechanisms are also present in 6LoWPAN protocol. These mechanisms allow IPv6 packets to be sent. They even help in receiving packets. The packets are sent and received through the use of IEEE 802.15.4 based networks.

EXAMPLE OF IOT

For better understanding of IoT, let's consider an example. A room consists of an air conditioner which the user wants to switch on automatically when the room temperature reaches 25° Celsius. For this purpose, the user connected the air conditioner with a temperature sensor. Using the temperature sensor, the exact temperature of the room can always be known. This sensor is then connected to the IoT and the IoT system is connected to the air conditioner. As soon as the temperature of the room reaches 25° Celsius, the sensor alerts the IoT system. Then this system triggers the air conditioner unit and turns it on. Figure 1 shows example of controlling temperature using IoT.

Let's consider another example of IoT. A tube light in a room needs to be switched on when sun light rays i.e., ultraviolet (UV) rays are not present. A sunlight sensor is used for this purpose. The sunlight sensor shall be connected to the tube light. Till the time the UV rays are present, the sunlight sensor will keep on sensing it and as soon as they are not present, the sensor shall indicate the IoT system. The IoT system shall then trigger the tube light to switch on. Figure 2 shows example of controlling tube light using IoT.

The IoT can greatly help humans in their day to day life chores also. Let us consider another example. A coffee machine is connected to internet and it is controlled through an application built in cell phone. Whenever the user is going to arrive, the user can alert the coffee machine to make and keep coffee ready for him/her.

IOT DEVICES

There are many different IoT devices present. They are providing great benefits to the users. Below re two such examples of IoT devices.

Figure 1. Example of controlling temperature using IoT

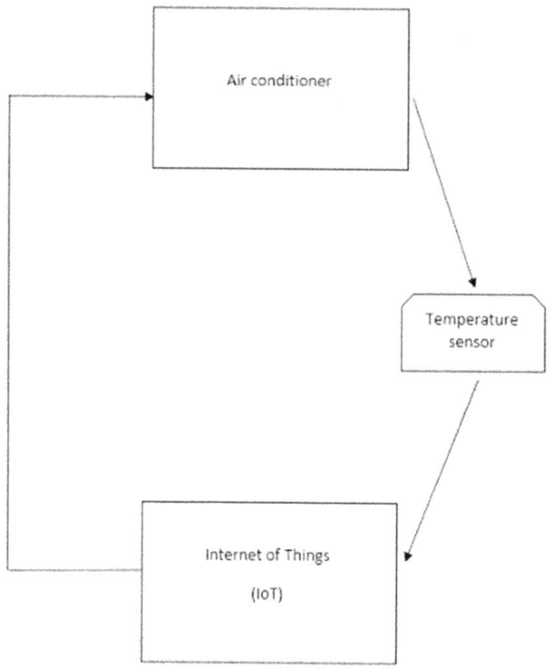

Figure 2. Example of controlling tube light using IoT.

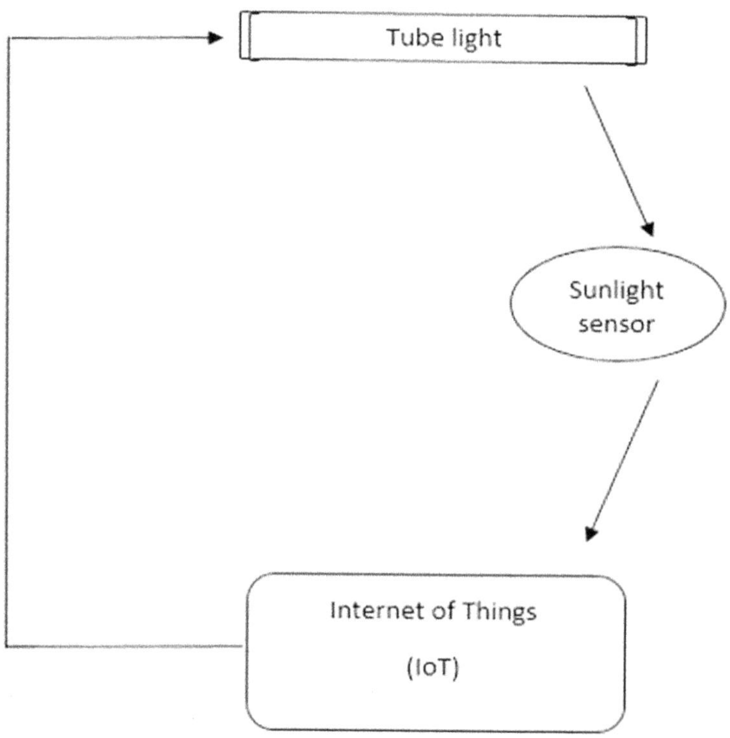

IoT device example 1

Amazon Echo

This device is such well-built and programmed that it recognizes human voice and follows their commands. This IoT device can perform various functions right from reading out important news from internet to playing our desired songs, etc.

IoT device example 2

August Doorbell Cam Pro

This IoT device ensures our house is safely locked. We can view who visits our home from the video recording feature of this device from any remote location. This lock has the ability to unlock on its own when the house owner arrives and lock the door when the house owner goes out. It also enables the house owner to provide digital keys to their relatives or guests.

ADVANTAGES OF IOT

There are enormous advantages associated with IoT. They are enlisted as follows:

1. IoT provides remote accessibility of devices and objects.
2. IoT also provides remote control of devices and objects connected.
3. IoT helps in automating devices.
4. IoT provides great security features to our devices and objects.
5. The workload of humans is either completely eliminated or greatly reduced by the introduction of IoT in different sectors.

CHALLENGES

The IoT is having major challenges related to security. The devices and the objects connected to internet are vulnerable. The data associated with the devices and objects as well as the devices and objects itself are at great risk as no one would want their data to be stolen and their devices and objects to be intercepted. Providing total security is quite necessary while using IoT technology. There are measures to keep the devices and objects secure while using this technology like use of digital keys, use of biometric features, etc. Despite of different security measures, our IoT connected devices and objects and its associated data is always under great threat. That is it why, to constantly keep on improving the security of IoT devices and objects and its data is a challenge for IoT.

OPPORTUNITIES

IoT can be introduced in various sectors like healthcare, agricultural, etc. IoT has the capability of tremendously changing the way any sector functions. IoT has a very positive influence on which ever sector it enters. It has great merits which hugely benefit the sector it enters. Hence, there are enormous opportunities with IoT.

INTRODUCTION TO HEALTHCARE SECTOR

No one in the world can deny from the fact that health is wealth. Health is such a factor that cannot and should not be neglected. The government of each and every country invests a lot of money every year in the development of the health care sector of their respective countries. All the people associated with the health care sector strive extremely hard almost every day round the year for the progress of this sector. Time to time new discoveries and inventions are made in health care sector for fighting against diseases and for preventing the cause of diseases and upgrading the health vigilance systems currently available and used.

NEED FOR AN INTELLIGENT HEALTH VIGILANCE SYSTEM

When any person is unwell and is admitted to the hospital, the health of that person requires continuous vigilance. Likewise, the health of any and every patient admitted to the hospital is in need of continuous vigilance and it becomes the responsibility of that particular hospital in which they are admitted, to provide this vigilance. Each hospital is acquainted with a certain amount of staff consisting of nurses, ward attendants, peons, doctors and security and administrative staff. The security and administrative staff of the hospital is not responsible for managing any vigilance work related to patient's health. The remaining staff consisting of doctors, ward attendants, nurses and peons work to the best of their ability but however it is not possible for them to be always present with the patient and by just being with the patient, the health related parameters of the patient cannot be known.

In order to keep a continuous tab of patient's health and to keep a check on different parameters related to patient's health like heart rate, etc. different machines are used. For example, when patients who have recently got a heart attack or have undergone some kind of heart related surgery or are suffering with serious heart problem like arrhythmia in which the patient's heart beats too fast or too slow or either the rhythm of heart is irregular, then the patient's in any of these cases are immediately connected or attached with a cardiac monitoring machine which performs vigilance of patient's heart rate and rhythm. Heart rate is basically the number of times our heart beats within a minute's duration. Heart rhythm is defined as the pattern of heartbeats i.e., the pattern in which our heart is beating.

Like cardiac monitoring machine, other machines are also present which are used for monitoring different parameters related to patient's health depending on the patient's illness. All such machines are a great bliss in keeping vigilance of the patient's health but to see the status of patient's health on the machine's monitor screen, a doctor or some other member of the hospital who can look after the patient's health in hospital like ward attendants or nurses or peons are always required near the machine connected to the patient. This is a major drawback as a person is always required to be present in front or near by the machine to which the patient is attached. Due to any reason if a person is not present near the machine and if there is some kind of sudden drastic change in the patient's health which is indicated by the machine, then the patient's health will suffer and this can be very dangerous. This arises a need to have an intelligent health vigilance system.

INTRODUCTION OF IoT IN HEALTHCARE SECTOR

Introduction of technology can help in health care sector in great way and can change the way heath care sector functions at present. Technology like IoT can be brought into use to keep vigilance of patient's health without the need for any person to be physically present in front or near the machine. Using IoT, vigilance of patient's health can be performed even from a remote location. This is one of the major advantages of using IoT in health care sector.

IoT can tremendously change the functionality of healthcare sector. With IoT a patient's health updates can easily be sent time to time to the patient's care taker which will be a great help and benefit for both, the care taker as well as the patient. IoT in true sense bridges the gap between the patient and their care taker. The doctor can also keep a tab on his/her patient's health.

IoT BASED VIRTUAL ASSISTANCE

Along with monitoring purpose, IoT can be also used in virtually assisting and reminding patients to take care of their health by taking medicines on time. It is many times seen that patients tend to forget to take their medicines on time. Also, many times the patients forget to take their medicines at all. This can have drastic ill effects. The patient's health can be affected adversely. This small forgetting habit can take a toll on the patients' health. Wen any patient is on a daily dosage of medicine for some kind of health issue like blood pressure, then the patient's body system gets used to that particular medicine.

A single day without intake of that particular medicine will affect the patients' health negatively as such kind of medicines are quite vital for the patient. It helps in maintaining and keeping the blood pressure at a required essential level. Likewise, the patients with high diabetes levels are required to take insulin injections on a daily basis. They cannot afford to miss out on their dose of insulin injection. They are even required to do blood test on a regular monthly basis so that they can keep a check on their blood sugar level.

There are many elderly patients who are severely ill or are not able to visit hospitals. IoT comes into role here. It helps by virtually keeping the patients connected to the doctor. The patient's real-time health parameters can be checked by the doctor from any location of the world. This shall be highly beneficial for the patients and the doctor as well. Both can easily overcome the barrier of distance by being virtually connected with each other.

ADVANTAGES OF INTRODUCING IoT IN HEALTHCARE

There are enormous advantages associated with introduction of IoT in healthcare. They are as follows:

1. Patient's health can be kept under continuous vigilance from any remote location.
2. The doctor and the patient's care taker can receive time to time updates regarding the patient's health.
3. The need for any person to be always physically present with the patient for vigilance of the patient's health is eradicated to quite a great extent.
4. IoT can provide virtual assistance to the patients.
5. A timely alert or update through IoT regarding the patient's health can help prevent any major mishap related to the patient's health.

CHALLENGES ASSOCIATED IN IMPLEMENTING IoT IN HEALTHCARE

Despite the various advantages the introduction of IoT brings in healthcare, there are certain challenges associated in implementing IoT in healthcare. They are as follows:

1. Ensuring the data always securely reaches the intended recipients.
2. Ensuring there is no data theft or tampering with data.
3. Keeping the system error free.
4. Making sure that timely updates in real-time are sent and received.

CASE STUDIES

Case Study 1

Gowrishankar S. et al. (2017) developed a heart rate and temperature monitoring system. The system consists of different modules. These modules are sensing module, main module and interaction module. The sensor module comprises of pulse sensor and temperature sensor. Main module essays the role of collecting information through sensors and storing it on cloud. The interaction module is for interacting with the system. There is a wireless –fidelity (Wi-Fi) module present in this developed system. All the information is uploaded at the IoT server using Wi-Fi module. Different test cases were considered for testing the developed system. These test cases included checking the heart rate of the patient when the patient is idle, when the patient is asleep, when the patient running and when the patient just gone through some abnormal heart rate. When the heart rate is not normal, i.e., it is not within a specified normal range, the liquid crystal display (LCD) displays a flash message saying heart rate alert and a buzzer raises a beep alarm. All the data from the sensors is uploaded on the IoT server in real-time and is accessible to the doctor from any part of the world. From this case study it is evident that IoT is greatly helpful in healthcare sector.

Case Study 2

R.N. Kirtana et al. (2017) developed an IoT based heart rate variability measure system. This system consists of a pulse sensor to check the heart rate of the patients. The microcontroller used in this system is Arduino Uno. Wi-Fi module is used for transmitting data to server. This system was tested on six different patients aged between three to sixty years. Out of these six patients, three were normotensive and three were hypertensive. The measured results are as in table 1. When any measured parameter is not in the normal range and is indicating some serious health issue then the doctor and care taker of the patient both are alerted through short message service (SMS). Based on the measured results, a graphical form of the results is also generated by the system which further helps the doctor in analyzing and comparing the health condition of the patient. From the measured results seen in Table 1, it is can be stated that except for the heart rate, all the other parameters of the hypertensive patients are less than the normotensive patients. The inter-beat interval (RR) of the normotensive patients is much more as compared to that of hypertensive patients. The intervals between the normal beats (NN) interval of normotensive patients is much higher than the hypertensive patients. The time domain measurement of heart rate variability (pNN) and the standard deviation of normal to normal intervals (SDNN) of normotensive patients is also greater than that of hypertensive patients. The results obtained from the developed system clearly show that the system is working efficiently and IoT can majorly benefit the healthcare sector.

Case Study 3

An intelligent pill box was developed by authors Juan Marcelo Parra et al. (2017). The intelligent pillbox is utilized as a storage device for storing pills, which contains an alarm system that can be programmed and an automatic opening and closing system. The pillbox also consists of an interactive and friendly user interface and a notification system that sends notification through GSM network. The pillbox alarm system would raise an alarm for each dose of medicine the particular patient using the pillbox is required to take. When the alarm system is activated, the pillbox automatically opens the door of the compartment where the required medicine to be consumed by patient at that particular time is kept. When the patient takes the medicine, infrared (IR) sensor indicates it to the microcontroller and then the door closes in some time. If incase the patient does not take the medicine then the IR sensor will not receive any input to indicate the microcontroller to close the door, so for such case, the pillbox will wait for a total of 10 mins after opening the compartment door. After 10 mins it will close the door as it did not receive any command from microcontroller. There is a liquid crystal display (LCD) which shows the information regarding the medicine dose that should be taken and it also shows the time at which it should be taken. Along with it, a short message service (SMS) will be sent to the patient reminding him/her to their take their medicine which also specifies the particular medicine dose to be taken as per the timing. In the

Table 1. Measured results.

Measured Parameters	Person under review	
	Normotensive	Hypertensive
Heart rate (beats/minute)	77.5	127.0
RR mean (ms)	789.0	579.5
NN50	97.0	61.0
PNN50%	43.5	27.5
SDNN	210.68	166.08

Source: R.N. Kirtana et al. (2017), IEEE International Conference on Computer, Communication and Signal Processing.

case when the patient does not take the medicine and IR sensor does not receive any input, the doctor or the care taker of the patient will be sent an SMS stating that the patient did not take their medicine. This system is proof enough to illustrate the miraculous use of IoT in healthcare sector.

FUTURE OF IoT IN HEALTHCARE

IoT can majorly change the current healthcare functioning scenario. IoT if promoted and encouraged well by the society, has a great potential of improving the facilities for the doctor as well as the patients. It shall not only assist the doctors and patients by keeping a vigilance on the patient's health but it shall also prove to be a boon for them both. Despite of having great vigilance facilities, there is a need felt for improving the existing facilities for the betterment of patients. IoT fits very well in this role by fulfilling this need. Hence, it can be rightly said that IoT indeed has a great future in healthcare.

CONCLUSION

The case studies above prove that IoT can very well be introduced and utilized in healthcare sector. The results of the heart rate and temperature monitoring system and the heart rate variability measure system proved to be efficient enough. The intelligent pillbox system is also very beneficial. It is a great virtual assistant for the patient. All the three above case studies showcased systems that proved the difference IoT can bring in healthcare sector for the betterment of mankind. Hence, it can be rightly stated that IoT has a great potential and should be encouraged to be used on a large scale in healthcare sector.

IoT is a technology which helps in connecting things and devices through internet. Despite of the physical distance between devices, it helps in maintaining a virtual connection between them. It has great a potential and due to its enormous benefits it can very well be included in healthcare sector. On the basis of the results seen in above case studies, it can be corrected said that IoT can ease out work in healthcare sector for everyone and act as a virtual assistant. It can help in keeping regular tab on patient's health parameters. Thus it can be concluded that IoT can very well help in making an intelligent health vigilance system.

REFERENCES

Ali & Ghazal. (2017). Real-time Heart Attack Mobile Detection Service (RHAMDS): An IoT Use Case for Software Defined Networks. *IEEE 30th Canadian Conference on Electrical and Computer Engineering (CCECE).*

Bhoomika & Muralidhara. (2015). Secured Smart Healthcare Monitoring System Based on IoT. *International Journal on Recent and Innovation Trends in Computing and Communication, 3*(7), 4958-4961.

Gowrishankar, S., & Prachita, M. (2017). IoT based heart attack detection, heart rate and temperature monitor. *International Journal of Computers and Applications, 170*(5), 26–30. doi:10.5120/ijca2017914840

Kirtana, R. N., & Lokeswari, Y. V. (2017). An IoT based remote HRV monitoring system for hypertensive patients. *IEEE International Conference on Computer, Communication and Signal Processing*. 10.1109/ICCCSP.2017.7944086

Li, C., Hu, X., & Zhang, L. (2017). The IoT-based heart disease monitoring system for pervasive healthcare service. *Procedia Computer Science*, *112*, 2328–2334. doi:10.1016/j.procs.2017.08.265

Parra, J. M., Valdez, W., Guevara, A., Cedillo, P., & Ortiz-Segarra, J. (2017). Intelligent Pillbox: Automatic and Programmable Assistive Technology Device. *Proceedings of the IASTED International Conference Biomedical Engineering*, 74-81. 10.2316/P.2017.852-051

Tjiharjadi, S., & Fajar, A. (2017). Human Heart Rate Detection Application. *International Conference on Soft Computing, Intelligent System and Information Technology (ICSIIT)*, 167-172.

KEY TERMS AND DEFINITIONS

Internet of Things (IoT): It is a technology used for connecting objects or devices together virtually irrespective of the physical distance between devices and objects.

Vigilance System: It is system required to keep a continuous tab of any particular thing or person.

Chapter 8
Economic and Commercial Aspects of IoT in Agriculture Digitization

Sunil Kumar
https://orcid.org/0000-0002-8348-1553
Amity University, Noida, India

ABSTRACT

Around the world, city, state, governments, and public sector groups are leading the way in taking the internet of thing (IoT) to existence and improving the lives of citizens everywhere. Currently 21 billion devices are connected. For 2025, the installed base of IoT devices is forecast to grow to almost 76 billion worldwide. At the same time, the global smart agriculture market size is expected to triple by 2025, reaching $15.3 billion (compared to being slightly over $5 billion back in 2016). Because the market is still developing, there is still ample opportunity for businesses willing to join in. Building IoT applications, devices, and products for future agriculture set you apart as an early adopter, and as such, help you pave the way to success. Wireless sensor network has an ability to solve many problems of real world (i.e., agriculture, harvest, farm monitoring, post-harvest, remote irrigation control, warehouse, livestock monitoring, cold chain monitoring) and improve the decision-making capability of government in response to natural disasters.

INTRODUCTION

Digitization of agriculture in India can play a very important role but harsh economic conditions make it difficult to implement. Only 7% of farmers have more than of 2 hectare land, where 80% of farmers have less than of 1 hectare land thus source of income is very low. In such conditions investment in technology is very difficult for Indian farmers.

The population in urban area is increasing rapidly and population growth will lead to more demand for food. This demand only fulfills when the production will increase in the agriculture field. In earlier days the farmers predict the rainfall by looking the sky but now days due to the global warming the

DOI: 10.4018/978-1-5225-9574-8.ch008

weather becomes unpredictable. Therefore we need technology to control the weather and predict the production in agriculture.

The discussion of IOT in Agriculture today is intriguing a result of numerous intonation focuses. The semiconductor and remote innovation is effectively accessible and at value focuses that would bode well from a ROI viewpoint of even a little time agriculturist when fabricating is done at creation scale. Keen cultivating dependent on IoT innovations will empower cultivators and agriculturists to lessen waste and improve efficiency (Babcock et al, 1998).

Utilization of innovation to agribusiness is a major subject. It can cover zones like miniaturized scale science, tissue culture, ranch motorization, Farm ERP frameworks, better seeds, better manures, Hydroponics frameworks, better additives, showcase revelation instruments, production network upgrades, better credit plans, sun based controlled homestead apparatuses and a large group of different things. Here we will concentrate mostly on use of hardware and PC in horticulture (Babcock et al, 1998).

Wireless sensor network has an ability to solve many problems of real world, i.e. agriculture, harvest, Farm monitoring, post-harvest, remote irrigation control, warehouse, livestock monitoring, cold chain monitoring and improve the decision making capability of government in response to the natural disasters. Yuktix is an IOT and M2M start-up working in the field of WSN and it provides solutions in the field of agriculture, air quality check and smart infrastructure and changing the lifestyle of the millions of people. Yuktix provides remote monitoring and control of automatic weather stations and air quality stations in many cities of India. Yuktix is working on weather project, transforming agriculture by crop disease prediction, air quality check and changing the lives wirelessly ("Food and Agriculture Organization", 2016). Some agriculture examples how WSN can impact the economy of any nation are given below:

Agriculture: Yuktix is working in transforming agriculture sector and providing solutions from agriculture fields to warehouse.

1. **Crop Disease Prevention:** The quantity of pests in crops can be projected by monitoring the variables like level of water, soil Temperature, growth of crops, humidity, leaf wetness, moisture, CO2 and temperature and therefore suitable quantities of pesticides could be sprayed at correct time. Using this system a farmer easily access the all required data and can predict the type of diseases in crops and take the necessary actions. This model can solve the problem of pesticide excess in agriculture produce.

2. **Smart Greenhouse:** Yuktix provides sensor based smart solution for climate control and provide the suitable environment for agriculture. It helps in the growth of the production of agriculture products by measuring temperature, humidity, soil temperature, moisture in environment, CO2, leaf wetness and soil pH etc. It controls the climate by measuring the variables inside a greenhouse and maintains the ideal conditions in agriculture. Yuktix developed many smart devices and provides interface with many type of sensors. It stores the data over the cloud and any authorize, authenticated farmer can monitor the environmental conditions from anywhere any time using the application on a mobile device.

Figure 1. Crop disease prevention system

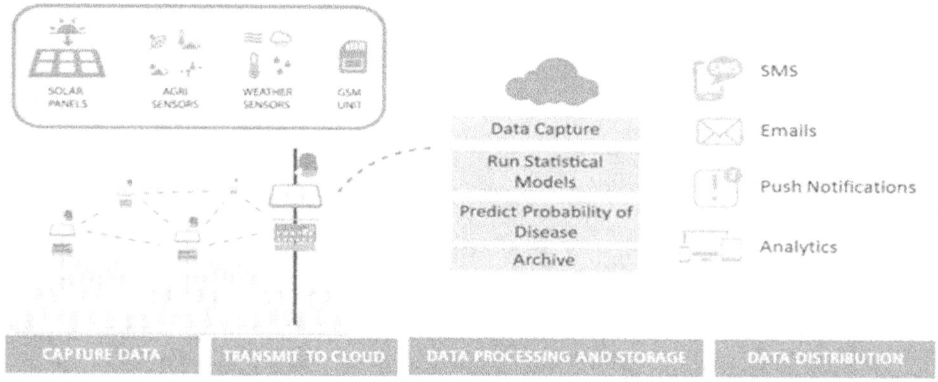

3. **Post-Harvest Monitoring and Remote Management:** A large part of the agriculture produce is stored to fulfill the requirements of tomorrow. Yuktix developed a system which controls the climate conditions and keeps the stored produce safe from any diseases.

Smart sensor based solutions monitor the growing pattern of any produce and take automatic defensive actions. These smart sensors deployed in the storehouses which are IoT enabled. The sensor nodes are exchanging their information across the storehouse and helping in improving the lifetime of agriculture products. These sensors communicating with gate way node using Zigbee technology and gateway node connected to controller which is controlling the storehouse as per the programming.

4. **Warehouse and Cold chain monitoring:** A large part of agriculture produce is stored to meet future demand and to keep it safe from changing weather patterns and diseases. The table below shows temperature and humidity ranges for products which are stored in cold storage. There are products that don't require cold chain to storage but they also need to be protected from the elements and kept under a controlled climate, e.g. rice and wheat. There is definitely an impact of environmental conditions on every produce.

Figure 2. Smart greenhouse system

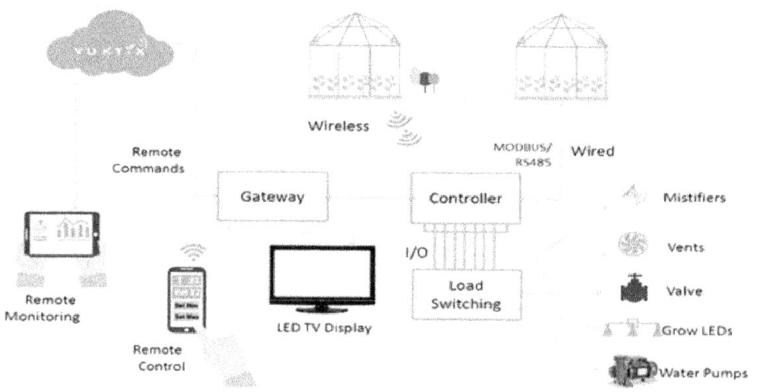

Figure 3. Post harvest monitoring and remote management system

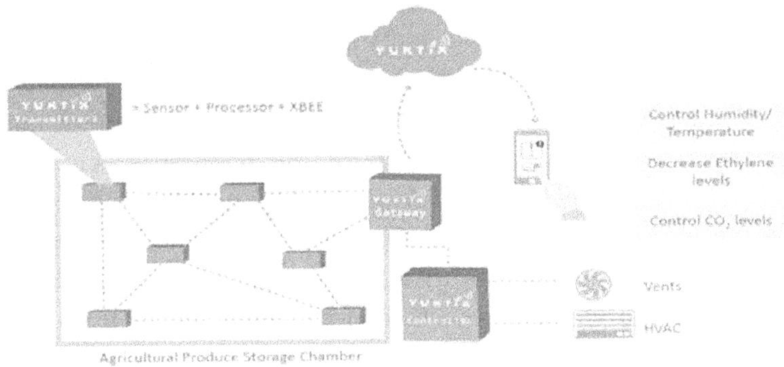

5. As per Government of India (GOI), The harvest and post-harvest losses run to the tune of Loss of INR 92, 651 Crores (more than 14 Billion US dollars). Most of post-harvest losses happen due to lack of monitoring infrastructure. There is no easy and cost effective way to monitor what is happening inside cold storage. We do not have enough cold storage with climate control facilities. Problem is that losses at post-harvest and storage stage are much severs than at the production stage. For example red chili should be stored 6 degree - 8 degree Celsius. Few days' deviation might not affect the quality but if deviation continues for few weeks or a month, it can result in quality reduction and hence export worthiness of the chili. Yuktix COLDSENSE allow you to set alerts using SensorDB and alerts are send to you on SMS or email as soon as a sensor reports deviation from the prescribed range. The cold sense data can be used for thermodynamic modelling of the warehouse which can then help improve the cooling efficiency and reduce the power bill of the cooling tower. That saving of power on cooling towers leads to direct savings (ROI) for the warehouse operators.
6. **Agricultural Drones:** Using drones for agriculture is a hot topic these days, and for good reason. These unmanned aerial vehicles (UAVs), as they are sometimes called, are rapidly becoming a core tool in a farmer's precision equipment mix. Today's farmers have to deal with increasingly complex concerns. Issues such as water – both quality and quantity, climate change, glyphosate-resistant weeds, soil quality, uncertain commodity prices, and increasing input prices to name a few (Babcock et al, 1998).

Growers are turning to high-tech tools, often under the banner of precision agriculture, to respond to and mitigate these and other concerns. Precision agriculture divides a field into zones that can be individually managed with a range of GPS-equipped precision machinery. Technology enables farmers to collect, store, combine and analyze the layers of data that drive precision nutrient and irrigation management. Using drones for agriculture is a hot topic these days, and for good reason. These unmanned aerial vehicles (UAVs), as they are sometimes called, are rapidly becoming a core tool in a farmer's precision equipment mix.

7. **Agriculture and its Livestock Monitoring:** Around 22.1 million populations depend upon livestock for their livelihood which contribute to 16% of the income of small farms household income. The

8.9% people of India living their lives from it. IoT plays a bigger role in many ways when it comes to agriculture and its monitoring (Abdullah et al, 2016).

8. **Remote Irrigation Control - Without sensor and with Sensor based feedback**
9. Without sensor and with Sensor based feedback. Irrigation plays an important role in Indian agriculture due to unpredictable climate changes and inefficient management of water resource resulting shrinking water labels, it will play a very important role in many aspects in the coming time. The water system is straightforwardly identified with the water preservation. On the off chance that we can do keen water system, at that point we can spare water. One more precedent is remote control of water system siphons. Numerous agriculturists live very far from the field and they need to achieve the field to kill ON/OFF the engines at odd hours for example late in the night and promptly in the first part of the day. This troublesome errand can be mechanized utilizing IoT, a cell phone and sensor criticism based framework (Harper et al, 2015).
10. **Cell Phone Based Control:** You can consider the number and press a catch and, your siphon would turn ON/OFF contingent upon the settings. You can control one engine or numerous engine with a similar gadget. KISAN RAJA is one of the outstanding Indian organization working here. Alternate organizations are Flybirdinnovation and Avanijal..
11. **Sensor Based Automatic Control of the Irrigation Line:** An automatic version of the previous solution can be done by measuring the soil moisture with the help of conductance based soil moisture sensors. Irrigation pumps can start depending on the actual soil moisture conditions ("Food and Agriculture Organization. Some companies which are active in this are Corpx, FarmX, Farmobile, Edyn, Hortau, tevatronic, Aquaspy. You can irrigate a wide area with just using a ", 2016). single pump and drip irrigation. Such solutions are very effective where there is a water shortage for the crops that need very accurate irrigation like Banana or cotton plants. Using latest communication technologies like LORA or Sigfox, a big and large network of wireless soil moisture sensors can be deployed over a wide area and water supply can be made to specific areas depending on the real time irrigation requirements, thus optimizing the usage of precious resource. In Gulf countries like Qatar or Saudi Arabia where there is an acute shortage of water, such solutions can prove to be very useful.

The Economic and Commercial aspects of Agriculture

The IoT has the power to transform agriculture in many aspects. The way IoT can change the farming is given below:

- Collection of real-time data, e.g. Temperature, water level, weather conditions, quality of soil, crop's disease and cattle's health. Collected data is useful to track the status of events in general as well as staff and tools efficiency, etc.
- Better resource control and low production risks. The IoT gives better prediction about your product production to sell. The ability to foresee the production gives you the choice of distribution. In early stage if farmers know about how much crops they are going to harvest, they can make assured that their product won't lie nearby unsold.
- Waste and cost asset the executives. On the off chance that agriculturists ready to perceive any peculiarities in the harvest development, any sickness or animals wellbeing, at that point ranchers will most likely dull the dangers of losing their yield by applying numerous specialized arrangements.

- Process Automation impacts on Business Growth. By utilizing brilliant gadgets, ranchers can computerize numerous procedures over your creation cycle, for example water system, treating, or bother control.
- Better Quality and Quantity of item. Accomplish better authority over the creation procedure and keep up higher measures of yield quality and development limit through computerization.
- As an outcome, these variables can in the long run lead to higher income. Presently that we have illustrated how IoT can be beneficially connected in the circle of horticulture.

The IoT is Set to Give Next Generation Farming

The digitization of horticulture is as of now ending up progressively ordinary for substantial landowner ranchers, and innovative cultivating is rapidly turning into the standard for little agriculturists as well, much gratitude goes to IoT and devise (Harper et al, 2015).

- Farmers have just started utilizing some cutting edge cultivating procedures and innovations so as to enhance the proficiency of their everyday work. For instance, sensors set in fields enable agriculturists to acquire point by point maps of both the geography and assets in the region, just as factors, for example, causticity and temperature of the dirt. They can likewise get to atmosphere estimates to foresee climate designs in the coming days and weeks.
- As a solid model, John Deere (composers hardware maker) as of now began interfacing its tractors to the Internet and has made a strategy to show information about ranchers' harvest yields. Besides, the organization is spearheading self-driving tractors, which would free up agriculturists to perform different errands and further increment effectiveness.
- Taking pictures from the satellite, IoT and WSN to watch and gather the information with the point of expanding the generation, diminishing the expense and keeping up the assets all assistance make exactness cultivating or accuracy horticulture conceivable.

Shrewd agribusiness and exactness cultivating are taking off, yet they could simply be the antecedents to much more noteworthy utilization of innovation in the cultivating scene.

- BI Intelligence predicts that IoT gadget establishments in the farming scene will increment from 30 million out of 2015 to 75 million out of 2020, for a compound yearly development rate of 20% (Abdullah et al, 2016).
- On Farm, which makes an associated ranch IoT stage, anticipates that the normal homestead should produce a normal of 4.1 million information focuses every day in 2050, up from 190,000 of every 2014. (Zhou et al, 2016).
- Furthermore, On Farm ran a few investigations and found that for the normal homestead, yield ascended by 1.75%, vitality costs dropped \$7 to \$13 per section of land, and water use for water system fell by 8%.

Return on Investment in Digitizing the Agriculture

Agribusiness is spine of any nation and India rank second in homestead yield, half of absolute workforce is in Agriculture and it includes 13.7% in India GDP. Still Agriculture is demographically the broadest economy division and assumes a noteworthy job in general financial texture of India. With expanding weight of populace, the weight ashore creating more yield have expanded thus as the weight on the normal assets. In situation like this, accuracy agribusiness assumes a critical job and when a similar exactness horticulture is executed in a restricted region with controller condition, that training is named as nursery development. So on the off chance that I need to clarify what Greenhouse development is in straightforward words, it is "development of vegetation in controlled condition, giving them fake condition that encourage their development". Variable like Temperature, mugginess, measure of light, CO2, soil dampness, supplements are estimated and controlled inside a nursery. By and large in Indian situation, the normal spending plan for a 1 hectare nursery is estimated 60 lakhs which incorporate foundation cost, ranch, stockpiling, work and so forth. Utilizing top of the line innovation is troublesome for any agriculturist thinking about this financial plan. Digitization of agribusiness is an answer that given you a chance to get to continuous information from anyplace on the planet on your telephone too on a web application and let you control neighborhood condition with a tick of the catch. Neighborhood show is likewise conceivable on a major LCD screen. So now the accompanying unavoidable issue is the reason? For what reason would someone even use sensor based splendid contraption for a nursery when everything is working fine with the physical work and age is extraordinary? Data reliably expect a crucial activity in each field. What happen now days in typical nursery, all the manual computerization is done subject to secret which lead to cut down creation, wastage of advantages like water and control, etc and this end up being a shortfall in net advantage!. Lets take a point of reference here of a 1 hectare nursery with chrysanthemum as the cash crop. Chrysanthemum are light sensitive plants and they create in different stages with various requirements like beats of light when blooms are coming up, high wetness when they are in early create stages (Quigley et al, 2009).

- **High Humidity:** When chrysanthemum are in their beginning time, appropriate measure of stickiness is required for the plants - called as "mums". Dampness range ought to be 80 - 95% for the entire day and the time of "mums" organize is some what 15 - 20 days. At the present time, moistness is controlled with mystery or clock mode for example turn on Mystifier after each 2 hour for 5 minute independent of continuous conditions. Additionally some time because of abundance water on the abandons, a portion of the plants pass on. For the most part, the yield here is 85 of 100 plants.
- **Right Amount of Light:** Chrysanthemum is a blossom use for enriching reason and is sent out to nations like Holland, Germany, Singapore, America(USA). The Main condition for fare is that the distance across of bloom ought to be 12-14 cm and this breadth must be achieved when beats of white light(Blue + red) is given for explicit interim of time and in explicit length of multi day.

How Digital Smart Greenhouse Work and how Effective It Is

- Advanced Smart nursery has different detecting units both wired and remote. With various sensors conveyed in various chambers, constant estimations of temperature, moistness, lux, CO2, soil dampness are recorded and send to the focal controller. Focal control the hardware like Mystifier(which thus control the mugginess keeping imperatives like power utilization, recurrence of mysting in thought.), Lights (for bloom bud development keeping limitations in thought). A neighborhood huge showcase is likewise used to demonstrate constant readings to the supervisor or proprietor of nursery. Similar qualities can likewise be gotten to from anyplace on the planet utilizing Smart telephone or App. Reaches like Humidity range or term of light heartbeat can likewise be changed utilizing the advanced mobile phone and manual supersede is additionally conceivable.

So now let calculate the ROI and some points are given below:

- Based on Guess work, work turn on Mystifier after ever 2 hour for 5 minute. That infers in 8 hour working range - 20 minutes. Propelled Smart nursery checks temperature, dampness and leaf wetness following 15 second and turn on mystifier after every 10 minute for 10 seconds. So in a 2 hour 2 minute and in a term of 8 hours 8 minutes. So a net speculation assets of 12 minutes. So a full scale assets of Rs 3.288 = ~Rs 1000 consistently. *(1 unit cost = Rs) at least 6 that, no estimation is done after those 8 hours. That will again cause some setback in progress.
- Production cycle is of 15-20 days. 2 Sunday there is no work and if you see, following 8 hour in day, no one turn on mystifier by any stretch of the creative ability. So now if 2 Sunday there is no work it infers 13.33% adversity. By and by expense of Chrysanthemum one cut stem is = Rs 3. Out of 100 plants - 13 fails horrendously = Rs 39 adversity and out of 1000, 130 plants kick the basin, that suggests Rs. 390 adversity in 15 days and Rs. 780 setback in a month .
- No of relentless light hours should not be more than 10.5 hours and dull hour should not be more than of 4 hours. If this progression is managed, blossoms will accomplish 12-14 cm in separation crosswise over and will meet all necessities for charge publicize. At present, one rose is sold at Rs 16(Maximum) and Rs 4(minimum). Typical cost is Rs 10. Likewise, at a comparable point, gerbera bloom is sold at Rs 1-4 for each sprout. So again lost Approximate Rs 6 for each blossom. That infers if 100 blooms are sold in close-by market rather than exchanging them, an overal shortage of Rs 600. Generally in a 3-month pretty much 5000 - 7000 blossoms are produced. This infers lost Rs 36,000 (tolerating 6,000 blooms in multi month). Directly in a year a hard and fast loss of Rs. 1,44,000.
- Convenience and progressing cautions - Even for an age head, visiting each chamber every day is a troublesome and tiring task. Besides, if a comparative age chief or nursery owner needs to visit 3-4 foundations in a city in multi day, it is for all intents and purposes close limitless. Automated Smart nursery offer impact to the boss/owner to get to consistent data, alerts from his nurseries on his phone and that too from wherever in world.

CONCLUSION

Therefore, the IoT applications in agrarian field are building it promising for landowners and agriculturists to gather genuine information of their items. Presently to satisfy the interest of future, each huge or little rancher needs to comprehend the capacities of IoT applications for digitizing the agribusiness by introducing the IoT gadgets to build viability and supportability in their creations. The interest for future in changing condition can be effectively met when ranchers acknowledge the IoT arrangements and execute them in a fruitful way.

REFERENCES

A. Abdullah, S. A. Enazi, and I. Damaj, "Agrisys: A smart and ubiquitous controlled-environment agriculture system," pp. 1–6, March 2016.

Harper, C., & Siller, M. (2015). OpenAG: A Globally Distributed Network of Food Computing. *IEEE Pervasive Computing*, *14*(4), 24–27. doi:10.1109/MPRV.2015.72

Moving from Uniform to Variable Fertilizer Rates on Iowa Corn. (1998). *Effects on Rates and Returns by Bruce A.* Babcock and Gregory R. Pautsch.

Quigley, M., Conley, K., Gerkey, B. P., Faust, J., Foote, T., Leibs, J., ... Ng, A. Y. "ROS: an open-source Robot Operating System," in *ICRA Workshop on Open Source Software*, 2009.

"The state of food and agriculture: Climate change agriculture and food security," Food and Agriculture Organization of the United Nations (FAO), Rome, Tech. Rep., 2016.

L. Zhou, N. Chen, Z. Chen, and C. Xing, "ROSCC: An Efficient Remote Sensing Observation-Sharing Method Based on Cloud Computing for Soil Moisture Mapping in Precision Agriculture," IEEE Journal of Selected Topics in Applied Earth Observations and Remote Sensing, vol. PP, no. 99, pp. 1–11, 2016.

Chapter 9
Integrating IOT-Commodity Cameras Through LoRaWAN:
An Architectural Implementation

Rajiv Pandey
https://orcid.org/0000-0003-3348-9888
Amity University, Lucknow, India

Shahnaz Fatima
Amity University, Lucknow, India

Shubham Asthana
Amity University, Lucknow, India

Ayush Kumar Rathore
Amity University, Lucknow, India

ABSTRACT

IoT devices and their applications are supporting humankind in almost all domains. This chapter explores LoRaWAN and proposes to integrate the commodity cameras installed at the security points of most of the gates to residential areas. LoRaWAN is a media access control (MAC) protocol for wide area networks and it has been opted for its functional and architectural scalability. The intelligent inputs are transferred from the commodity cameras to the data concentrators (DC), on edge-based computing, the DC can transfer this input to fog, cloud, remote servers for machine learning integrations. This chapter demonstrates the basic architectural framework of the said implementation. However, the detailed implementation and prototype is beyond the scope of this chapter. The chapter has however demonstrated the architecture. The features of the commodity cameras have been listed that can serve as the feed to the concentrators that shall enable alarm generations at the local and remote policing sites.

DOI: 10.4018/978-1-5225-9574-8.ch009

INTRODUCTION

This chapter highlights the points related to integrating the small existing surveillance systems in ordered to obtain a unified and autonomous system and the steps involved in this process. The chapter also presents the use of python programming language as a tool for implementation of the same and its features shall also be highlighted. The Python programming language is a versatile programming language and is commonly used for application developments and data analytics and visualization purposes with active support of libraries like CV2 or OpenCV for image capture and analysis also for user friendly graphical environment developments.

Internet of Things (IoT)

In modern age, Internet of Things (IoT) is emerging as a very promising technology with capabilities to transform how we live in the environment which is surrounded by IoT devices. The Internet of things has a large impact on the consumers and the market. IoT has become a significant topic in places like industries, policies, government, and college research topics. Many of the companies are and will be investing in research projects and owned to that various projects are in the current market scenario, like the home automation system, fitness wearables, traffic congestion monitoring system, network enabled health monitoring system which make use of the sensors to provide the services. Many disabled and elderly people will be benefitted by these emerging technologies which will improve the quality and individuality of the living. In a future scope the large chunks of data collected by the sensors across the surroundings will help in transforming cities in to smart cities by which we move closer building an energy efficient city. This also enable us to elevate areas like the agricultural industry, smart education, industries, health and medical, transportation.

According to RFC 7452, "The Internet of Things (IoT) is a framework in which all things have a representation and a presence in the Internet. More specifically, the Internet of Things aims at offering new applications and services bridging the physical and virtual worlds, in which Machine-to-Machine (M2M) communications represents the baseline communication that enables the interactions between Things and applications in the cloud.".

An operative point of view suggests us to contemplate over how the IoT devices will get their spark in terms of their currently used technical models. So, on March 2015, IAB (Internet Architecture Board) released a document as a guide for networking smart objects [RFC 7452], which gives a full description of the models which will be used in abolishing the barriers in connecting IoT devices.

- Device-to-Device Communications
- Device-to-Cloud Communications
- Device-to-Gateway Model
- Back-End Data-Sharing Model

The above-mentioned communication models demonstrate the developing way to be used to facilitate current devices to connect. The networking of these devices is greatly influenced by the type of models they are using and there open versus proprietary landscape. Device-to-gateway model as its primary feature is its facility to overcome the restrictions in connecting IoT devices which are proprietary and opensource making the data exchanging and open standards its key contemplation.

From a wider opinion, the interconnection of devices with technologies and trends is only possible because of the following features:

1. **Pervasive Connectivity**: The wireless services provided are enabling us with omnipresent low cost highspeed connectivity. Being a licensed and unlicensed service.
2. **Adoption of IP's**: All devices connected todays has a unique ip assigned with it which enables a device to be omnipresent as IP is now a dominant global standard.
3. **Computing Finances**: Moore's law gives us the technology for low powered and great computing power which is developed in and drove by research, investment.
4. **Diminishment**: Owing to computing finances the diminishment of the sensors with greater computing power and communication led to various creation of inexpensive devices which are the base of many IoT application.
5. **Data Analytics**: As the data is rising due to the miniaturization of the devices new algorithms are incorporated in cloud which enables us to analyse, correlate, and aggregate the data which helps us to abstract it for knowledge and useful information.
6. **Cloud Computing**: Comes with fog computing which allows these devices and sensors to use the computing power with powerful backend analysis and control and help us interact, process, manage and store data.

It is a huge network of interconnected objects that can increase the availability of information dramatically. To achieve greater results and performance, the advancement of technologies such as artificial intelligence systems that can communicate and share data can be used in any industry. In almost every corner of the world, IoT is likely to transform companies and organizations because it has vast applications in our day - to - day activities. Deployment of IoT is challenging. Apart from IT teams, several operational teams and business units must be involved in an effective IoT implementation.

Proposed Idea

The perspective for writing this chapter is to find an effective method to integrate the now scattered security systems. The prime focus for this unification would be on the surveillance systems incorporating CCTV (Close Circuit Television) cameras, referred as commodity eyes in this chapter.

The objective for this project is: Distributed parallel processing by servers installed at various locations around the city, an ideal system redundancy and fault tolerance that could withstand any danger.

The current scenario for law enforcement requires a great number of manpower and other resources like vehicular fuel just for patrolling purposes. On the other hand, the civil areas under the patrolling supervision are already been guarded by personal and/or private security systems, generally incorporating security via cameras. The element lacking in this current design is a proper integration of these individual arrangements (Han et al, 2013).

The existing system lacks in efficiency of communication between police, society and their cameras, hence a proper integration has to set up in order to make the cities a lot safer. This could also be seen as another small step towards achieving the concept of smart city, particularly by providing a smart surveillance system. These individual systems, of society as described earlier, would act like as discrete points of data collection in the unified pan city architecture. After effective integration, the proposed design shall provide with a single or replicated point of observation to monitor an area as huge as a metro city.

A proper AI (Artificial Intelligence) algorithm shall be then applied to this to further automate the alarm generation and warning system to lower the human intervention even further.

This also solves the current methodologies substantial problem. The present structure lacks a real time warning generation system; this makes the surveillance only a passive observer and a proof collector to be used during legal trials (Lee et al, 2013).

For making a city, security smarter includes an up gradation or modification of conventional system. The conventional system as said in above text, needs to overcome the lacking in efficiency of communication between police, society and their cameras. There are two ways that can be implemented to achieve the goal of smart security,

- Building a new architecture totally financed by government or a PPP model.
- Enlightening the citizen about smart security and integrating their existing architecture with slight modifications. This makes project more cost effective and more reliable by ensuring public partnership.

The arrangements which the user or citizen is providing for the project are:

1. A security camera with transmitter (internal or external) to connect to internet or separate frequency.
2. Getting the camera's IP address registered for online monitoring.

Government provided support can be through the under mentioned framework facilitation.

1. A radio tower or a receiver/transmitter to an existing tower for transmission. A control room for 24 x 7 monitoring.
2. Cloud storage for further analysis of gathered data (Big Data).
3. Team of IT professional for big data analysis and devising monitoring software for further reducing the manpower requirements, which is to develop a suitable AI (Artificial Intelligence) Algorithm.

COMMODITY DEVICES

Commodity Cameras

From a perspective of IOT camera comes under the controller and devices layer. Camera is a sensing device which senses the environment and with the help of various components it captures the image of the environment and allow us to use this image for future. Having this kind of feature the makers of camera designed various models of cameras which can be used as a surveillance tool. In this chapter as we are using camera as a commodity, so it is a crucial part of our paper with its components.

In this section we deliberate about:

1. CCTV cameras,
2. Different type of CCTV camera,
3. DVR,
4. NVR,

5. Ip Networked cameras,
6. Data Compression.

CCTV(closed circuit television) cameras first we will talk about what a LAN is as cctv cameras using their own closed network to transmit the video feed rather sending it from an internet channel. CCTV itself stands for closed circuit television. LAN stands for local area network which means that the network is limited and by this, the author intends to mean it is bounded to be within a building or an office or to a limited geographical are but not connected to internet. The place where the cctv cameras are installed a cable connected with the cameras come towards dvr which manages different camera feed (Franklin et al, 2013).

There is a lot diverse type of cameras present in the market which can be used as a commodity like:

1. CCTV dome camera
2. CCTV wireless camera
3. CCTV bullet camera
4. CCTV C-mount camera
5. CCTV Time camera
6. CCTV night vision/infrared camera
7. CCTV network camera

Etc.

CCTV Dome Cameras: As there name suggest it has a body casing which seems like a dome as these cameras are used in places like houses, and markets where the camera needs to undetectable as to where the direction of camera is at a given time making this camera discreet but visible. These cameras have features like zoom, pan effect, tilt effect. They are used to cover a wide area where the camera is limited to one piece and where there is less light

CCTV wireless Camera: This camera has a network chip which act as a Wi-Fi receiver and this is used to send videos using the Wi-Fi. The main advantage of having this setup is that we get the quick installation as there are no cables, all the transmission takes place wirelessly.

CCTV Bullet Camera: This camera looks like a sniper rifle bullet as the camera case have a long, cylindrical shape. This camera do not have any user intervened control like zoom/pan/tilt control. The image it captures is of a fixed location pointing at a particular area. The camera can be mostly used for an indoors but have a flexible outdoor use as the case provide protection from dust, dirt, rain, hails etc.

CCTV C-Mount Camera: Like any other camera it has features to be used in an indoor/outdoor situation the only difference is that it has a detachable lens which fit to different applications. Normal CCTV cameras can cover range of capture till 40ft but the CCTV cameras can capture more that 40ft.

CCTV Time Camera: This camera has a unique feature to capture images in any condition possible these cameras have a very good camera lens which can handle glare, direct sunlight, reflections and strong backlight. They are mainly made for outdoor application as it has capability for working in both day and night time which has varying light conditions.

CCTV Infrared Camera: These cameras have an infrared illuminator which has capability to see images in complete dark conditions and ideal or a place where lighting is poor or zero.

CCTV Network Camera: These cameras sends/transmits the video feed over internet using both wired as well as wireless medium, but to match up with the speed of transmission it sends a compressed form of video to save the bandwidth for other cameras. The main feature is that is requires a separate cable for power.

DVR: It is device which projects the video feed captured by a camera to a connected device like a TV or sends it over a network. The video feed captured is converted into a video format that can easily be recorded in a HDD, USB drive, SD memory card, SSD or other mass storage device on the network. The term includes set-top boxes with direct to disk recording, portable media players and TV gateways with recording capability, and digital camcorders. Personal computers are often connected to and used as DVRs for video capture; in such cases the video capture application software is an integral part of the DVR.

Many DVRs are classified as consumer electronic devices such devices may alternatively be referred to as personal video recorders (PVRs). Video images of MP3, MP3 and AC3 are typically recorded and played on MPRs, VOB, MPEG-4 Part 2, MPEG-2.mpg, MPEG-2.TS, and ISO. It is also possible to display JPEG and PNG images and to play MP3 and Ogg music files. Certain devices can be updated in new formats to play and record. DVRs generally record for copy protection in proprietary file systems even when some file systems can use FAT. Standard TV recordings are typically 480p / i/576p / I whilst HDTV is normally 720p/1080i.

A DVR nowadays provide functions like searches of video feeds by camera/date/time. Elasticity over quality and FPS in now in our hands. Due to their intelligence, storage can easily be optimized for to store footage for future purpose ("Reclus et al, 2009).

Hardware Features

Here are the features associated with DVRs:

- Designed for PC/TV's or where possible mounted.
- Inputs can be single or from multiple sources. The DVRs now made are fully functional to handle analog signal provided by cables such as twisted pair, coaxial cable or optical fiber cable. Inputs are assigned in a power of 2 like 1, 2,4,8,16,32 and so on. Multiple DVRs can be configured to connect together using a LAN.
- Switches, multiplexers and other display devices can make use of the video output given by the DVR directly as the output given by the DVR corresponds to the same video signal input.
- Secure output to display devices.
- DVR consists of various buttons and indicators which be used to fully control the DVR from outside.
- Today the DVRs are consistent with the networks and are made to send/receive video feeds and also record them when needed
- Control devices can be attached to the DVRs like a Mouse and Keyboard.
- The features like pan/tilt//zoom can be achieved using a single connector.
- Hubs and DVD writer provided for writing video files into them for archiving process.
- Includes connection to external storage media.
- Includes triggered alarm event for one alarm per video signal.
- Alarm evens to be analysed by various internal features like motion detection and other image sensing.

Software features

Here are some software features associated with DVRs:

- Selection of fps is given to user not the software to each of input differently giving the user the flexibility to set the fps more for important cameras and less to less important cameras enhancing security.
- User selected image resolution to each of the input differently giving the user to program a triggered event for DVR to change resolution on an alarm for clear picture of the detected event.
- Data compression methods used decides the speed of transfer of the video feed over the internet.
- Motion detection a new technology where the motion is detected by using a frame change in the video feed. It is like a camera taking a picture of a still car and it generates an alarm when the car starts moving.
- User Selected motion detection system on each input differently giving the user full power to use motion detection on a particular camera. This feature of software make an efficient chain of processing power.
- Triggered selection of an object which is selected by user. Like for a video feed a user selected a part which is unacceptable next time the alarm will be generated on the basis of this selected criteria.
- In case of an event which is unacceptable the DVR directs the video feed to the selected video display device.
- All the video feeds are stamped with date and time.
- Logs of the triggered event are kept accordingly.
- Searches related to triggered event.
- Various sound recording channel.
- Feature of making cabinets/archives of data.

NVR SECURITY SYSTEM – PROS AND CONS

NVR security camera systems incorporate the newest technology to provide an enhanced, feature-rich security system. Also known as POE security camera systems, NVR based systems are more flexible and complex than DVR systems.The detailed comparison is covered in the following paragraphs (Lee et al, 2013).

Camera Type – IP Camera

As NVR systems process the camera video data instead of the recording device, NVR cameras are much more robust than their DVR counterparts. NVR systems use standalone IP cameras that capture images. IP cameras each have a chipset to process the video data that is then passed to a recording device. Unlike analog cameras, IP cameras can generally record and send both audio and video. The more potent IP-based hardware also enables smart and video analytics, including face recognition, to be enhanced.

Cable – Ethernet

NVR systems connect the camera to the recording machine, like DVR systems. But it's totally different how the camera is connected to the recorder. NVR systems are transferring data using standard Ethernet cables, for example cat5e and cat6. Professional installers prefer Ethernet cables based on the number of benefits compared to coaxial cables:

Ethernet cable powers the camera with Power over Ethernet (POE), thus eliminating a need for messy splitters like a DVR system, which will take the camera with a cable running to record video, audio, and power the camera. Ethernet cable tends to be easier to route and terminate because it is thinner and has a smaller connector allowing for less drilling.

Ethernet is cheaper than coaxial cable and much better available, making it more accessible and affordable for cable replacements or system extensions. Many modern houses and companies are wired for Ethernet and facilitate the installation. Added value is that every camera on the system can transmit audio because Ethernet is able to natively send audio data. No cables between each camera and the recording device are required. Their wireless network needs to be the same. Installation is easier and cleaner because multiple cables are not necessary.

Despite a shorter max Ethernet cable length, 328ft or 100m, network switches can be used to extend total distance without impacting image quality.

Recorder

In contrast to a DVR system, the recorder does not process video data in an NVR system. Before the camera is transmitted, this step is completed. NVR recorders are used only to store and view footage.

System Flexibility

Inherently, NVR systems are more flexible because security cameras do not necessarily have to be directly connected to the recorder physically. IP cameras must instead only be on the same network. As such, cameras around the world can be viewed as a comprehensive system on the same network that connects to your NVR.

Image and Audio Quality

Since NVR recorders receive a pure digital signal from the cameras, the quality of the video is better than a DVR with the same resolution. Furthermore, all cameras with microphones could record audio to the NVR as Ethernet cables carry audio. Both systems record and rely on video data. Although DVR systems ' video quality lagged behind comparable NVR systems in the past, this gap is now significantly reduced. The differences between DVR and NVR systems depend on the cost, the transmission of the data and the type of cameras. NVR systems tend to have better image quality, easier installation, greater flexibility and native audio support on every camera that has a sound system/microphone (Han et al, 2012).

Data Compression

Without digital compression, digital video surveillance would not be practicable. Digitizing high-resolution video generates a great amount of fast data, which is not easy to send over wide distance, but also requires large data vaults to store. Several video compression techniques helps to rectify this kind of problems. JPEG (MJPEG), MPEG-4 and H.264 are 3 of the several methods used for compressing video feed surveillance. Firstly all DVR used JPEG (MJPEG) which compress the feed image by image. This method uses the least CPU power but more transmission power.

MPEG-4 is now used as an efficient method of video compression it requires an ongoing processing of frames. A lot of CPU preprocessing is needed, but today it is far easier with new processors. H.264 is an ITU standard, MPEG-4 standard. It offers the best quality-bandwidth transmission. H.264 provides the basic MPEG-4 compression for the matching video quality approximately twice, which means that the video quality is untouched but the bitrate is changed. Nowadays most compression techniques uses MPEG-4/H.264, which is roughly the same compression but now H.264 is widely used.

The transmission bandwidth required for compressed transmission of video feed depends on all the above mentioned physiognomies, including format and Frame per second or fps. This is then packaged in TCP / IP. The digitized and compressed video is usually compressed in a frame which uses Real-time Transport Protocol (RTP), which is further compressed in User Datagram Protocol (UDP). UDP is then molded in IP, usually in IPv4, but later in IPv6. The resulted transmission packets are then usually further packed in standard Ethernet frames for use with CAT5e/6 or wireless ("InfoSec", n.d).

Peak videos have been saved in the past on a videotape. Most VCRs are using a two-hourT-120 VHS cassette. The slower fps causes the increase in video significantly. Numerous VCRs existed for many systems. Moreover, every system had different processes regarding how much and for how long a video is to be archived. Thanks to the DVR introduction, now no new systems use VCRs.

Cable and IP television (IPTV) systems are commonly used for DVRs stored on a large hard drive. Video compression makes it easy to store enormous amounts of video. Special DVR systems for surveillance are common. This includes inputs of analog camera that are digitized and compressed before storage by analog to digital converters (ADCs). Other inputs are supplied by IP cameras.

The standard DVR is a multi-channeled time-lapse video and web server device. There are four, eight, 16, and 32 input channels in a common box. You can change or remotely access the frame rate, record, or playback function. The majority have a hard drive of at least 500 GB, although there are frequent options of 750 GB and 1 TB. The most frequently used compression is H.264. A DVD burner is provided for some units (Han et al, 2013).

The FPS of a DVR, typically 120 fps maximum, and is a key factor to consider. This rate needs to be split between the cameras. So after splitting each can record at 30 fps with four cameras. However, if more cameras are used, each frame rate must decrease. It turns out that today, several cameras only record with a rate of 3 to 5 fps.

The NVR or network recording system uses RAID as the primary storage medium to store compressed videos and now the scenario is that all the present DVR are converted to Hybrid DVR/NVR systems.

Output Data Format

As described in preceding sections commodity cameras getting incorporated would be of three broad forms analogue, digital and network capable. Currently this section would focus on digital cameras however some of the content may also apply for the network capable cameras.

Digital Video Recorders began to replace older Analog CCTV recorders about 15 years ago, and they promised a lot: better quality, ease of use and reliability. One important feature was lost in the change-over: a standard data format. With Digital Video Recorders-DVR file system the storage mechanism and encoding techniques for the video data was to be resolved by the manufacturers. Owing the absence of a universal regulation or globally accepted industry standards, all the manufacturer started developing their own file formats and codec the need for media players were also obliged individually. The result was thousands of proprietary file format and ever growing codec (Camps-Mur eta l, 2013).

The letters or numbers after the final '.' in the video file name are called the file extension, these extensions are the key to identifying the file format, further to anticipate the possible codec used. If the file ends in 'AVI' the next step is to determine what codec the file requires. If it is 'mpeg', 'mpg', 'flv' or one of the other open formats, the file will likely play in a standard movie player such as VLC or Windows Media Player. If it is none of these, the file format is most likely closed and will only play in the manufacturers own software.

AVI is the most common file format in CCTV, but within AVI, most manufactures use their own, non-standard, codecs. The AVI files have their codec written into the file header. It is possible to look at the codec by opening the file in HxD, that is a HEX viewer/ editor (look at byte position 188). to make it more simple there are free tools available online. There are more than 850 different codec available, and that is only for AVI file system. Manufacturers provide with the codec they are using for the prducts. So the user can download a specific codec and install it to any existing media player to run the video coded using this codec.

Relevant Protocol

LoRaWAN and Evolution of LoRaWAN

In preceding sections the cameras and the output formats were described, in this section the focus shall be on the effective transmission of these sensor data, that is the video feeds to their respective actuators. This section shall focus on LoRaWAN, as it is relevant and low- cost protocol for this purpose. Low Power Wide Area Network or a simple acronym LPWAN is a network technology standard for interconnecting devices that have a critical power constraint like a batter with a single charge cycle of years or may be even decades, besides this but on the other hand these devices do not have much of bandwidth requirements. So these are low-bandwidth, battery operated, long communication range devices that requires a very low baud rate. Some common implementation of this technology is pressure sensors on hilly areas or soil dampness and moisture sensors in framing fields.

The figure above, Fig-1 is a reference architecture of LoRaWAN. In this architecture, left most devices are referred to as end Nodes or end devices- these nodes are capable of low power communication by the virtue of microsystems embedded. The next stage of architecture, as also labeled is Concentrator layer, these devices are placed to receive the data being transmitted by the end nodes and further to communicate with the Network servers. Then there are application servers and Network servers to monitor, record and

Figure 1. LoRaWAN Architecture

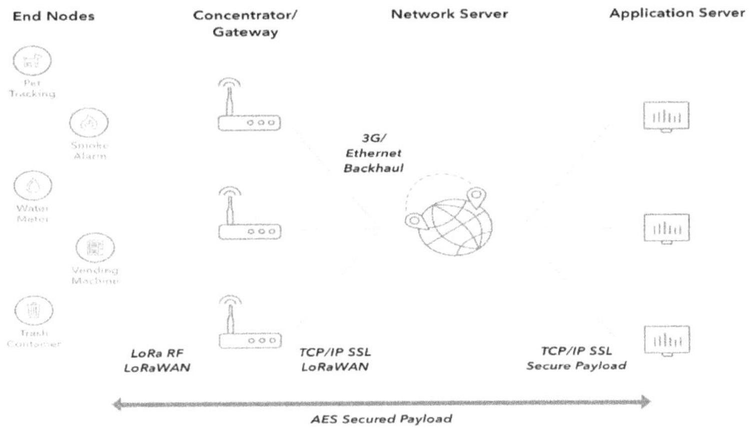

generate responses. This architecture shall be redesigned and modified in the subsequent sections to fit in with the architectural requirements of the proposed project (Asthana et al, 20180.

LoRa is a physical layer implementation of above described LPWAN it is a technology which utilizes wireless frequency modulation (“LoRa”, n.d.). LoRa is proprietary digital wireless data communication system which was developed by France based company, Cycleo of Grenoble and was latter acquired by chip manufacturer Semtech in 2012 (“LoRa”, n.d.). As described by them LoRa is a technique of modulation and transmission of data by means of electromagnetic waves i.e. radio signals. This method of transmission was called Chirp Spread Spectrum of encoding data into frequency modulations named chirps. This data transmission approach was quite popular in military and space communication for decades for its aforementioned reasons. LoRa operated under a license free sub gigahertz radio frequency band like 169 MHz but the most popular ones are 868 MHz in the EU (ITU region 1) and 915 MHz in American region(ITU Region 2). Though India falls into ITU 1 but the unlicensed frequency ranges from 865 MHz to 867 MHz. Following this confusion LoR IOT, a LoRaWAN offering start-up has been operating and providing LoRaWAN services at a frequency of 865.7 MHz in India. Though the claimed range of LoRa is said to be in tens of kilometer (10-15 Kms) however in a typical urban environments many factors cast their effects and the practical range is expected to be somewhere between 3 or 4 kilometers whereas in rural scenarios the network could have a coverage area of 5 to 7 kms. Some common factors that affect the realistic range of LoRa are network and node-gateway positions i.e. placement of gateways being indoors/outdoors, the antenna used and its performance in terms of gain ration, Tx power mode also the payload of the message being transmitted and various other environmental factors like trees and/or buildings. More information can be read at (“The Things Network”, 2017).

While LoRa is a physical layer implementation its counterpart LoRaWAN defines the communication protocol on network layer. The protocol and network architecture for the network defined by LoRaWAN plays a vital role in determining the battery life of the connected nodes and various networking factors like QoS (Quality of service), network capacity, security and implicational capabilities of network. The LoRaWAN architecture and protocol directly influence the battery life of the nodes. For achieving a better and longer battery life the devices are programmed to go to sleep mode when there is no communication ongoing and are event triggered to wake and transmit/ generate alarm. LoRaWAN uses the

Figure 2. Proposed architecture

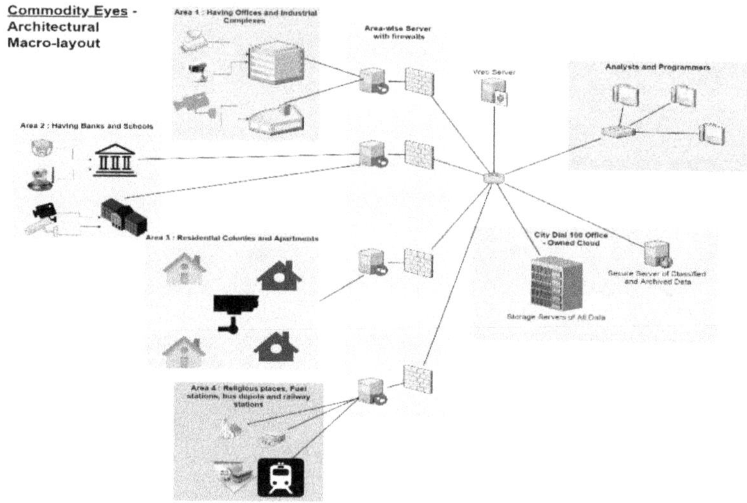

star topology for end note connection as it is advantageous for increasing the battery life in scenarios of long range connectivity. Basic LoRaWAN / LoRa network consist of three elements or layers with an addition of Application server as the forth layer. Following are the major four components:

1. End nodes/ End points
2. Gateway or Concentrators
3. Network server
4. Application server

PROPOSED FRAMEWORK AND INTEGRATION OF COMMODITY EYES

Architecture: Layers Described

In reference to fig-1 the modified and redesigned architecture is reproduced below to suit the operational of requirements of the proposed model for data capturing and the follow up processes like alarm generation and reporting.The architecture as is divided into 3 broad stages stage1 incorporates the commodity camera and concentrators and the connection between respective pairs. Stage2 has Artificial Intelligent systems to analyze the real time feeds for suitable response and alarm generation also firewall to send the data worth collecting and storage to next stage. Final stage3 constitutes of datacenters for storing the data for regular purposes and to fetch data for ML Algorithm modifications and upgrades. This stage shall also have a archives server to store the critical data that could be substantial for some future references.

Following is an reference schema to represent the connectivity between the end device, sensors(Cameras in this case), the application server as depicted in the block diagram .

The popular Wi-Fi network is the best possible way to connect and implement in the given networking requirements, but the high power requirements of wifi limits its use in IoT set-up. On the other hand

Figure 3. IoT system composition based on Wi-Fi & LoRa

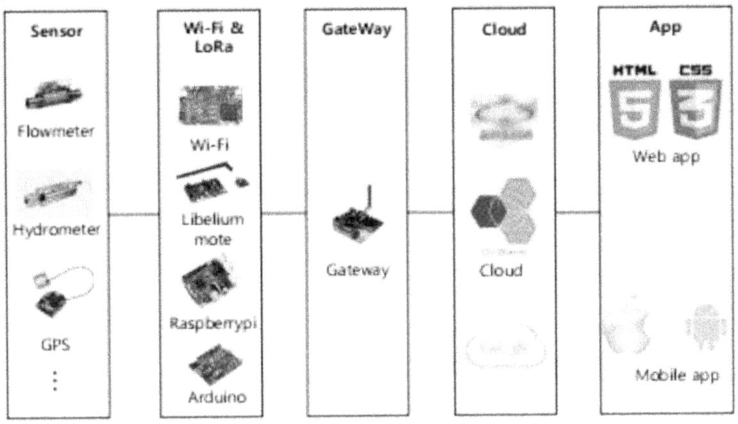

LoRa with the low power consumption and long range is by far the best fit for this project.Having said that LoRa's limited in its bandwidth rates which then can be solved by combining LoRa and WiFi setup.

Figure 1 depicts the IoT framework organizational dependency on WiFi and LoRa to offer different features. The Wi-Fi and LoRa module are associated with different gadgets in the architecture, here the LoRa module is giving LoRaWAN and the Elix board is giving Wi-Fi. The gadgets giving LoRa is made out of the handset(transmitter and receiver) which is utilizing sx-1272 chipset created by Semtech and Waspmote, Raspberry Pi and Arduino. The gateway is the module to interface the collected information, gathered from LoRa module to Internet. To save this information/ data collected the data center described in other section shall be used which shall be either connected via this wireless system or wired network.

The programmatic and logical algorithms in this project are made out of sensor data division that is gathering data from each sensor, Wi-Fi handler modules that are generating and transmitting the video feeds or video buffers via the Wi-Fi network, the LoRa controller or handler node that is responsible for the transmission of the captured video data through LoRa module, the power consumptions in these operations and data validation, that is checking whether Wi-Fi or LoRa to be used, depending upon data to be transmitted, like the isAlive() signals are short length signals and so for cameras to send signals of them being working fine LoRaWAN shall be used, however when actual video feeds are to be transmitted Wi-Fi shall be used. The typical advantages for this arrangement is that it is optimizing both power consumption and bandwidth usage.

Depending on the criticality of the event there would be 3 levels of alarm generation and response generation,

The first response or the Level-1 response (here) shall be a notification in form of messages like the IRC to the respective property managers like owners or society secretaries. For a more Moderate level event the local authorities like public safety departments, fire departments or so shall be informed of the situation along with the person registered against the turf of event occurrence. The most critical events would require utmost concern and so the real time feeds shall be transmitted to the higher authorities for monitoring and swift decision making. Further the next level of alarm shall be for more severe incidents which require a moderate degree of response, for instance, alerting the local authorities like public safety departments (Police and/or Fire). Finally, the most critical that is the severe events would require utmost concern and real time reporting of the matter to the concern departments as well as to the top brass of the departments for prompt decision making and so the real time feeds shall be transmitted to the higher authorities for monitoring and swift decision making.

On later stages after proper trying and testing of the system on ground, the phase 2 shall incorporate packages to be implemented on road safety issues also to revolutionaries and further automate the working of city traffic department.

Central nodes or concentrators shall be equipped with LoRaWAN protocol to effectively communicate with the other elements in the network. Like the response and alert system would be working on this protocol to receive and transmit

Cameras as Sensors

The cameras are, and their features are well described in the previous sections, and hence this section shall concentrate on the concentrators and their connectivity. The cameras shall have a unique ID or UID these UID(s) would act as a primary key identifier to have a unique identification of record in the central database of cameras. These records will provide the system to obtain the information about the locality, also the information regarding the person or groups to be contacted for level-1 notification generation.

While designing and devising an integrated architecture of this system it was discovered that some commodity cameras lack the ability to effectively transmit the video feeds over the web

Actuating Devices/ Concentrators Nodes

Concentrator nodes acts as gateways in the proposed architecture of this paper. The location of installation of these actuators is at stage 2, named as area wise server with firewall these actuators would be in two layers. The first layer shall be equipped with server-side scripting and function as a central hub (area wise) to collect video feeds transmitted by the commodity cameras from their respective area and domains, for example an actuator acting as a concentrator of school camera feeds in a particular shall only collect the feeds coming from the school transmitters, for the residential commodity camera feeds from the same area a different actuator shall act as a concentrator. These concentrators will then analyze the feeds and send the details to the next concentrator, in case any alarm is generated. The higher-level actuator shall be connected to the application server, the level of action to be taken will be decided at this actuator and the IRC messages or other actions shall be initialized at this stage by the utilization of the application server in connection with this actuator node. These gateways / actuators are serving as a bridge in this architecture between the commodity eye (cameras) and the core network, which includes the database servers, Data analysts and the actual Artificially intelligent processors.

The Events above level-2 on above described scale needs to be stored for a short duration for the authorities and the AI programmers to analyze the feed for upgrade and Modification of the ML algorithms applied. These would be stored on the data server/ datacenter shown in stage-3 of the layout architecture. These datacenters could be located in a single facility or utilized as a distributed big data storage mechanism. Data coming to these storage facilities are voluminous and at a great velocity, hence auto deletion of the unnecessary and old data is important. Yet in some cases

Figure 4. Importing APIs in Python

```
import os
import cv2
import face_recognition as fr
```

Data Analysis

Only collection of the data does not yield any result until it is passed through a decision making process or algorithm. Any algorithm which is a part of Machine Learning takes a whole lot of data as an input and learns from it. The data which is accumulated is abstracted using the algorithm. Scenario like, identifying a particular face in a video feed or normally from an image can be achieved by such Machine Learning algorithms, so let's talk about face. Humans are intelligent to identify face because of their eyes but when it comes to computer vision the computer can only understand number values so to make him understand we take help from the algorithm like face-recognition which convert the human face into featured vector and which I will be using for this chapter.

A face featured vector looks like an array [(23.1, 15.8, 255, 224, 189, 5.2, 4.4)].The conversion of a face image in to a featured vector is known as encoding. Now, once we have encoded each image into a feature vector, the problem becomes much simpler. Clearly, when we have 2 faces (images) that represent the same person, the feature vectors derived will be quite similar. Put it the other way, the "distance" between the 2 feature vectors will be quite small and by this we recognize the face.

Now that we have a basic understanding of how Face Recognition works, let's build the Face Recognition algorithm using some of the well-known Python libraries. I will be using Opencv and face-recognition to make the algorithm.

OpenCV or open source computer vision is a library made in python which is used to provide an eye to computer as it contains predefined methods to detect cars, faces, trees, and objects its core functionality include Image Processing, Video Analysis, Camera Calibration, 2dfeature framework, object detection high level gui, and video I/O.

Implementation aspects of the above algorithm is deliberated as below:

```
import os
import cv2
import face_recognition as fr
```

Initial statement for any python code, import is used for including a library which is already made by someone os is library which provide function related to operating system work like execution of program, getting system time, making directory, reading a file and so on. cv2 as explained earlier is OpenCV library due to its extensive update it has two versions with are cv and cv2. Face_recognition is the library which is made over openCv for recognizing faces. It has many straightforward function names which can be used for easy build up. Here the fr is object of library face_recognition and this type of standard is used to shorthand the long library names.

```
directory = "/home/ayush/FACERECOG/img/"
```

This line defines the path to the image folder where the named images of know persons are kept. This image folder is used by the algorithm to check for faces in a video feed or an image frame.

```
persons = []
```

Figure 5. Defining variables

```
directory = "/home/ayush/FACERECOG/img/"
camera = cv2.VideoCapture(0)
persons = []
frame = []
small_frame = []
face_encodings = []
face_locations = []
person_encodings = []
person_names = []
```

```
frame = []
small_frame = []
face_encodings = []
face_locations = []
person_encodings = []
person_names = []
```

In python arrays are defined using [] the brackets, these defined arrays are not limited to one dimension but can be used an object list, vectors, associative array and multidimensional array as well. Here the persons array is defined to carry the list of persons created using the path given above by the class which will be explained later. The frame is used to reference different image frames to check. The small_frame is used to store the ¼ of the real frame to save cpu power in checking a big image as how small an image is the feature vector can be obtained using the ration formulae. The face_encodings is defined to store the featured vector array of each face captured in a given input. The face_locations is defined to store the position of the faces in the image in a four dimensional array, and this array can be further used for drawing purpose. The person_encodings is defined to store the encodings of all the persons created and stored in person array. The person_names is defined to store the names of the person created, and these names are the filenames which actually the name of person in the image.

```
camera = cv2.VideoCapture(0)
```

The camera is defined to capture an instance of the Opencv which carries out the video i/o and as its used on a local computer and the only video feed is through the webcam and thus 0 is passed as an argument.

```
class Person(object):
def __init__(self, file, name, path):
super(Person, self).__init__()
self.path = path
self.name = name
self.fname = file
self.img = fr.load_image_file(path)
self.encoding = fr.face_encodings(self.img)[0]
self.counter = 0
```

The above code snippet defines a person class, we already know about what a class is? It is a group of objects sharing common characteristics and behavior and also an object is an identifiable entity with

Figure 6. Making a person class

```
directory = "/home/ayush/FACERECOG/img/"
camera = cv2.VideoCapture(0)
persons = []
frame = []
small_frame = []
face_encodings = []
face_locations = []
person_encodings = []
person_names = []
```

some unique characteristics and behavior. Like for case of an apple it is the object of class fruits which is unique from rest of the fruits. Here the persons are categorized by the image files their names and their face encodings. The self keyword is same as the this keyword we often use in java to point the current object. To shows how my code works I have used counter variable which will keep the account of how many times the similar face appeared in a video feed.

```
def makePeople():
for root, dirs, files in os.walk(directory):
for file in files:
persons.append(Person(file,os.path.splitext(os.path.basename(file))[0],directory+file))
for person in persons:
person_encodings.append(person.encoding)
person_names.append(person.name)
```

The function makePeople is the main function which is making the persons actually, using the os library in a for loop it loop around all the images we kept in the img directory and after doing so it appends a Person object in persons array. For the sake of illustration I have defined the arrays person_encoding and person_names. However future scope of this code may only use the person array.

```
def increaseCount(name):
for person in persons:
if name==person.name:
person.counter = person.counter + 1
```

The function increaseCount is defined to increase the counter variable which is the backbone of the code as it shows how many times the person appeared in an input.

Figure 7. Depicting how we are making a list of person class object

```
class Person(object):
    def __init__(self, file, name, path):
        super(Person, self).__init__()
        self.path = path
        self.name = name
        self.fname = file
        self.img = fr.load_image_file(path)
        self.encoding = fr.face_encodings(self.img)[0]
        self.counter = 0
```

Figure 8. Increase the counter variable by one every time a similar face appear

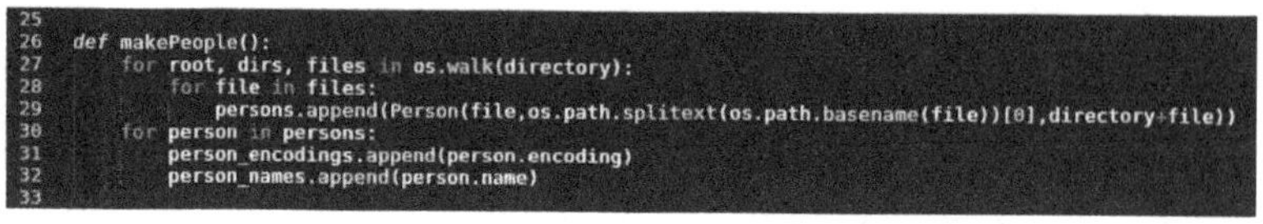

```
def makePeople():
    for root, dirs, files in os.walk(directory):
        for file in files:
            persons.append(Person(file,os.path.splitext(os.path.basename(file))[0],directory+file))
    for person in persons:
        person_encodings.append(person.encoding)
        person_names.append(person.name)
```

```
def detectFaceFromCamera():
process_this_frame = True
while True:
ret, frame = camera.read()
small_frame = cv2.resize(frame, (0, 0), fx=0.25, fy=0.25)
rgb_frame = small_frame[:,:,::-1]
if process_this_frame:
face_locations = fr.face_locations(rgb_frame)
if len(face_locations)>0:
face_encodings = fr.face_encodings(rgb_frame, face_locations)
for face_encoding in face_encodings:
results=fr.compare_faces(person_encodings,face_encoding,tolerance=0.45)
for x in range(len(results)):
if results[x] == True:
increaseAppearCount(person_names[x])
process_this_frame = not process_this_frame
cv2.imshow('Video', frame)
if cv2.waitKey(1) & 0xFF == ord('q'):
break
```

The detectFaceFromCamera is function which detect the face using the webcam feed. Firstly to fasten the task we define a variable process_this_frame which will skip the alternate frame to speed up search. Then in while loop we define ret and frame which stores the Boolean valid and single image frame respectively.

Then comes the small_frame which keeps the ¼ size of the actual frame to increase processing efficiency as a limited amount of image resize will not make a difference as feature vector is calculated after ratio equalization. Then comes the rgb_frame as cv2 captures CYMK it has to be converted to RBG format so that it can be used by any of the modules in face_recognition. Now to further save the memory and processing power we will first confirm if the feed contains a face or not so in above condition only

Figure 9. Backbone of this code it capture frame and analyze it further

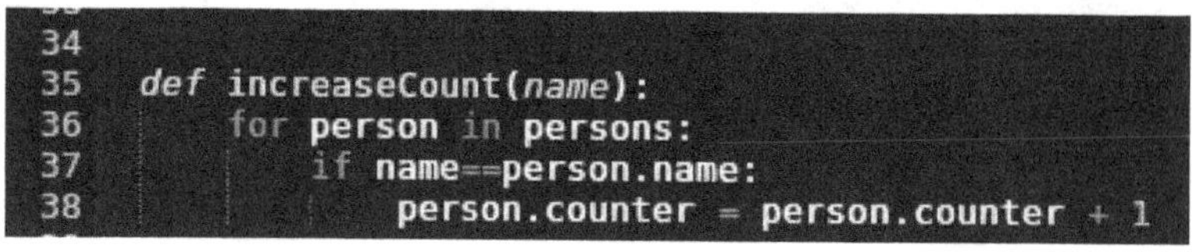

```
def increaseCount(name):
    for person in persons:
        if name==person.name:
            person.counter = person.counter + 1
```

Figure 10. Function call.

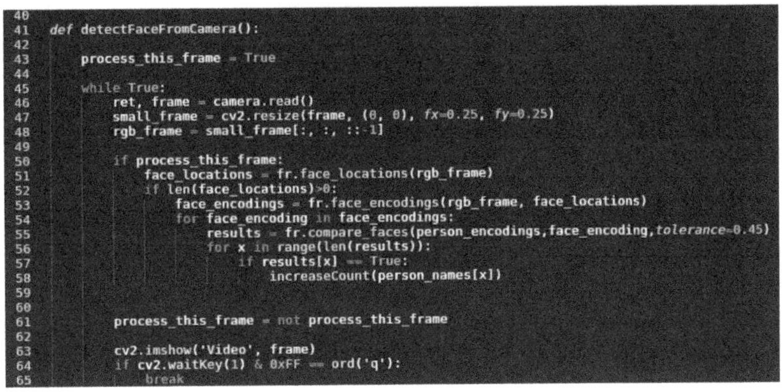

```
def detectFaceFromCamera():

    process_this_frame = True

    while True:
        ret, frame = camera.read()
        small_frame = cv2.resize(frame, (0, 0), fx=0.25, fy=0.25)
        rgb_frame = small_frame[:, :, ::-1]

        if process_this_frame:
            face_locations = fr.face_locations(rgb_frame)
            if len(face_locations)>0:
                face_encodings = fr.face_encodings(rgb_frame, face_locations)
                for face_encoding in face_encodings:
                    results = fr.compare_faces(person_encodings,face_encoding,tolerance=0.45)
                    for x in range(len(results)):
                        if results[x] == True:
                            increaseCount(person_names[x])


        process_this_frame = not process_this_frame

        cv2.imshow('Video', frame)
        if cv2.waitKey(1) & 0xFF == ord('q'):
            break
```

after getting a face or 2 the face_location gets increased to 2 so now further processing can be done. Then comes the encoding of the face location and then comparison with the input feed.

After completion of the comparison the matched person name is passed to increaseCount which uses this to increase the counter variable thus proving our recognition works.

The code statement 69 and 70 are the function calls to makePeople() and detectFaceFromCamera().

for person in persons:

print("Person:"+" "+person.name+" in camera for "+str(person.counter)+" times!")

The below screen shot is the result about a person. To view the results we have to run the script and after that we can see result like this:

CONCLUSION AND FUTURE RESEARCH DIRECTIONS

This chapter provides a framework to integrate commodity cameras (IoT devices) for the purpose of effective policing and overcoming the regular patrolling and thereby saving the huge amount of human intervention. The end devices shall enable data capture based on the rule setting to raise alarm. The captured data is being streamed to the data concentrators for effective decision making and raising alarm

Figure 11. Result of the code executed.

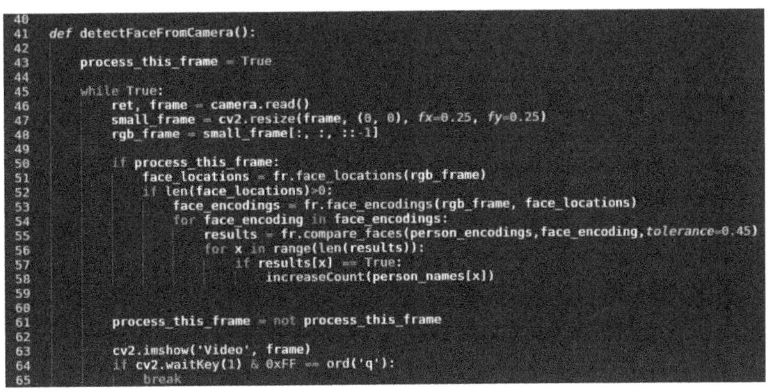


```
def detectFaceFromCamera():

    process_this_frame = True

    while True:
        ret, frame = camera.read()
        small_frame = cv2.resize(frame, (0, 0), fx=0.25, fy=0.25)
        rgb_frame = small_frame[:, :, ::-1]

        if process_this_frame:
            face_locations = fr.face_locations(rgb_frame)
            if len(face_locations)>0:
                face_encodings = fr.face_encodings(rgb_frame, face_locations)
                for face_encoding in face_encodings:
                    results = fr.compare_faces(person_encodings,face_encoding,tolerance=0.45)
                    for x in range(len(results)):
                        if results[x] == True:
                            increaseCount(person_names[x])


        process_this_frame = not process_this_frame

        cv2.imshow('Video', frame)
        if cv2.waitKey(1) & 0xFF == ord('q'):
            break
```

in the local police stations and forwarding the feed to cloud data centers for further monitoring on need basis. The detailed architecture would need a protocol level integration/depiction. The model can be customized on the need basis selecting the protocols at each level.

The description of detailed LoRaWAN implementation is beyond the scope of this chapter however the authors may be contacted to explore the integration at the protocol level. An effort has been made to demonstrate the framework and display the initial coding to configure the commodity IOT cameras

REFERENCES

Asthana & Pandey. (2018). Securing IoT: Threats and Vulnerabilities. IJCA Proceedings on International Information Security Conference, 2017(1), 12-16.

Camps-Mur, D., Garcia-Saavedra, A., & Serrano, P. (2013). Device-to-Device Communications with WiFi-Direct: Overview and Experimentation. *IEEE Wireless Communications*, *20*(3), 96–104. doi:10.1109/MWC.2013.6549288

Franklin, M. (2013). The Berkeley Data Analytic Stack. *IEEE International Conference on Big Data.*

Han, H., & Xu, H. (2012). Interactive Charging Strategy of Electric Vehicles connected in Smart Grids. *7th International Power Electronics and Motion Control Conference*, 2099-2103.

Han, J., Haihong, E., Le, G., & Du, J. (2011). Survey on NoSQL database. *6th International Conference on Pervasive Computing and Applications (ICPCA)*, 363-66.

IEEE Trial-Use Standard for Wireless Access in Vehicular Environments – Security Services for Applications and Management Messages. (2006). IEEE Std 1609.2TM.

Infosec. (n.d.). Internet of Things: How Much are We Exposed to Cyber Threats? Retrieved from http://resources.infosecinstitute.com/internet-thingsmuch-exposed-cyber-threats/

Lee, J. (2013). SAP HANA Distributed In-Memory Database Systems: *Transaction, Session and Metadata Management. ICDE Conference*, 1165 – 1173.

LoRa. (n.d.). In Wikipedia. Retrieved from https://en.wikipedia.org/wiki/LoRa

Park, J. K., & Han, K. S. (2018). Intensity Classification Background Model Based on the Tracing Scheme for Deep Learning Based CCTV Pedestrian Detection. 2018 IEEE 9th International Conference on Mechanical and Intelligent Manufacturing Technologies. doi:10.1109/ICMIMT.2018.8340453

Reclus, F., & Drouard, K. (2009). Geofencing for Fleet and Freight Management. 9th International Conference on Intelligent Transport Systems Telecommunications (ITST), 353-356. doi:10.1109/ITST.2009.5399328

RFC 7452. (2015). Architectural Considerations in Smart Object Networking. Retrieved from https://tools.ietf.org/html/rfc7452

Shvachko, K., & Kuang, H. (2010). The Hadoop Distributed File System. *26th IEEE Symposium on Mass Storage Systems and Technologies*, 1-10.

The Things Network. (2017). Ground breaking world record! LoRaWAN packet received at 702 km (436 miles) distance. Retrieved from: https://www.thethingsnetwork.org/article/ground-breaking-world-record-lorawan-packet-received-at-702-km-436-miles-distance"

KEY TERMS AND DEFINITIONS

2dfeature Framework: This subclass of openCV contains various functions that are common to all feature detection and descriptor extraction classes.

Chirps: A chirp is a signal in which the frequency increases (up-chirp) or decreases (down-chirp) with time.

Face Recognition: Face Recognition is a recognition technique used to detect faces of individuals whose images saved in the data set.

Machine-to-Machine: M2M describes the interaction of billions of devices and machines that are connected to the internet and to each other.

QoS (Quality of Service): Quality of service (QoS) technologies used in the electronic or telephone networking business typically assists in optimizing network traffic management in order to improve the experience of network users.

Smart Home: In IoT-enabled Smart Home environment various things such as lighting, home appliances, computers, security camera etc. all are connected to the Internet and allowing user to monitor and control things regardless of time and location constraint.

Ubiquitous: Ubiquitous computing (also called pervasive computing) is the growing trend towards embedding microprocessors in everyday objects so they can communicate information and make intelligent actions possible.

Chapter 10
Implementation of IoT in Healthcare

Rashima Mahajan
Manav Rachna International Institute of Research and Studies, Faridabad, India

Pragya Gupta
SRM Institute of Science and Technology, Chennai, India

ABSTRACT

The progressive research in the field of internet of things provides a platform to develop high performance and robust automated systems to control external devices via internet data transfer and cloud computing. The present emerging IoT research including user-friendly and easily-wearable sensors and signal acquisition techniques have made it possible to expand the IoT application areas towards healthcare sector. This chapter aims at providing a rationale behind development of IoT applications in healthcare, architecture details of internet of healthcare things (IoHT), and highlights a step-by-step development of IoT-based heart rate measurement and monitoring system using Arduino. The developed module has been advanced to transmit data over the internet on the ThingSpeak channel to allow remote monitoring in real time. This may help to improve/restore useful life among cardiac patients via real-time monitoring through remote locations.

INTRODUCTION

Increasing incidences of chronic disease, sudden heart failures and degrading health care services in the rural and remote areas generate a need for a system of continuous health monitoring and real time transmission of patient's biomedical data to the experts. Mobile and Internet of things (IoT) technologies have overwhelming superiority in solving the problem of patients care as they can expand the availability of medical resources greatly and is termed at Internet of healthcare things (IoHT). The Internet of Things is just not limited to conventional gadgets/devices like laptops, tablets, smart phones and desktops; however, it possess the capacity to connect a diverse and vast set of devices utilizing embedded based automation. All the connected devices are made to communicate through a huge Internet connectivity and thus even

DOI: 10.4018/978-1-5225-9574-8.ch010

making remote monitoring possible. The present emerging IoT research including user friendly & easily wearable sensors and signal acquisition techniques have made possible to expand the IoT application areas towards healthcare sector. This chapter aims at providing a rationale behind development of IoT applications in healthcare and architecture details of Internet of healthcare things (IoHT) via suitable project implementation for heart rate measurement.

IoT is a technique to connect all the sensors and other devices to the internet and will form a global infrastructure for the networked physical objects that are used for data transmission between a transmission device and remote end via cloud network. The Internet of Things is progressively used in automation of devices via Internet for enhanced and efficient societal relevant applications which can hugely impact the human lives. An IoT network is constituted of variety of physical objects connected together for transportation of information in real time. The interconnected devices are embedded with distinct sensors to acquire the information at designated intervals of time. The recorded information is monitored, processed and analyzed for relevant control action or further decision making. As per present estimates, the today's Internet of things network can accommodate even connection of more than twelve billion external devices. It is being forecasted by International Data Corporation that there will be twenty six times more connected devices than humans by the end of this decade. In this IoT innovation driven period where a whole world will be administered by the web, it will be crucial to add and update web based automated healthcare frameworks for regular monitoring of subjects.

IOT IN HEALTHCARE: NEED AND CHALLENGES

Earlier the patients were required to visit doctors personally as per physician's availability or they need to rely on tele/text communications. It was not possible to continuously monitor the patient's health via remote access. The Internet of things platform have revolutionized the healthcare sector by making live remote monitoring possible in real time. It has undoubtedly transformed the healthcare sector by reducing the need to visit hospitals physically, decreasing the stay time at hospitals, timely disease diagnosis and management at an early stage in real time, minimizing healthcare logistics and thus improving overall treatment solutions for patients particularly aged population, their families, doctors and hospitals.

According to well known saying *Health is wealth*, it is the need of an hour to utilize innovation for sustainable and improved healthcare facilities. The researchers are developing and implementing more efficient IoT frameworks to provide early diagnosis facilities and immediate medical aid. These are designed to acquire subject's health parameters through embedded sensors and transmit it over the cloud to be monitored by approved clients. The whole IoT network includes sensors to acquire initial data, wireless networks to transmit and receive data to and from the cloud, respectively.

Let us imagine a situation where most of the patient's body parameters such as blood pressure, pulse rate, brain signals, muscle signals, and other vital parameters are acquired by the sensors embedded to his/her body. The subject does not even need to visit hospitals for body checkups and subsequent treatments. Rather he can even get his all the medical examinations at home itself with the implementation of portable medical equipments, sensors etc. The physicians can remotely monitor their body parameters and can appropriately guide them as per the received data at regular intervals. These advancements can definitely lead subjects towards a quality life.

The sensors in IoT network acquire human health parameters (blood pressure, pulse rate, blood sugar level etc). The acquired sensor data is processed by the interfaced microcontroller and the digitized data is used to trigger the actuators for necessary actions via implementation of cloud computing. Thus, the sensors and actuators are used automatically for acquiring signals and controlling external devices, respectively. This is implemented by Internet data transfer and cloud computing.

The IoT for medical applications is also not free from challenges.

- With billions of devices being connected together, it is very crucial to make sure that the information is secure.
- Another major concern is how to deal with massive amount of data generated through a network of connected devices. A method is required to be devised to handle, analyze and interpret the generated data.

INTERNET OF THINGS: LAYERED ARCHITECTURE

The development of an IoT based system involves multi layered architecture (Sethi, & Sarangi, 2017) as shown in Figure 1. It includes six layers namely coding layer, perception layer, network layer, middleware layer, application layer and business layer. The coding layer provides information about the entities involved in the system. Each entity is assigned a unique code at this layer. The objective is to identify the object or sensor used for acquiring primary information. This code may be some unique object code, Internet protocol version six ID, media access control address ID or some custom ID. It is the responsibility of coding layer to collect all the Meta information related to the devices taking part in particular IoT operation.

The next layer is known as 'Perception layer', the task of which is to collect information from all the connected sensors such as IR sensor, RFID tags, biomedical sensors, environment sensors etc. It could sense the physical condition such as height, heart rate, temperature, humidity, location etc or even sense the presence of any smart object in the range. The acquired sensor data is provided to the next level known as the 'Network layer' in digitized form. The transmission technique such as ethernet, Wi-Fi (Wireless-Fidelity), bluetooth etc. can be used to transmit the sensor data. It is the responsibility of network layer to provide interfaces for connectivity to other smart devices, network gateways/routers and servers. The data is attained from network layer and processed by the middleware layer using technologies like cloud computing and edge/fog computing. It is also known as processing layer and incorporates sub modules for storage, processing and analyzing a wide set of data. The intelligent action is planned and automated at this layer only. The real realization of IoT applications on the processed information is done in the 'Application layer'. It helps in developing application IoT networks to deliver specific services including smart homes, smart grid, smart healthcare and smart cities to the user. The various services & applications of IoT, privacy issues of user are managed and profitable business models are developed at the 'Business layer'.

Internet of Things in Healthcare: Framework

The framework of Internet of things in healthcare is depicted in Figure 2. An IoT device is comprised of sensors (wired or wireless) to collect the medical data, actuators to trigger an instant action, microcontrollers to process acquired data and a set of interfaces for transmission of data to fog or cloud data center for further analysis. It is primarily responsible for sensor data aggregation, analog to digital conversion,

Figure 1. Layered architecture of IoT

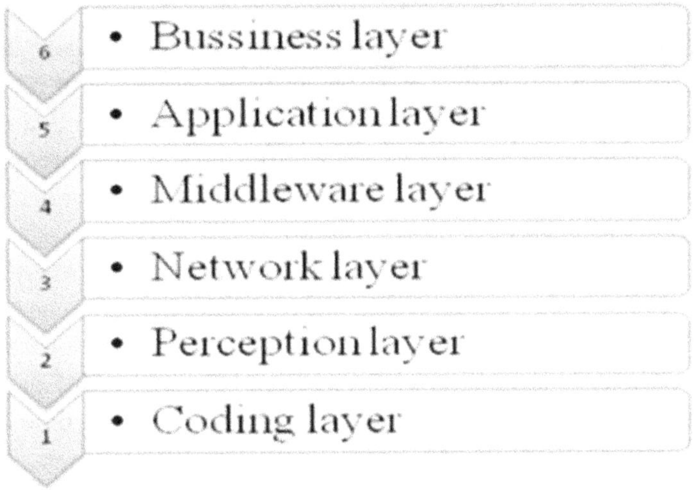

processing of digital data, measurements associated to trigger certain control actions and transmission of data for subsequent processing/ analysis/ storage at fog layer and cloud layer.

A variety of medical sensors can be embedded on IoT device itself or can be connected to IoT device externally to measure distinct biomedical parameters by recording electrical impulses or variable parameters from target human body. The medical sensors are installed at user/ patient end to aid regular and real time visualization of patient's health parameters without any physical assistance even at remote locations (McGrath and Scanaill, 2013). A number of wearable medical sensors in the form of wrist bands, body patches, scalp electrodes, smart jackets and fitness trackers are available these days to measure vital body parameters including heart rate, skin resistance, body temperature, brain activity, blood pressure, blood glucose levels etc (Mora et al., 2017; Haghi et al., 2017).

The microcontrollers are used to process the incoming sensor data stream to convert raw data into meaningful information. The acquired data in IoT systems is processed and analyzed primarily to remove redundant and non-redundant information using a set of hardware and software platforms. It may include Arduino, Raspberry Pi, Intel Galileo etc as hardware platforms and a set of operating systems including Android, Lite OS, RIoT, Contiki etc act as software platforms for IoT implementation.

Actuators are attached to IoT devices to take instant action on the external devices which are very near to the IoT device. For example, ON and OFF instants of any appliance can be controlled by using a relay switch through a set of control commands. Communication modules consist of a set of interfaces for connectivity to transmit the acquired sensor data to a set of external devices, to cloud center or to fog layer for required processing and analysis of data. IoT devices may send and receive data through IP or non-IP networks. The IP networks are used for longer distance communication across wide area networks (WANs) and non-IP networks are used for communication across local area networks (LANs) and personal area networks (PANs). The technologies which are used to facilitate communication across IoT networks may include Ethernet, bluetooth, Zigbee, near field communication (NFC), Wi-Fi, LoRa (long range communication) etc. The non-IP set of networks, however, consume less power and their cost of implementation is comparatively very less.

A massive amount of sensor data is generated and transmitted to on-demand cloud data centers for subsequent data handling operations. The operations include everything from creating databases by managing the received data, processing of data, analysis of data to draw meaningful conclusions and actuating control actions to trigger targeted external devices (Mohammed et al, 2014). The transmission and processing of huge amount of sensor data to cloud may lead to loss of real timelines required in internet of things implementation. The major reasons are availability of limited bandwidth, availability of limited storage space and limited energy resources to fulfill power requirements. Therefore, a need is there to reduce latency (delay) introduced due to the time taken by sensor data to travel to cloud data center plus time taken for data analysis at cloud in addition to time taken by subsequent encoded command or action to reach to the intended device or user from the cloud. Further, if each and every byte of sensor data generated by sensors integrated in IoT devices are intended to be transmitted to the cloud layer for data management and further analysis, this may lead to consumption of whole available bandwidth. The solution provided is to extend the cloud service at the edge of IoT devices (Shi et al., 2016). This is implemented by introducing a fog layer for preprocessing and real time analysis of sensor data at the device side and is also termed as edge or fog computing (Stojmenovic, 2014). The sensor data is preprocessed and analyzed at an intermediate fog layer to save a storage space requirement at cloud data center, save bandwidth for data transmission, reduce latency and thus, enables a quick decision making for critically time sensitive dataset (e.g. surveillance applications, healthcare applications). The communication with cloud data center through fog layer is also permissible for historical data analysis. The less time sensitive dataset (which does not involve real time applications) can be sent to cloud for subsequent signal processing and storage.

CASE STUDY: IOT IN HEALTHCARE OUTLINE

This chapter aims at building a prototype for future diagnosis at hospitals with the help of the Arduino, ESP-8266(Wi-Fi module) and a Heart Rate Sensor, which measures the Heart Beat per minute. The main idea behind this prototype design is to demonstrate the application of IoT in healthcare. This may help in diagnosing patients even at remote locations and reduce diagnosis time at the hospitals followed by required treatment to be provided to the patients as soon as possible. The early diagnosis of patients waiting outside in the queue or at any remote location equipped with body parameter measuring device can be done and data can be sent to the doctor's laptop/phone. The patient can be directly treated according to the diagnostic data acquired and stored via IoT implementation. The other way around it can also be used for post treatment remote monitoring in real time.

In this chapter, a heart rate identification system using Arduino has been developed. The pulse sensor is integrated to acquire heart beat readings of a subject and the recordings in beats per minute (BMP) are made to display on the LCD interfaced to it. For remote monitoring of readings across the world, the acquired heart rate recordings are transmitted to remote server (ThingSpeak for present project) via interfaced Wi-Fi module. ThingSpeak is a free cloud server and is frequently used to display acquired dataset online. The displayed data can be accessed through ThingSpeak at any instant from any place.

The architectural and implementation details of Internet of healthcare things have been explored in detail. Thus, IoHT is redefining medical care facilities and opportunities by providing improved care and treatment even at reduced costs. As per records from MarketResearch.com, the IoHT market is expected to hit $117 billion by the end of this decade (https://www.marketresearch.com/BIS-Research-v4011/

Figure 2. Framework of IoT in healthcare

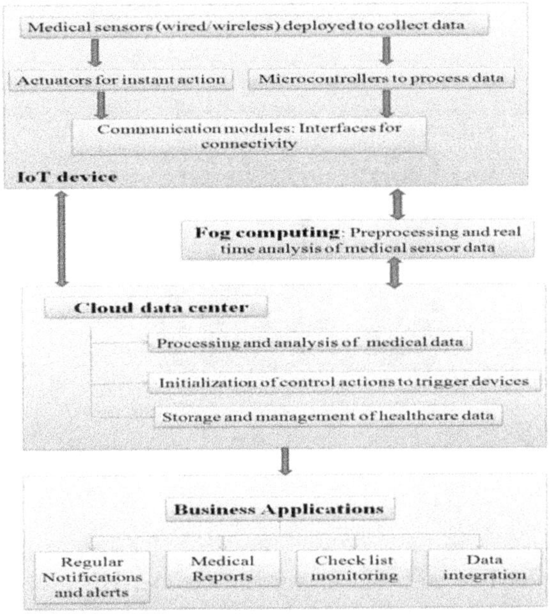

Global-Sensors-Internet-Things-IoT-10703194/). This will definitely lead to impart quality medical care to patients even at remote locations with minimal infrastructure requirement (Ungurean and Brezulianu, 2017).

DESIGN CONSIDERATIONS

The basic design for implementation of healthcare in IoT to record and analyze heart rate in real time incorporates a set of IoT device consisting of pulse rate sensor and arduino microcontroller board; connectivity module for sensor data transmission and selected cloud data center for data storage/ visualization.

Arduino Platform

Arduino is open source programmable board with inbuilt microcontroller and the software integrated development environment (IDE) (Louis, 2016). A variety of sensors and actuators can be interfaced to this microcontroller board. The actuators can directly be controlled at device edge or the acquired sensor data can be transmitted to cloud data centers for further storage and analytics. The physical appearance of an arduino board is shown in Figure 3.

It is capable to accept both analog and digital signals as input and provide the processed digital output. All the hardware required to load a software program into the microcontroller board is integrated on the same board itself without any further requirement of extra hardware for arduino systems. However, multiple interfacing ICs are required externally to interface and load the respective program into conventional 8051 and PIC microcontroller series. There are four types of arduino boards available on the

Figure 3. ARDUINO-UNO Board (Modified and reproduced original figure from https://datasheet.octopart.com/A000066-Arduino-datasheet-38879526.pdf)

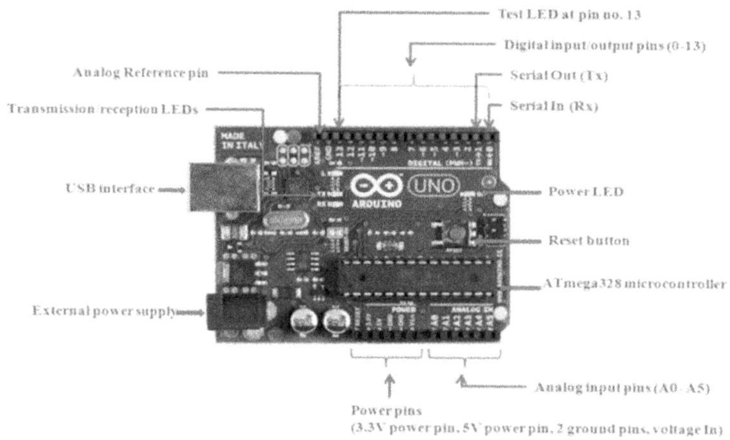

basis of type of microcontroller IC used. These microcontroller ICs are ATmega328 microcontroller, ATmega32u4 series microcontroller, ATmega2560 series microcontroller and ATmega91SAM3X8E series microcontroller. An arduino board operates at five volts voltage with clock frequency of sixteen hertz. It has fourteen digital input-output pins (0 to 13), six analog input pins (A0 to A5) to receive analog inputs, six PWM (pulse width modulation) pins, atleast one UART (universal asynchronous receiver and transmitter) pin for serial communication between arduino board and connected devices and one USB interface. Out of fourteen digital input-output pins, two pins are reserved for serial data transmission and reception. Apart from these pins, it also contains inbuilt voltage regulator, LED for power indication, transmission-reception LED indicator and ground pins. A reset button is used to rest the microcontroller. It contains three categories of memories on chip which includes flash memory of thirty two kilobytes, static RAM of two kilobytes and electrically erasable programmable ROM of one kilobyte. Arduino implementation has been chosen because of the following advantages on conventional bread board circuits or implementation on other microcontrollers.

The various advantages of using Arduino platform, which answers to the question "Why Arduino" are:

- **Cost-Effectiveness:** Programmable ATMEGA series based microcontroller boards are very cost effective solutions as compared to rest of the microcontrollers such as 8051, ARM, PIC series etc. The arduino boards can be assembled individually and these also come as pre assembled boards with distinct microcontroller ICs. The cost in both cases is not even more than fifty USD.
- **Multi-Platform Compatibility:** It offers multiple platform compatibility. The arduino IDE is compatible with multiple operating systems including Linux, Windows and Macintosh OSX. However, compatibility of most of the other microcontroller ICs is limited to Windows operating system only.
- **User-Friendly Programming Platform:** The arduino IDE provide an extremely simple, understandable and flexible programming platform to accommodate multiple complex processing problems. The arduino provides a user friendly platform to develop prototype of designed frameworks. An open source IDE and easily available programming support/ tutorials make it very much con-

venient to develop a working model of innovative ideas. For example, it requires the simplest of coding lines to connect a LCD (liquid crystal display) to display distinct instructions/ messages in just few minutes.

- **Open Source Programming Software:** Arduino IDE software is open source i.e. it is freely available to use and to be modified also. This pre-wired board comes with a large set of open source coding libraries to convert an idea into prototype and subsequently test the validity of designed model. The libraries can be expanded by developing and adding more modules and functions using C-programming language. The AVR code in C language can be directly added into arduino based software programs. The individual and more productive versions of modules can be added by experienced software programmers to improve its outcome. It saves time by letting the designer concentrate on testing testing of project idea instead of wasting time on developing peripheral circuitry and writing huge lines of codes. The saved time can be utilized to develop compact circuitry and optimizing the code, once prototype using arduino has been tested with desired output. This leads to conversion of developed prototype into end user product in stipulated time period. Therefore, an already wired and multipurpose arduino platform aids in saving users from troubles involved in designing printed circuit boards and excessive wiring.
- **Wide and Active User Community:** A wide community of embedded circuit developers is using arduino for building prototypes. Intended users can easily have their account on different software development community platforms including GitHub and can share/ develop software codes by helping one another. The signed users can publish full code information and hardware description on GitHub, a great resource for new users.

Pulse Rate Sensor

A three-pin pulse rate finger sensor has been selected to measure number of heart beats in a minute as shown in figure 4. The three pins are named as ground, supply voltage (3Vor 5V) and data pin. It also contains an inbuilt circuitry for amplification of sensed signal and noise rejection module to obtain a signal with better signal to noise ratio. This sensor operates on three to five volts of supply voltage and its measurement range is twenty to one hundred forty beats per minute. The sensor is based on principle of photoplethysmography (Castaneda et al., 2018). It contains an infrared LED to act as a source of light and a photo detector in the form of light dependent resistor (LDR) to receive reflected light variations. The varying intensity of reflected light through blood vessels of finger because of varying volume of blood measure the heart rate (Saquib et al., 2015). When a finger tip is placed on the top of pulse rate sensor, the heart beat is measured by recording variations in intensity of reflected light due to blood volume variations in capillary blood vessels of finger. Whenever there is a heartbeat, the volume of capillary blood vessels tends to increase and thus, the intensity of reflected light will be less. However, during the instants between two consecutive beats, the intensity of reflected light tends to be high due to less volume of capillary blood vessels. These variations in transmitted and reflected light are captured to measure the number of beats per minute (BPM) by interfacing pulse rate sensor with arduino microcontroller board.

Liquid Crystal Display (LCD)

The acquired heart rate data is processed by arduino and corresponding digital output is displayed on LCD screen. A sixteen by two LCD has been interfaced with arduino to continuously display the beat

Figure 4. Pulse rate sensor (Modified and reproduced original figure from https://components101.com/sensors/pulse-sensor)

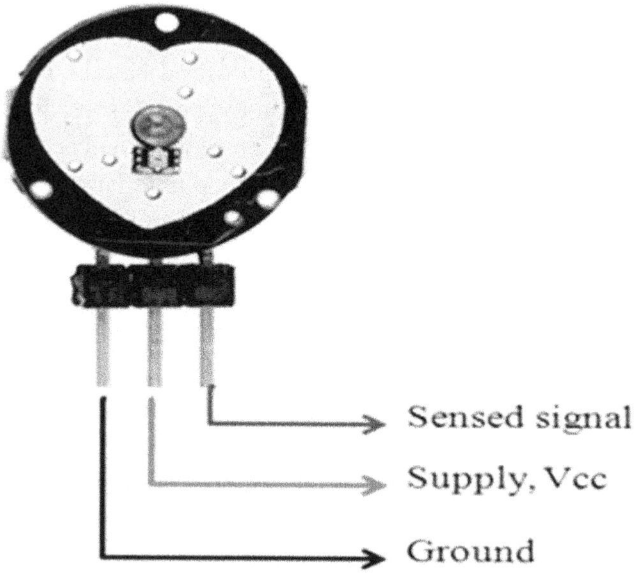

per minute readings. Here, sixteen indicates the maximum number of characters that can be displayed horizontally and two indicates the number of rows. The pin diagram of selected LCD is shown in Figure 5. It operates at a voltage ranging from 4.7 volts to 5.3 volts. It is a sixteen pin package. The pin Vss is a ground pin, V_{dd} is supply voltage pin for LCD and Vcc pin is used to vary the contrast/brightness of LCD screen by connecting it to mid terminal of the potentiometer. This pin is grounded to get maximum contrast. The next pin RS is known as register select pin and is connected to ATmega 328 microcontroller. It is set to one to use LCD as data register i.e. to display the stored and received data from microcontroller. This pin is set to zero to use LCD as command register. In this mode, LCD receives instructions as predefined hexadecimal codes to perform certain actions including clear screen, set cursor position, and initialize LCD etc. The fifth pin RW is provided to read and write the received data. To read the data, this pin is set high and if the data is to be written on LCD screen, it is grounded. The next pin is enable (E) which is used to enable writing operation in the registers. The combination of these three pins RS, RW and E is termed as control bus. An eight bit data bus (D0 to D7) is interfaced with data pins of ATmega328 microcontroller to transmit/receive eight bit data.

Wi-Fi Module

The arduino microcontroller module is connected to Internet network by interfacing it with a low cost wireless shield ESP8266 as shown in figure 6. It is very suited and widely used for implementation of Internet of things based end point embedded applications. This module operates at a voltage ranging from three volts to 3.6 volts and provides a 2.4 gega hertz of Wi-Fi data transfer rate. The internetworking is provided by using version four type network protocol (IPv4). The RST pin is used to reset the

Figure 5. Liquid crystal display (Modified and reproduced original figure from https://www.elprocus.com)

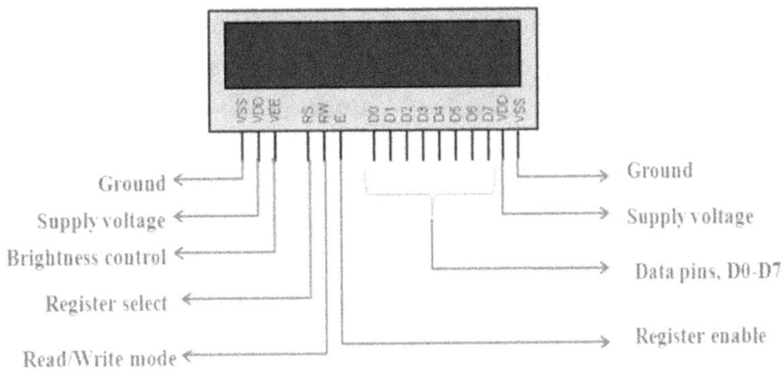

Wi-Fi module and is of active low type. The module is enables by sending a high signal at enable pin (EN). The data can be serially transmitted and received via pins TX and RX, respectively. It is provided with general purpose input output pins labeled as GPIO 0 and GPIO 2. These are used to select a boot (program) mode or normal working mode.

Few applications of ESP-8266 in IoT are listed below:

- Internet of things is a network of embedded and electronics devices. In this network a minor computer is directly indulged to control the household appliances such as bulbs, fans, fridges and microwave oven etc. All these home appliances can be effortlessly controlled by using the Wi-Fi ESP-8266 module and this system is known as home automation system.
- It is used in security cameras for monitoring and check the cameras position at any moment of a day.
- It is used in robotics arms for controlling the position of arms using the Wi-Fi network.
- It is used in temperature sensor for controlling the temperature automatically.
- It is used in quad copters and drone attacks for checking the target through Wi-Fi.

Figure 6. Wi-Fi shield- ESP8266 (http://esp8266.net/)

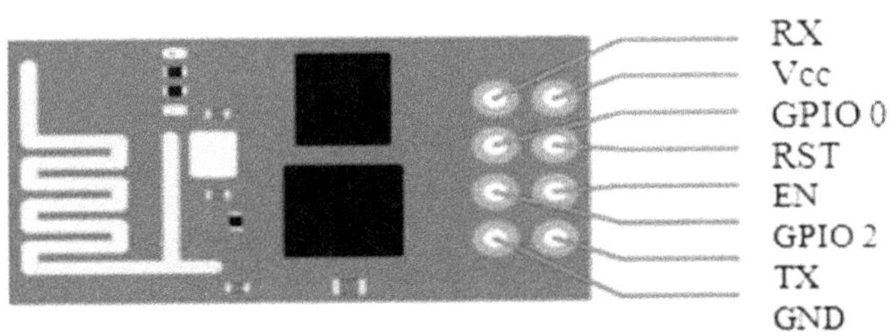

METHODOLOGY AND RESULTS

The design set up to record beats per minute via internet of things implementation is shown in Figure 7. The whole system consists of pulse rate sensor to acquire heart rate readings, arduino microcontroller board with ATmega 328 to process the incoming data, LCD to display the acquired data and wi-fi transceiver module to transmit data over internet for subsequent processing and analysis of sensed data to initiate designed actions. This section mainly includes the major technical specification and interfacing of various modules with each other. Further details of each module with another are explained with supportive techniques.

The entire methodology has been divided into two modules. The first module implemented the making of 'Pulse Rate Sensor' and displaying the results on interfaced LCD screen. This module has been then advanced to send the sensed heart rate data over the internet on the 'ThingSpeak' channel.

Module 1: Sending Pulse Sensor Data on the LCD

This module consists of two sub-modules. First involves interfacing of pulse rate sensor with arduinoUno board followed by second sub module involving interfacing of arduinoUno board with LCD.

At first the pulse sensor is connected with arduinoUno board to receive analog sensor data. The supply pin of sensor (refer figure 4) is connected to power pin of arduinoUno, providing voltage of five volts (refer figure 3). The ground pins of both sensor and arduinoUno board are connected with each other. An analog heart rate data from pulse rate sensor is received at analog pin A0 of arduinoUno board. An external LED is connected at pin number 13 of arduinoUno board to indicate the instances of received heart beats.

The arduinoUno board is now connected with LCD to display received heart rate data. A very first pin V_{EE} is connected to ground, second pin V_{DD} is connected to get supply of five volts. The third pin V_{SS} is connected to the mid pin of a ten kilo ohms potentiometer. The rest two pins of potentiometer are connected to supply voltage (five volt) and ground. Contrast and brightness of LCD screen is controlled by varying the connected potentiometer's resistance. A pin number four (RS-register select) from control bus pins of LCD is connected with pin twelve of arduino to operate LCD as data register or program register. Pin number five (RW) of LCD is connected to ground pin of arduinoUno board. Another control pin (enable) of LCD is connected with pin number eleven of arduinoUno. The data pins eleven (D4), twelve (D5), thirteen (D6) and fourteen (D7) are connected to pins five, four, three, two of arduinoUno, respectively. Pins fifteen and sixteen are connected to Vcc via a 220 ohm resistor and ground.

Figure 7. Design set up to record beats per minute

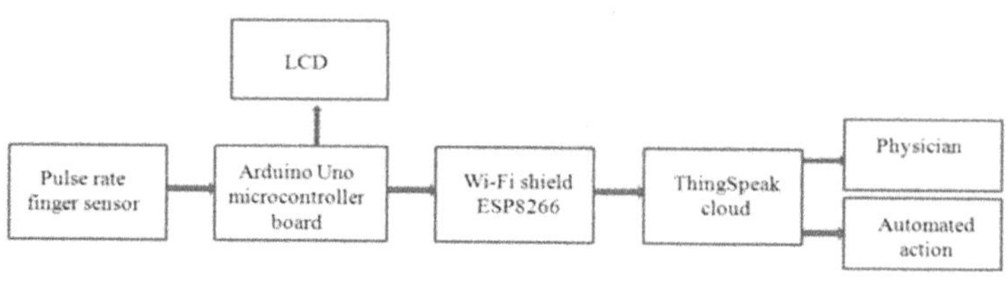

Module 2: Interfacing of Model with Wi-Fi Shield ESP8266

After successfully sending the data on LCD screen, the model has been interfaced with an ESP8266 module to initiate data transfer from connected pulse rate sensor to the ThingSpeak channel. The arduino module is connected with wireless transceiver ESP8266 to transmit data over internetwork for demonstration of implementation of IoT in healthcare. The pins Vcc and RST (refer figure 6) of Wi-Fi shield are connected with supply pins (3.3 volts) of arduinoUno board. Another pin RX requires a voltage of 3.3 volts to operate. This is provided by designing a potential divider network consisting of three series resistors to convert five volts into required 3.3 volts. Now, another pin TX is connected with pin number nine of arduiniUno board and pin RX with pin number ten of arduino via a potential dividing network consisting of series combination of resistors.

The complete circuit diagram has been developed using online tool FRITZING as shown in figure 8 and experiment setup for the same is shown in figure 9.

Sending Data to Cloud

The processed heart rate data from arduinUno board is transmitted to ThingSpeak cloud for real time data analysis via wi-fi shield ESP8266. ThingSpeak is an open source cloud data center that allows reception and storage of sensor data, thereby, allowing implementation of IoT in healthcare. It shows great compatibility while receiving data from multiple embedded circuit modules like arduinoUno, Raspberry-Pi, Beaglebone black etc and even from software platforms including MATLAB workspace. The data from multiple edge sensor nodes can be captured and stored for time dependent data analysis and processing to initiate certain actions for external device control. A very first step to use ThingSpeak as cloud data center is to create a channel on ThingSpeak platform. Each channel has its unique data, location and status field. Once the channel is created, data can be written onto the created channel for subsequent processing, analysis, visualization and storage. The four step workflow to work with ThingSpeak cloud is:

- Create a channel and name it as 'Pulse rate channel'
- Collect sensor data

Figure 8. Circuit diagram developed using FRITZING.

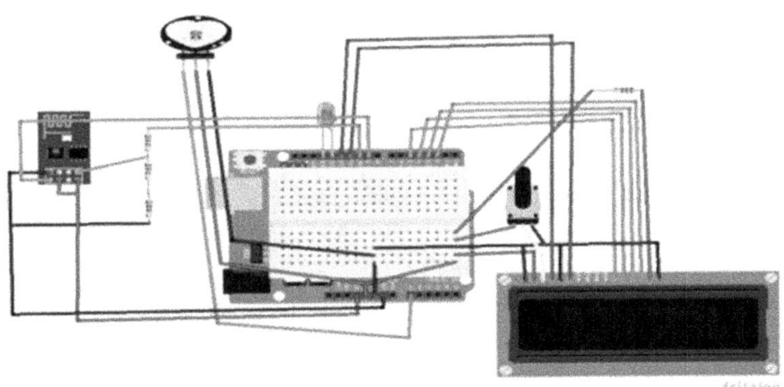

Figure 9. Experimental set-up using ESP8266 and Pulse Rate Sensor (SEN11574)

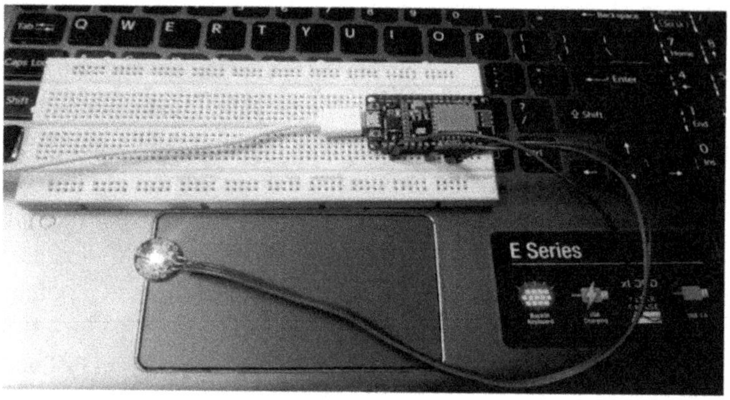

- Analyze and visualize the sensor data
- Act on the received data using any of several Apps

The created channel on ThingSpeak cloud and pulse rate data received in real time is shown in figure 10. To make all these entities work together, coding in arduinoUno IDE is developed. The code files written in arduino IDE are saved as .ino extension files. Any arduino code has two functions void setup() and void loop(). A void setup() function is called only once at the start of arduinoUno functioning and it is executed only a single time for whole program. This function includes the initialization of each input and output pin to be used in particular project. Further, the initialization of rest of the peripherals including LCD, wi-fi shield etc is also done once within this function. On the other end, void loop () function contains the coding lines which are to be executed again and again in continuous manner. One can also produce delay in displaying the successive pulse rate readings on LCD by writing a code within void loop () function. The code code developed for this project is provided in Appendix I.

Figure 10. The 'Thingspeak' channel and data received in real time

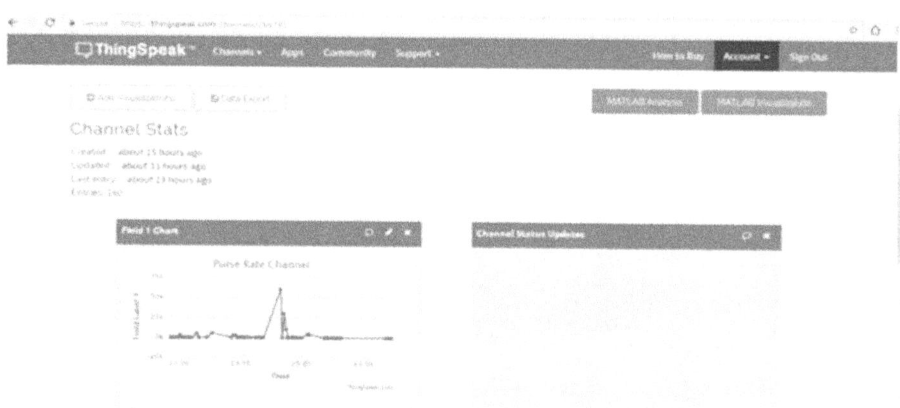

The various observations made using the developed circuit was recorded from cloud channel 'pulse rate channel' in the following observation tables. Samples of the observation are recorded as tabulated in table 1(while sitting ideally) and table 2 (while walking).

It can be observed that the high heart rate is captured when a person is moving and lower values have been captured when a person is sitting ideally. This heart rate data from cloud can be visualized directly by physician to decide for the further treatment instructions. It can also be used to initiate some automated action. For example, if heart is under stress as captured through heart rate sensor and visualized over cloud. A machine learning algorithm can be integrated to play some cheerful music in this particular scenario or it can automatically dial a number to family members.

Therefore, IoT in the healthcare has 4-stage architecture in general. All the stages are connected in such a way that the sensor acquired at first stage will be processed at next stage followed by a stage involved in transfer of data to cloud data centers. The very first stage involves deployment of individual or multiple biomedical sensors to capture the data. The acquired sensor data is analog in nature, which is converted into its digital equivalent at stage two for subsequent processing of data. This digitized sensor data is transmitted to fog/ cloud data centers at stage three via Ethernet or Wi-Fi connecting modules. The stage four constitutes the analysis and management of received dataset at cloud centers for subsequent decision making.

IOT IN HEALTHCARE: APPLICATION SCENARIOS

Internet of things in healthcare domain has opened up numerous opportunities starting from cost effective medical aid, reduced diagnostic times, and improved treatment records of patients, reachable

Table 1. When sitting ideally, the heart beat is measured

Entry Id	Time	Pulse Rate Recorded (In BPM)
1	**2019-01-27 17:15:08 UTC**	**62**
2	**2019-01-27 17:15:27 UTC**	**69**
3	**2019-01-27 17:15:45 UTC**	**49**
4	**2019-01-27 17:16:04 UTC**	**73**
5	**2019-01-27 17:16:24 UTC**	**83**
6	**2019-01-27 17:16:43 UTC**	**82**
7	**2019-01-27 17:17:01 UTC**	**82**
8	**2019-01-27 17:17:20 UTC**	**81**
9	**2019-01-27 17:18:09 UTC**	**83**
10	**2019-01-27 17:18:27 UTC**	**72**
11	**2019-01-27 17:18:47 UTC**	**74**
12	**2019-01-27 17:19:07 UTC**	**75**
13	**2019-01-27 17:19:26 UTC**	**88**
14	**2019-01-27 17:19:45 UTC**	**87**
15	**2019-01-27 17:20:04 UTC**	**87**

Table 2. While walking, the heart beat is measured

Entry Id	Time	Pulse Rate Recorded (In BPM)
1	**2019-01-27 18:15:08 UTC**	**92**
2	**2019-01-27 18:15:27 UTC**	**94**
3	**2019-01-27 18:15:45 UTC**	**89**
4	**2019-01-27 18:16:04 UTC**	**93**
5	**2019-01-27 18:16:24 UTC**	**88**
6	**2019-01-27 18:16:43 UTC**	**102**
7	**2019-01-27 18:17:01 UTC**	**100**
8	**2019-01-27 18:17:20 UTC**	**97**
9	**2019-01-27 18:18:09 UTC**	**93**
10	**2019-01-27 18:18:27 UTC**	**97**
11	**2019-01-27 18:18:47 UTC**	**101**
12	**2019-01-27 18:19:07 UTC**	**95**
13	**2019-01-27 18:19:26 UTC**	**98**
14	**2019-01-27 18:19:45 UTC**	**97**
15	**2019-01-27 18:20:04 UTC**	**97**

medical facilities, improved physician availability and many more (Islam et al., 2015). The healthcare specific IoT based embedded devices possess the potential to completely revolutionize this sector by providing healthcare assistance even at remote locations. In healthcare, the implementation of internet of things include applications at patient's end, physician's (healthcare provider's) end and medical center's (hospitals) end.

- **IoT for Patients**: At patient end, IoT can be implemented by designing IoT devices integrated with wirelessly connected wearable medical sensors like wrist bands, finger pulse rate sensors, blood sugar level glucometers, neural activity measuring sensors, blood pressure sensors etc. It may designed with a capability to impart individual attentions by reminding patients about their appointments, calorie check level alerts, pulse rate variation notifications, blood pressure fluctuation alerts followed by many more such scenarios. These complete IoT scenarios in healthcare possess the capability to aid constant health monitoring of elderly and differently-abled subjects, who otherwise are unable to report physically. Any variation in normalized health parameters being monitored continuously will definitely send an alert notification to respective guardians and physicians through internet of things implementation. Health monitoring at remote location is an important and the most needful application of IoT. The timely monitoring and notification via fitted sensors help in providing an immediate medical attention required to save people lives.
- **IoT for Physicians:** At physician end, the wirelessly connected and embedded biomedical sensors enable the physicians to keep record of patient's health parameters in more precise and effective manner even at remote locations. The implementation of IoT in healthcare enables medical practitioners to monitor their patients in real time. This can lead to imparting an immediate medical

aid and attention to patient needs proactively even if patients are physically not present at medical centers.

- **IoT for Hospital Resource Management:** The implementation of internet of things in healthcare even helps in enabling the efficient resource allocation and management to hospitals. Live location of distinct biomedical sensing and monitoring equipments may also be known if these are equipped with related sensors and IoT devices. It can also provide updates on availability of medical staff (with respective unique IDs and location) in real time through connected sensors. Further, an automated management of physical assets and conditions including medical inventory control, temperature monitoring and control, oxygen level control, lighting control, wheel chair management, hygiene monitoring etc has become also possible due to implementation of internet of things in healthcare. A whole connected arrangement can even predict the number of patients visiting the particular/ physician and forecast the respective demand in real time.

The progressive adaptability of internet of things based healthcare solutions in medical centers is driving the need for more innovative solutions. This can lead to provide timely medical assistance, efficient management of biomedical equipments, cost effective healthcare solutions and generation of extended revenue gains. A variety of scalable solutions have already been provided by the healthcare industry. Few of these solutions include a Philips company product named as Medical Dispensing Unit which ensures that there will be no miss of any dose by patients, Niox-Mino which is a product of Aerocrine company and is used for automated monitoring of asthma patients and Uro-Sense which is a product of Future path medical company to monitor body temperature and urine output of catheterized subjects. A very useful GPS-SmartSole product is launched to track Dementia patients by wearing a shoe tracking device. This shows an enormous scope for growth of internet of things based technology in healthcare sector.

CONCLUSION

A remarkable acceleration in implementation of Internet of things for healthcare sector has been witnessed due to progressive advancements in biomedical sensor technologies, improved computation performance of microcontrollers with reduced pricing, enhanced cloud infrastructure availability and application development platforms. The medical sensor technologies include a set of wearable devices, body implanted devices and telehealth monitoring devices. This integration and internetworking of human beings with smart physical systems may result in improved utilization of medical resources while imparting more independent life to the elderly and people with severe medical conditions. The database management application may be used to create and maintain electronic health records having minute body parameter details of patients. The Arduino platform is a fad these days and has an immense potential of developing and testing high level prototypes in the same. The developed module has been advanced to transmit data over the Internet on the ThingSpeak channel to allow remote monitoring in real time. This may help to improve / restore useful life among cardiac patients via real time monitoring through remote locations.

An efficient implementation of Internet of things in healthcare is a need of an hour. Such hospitals and diagnostic centers aim towards early and remote diagnosis of the patients which also tend to reduce the diagnosis time. This may help physicians to decide treatment protocols at early stages also. This domain mainly finds it use in the field of medical diagnosis, in a country like India where rural population and remote areas plays an important role. It is important to have an effective and on-time diagnosis system.

One can make the whole diagnosis process better by introducing such ideas in healthcare domain. The multiple sensors like body temperature sensor, neural signal acquisition sensor, blood pressure sensor etc can be deployed with each patient to record full body parameters.

REFERENCES

Arduino UNO. (n.d.). Retrieved from https://datasheet.octopart.com/A000066-Arduino-datasheet-38879526.pdf

Castaneda, D., Esparza, A., & Ghamari, M. (2018). A review on wearable photoplethysmography sensors and their potential future applications in health care. *Int J Biosen Bioelectron*, *4*(4), 195–202. doi:10.15406/ijbsbe.2018.04.00125 PMID:30906922

Global Sensors in Internet of Things (IoT) Devices Market, Analysis & Forecast: 2016 to 2022; (Focus on Pressure, Temperature, Light, Chemical, & Motion Sensors; and Applications in Healthcare, Manufacturing, Retail, & Transportation. (n.d.). Retrieved from https://www.marketresearch.com/BIS-Research-v4011/Global-Sensors-Internet-Things-IoT-10703194/

Haghi, M., Thurow, K., & Stoll, R. (2017). Wearable Devices in Medical Internet of Things: Scientific Research and Commercially Available Devices. *Healthcare Informatics Research*, *23*(1), 4–15. doi:10.4258/hir.2017.23.1.4 PMID:28261526

Islam, S. M. R., Kwak, D., Kabir, M. H., Hossain, M., & Kwak, S. (2015). The Internet Of Things For Health Care: A Comprehensive Survey. *IEEE Access: Practical Innovations, Open Solutions*, *3*, 678–708. doi:10.1109/ACCESS.2015.2437951

Liquid Crystal Display. (n.d.). Retrieved from https://www.elprocus.com

Louis, L. (2016). Working Principle of Arduino And Using It As A Tool For Study And Research. *International Journal of Control, Automation Communication and Systems*, *1*(2), 21. doi:10.5121/ijcacs.2016.1203

McGrath, M. J., & Scanaill, C. N. (2013). Body-worn, ambient, and consumer sensing for health applications. In Sensor Technologies, (pp. 181–216). Springer.

Mohammed, J., Lung, C. H., Ocneanu, A., Thakral, A., Jones, C., & Adler, A. (2014). Internet of Things: Remote Patient Monitoring Using Web Services and Cloud Computing. *IEE International Conference on Internet of Things (iThings)*, 256-263. 10.1109/iThings.2014.45

Mora, H., Gil, D., Terol, R. M., Azorín, M., & Szymanski, J. (2017). An IoT-Based Computational Framework for Healthcare Monitoring in Mobile Environments. *Sensors (Basel)*, *17*(10), 2302. doi:10.339017102302 PMID:28994743

Pulse Sensor. (n.d.). Retrieved from https://components101.com/sensors/pulse-sensor

Saquib, N., Papon, M. T. I., Ahmad, I., & Rahman, A. (2015). Measurement of heart rate using photoplethysmography. *Networking Systems and Security (NSysS) 2015 International Conference*, 1-6.

Sethi, P., & Sarangi, S. R. (2017). Internet of Things: Architectures, Protocols, and Applications. *Journal of Electrical and Computer Engineering*. doi:10.1155/2017/9324035

Shi, W., Cao, J., Zhang, Q., Li, Y., & Xu, L. (2016). Edge Computing: Vision and Challenges. *IEEE Internet Of Things Journal*, *3*(5), 637–646. doi:10.1109/JIOT.2016.2579198

Stojmenovic, I. (2019). Fog computing: a cloud to the ground support for smart things and machine-to-machine networks. *Proceedings of the Australasian Telecommunication Networks and Applications Conference (ATNAC '14)*, 117–122.

The Internet of Things with ESP8266. (n.d.). Retrieved from http://esp8266.net/

Ungurean, O., & Brezulianu, A. (2017). An Internet of Things Framework for Remote Monitoring of the HealthCare Parameters. *Advances in Electrical and Computer Engineering*, *17*(2), 11–16. doi:10.4316/AECE.2017.02002

ADDITIONAL READING

Augustin, A., Yi, J., Clausen, T., & Townsley, W. M. (2016). A study of LoRa: Long range & low power networks for the Internet of Things. *Sensors (Basel)*, *16*(9), 1466. doi:10.339016091466 PMID:27618064

Baker, S. B., Xiang, W., & Atkinson, I. (2017). Internet of Things for smart healthcare: Technologies challenges and opportunities. *IEEE Access: Practical Innovations, Open Solutions*, *5*, 26521–26544. doi:10.1109/ACCESS.2017.2775180

Firouzi, F., Rahmani, A. M., Mankodiya, K., Badaroglu, M., Merrett, G. V., Wong, P., & Farahani, B. (2018). Internet-of-Things and big data for smarter healthcare: From device to architecture applications and analytics. *Future Generation Computer Systems*, *78*, 583–586. doi:10.1016/j.future.2017.09.016

Kumar, T., Ramani, V., Ahmad, I., Braeken, A., Harjula, E., & Ylianttila, M. (2018). Blockchain Utilization in Healthcare: Key Requirements and Challenges. *e-Health Networking Applications and Services (Healthcom) 2018 IEEE 20th International Conference on*, 1-7.

Muhammed, T., Mehmood, R., Albeshri, A., & Katib, I. (2018). UbeHealth: A Personalized Ubiquitous Cloud and Edge-Enabled Networked Healthcare System for Smart Cities. *Access IEEE*, *6*, 32258–32285. doi:10.1109/ACCESS.2018.2846609

Ramesh, D., Suraj, P., & Saini, L. (2016). Big data analytics in healthcare: A survey approach. *Proc. Int. Conf. Microelectron. Comput. Commun. (MicroCom)*, 1-6, Jan. 10.1109/MicroCom.2016.7522520

APPENDIX

Code

```
#include <SoftwareSerial.h>
#define DEBUG true
SoftwareSerialesp8266(9,10);
#include <LiquidCrystal.h>
#include <stdlib.h>
LiquidCrystallcd(12,11,5,4,3,2);
#define SSID "sk" // "SSID-WiFiname"
#define PASS "abcd1234" // "password"
#define IP "184.106.153.149" // thingspeak.com ip
String msg = "GET /update?key=GLE4446OJTSA7NF4";
//change it with your api key like "GET /update?key=Your Api Key"
//Variables
float temp;
int hum;
String tempC;
int error;
88-98-9int pulsePin = 0;
// Pulse Sensor purple wire connected to analog pin
intblinkPin = 13; // pin to blink led at each beat
intfadePin = 5;
intfadeRate = 0;
// Volatile Variables, used in the interrupt service routine!
volatileint BPM;
// int that holds raw Analog in 0. updated every 2mS
volatileint Signal; // holds the incoming raw data
volatileint IBI = 600;
// int that holds the time interval between beats! Must be seeded!
volatileboolean Pulse = false;
// "True" when heartbeat is detected. "False" when not a "live beat".
volatileboolean QS = false;
// becomes true when Arduino finds a beat.
// Regards Serial OutPut -- Set This Up to your needs
staticbooleanserialVisual = true;
// Set to 'false' by Default. Re-set to 'true' to see Arduino Serial Monitor ASCII Visual Pulse
volatileint rate[10]; // array to hold last ten IBI values
volatile unsigned long sampleCounter = 0;
// used to determine pulse timing
volatile unsigned long lastBeatTime = 0; // used to find IBI
```

```
volatileint P =512; // used to find peak in pulse wave, seeded
volatileint T = 512; // used to find trough in pulse wave, seeded
volatileint thresh = 525; // used to find instant moment of heart beat, seeded
volatileint amp = 100; // used to hold amplitude of pulse waveform, seeded
volatilebooleanfirstBeat = true;
// used to seed rate array so we startup with reasonable BPM
volatilebooleansecondBeat = false;
// used to seed rate array so we startup with reasonable BPM
void setup()
{
lcd.begin(16, 2);
lcd.print("circuitdigest.com");
delay(100);
lcd.setCursor(0,1);
lcd.print("Connecting...");
Serial.begin(9600); //or use default 115200.
esp8266.begin(9600);
Serial.println("AT");
esp8266.println("AT");
delay(1000);
if(esp8266.find("OK")){
connectWiFi();
Serial.println("wificonnect");
}
interruptSetup();
}
void loop(){
lcd.clear();
start: //label
error=0;
lcd.setCursor(0, 0);
lcd.print("BPM = ");
lcd.print(BPM/2);
Serial.println(BPM/2);
delay (100);
lcd.setCursor(0, 1); // set the cursor to column 0, line 2
delay(1000);
updatebeat();
//Resend if transmission is not completed
if (error==1){
goto start; //go to label "start"
}
delay(1000);
```

```
exit(1);
}
voidupdatebeat(){
String cmd = "AT+CIPSTART=\"TCP\",\"";
cmd += IP;
cmd += "\",80";
Serial.println(cmd);
esp8266.println(cmd);
delay(2000);
if(esp8266.find("Error")){
return;
}
cmd = msg ;
cmd += "&field1=";
cmd += BPM;
cmd += "\r\n";
Serial.print("AT+CIPSEND=");
esp8266.print("AT+CIPSEND=");
Serial.println(cmd.length());
esp8266.println(cmd.length());
if(esp8266.find(">")){
Serial.print(cmd);
esp8266.print(cmd);
}
else{
Serial.println("AT+CIPCLOSE");
esp8266.println("AT+CIPCLOSE");
//Resend...
error=1;
}
}
booleanconnectWiFi(){
Serial.println("AT+CWMODE=1");
esp8266.println("AT+CWMODE=1");
delay(2000);
String cmd="AT+CWJAP=\"";
cmd+=SSID;
cmd+="\",\"";
cmd+=PASS;
cmd+="\"";
Serial.println(cmd);
esp8266.println(cmd);
delay(5000);
```

```
if(esp8266.find("OK")){
Serial.println("OK");
return true;
}else{
return false;
}
}
voidinterruptSetup(){
TCCR2A = 0x02; // DISABLE PWM ON DIGITAL PINS 3 AND 11, AND GO INTO CTC MODE
TCCR2B = 0x06; // DON'T FORCE COMPARE, 256 PRESCALER
OCR2A = 0X7C; // SET THE TOP OF THE COUNT TO 124 FOR 500Hz SAMPLE RATE
TIMSK2 = 0x02; // ENABLE INTERRUPT ON MATCH BETWEEN TIMER2 AND OCR2A
sei(); // MAKE SURE GLOBAL INTERRUPTS ARE ENABLED
}
ISR(TIMER2_COMPA_vect){ // triggered when Timer2 counts to 124
cli(); // disable interrupts while we do this
Signal = analogRead(pulsePin); // read the Pulse Sensor
sampleCounter += 2; // keep track of the time in mS
int N = sampleCounter - lastBeatTime;
// monitor the time since the last beat to avoid noise
// find the peak and trough of the pulse wave
if(Signal < thresh && N > (IBI/5)*3){
// avoid dichrotic noise by waiting 3/5 of last IBI
if (Signal < T){ // T is the trough
T = Signal; // keep track of lowest point in pulse wave
}
}
if(Signal > thresh && Signal > P){ // thresh condition helps avoid noise
P = Signal; // P is the peak
} // keep track of highest point in pulse wave
// NOW IT'S TIME TO LOOK FOR THE HEART BEAT
// signal surges up in value every time there is a pulse
if (N > 250){ // avoid high frequency noise
if ((Signal > thresh) && (Pulse == false) && (N > (IBI/5)*3)){
Pulse = true; // set the Pulse flag when there is a pulse
digitalWrite(blinkPin,HIGH); // turn on pin 13 LED
IBI = sampleCounter - lastBeatTime; // time between beats in mS
lastBeatTime = sampleCounter; // keep track of time for next pulse
if(secondBeat){ // if this is the second beat
secondBeat = false; // clear secondBeat flag
for(int i=0; i<=9; i++){ // seed the running total to get a realistic BPM at startup
rate[i] = IBI;
}
```

```
}
if(firstBeat){ // if it's the first time beat is found
firstBeat = false; // clear firstBeat flag
secondBeat = true; // set the second beat flag
sei(); // enable interrupts again
return; // IBI value is unreliable so discard it
}
wordrunningTotal = 0; // clear the runningTotal variable
for(int i=0; i<=8; i++){ // shift data in the rate array
rate[i] = rate[i+1]; // and drop the oldest IBI value
runningTotal += rate[i]; // add up the 9 oldest IBI values
}
rate[9] = IBI; // add the latest IBI to the rate array
runningTotal += rate[9]; // add the latest IBI to runningTotal
runningTotal /= 10; // average the last 10 IBI values
BPM = 60000/runningTotal;
// how many beats can fit into a minute? that's BPM!
QS = true; // set Quantified Self flag
// QS FLAG IS NOT CLEARED INSIDE THIS ISR
}
}
if (Signal < thresh && Pulse == true){ // when the values are going down, the beat is over
digitalWrite(blinkPin,LOW); // turn off pin 13 LED
Pulse = false; // reset the Pulse flag so we can do it again
amp = P - T; // get amplitude of the pulse wave
thresh = amp/2 + T; // set thresh at 50% of the amplitude
P = thresh; // reset these for next time
T = thresh;
}
if (N > 2500){ // if 2.5 seconds go by without a beat
thresh = 512; // set thresh default
P = 512; // set P default
T = 512; // set T default
lastBeatTime = sampleCounter; // bring the lastBeatTime up to date
firstBeat = true; // set these to avoid noise
secondBeat = false; // when we get the heartbeat back
}
sei();
// enable interrupts when you are done!
}// end isr
```

Chapter 11
Internet of Things and Location-Based Services

Shaveta Bhatia
Manav Rachna International Institute of Research and Studies, Faridabad, India

ABSTRACT

The progressive rise of mobile computing devices and wireless networks have created a lot of interest in location-based systems and services. The involvement of internet applications in almost every field has changed our lives. Location-based services are the services provided to mobile users according to their geographic locations. Each user wants to get the service according to his/her own interest. The general user's actions in location-based services are locating, searching, navigating, identification, and checking. The location identification has now become a critical attribute. Today, internet of things in the field of location-based services (LBS) provide services to the mobile users by explore the location depending on the geographical coordinates for their valuable needs. Mobile phones that are equipped with new technologies and supported by the presence and development of broadband mobile data networks have created new opportunities for the processing of location-based applications.

INTERNET OF THINGS AND LOCATION BASED SERVICES

In today's technological world Location Based Services (LBS) is applicable in the field of Internet of thing (IOT) where services are being provided to the cell phone users depending upon the geographical locations in terms of cardinal axis. At present mobile phones are very well equipped with inbuild features that support new technologies with the ease of advancement in the field of broadband connectivity that provides connectivity and faster rate of data transfer and creates several outstanding opportunities for location-based services usually geographical based information about the device location and persons carrying that device can be retrieved using the users account information. Location Based Services provide valuable information in terms of mobile computing and location tracking

In terms of next generation technology, the advancement in Internet of Things (IOT) plays a crucial and potential role in terms of location-based services. To trace location of any person or device location-based services is essential to relate themselves with geographical axis. As mobile phones are easily affordable

DOI: 10.4018/978-1-5225-9574-8.ch011

to every person due to lower cost and higher features it has increase the level of advancement in terms of features that provide multiple advanced features such as conversation in different languages using voice recognition, social networking like Facebook, WhatsApp, Viber etc. Technological advancement in Geographical Positioning System(GPS) provides essential geographical information like if anybody's child is carrying mobile phones with him/her having internet connectivity and he/she is out of home and if met with any kind of trouble then one can search him/her using GPS asap is running into his/her devices with which exact location can be traced and can be easily reached to that location without any delay to save your child.

These services provide various innovative ideas and support to the users using smartphones. The general functions in location-based services are locating, searching, navigation, identification and verification. Locating means where the user himself is. Searching means where the other person is. Navigation means to know the various routes. Identification means to ask the properties of a particular location. Verification means to check the accuracy of location by identifying the events happening nearby the particular location. The popularity of the mobile operators like Airtel, Vodaphone, Bharat Sanchar Nigam limited (BSNL) based on its capability the implementation of wireless network is increasing day by day at a faster rate.

The general classification LBS is based on the concept of person-oriented and device-oriented services. The person-oriented services are user-oriented services having main idea is to get the position of a person to avail the services easily. He person-oriented services have basically two application designs i.e. push services and pull services. The push services are those services with which user receives the information of his or her where about without requesting for it for example advertising whereas pull services are those services where user actively requests for the information through the application for example "where is the nearest school" and "friend finder". The device-oriented location-based services are those services which are external to the user where focus is not only to search a person but also to search different objects in different areas.

The major positioning method of location-based services is GNSS that uses trilateration technique. One of the GNSS is NAVSTAR Global Positioning System (El-Rabbany et al, 2002)., developed by U.S Department of Defense, which is widely used these days for mobile phones. This system acts as a service provider that helps to retrieve longitude and latitude of a location.

Model of Location Based Services

Location-based services consists of three tier architecture. The three layers of this architecture are Positioning layer, Middleware layer and an Application layer (Gleason et al, 2009).. The architecture is shown in figure 1. below. The function of first layer is to calculate the exact location of a device using position detection techniques having availability of geospatial data stored in GIS.The position detection techniques include handset-based technologies like global positioning system, network-based technologies like cell-identification or cell-identification with time advance and other hybrid technologies like E-OTD in combination with global positioning system.

The Geographical Information System (GIS) helps to translate the raw information retrieved by the positioning layer based on network into geographic information (that is in terms of longitude and latitude). This geographical information is then passed to the middleware layer or sent directly to the application layer which request for the service through location gateway. The application layeris basically a client who is using all the services and requests for the location information like a request to search a friend,

Figure 1. General model of location-based services

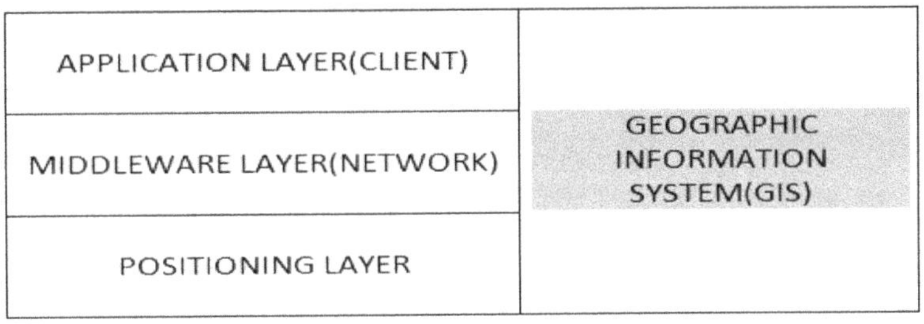

to search the nearest hotel, to track a child and so on. The middle layer is between the positioning layer and application layer, and its job is to reduce the complexity of integration of various services without intervention of operator and its network so that it can control all the services executed in the network. Therefore, it saves the time and cost of network operators and application providers. The applications can be integrated with and without middle layer (middleware) as shown in figure 2. In some of the cases like advertising middleware is not used. The advertising companies buy an access to the mobile phones of different companies while entering in particular areas and then make regular contacts with those mobile phones by sending push messages. For example, entering the city Jaipur, one will get the messages containing the information of Jaipur whether one has requested for it or not. The other example of service, where middleware is not used, is fleet management, where the assistance companies offer the automatic mobile based positioning services to their customers for emergency purposes, by purchasing the rights from operator to access the data of those customers. But the operators generally hesitate to give rights to the companies due to privacy concerns. Here comes the role of middle layer (middleware), which allows the companies to access the rights and takes the overall control for privacy and authenticity. Thus, the location middleware is used to manage the interoperability between various networks for giving the exact information about the locations.The various protocols like Signaling System 7 (SS7) and Wireless Application Protocol gateway available for mobile communication are integrated through middle layer so that users can easily access the location-based services on their mobile phones.

The various application areas of location-based services with and without use of middleware are:

1. **Proximity Searches or Point of Interest:** The main example of this area is yellow pages also called concierge services. These are the services where the users can locate their homes or the areas nearby their homes. These servicesare helpful inorder to answer a questions like, "where is the railway station?" and it isusually used for business purposes. These services are both pull and push services with and without use of middleware according to the requirement.

2. **Dynamic Data Information:** It is connected with emergent services for example, when we are moving towards Delhi and due to Jam or an accident, there is need of changing of route, the dynamic data information help the user. The location-based service provides the information in this emergency with the help of middleware.
3. **Neighborhood Search by Proximity:** There is role of middleware in these pull services, where the user request for the location of his child, friend or any other person.

Figure 2. Location based application integration with or without middleware

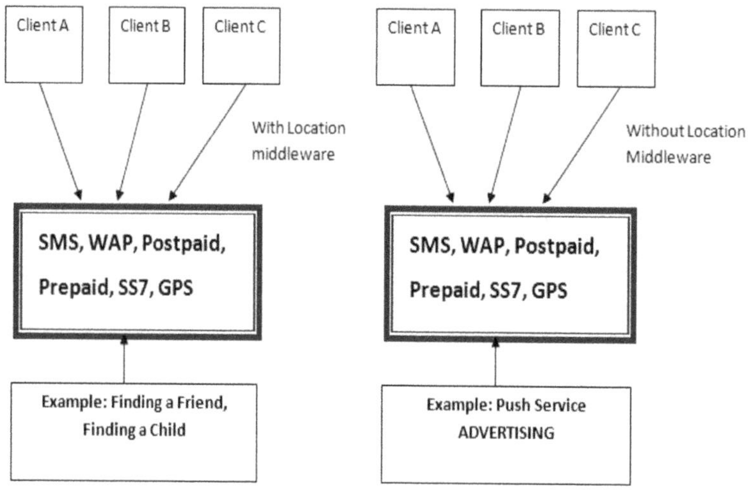

Components of Location Based Services

If any user is willing to use location-based services then users requirements are components like mobile device, positioning system, content provider, communication network and service provider are necessary. The figure 3 illustrates the various components.

Mobile Device

This is a major tool for the user who requests for the detailed information. The user request from his mobile phone and the results are retrieved through text or audio or through maps on mobile phone. The other devices for requesting the information are e-Laptops, desktops, tabs etc.

Figure 3. Components of location based services

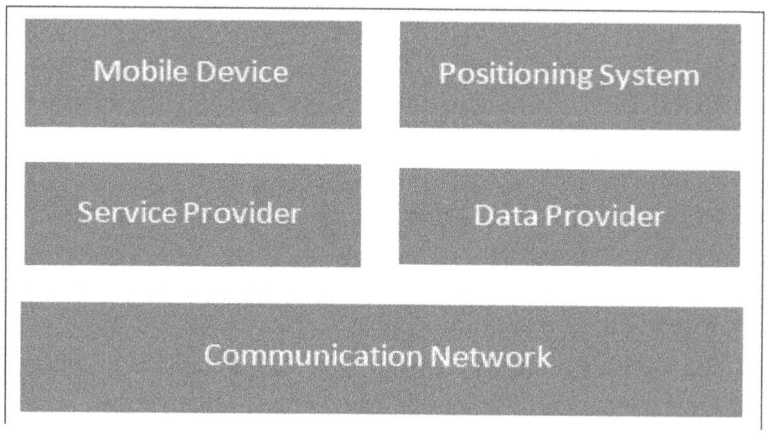

Communication Network

This is a communication layer that links and the data of mobile users and are sent to the service provider and received the acknowledgment containing the required information revert back to the user using a mobile.

Positioning System

It is a mobile communication network through which position of the user can be retrieved. The examples of various positioning systems are global positioning system, sensors or radio beacons, active badges etc.

Service Provider

The service provider responds to different services to the mobile user and is also responsible for the proper processing of services requested by users. The different services provided by providers are internet facility, user address, user International Mobile Equipment Identity (IMEI) number, positioning and other interesting information about user. Various location-based applications that are provided by telecommunications are NTT DoCoMo, Telecom, Vodaphone, AT&T or other specified companies that provide their services to user connected to different telecom networks (Steiniger, n.d.).

Data Provider

The service providers do not save the information by own. But they take the collected information, geographical data and positioning information from data providers and send them to user after processing. The different data providers who store the information are directory-based services, gateway-based services, location-based utility services, presentation-based services and router-basedservices (Gyeyoung et al, 2012). The various directory services available are yellow pages, google, yahoo, electronic libraries like Wikipedia, personal websites like school website, university websites like Manav Rachna International University, patient information from the hospitals, student's information from the college. The gateway services include various positioning providers like global positioning system, cell identification and so on. The examples of location utility services are the data collection of a society for which the location application is to be developed. The presentation services include the providers providing data about monuments and museums etc.

Classification of Location-based Services

The location-based Services are classified in the following categories according to their functionalities.

Single Target and Multi Target Location Based Services

In a single target location-based services, the position of single mobile user is targeted on map and in case of multi-target location-based services, the position of various mobile users is targeted simultaneously on the map. The multi targeted location-based services are also called zone services, where a person can request for the location of different users at same time.

Reactive and Proactive Location-Based Services

Reactive Location based Services are those where location is tracked on the request from the user. These are also called location tracking services, where user and requester are different persons. The requester wants to know the location of another user. For example, parents want to know where their children are. Proactive Location based Services are those services where information is triggered on the occurrence of prespecified events such as entering, approaching and leaving any city or monument, the information of the particular place is automatically notified to the user. These are also called position aware services where user and requester are same; a user can get his own location with the required information. For example, user request to get route information from Faridabad to Jaipur.

Central and Peer to Peer Location Based Services

Peer to Peer Services are self-organized services like location based gaming where the position of each other is exchanged directly between mobile users. The central location based services like advertising are those where the service is provided and managed by the application server or central location.

Location-based Services that are Based on Indoor and Outdoor Positioning

Outdoor Location-Based Services are those which are available in open geographical areas with the help of satellites. The Indoor Location-based Services are those services that are available inside the buildings using various indoor techniques like Wireless Fidelity (WI-FI), sensors etc.

Geographic Information System (GIS) and Location Based Services (LBS)

Public is the main source for which location based services has been designed.Thus geographical information system is a traditional professional system designed for experienced users with wide variety of functionality whereas location based services are developed with limited services for non-professional users in mobile computing environment. GIS is defined as "Geographic Information System is a computer based system which is used to digitally reproduce and analyze the feature present on earth surface and the events that take place on it" (Sahoo et al, 2014). The geographical information system either contains geographical references like latitudes and longitudes or national grid coordinates like postal addresses, forest identifier, road name etc. whereas location-based services is the integration of three services namely communication and handheld devices, internet and geographic information system.

Standard Organizations of Location Based Services

The various standard organizations of the location based services are as given below:

Open Geospatial Consortium (OGC)

An Open Geospatial Consortium is an International Standard Organization that leads to the development of various standard for location-based services. The standards of an OGC are documents containing interfaces and encodings which are used by software developers to build open interfaces and to encode their products and services. These standards are freely available to public. For example, OGC Web Services (OWS) are the standards used for World Wide Web (WWW) applications ("Open Geospatial", n.d.).

Open Mobile Alliance (OMA)

This is a main leading industry forum formed by various world leading mobile operators, information technology companies, network suppliers and service providers in 2002. The main purpose of forming this alliance is to create interoperable services by providing open specifications. These services can work for any bearer network across all geographical boundaries. The alliance also provides protocols and Applications Programming Interfaces (APIs)for exchange of information of location between actors.

Internet Engineering Task Force (IETF) and World Wide Web Consortium (W3C)

Aim of IETF is to provide good and faster internet facility with special new protocols for integration of collected location data into Internet or web-based applications. The aim of W3C provides open development platform for application development ("W3C", n.d).

POSITIONING METHODS

Several positioning systems have pros and cons but a single positioning system can never fulfill all the requirements location based services. The satellite-based positioning systems like GPS is capable of achieving high coverage in outdoors, but they are lacking in indoors (Roth et al, 2004). Positioning in the indoor positioning systems is limited to buildings only with high cost. The analysis of location is very important in areas like robotics, land surveying, aviation, robotics, person, object tracking and gaming based on location. There are different positioning system that provides the location information with many characteristics features summarized as:

1. A coordinate systems (including latitude, longitude, and altitude) and positioning system to determine the location.
2. There is inaccuracy in the results of positioning system due to technological reasons and environmental conditions.
3. The users are mostly interested in the knowing the location of their nears and dears having semantic information with their geographic coordinates. The semantic information of locations is always a powerful representation for any location based service. For example, rather than the values of longitude and latitude, it is more useful to get the result like "You are in college".

(d)Besides the location, the other spatial information is also needed by the user. for example, in navigation services, the users orientation is required. Orientation in the space is specified by three angles i.e. roll, pitch and yaw (Roth et al, 2004). Most of the positioning systems calculate the direction when a child moves.The Yaw angle is determined by the various tools like electronic compasses.

Generally the positioning methods are classified in two main groups:-

1. Network Based Positioning

2. Device Based Positioning

The general classification of positioning methods is based on two main groups: Network-based Positioning and Device based Positioning. In the case of Network-based Positioning, the location of user is tracked on the basis of base stations network as shown in figure 4. In a Device-based positioning system, the location is calculated by device itself and is dependent on the signals received from the base stations as shown figure 5. The example of device positioning is Global Positioning System.

Positioning Techniques

The basic positioning techniques which are used in combination to get the location information are as follows:-

Cell Identification

The cellular network is divided into cells and each cell is associated with one base station. When the user moves to the particular area of base station, the cell identification technique determines the serving cell of that network according to the signal strength and gives the position of a mobile nearby that base station. Sending location of base stations are shown in figure 6. The cell-id is the identifier of the serving base station and position is defined in a circle around the serving base station and the accuracy of the location depends upon the size of cells (Neun et al, n.d.). The accuracy is about 30-35 kms for 2G networks and little bit less for 3G networks.

Time Difference of Arrival (TDOA)

This technique will determine the relative positions of the mobile transmitter by examining the time difference at which the signals from the mobile device reaches at different base stations. There must be atleast three base stations and each base station is connected to carrier network equipment or TDOA equipment which is used to calculate the position of mobile device.This technique provides more accuracy than the above method. The diagram of this technique is shown in figure 7.The System called global positioning system uses this technique to know the position but it works for outdoors not for indoors.

Angle of Arrival (AOA)

In this technique the position of the targeted object can be fetched by intersection of several pairs of angle direction lines (Ballazhi et al, 2012) as shown in figure 8. The angles of signals reached from the target device to different base stations are calculated. Basically this technique is mostly used for direction finding. There must be at least two reference points to get the 2D location of the target object. The limitation is that there is no synchronization between the measuring points and the technique needs complex antennas at each site.

Figure 4. Network based positioning

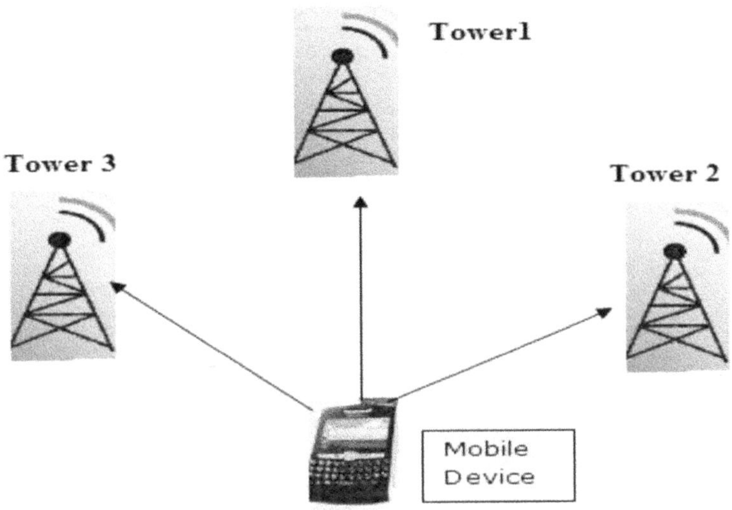

Enhanced Observed Time Difference(E-OTD)

It is a network center-based technique applied only for GPRS and GSM links. In this technique the technique the location measurement units are placed at various cell sites in a wide area and targeted mobile device must enabled with E-OTD software as shown in figure .9. Now, when the signals from atleast three base stations reach the mobile device, their times are recorded and the signals from base station to Location Measurement Unit (LMU) are also recorded. The time difference of arrival of both mobile device and LMU is calculated and the position of the mobile device is calculated. This technique provides good accuracy.

Figure 5. Device based positioning

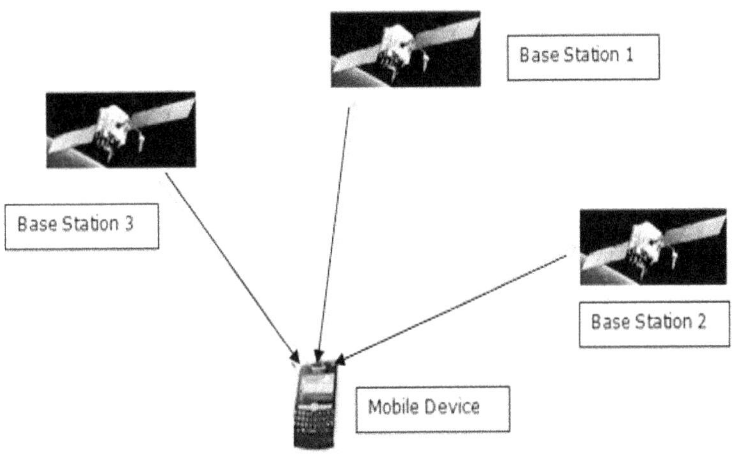

Figure 6. Cell identification

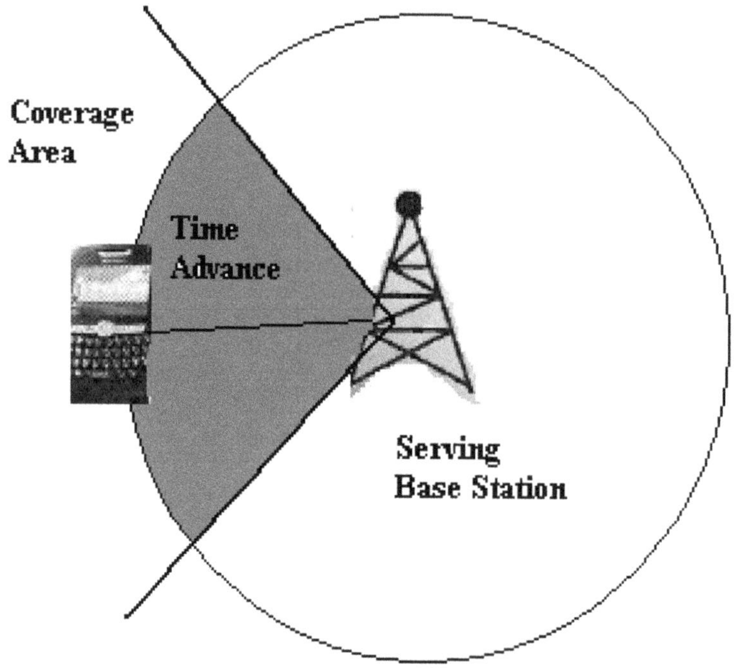

INDOOR TECHNIQUES

Infrared

Infrared Sensors can be placed inside the ceilings of any buildings and can be used to detect a person wearing a device called badge that periodically emits an ID code via an infrared transmitter. The location of the user is determined through sensors by detecting a badge identification of a wearer. This system does not provide higher accuracy and does not work when device will be in user's pocket.

Ultrasound

Ultrasound Transmitters are known as beacons that send signals to receiver and allows device to calculate its location on proximity (Steinfield et al, 2004).The ultrasound-based systems provide high precision results as the speed of sound is less than the speed of radio signals.

Radio frequency Identification Device

These are the wireless radio electromagnetic fields which are used to identify and track the tags attached to various objects ("Radio Frequency Identification", n.d.). These are used for navigational purpose by transferring the location information of the objects to the receiver by using the combination of angle of arrival along with time difference of arrival techniques.

Figure 7. Time difference of arrival

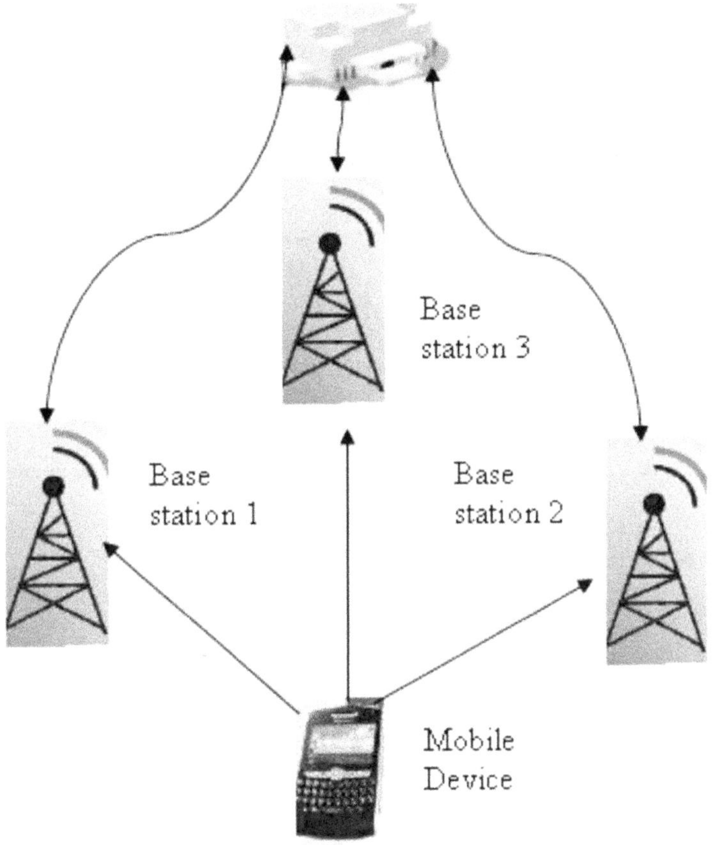

Wireless LAN (WI-FI)

These are used to provide the internet accessibility on mobiles, laptops and desktops which are equipped with wireless LAN cards. The system provides the information of location based on the serving base station roughly in the range of 50 meters.

Bluetooth

Bluetooth is a short range adhoc network used for personal data networks roughly in 10 meters of range. These systems can be used for Proximity-based location services when Bluetooth device comes within the range of service point (Steinfield et al, 2004).

Figure 8. Angle of arrival

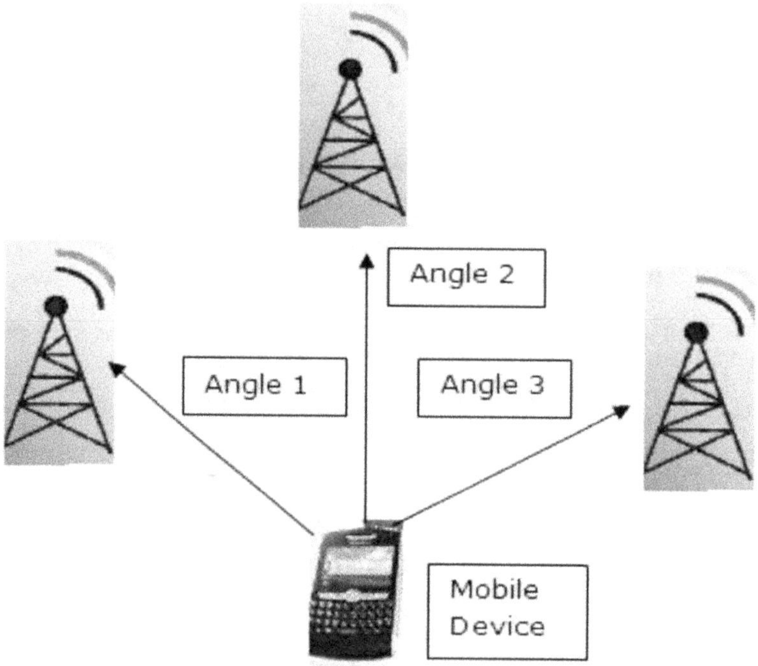

Figure 9. Enhanced observed time difference

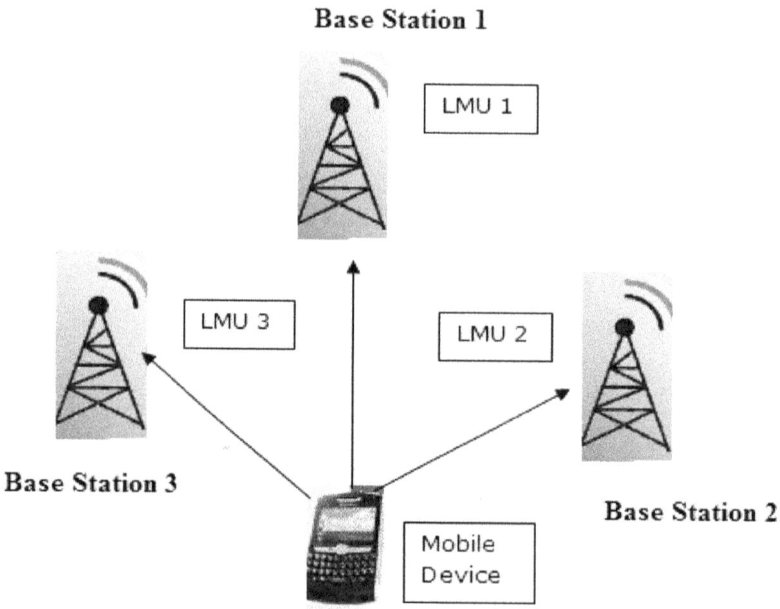

POSITIONING ALGORITHMS

Triangulation

The process of triangulation is based on an angle of arrival technique and the position of device is calculated by computing the angles of the received signals. The position of device is located at the fixed point of triangle.

Trilateration

The process of trilateration is based on two techniques, time of arrival and time difference of arrival where the position of a device is calculated on the basis of distances of received signals.

Proximity

This system provides the information of location on the basis of dense grid of antennas, having a well-known position (Hopper et al, 1992). The signal strength is less if the mobile is targeted by single antenna and the signal strength is higher if the mobile is targeted with more than one antenna.

GLOBAL NAVIGATION SATELLITE SYSTEM (GNSS)

Global Navigation Satellite System are constellations of satellites that are being designed to provide positioning and timing information for users on earth (Gleason et al, 2009). The various GNSS available are:-

Global Positioning System

It is a most widely used operational GPS.It was designed by U.S department of defense. It consists of 24 satellites in approximate 12-hour orbits. The system is widely used for military, public and private applications.

GLONASS

The Russian Federation GLONASS(GlobalnayaNavigationnayaSputnikovaya Sistema) (Krebs et al, 2014).] is another fully operational GNSS.It consists of 24 satellites in 1996 but reduced to 8 to 10 satellites in 2001.It is operated by Russian Aerospace Defense forces. It has limited demand and availability.

Galileo

The other GNSS is European Union's "Galileo Constellation". It was first launched in 1995 and expected to be operational at the end of 2013 (Gleason et al, 2009).This project was names after Italian Astronomer "Galileo Galiler". The main purpose of this project is to provide positioning with high precision for European Nations.

GPS Aided GEO Augmented Navigation (GAGAN)

It is an Indian regional Satellite Based Augmentation System (SBAS).This is jointly executed by an Airports Authority of India (AAI) and Indian Space Research Organization (ISRO) for the purpose of stramline satellite based navigation across the boundaries and controlling air traffic ("ISRO et al, 2013).

Indian Regional Navigation Satellite System (IRNSS)

This is an autonomous regional satellite navigation system that is developed by an Indian Space Research Organization (ISRO) under the complete control if Indian government.

GOOGLE MAPS

The maps are defined as the graphical representation of space and can also be seen as communication and information system conveyed by images (Drakopoulou et al, 2010). These days, google maps are used in every application area of location based services. We can say that the google maps are ways of organizing world's information spatially". In 1994, when google maps were introduced, they were only supported by Internet Explorer and Mozilla and now days they were supported by all browsers. The google maps are mostly used in wide area of location based services. For example,the users can create a social imaginary world while playing games based on the location. The maps display the information in the form of spatial data where spatial data is a combination of raster and vector data model and is based on geographical information system. In case of vector model, the data is represented in the form of geographical locations with the points and lines without any conversion or generalization whereas in case of raster data model, the location is represented according to the pixel value and position of cell generally used by optical scanners and CCTV cameras.The maps are represented in the form of vector form of spatial data as it is easy to understand and manipulate because working with pixels is very complex process. The example of spatial data like the map shows the streets, homes various monuments and important places etc.

EVOLUTION OF LOCATION BASED SERVICES

The first generation of location based services originating during 1997-2001, where the position of mobile device is determined through Signaling system 7(SS7) link .The first generation was according to the link services like Mobile Switching Centre (MSC) and GSM, Universal Mobile Telecommunication System (UTMS) to Gateway Mobile Location Centre (GMLC) provided to the operator to create a connection by operation available to MPC/GMLC and use applicable program. However, this generation faced lack of standard, lack of unity by the operator to provide services to user. The users were getting services at very higher prices and lack of distribution. For solving these problems, middleware and geoserver were included in the architecture to increase its functionality. The second generation of location based services includes two main standards namely Mobile Location Protocol (MLP) and Open Location Services (OpenLs) where mobile location protocol is used as a connection between the servers and open location services provides interface that allows companies to enable location based applications. The location based applications like location based games, tracking a child, neighbor and emergency services where the identification of location is an important factor are most emerging segments of the mobile industry.

WIRELESS STANDARDS FOR THE APPLICATION

Global System for Mobile Communication and CDMA

The construction of mobile communication system is due to the availability of mobile services provided to the users through network operators. For any mobile communication system, the following points are of great importance:

1. The system provides full coverage to any mobile phone whether a phone is in home area or in roaming area.
2. The phone calls provide a fixed network calls without any interruption or disconnectivity.

The first generation of communication system does not cover the above requirements and that leads to the development of Global System for Mobile communication (GSM) in 1982 standardized by the European Telecommunications Standards Institute (ETSI) (Mouly et al, 1992; Karl et al, 2004). The development of GSM network results in the design of cellular networks where various mobile users are connected through different base stations in order to get max coverage and high speed. The position is determined according the coverage area of the base station. The lesser number of cells leads in cheaper infrastructure with constraint of accuracy whereas the large cells results in higher accuracy but more expensive. Therefore, heterogeneous cell structure is used where large number of small cells are used in urban areas with many base stations to provide high accuracy and less number of large cells are used in rural areas according to signal's range. The GSM network works in combination of both fixed networks (circuit switched data) and wireless networks. The mobile phones having GSM networks have very good features as GSM carriers put customer information on removable SIM card. Therefore it is easy to transfer information by swapping sim cards. The service providers like Airtel and Idea uses GSM networks. Since the accuracy of GSM networks depends on its cell size, therefore to increase the accuracy of positioning, the Cell-id technique is combined with time advance to recognize the position. These recognized position information are delivered to the system provider. Since high level accuracy is not required, in most of cases it is kept in diameter of GSM cell in between 2km-20 km. The company named Ericsson developed a system called the Mobile Positioning System (MPS) in order to get the exact position of various locations in rural and urban areas. After making few changes in the current infrastructure of the towers.The following tasks can be performed with MPS:

1. The mobile user can query for its nearby places like nearest restaurant and Railway Station, Bus stops etc.
2. Users can also request for the location of other users using the mobile.

A GSM terminal can be integrated in various vehicles that helps the owner to locate his stolen vehicle, also it is helpful for transportation companies to locate their vehicles and to find the routes if a person is entering towards new place. The 3G standard used by GSM carrier is WCDMA which provides high speed circuit switched data at the data rates of 57.6 Kbits/second. For using WCDMA, the user has to switch to new smartphones which are costlier. When a user is using mobile networks on his phone, he has a choice to select either 2G or 3G networks as shown in figure10. In this figure, the GSM is a 2G network and WCDMA is a 3G network, user has a choice to switch to any network or put it in an auto mode where network is automatically available. The components of GSM network and its working is shown in Figure 11.The GSM network works in combination of Mobile Switching Center(MSC), Base Transmitter Station(BTS), Operation and Support System(OSS), International Switching Centers(ISC) and Gateway Mobile Switching centre(GMSC). The main purpose of the mobile switching center is to manage the complete working of GSM and also named as brain of various networks.. The other Components of GSM network are given below.

Base Station Subsystem

The Base Station Subsystem (BSS) is defined as "The section of a traditional cellular telephone network which is responsible for handling traffic and signaling between mobile phone and network switching subsystem ". BSS is divided into two sections:

1. Base Transceiver Station (BTS) is a station that communicates with all mobile users comes in the particular range
2. Base Station Controller (BSC) controls all BTS in the network.

Network Switching Subsystem (NSS)

This system is basically responsible for connecting or disconnecting various connections between different mobile users. The database of all the mobile stations is maintained by this system. The database includes Home Location Register (HLR), a database that stores permanent the information of subscrib-

Figure 10. Snapshot of selection of network by user

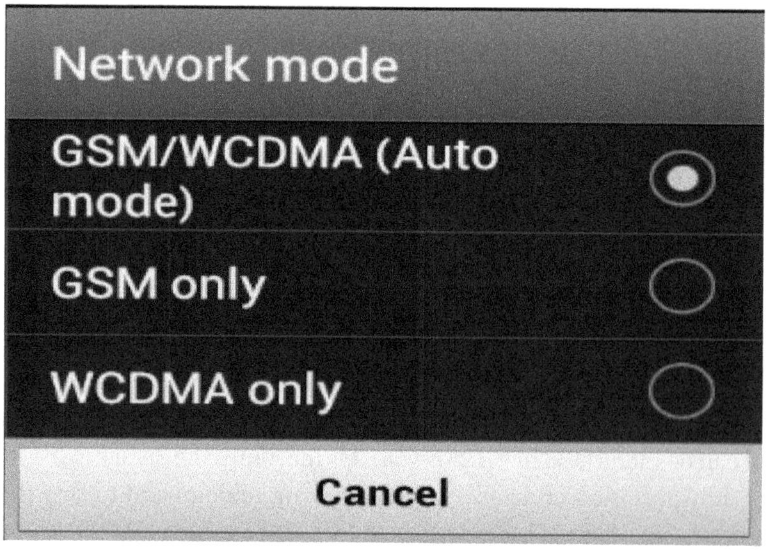

Figure 11. Architecture of global system for mobile communication

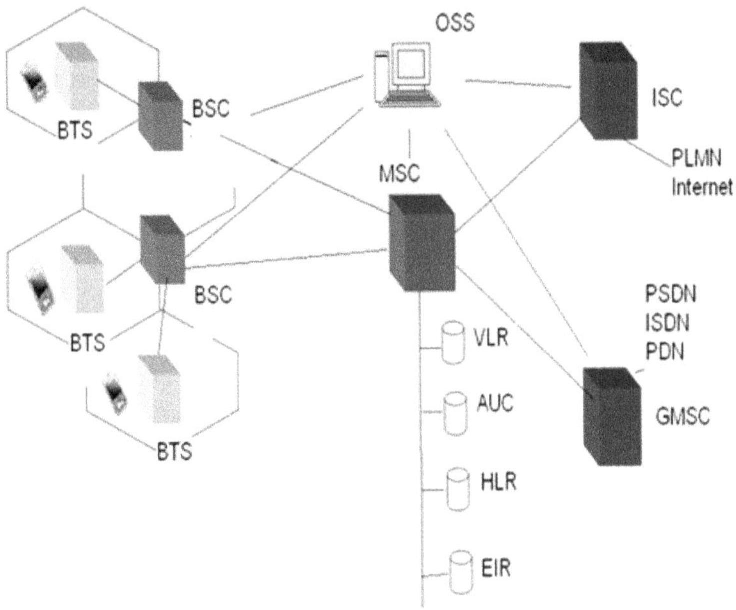

ers (local) such as location information, activity status and Service profiles. Visitor Location Register (VLR) that stores information of roaming users, It is a database used to store temporary information about the subscribers (Ghribi et al, 2000). The function of Mobile Switching Center (MSC) is to control the various functions of the network. The authentication center provides the necessary parameters to the MSC to perform authentication procedure .Authentication Register (AU) stores information of authentications provided to various users and equipment identity register that stores information of international mobile equipment identity numbers. The Equipment Identity Register (EIR) is a register or database containing the details of mobile equipments. When the mobile user enters in roaming area, the information of that mobile user is transferred from HLR to VLR so that the calls can be easily processed without intervention of HLR.

OPERATION AND SUPPORT SYSTEM (OSS)

General Packet Radio Service(Non 3G Internet on Mobile Phone)

General Packet Radio Service also called 2.5G is one of the major development in GSM standard with packet switched network in order to provide high speed (35 kbps) data rates to mobile users ("GSM", 2011). With the help of packet switching, any mobile user can share data independently without set up a connection regardless of availability of network. There are 12 multislot classes in GSM which are also associated with GPRS that contains large numbers of uplinks and also downlinks for every mobile user (Newton et al, 2009; Penlikousis et al, 2005).

GPRS Services

The two categories of GPRS services are: one is Point to Point (PTP) and second is Point to Multipoint (PTM), where PTP services are the connection oriented services provided for IP based applications having X.25 protocol.

1. These services are used for bursty and interactive applications.
2. Point to Point services are used for sending multiple packets are sent over a single destination for example telnet applications.
3. The Connectionless Point to Point services having no direct connection between the source and destination.
4. The Multipoint services for creating and maintaining weather and traffic reports.
5. The multipoint group services for conferencing in real time with availability of multimedia.

Figure 12 describes the working of GPRS with its components. The working of various components of GPRS is as follows:-

GPRS Terminals

GPRS is considered as one of the service of GSM which shares the resources and works interrelated. There are three types of terminals:

Class A Mobile Station A mobile station that works on both the connections having circuit switched networks and the connections having packet switched networks.

Class B Mobile Station A mobile station that connects GSM along with GPRS whenever there is a need of GPRS data transaction for resuming the incoming calls.

Class C Mobile Station It is a mobile Station that allows the mobile users to access the services in one by one order at a given time.

The mobile station contains a mobile terminal also, called handset, and a terminal equipment like Laptop or computer.

Figure 12. Architecture of GPRS

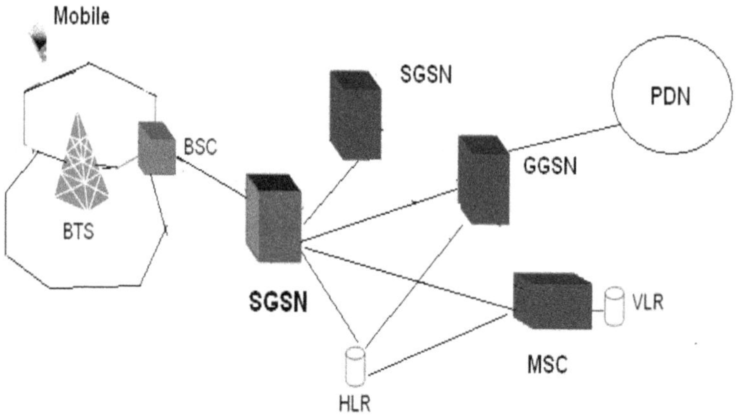

GPRS Base Station Controller

The Base Station Controller works through Base Station Subsystem. There is not much difference between the BSS of GSM and GPRS BSS because the frequency bands, hopping techniques, radio modulation, burst structure of both are same. The only difference is the addition of Packet Control Unit (PCU) in GPRS BSS for supporting management of packets. The connections in GPRS are established and released over an air interface.

GPRS Network Switching System (GPRS NSS)

The GPRS NSS acts as a network between the mobile customers and data networks. There is an addition of GPRS Support Node (GSN) that can act as a Serving GSN (SGSN) or it may work as Gateway GSN (GGSN) to support different services. The SGSN is responsible to keep tracking of various mobile stations and controlling their services. SGSN also provides the interface to the various outside networks.. The example of GPRS activities is an Internet Service Provider that provides accessibility of internet to various mobile users with other services.

Enhanced Data Rates for Global Evolution (EDGE)

EDGE is 2.7 generation network which is also called Enhanced GPRS or EGPRS,a data system(also named as Pre-3G radio technology) used in the GSM networks that allows improved data transmission rates which are three times faster than GPRS ("Enhanced Data Rate", n.d.). The EDGE works for any network which is deployed along with GPRS, with the upgradation of transceiver units and base stations to compatible with edge. Therefore, if network does not support the edge, the phone will automatically revert to GPRS. The EDGE provides the higher data rate of 384kbits/sec .

Universal Mobile Telecommunication System (UMTS)

UMTS is developed to provide a uniform network for providing multimedia in various communication using mobiles which covers the limitation of communication systems of mobiles. The UMTS integrates the technologies like GSM with wired networks and their backbone networks of base stations for providing easy accessibility and higher rates for users. UMTS is built on GSM and GPRS technology. In the air interface of this system a new BSS technology based on Wideband CDMA is added making it a third generation wireless standard that allows both voice and data offering data speed of up to 384 kbps. The architecture of UMTS is shown in Figure 13.The architecture of UMTS is similar to GPRS with the addition of first element called Radio Network Controller (RNC) and second element, called Node B.The function of RNC is similar to the function of Base Station controller of GPRS and the major functionality of Node B is similar to the function of Base transceiver station of GPRS network.

The architecture of UMTS is basically divided into major two domains: GSM circuit switched and GSM packet switched. Various other elements are home location register, visitor location register, authentication and so on. The other major domain of UMTS is UTRAN (UMTS Terrestrial Radio Access Network) is used to provide air interface to the device through WCDMA technology. The Radio Network Controller ("High Speed Downlink", n.d.) is a an active element of UTRAN which is responsible for Mobility Management and Radio Resource Management.

Figure 13. Architecture of UMTS

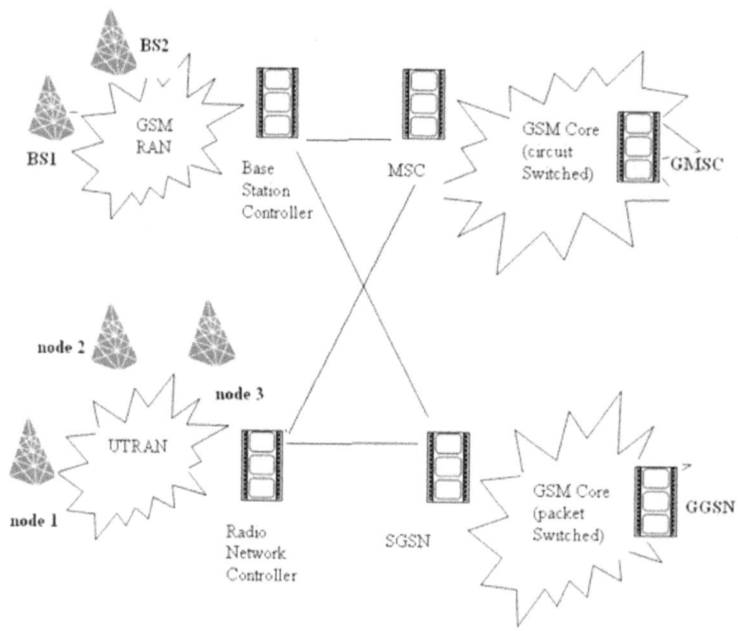

High Speed Downlink Packet Access (HSDPA)

The evolution of Global System for Mobile Communication towards the higher data rates of GSM leads to an effective cellular technology available to customers worldwide. This is also called 3G+ networks with higher data rates based on Universal Mobile Telecommunications System . The data service providers deployed Universal Mobile Telecommunication System for providing higher data services with downlink data transfer speed of 42.3 mbits/s to customers. It supports packet data service that helps end users for better results for downlink applications with shorter connections and shorter response times. HSDPA provides full potential to UMTS for Broadband Services and HSDPA is fully backward compatible with UMTS .

Long Term Evolution (LTE)

The 3rd GPP Committee is now working on UMTS Long Term Evolution. The long term evolution is also called as 4G LTE and Evolved terrestrial Radio Access (E-UTRA) which is part of release 8 of GPP specification, an access part of Evolved Packet System(EPS) and a standard that provides higher data rates .The evolution path from GSM to LTE is shown in Figure 14.

This is a next step towards 4G mobile phones, where 3GPP family of cellular systems include GSM, GPRS, UMTS and EPS leads to LTE, where LTE enables more speed, low latency and more compelling mobile environment. GSM has circuit switched network and developed for real time services but with low data rates. The next step is GPRS, where IP based packet switched network is used to provide high speed internet facility. The next step UMTS provides higher data rates for real time services using

Figure 14. Long term evolution

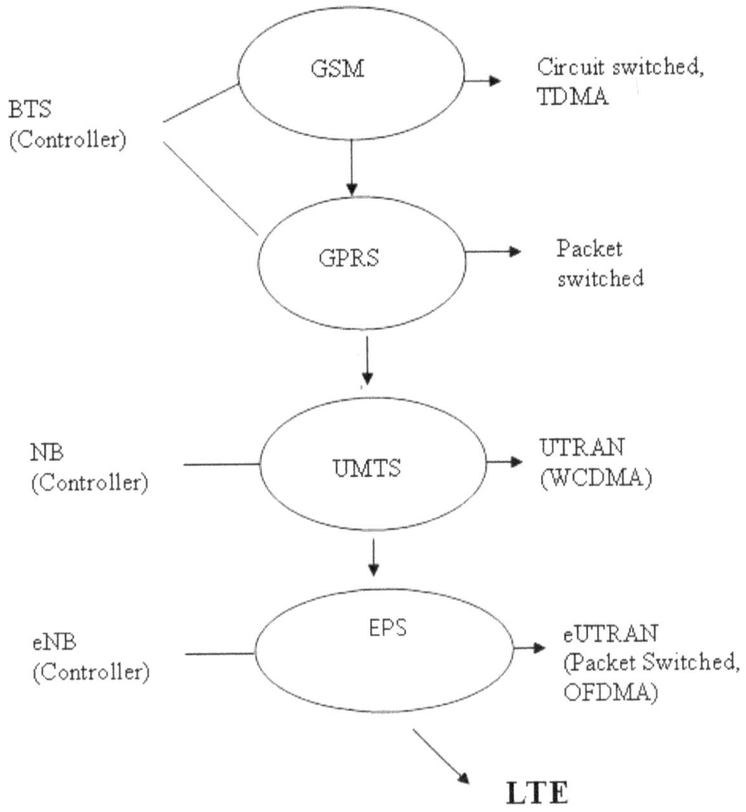

WDCMA technology. The last step is based on packet switching system with Orthogonal Frequency Division Multiple Access (OFDMA) technology for both uplinks and downlinks of frequency. Frequency Division Multiple Access (SC-FDMA) for uplink that provides powerful packet optimized mobile radio systems with uplink data speed of 50 Mbps and downlink data speed of 100 Mbps .The SC-FDMA allows multiple users to share single communication resource. Further, LTE is based on EPS which provides a long term growth to mobile operators that provides value added services to their customers. The Evolved Node B(eNB) is the main element of E-UTRA which in turn is a part of Long Term Evolution, a hardware that is connected to mobile phone network In case of UMTS, NodeB (NB) is controlled by Radio Network Controller but in LTE, there is no separate controller which results in reducing response times.

The main characteristics of LTE are:

1. Enhanced air interface, where all radio access networks are based on Orthogonal Frequency Division Multiplexing and provides higher data rates.
2. High Spectral efficiency, where it allows mobile operators to join more and more customers with reduced cost of delivery.
3. Flexibility radio planning, where it provides optimal performance in cell size of 5 km and effective performance in cell size on 30 km.

4. Reduced Latency with reduced round trip time for response to user which also provides high quality, interactive real time services.

Applications

Location based services are extensively used in various application areas. The few of examples are as follows:

A novel incremental Dijkastra's parallel System to analyze the applications of location based services for categories such as searching a nearest restaurant, nearest school, nearest gas Station etc.

Interactive Travel Assistant application that help user to record and share the wonderful experiences of their journey as per the interest of the user. The user can use and update its multimedia resources in journey. The application is implemented using GPS Technology. User can automatically get images, videos and comments on their PDA's according to their current location.

Location-based business marketing that helps user to do their shopping using their mobile phones where user can get all the information regarding the details of products, discount offers and so on of various shopping centers without moving here and there.

Cell based User Locating System where services offered by this system include navigation and routing, finding point of interest and geocoding which means mapping the user to an actual geographic location.

Tracking of trains and vehicles for avoiding accidents and checking the track of goods send by the sender and the route followed by the vehicles. Also to keep track of source and destination along with their time logs.

Road Management and maintenance system where drivers can send the details of the particular location, where problem exists to the management in order to avoid accidents.

An android based distribution application called LogLife, where the record of all the places which are visited by the tourists are recorded. The application also shows the locations on the map. The data stored in the log is in the form of addresses. To convert the coordinates into addresses, the google server is used with appropriate rights and reverse geocoding process is evaluated for converting coordinates into addresses.

Remote Patient Management System that can be used to provide the services to the patients using location based Services. The system is implemented using Assisted GPS in the mobiles for emergency circumstances. The Patients wearing various wearable devices through which the location of the patients can be recorded when they are not mentally or physically fit and they can also provide the treatments at their nearby areas.

Virtual world where Second Life 3D world where every player plays with a virtual objects which are just similar to a real world. Anyone can use its artistic talent in this game and can share their talents with their friends. It is a virtual world, where avatars play like a real person and do things like we do in real life. It is multiplayer game that can socialize their world and also make group activities.

Cell based User Locating System is used for finding the location of the mobile user from a network independent brokers' point of view. The services offered by this system include navigation and routing, finding point of interest and geocoding which means mapping the user to an actual geographic location. The system locates the user in coverage area of base transceiver station using signal parameters

Cricket Location Support System in 2000 which is also called location dependent application. This system runs on mobile phone to know the physical locations used by listeners. The application consists of nodes to analyze the beacons throughout the building. A beacon is a device that is attached with objects to determine the position. The listener is a small device that listens the messages from the beacons and advertises them in the space.

SUMMARY

The main aspects of the latest technologies in the current era is to know the location of your nears and dears for the purpose of their safety as safety is the main concern these days. Safety of elders, safety of kids, safety of women, and these all aspects can be covered by the various sensor technologies combined with location based services. The availability of 5G networks, smart computing, wireless fidilety, artificial intelligence and GSM in the era of internet of things makes life easier. Internet of Services are jointly used in the area of location based services for tracking mobile phones, tracking vehicles, navigation purposes, healthcare and in many other areas. Safety of family member and timely information regarding the places is always a very important aspect of location based services using GSM in internet of things.

REFERENCES

W3C Standards. (n.d.). Available at http://www.w3.org/standards/

Ahmad. (2011). Smart Metering and Home Automation Solutions for the Next Decade. *International Conference on Emerging Trends in Networks and Computer Communications*, 200-204.

Air Navigation Services-India. (n.d.). *GAGAN-GPS-Aided Geo-Augmented Navigation*. Available at http://www.aaians.org/?q=buzz-words/gagan-gps-aided-geo-augmented-navigation

Arena, G. S. M. (n.d.). *General Packet Radio Service*. Available at http://www.gsmarena.com/glossary.php3?term=gprs

Arena, G. S. M. (n.d.). *Enhanced Data for Global Evolution*. Available at http://www.gsmarena.com/glossary.php3?term=edge

Ballazhi. (2012). *Wireless Indoor Positioning Techniques*. Comminications System Seminar.

Calduwel Newton, P. (2009). A Quality of Service Strategy to Select Coding Schemes in General Packet Radio System. *International Journal of Advance Science and Technology*, *7*, 1–10.

Drakopoulou, S. (2010). A Movement of Experimentation: Spatial Practice and Representation of Space as narrative Elements in Location based Games. *Journal of Media Geography*, *5A*, 73–76.

El-Rabbany, A. (2002). *Introduction to GPS: The Global Positioning System*. Artech House Mobile Communication Series.

Gleason & Gziabher. (2009). *GNSS Applications and Methods*. Norwood: Artech House.

Gyeyoung & Yim. (2012). A Review of the Techniques for Indoor Location based Service. *International Journal of Grid and Distributed Computing*, *5*(1), 1–22.

ISRO. (2013). *Indian SBAS System-GAGAN*. GNSS. Asia India-Industry workshop.

Karl, H. (2004). *Data transmission in Mobile Communication Systems*. Location-Based Services. doi:10.1016/B978-155860929-7/50009-1

Krebs. (n.d.). *Gunter's Space Page Site*. Available at http://space.skyrocket.de/doc_sdat/uragan-k1.htm

Lin, Y.-B., Haung, Y.-R., Chen, Y.-K., & Chlamtac, I. (2001). Mobility Management: From GPRS to UMTS. *Wireless Communications and Mobile Computing, 1*(4), 339–360. doi:10.1002/wcm.27

Mouly & Bernadette. (1992). *The GSM System for Mobile Communication*. Academic Press.

Neun, M. (n.d.). *Techniques for LBS Cryptography*. Available at http://www.e-cartouche.ch/content_reg/cartouche/LBStech/en/text/LBStech.pdf

Open Geospatial Consortium. (n.d.). *Internet source*. Available at http://www.opengeospatial.org

Penlikousis. (2005). Wireless Data Networks. *Internet Protocol Journal, 8*(1).

Project Coordination Group. (2013). *The Third Generation Partnership Project Working Procedures* (Version 15.0). Available at http://www.3gpp.org

Ratti, C., Frenchman, D., Pulselli, R. M., & Williams, S. (2006). Mobile Landscapes: Using Location Data from Cell Phones for Urban Analysis. *Environment and Planning. B, Planning & Design, 33*(5), 727–748. doi:10.1068/b32047

Renka. (n.d.). Location based Services on Mobile Ecommerce. *International Journal of Computer Science and Information Technologies, 3*(1), 3147-3151.

Rohde & Schwarz. (2007). Headed Towards Next Mobile Radio Generation with UMTS LTE. *News on Mobile Radio Signal Generators/Signal Analyzers, 193*. Available at www.rohde-schwarz.com

Roth, J. (2004). Data collection. In Location-Based Services. Morgan Kaufmann.

Roth. (2004). Data collection. In *Location-Based Services*. Morgan Kaufmann.

Sahoo, P. M. (n.d.). *Introduction to Geographic Information System (GIS) – IASRI*. Retrieved from www.iasri.res.in/ebook/gis_ta/m3_1_intogis.pdf

Schiller & Voisard. (2004). *Location based Services*. Morgan Kaufmann Publishers.

Steinfield, C. (2004). The Development of Location based Services in Mobile Commerce. E-Life after Dot Com Bust, 4, 177-197.

Steiniger, S. (2012). *Foundation of LBS*. Lecture notes on LBS, Version 1.0. Available at http://www.e-cartouche.ch/content_reg/cartouche/LBSbasics/en/text/LBSbasics.pdf

Want, R., Hopper, A., Falcão, V., & Gibbons, J. (1992). The Active Badge Location Based System. *ACM Transactions on Information Systems, 10*(1), 91–102. doi:10.1145/128756.128759

Wikipedia. (n.d.). *Radio Frequency Identification*. Available at http://en.wikipedia.org/wiki/Radio-frequency_identification

Wikipedia. (n.d.). *Enhanced Data Rate for GSM Evolution*. Available at http://en.wikipedia.org/wiki/Enhanced_Data_Rates_for_GSM_Evolution

Wikipedia. (n.d.). *Radio Network Controller*. Available at http://en.wikipedia.org/wiki/Radio_Network_Controller#RNC_Roles

Wikipedia. (n.d.). *High Speed Downlink Packet Access*. Available at http://en.wikipedia.org/wiki/High-Speed_Downlink_Packet_Access

Chapter 12
Introduction to IoT Technologies and Its Applications

Neeta Yadav
KNIT Sultanpur, India

Neelendra Badal
KNIT Sultanpur, India

ABSTRACT

The internet of things is everywhere. It is an emerging and wide scope of study. Every day there is a new addition to this technology. It is simply defined as a network over which a collection of things communicate with each other using internet as a medium of communication. It is also sharing information with each other for the ease, welfare, and convenience of human beings. It makes human life easier. About one trillion devices are expected to connect with this technology in the upcoming future (most probably in 2025). Each and every year its popularity increases. It makes smart health; provides more safety, comfort, and convenience; and also increases wisdom. IoT is one of the platforms of today's smart city and smart energy management systems. This chapter explores IoT technologies and applications.

INTRODUCTION

This chapter covers technologies of IoT, development tools (software, middleware, and hardware), architecture, issues, and various applications of IoT. This chapter basically focuses on IoT technologies that are RFID, ZigBee, WiFi, sensors, and smartphones and its various applications like smart energy, smart home, smart farming, smart healthcare, etc. In this chapter, the internet of things is fully covered from each and every perspective. This chapter provides each and every detail regarding IoT and also its recent achievements.

The main motive of writing this chapter is to provide quality content to the beginner as well as those who are going to do something new in this field. It may act as a tutorial to them and also provide depth

DOI: 10.4018/978-1-5225-9574-8.ch012

knowledge of IoT technologies and applications. Our main concern is to provide quality content to the reader and target audience (Students, Researcher, and Project developer). For starting any research, project related to IoT, it's very necessary to have basic knowledge and information for carrying out quality work, so this chapter covers almost all topics related to IoT. This chapter is explained in an easy manner and written in simple English language so it is approachable, understandable, and acceptable to everyone.

Nowadays, the internet is not only connecting people, but it is also connecting a thing that's why it is called the internet of things. IoT is based on various technology and it has various applications like smart city, smart computing, and smart health, etc. in short this technology adds smartness to various things. Its adoption is growing very fastly in a day to day life. In the coming future, it may be renamed as the internet of everything. The rest of the things are discussed in detail inside the chapter.

Internet of Things

The Word "*Internet of Things*" first came into existence by Kevin Ashton (Cruz et al. 2018), and (Wu et al.2017). He Coined this term. In simple words, the Internet of Things is basically an ecosystem of connected physical objects.

It is a network of physical objects that are interconnected to each other. As everyone knows data, images, books, commerce, records, etc. are created for the people, by the people, and about people. So if everything is connected to the internet, human life becomes easier. In figure 1, the internet of things is defined.

When things have the ability to sense and they start communicating with each other for taking appropriate decision and giving an alert about change in the environment are defined as the internet of things. With the help of sensor internet of things are able to collect and exchange data. Things interact and collaborate with other things with the help of a sensor. This technology adds smartness in a day to day life. It is very necessary for human in the modern world. IoT is everywhere and it has the properties of anytime, anyplace, and anything connection. Five things are required for IoT that makes it work that is **"Things, IP Protocol, Internet, Sensor, and Mobile".**

Why IoT is needed?

IoT is needed because of the following reason:

- For the dynamic control of industry and daily life.
- To stop the wastage of resource and utilize the resources in an improved way.
- It is responsible for establishing a better relationship between human and nature.
- For universal transport and internetworking purposes.
- For accessibility and usability purposes.

Figure 1.

- For technologies integrator.

In Figure 2, the properties of IoT are discussed.

IoT provides anytime connection whether it is daytime or night or on move. It also has the property of anything connection, for example, it provides the connection between PC, Human 2 Human and Human to Thing. And the most important property of IoT is anyplace connection whether indoor or outdoor. These three properties of IoT make it more successful and acceptable in day to day life.

HISTORY OF INTERNET OF THINGS

Nowadays, the internet not only connects people throughout the globe but also things around us with the help of sensors. It is a very emerging technology, everyday enhancement, research, and innovation is done in this. It is a very vast area and humans have touched only a few percentages of this field a lot of research work still waiting. The internet is not a very old concept. In 1974 TCP/IP structure came into Knowledge of everyone means this concept emerged in 1974. The first online website was created in 1991. In a very few time, internet spread all over the world. By 1995 more websites and system came online. The term IoT is not a very new term it is used most of the time earlier. "The first look at the Internet of things arguably came from Nicola Tesla in 1926" (Firner et al. 2011, p. 337), and (Somayya et al.2015, p. 164). Real IoT is touched by Mark Weiser in 1998, he developed a water fountain that was very wonderful and everyone was mesmerized after seeing it. In 1999, term the Internet of Things coined

Figure 2.

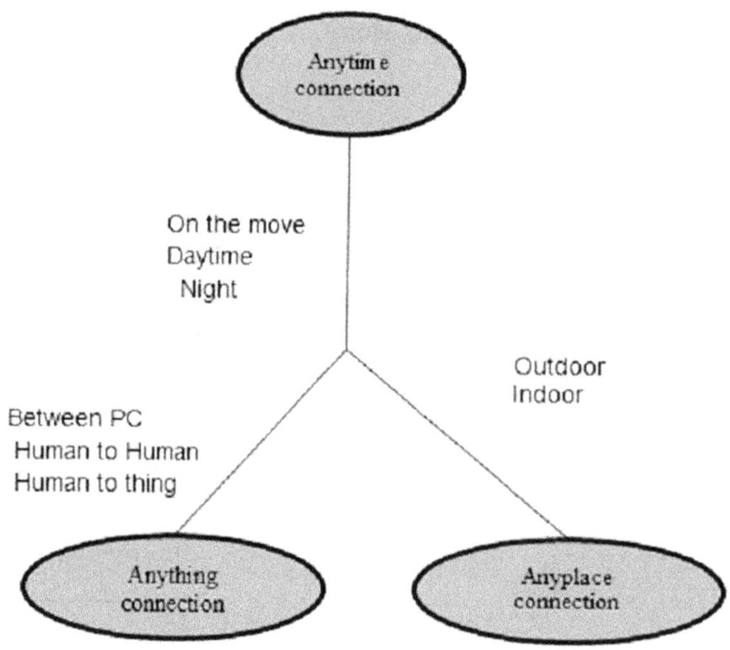

by Kevin Ashton. "IoT continued to grow, to evolve and projections are bright for new technology". In table 1, the zest of IoT history is shown are as follows:

LAYERS OF ARCHITECTURE

Nowadays, IoT is a very wide scope of the study and it is very emerging and trending technology which covers almost the whole globe in upcoming future definitely it will be renamed as IoE that is the internet of everything. Anyone who wants to work or know about IoT it is very necessary to understand its architecture firstly. There are six layers of "Internet of Things" that is Device layer, Service Provider layer, Infrastructure Services layer, IoT Services layer, IoT Development Services layer, and IoT Application layer. In Figure 3, the architecture of IoT is depicted.

Device Layer

This is the first layer of the internet of things and it is the main source of data. Devices like ambulance, mobile, refrigerator, etc. sensing devices can gather data and transmit them at regular interval. These devices interact with the intelligent gateway for further operation.

Electronic devices generally are the small, memory-constrained, and often battery-operated with onboard sensors and actuators. These devices work independently or are embedded as part of bigger machinery for sensing and control. There are three main capabilities of IoT devices which are as follows:

- The ability of sensing and recording data.
- The ability to perform light computing
- The ability to connect to a network and communicating with the data

Examples of these devices include fitness trackers, agricultural soil moisture sensors, medical sensors for measuring blood glucose levels and more. There is a huge competition between startup companies and established companies to come up with newer and newer sensors, actuators and devices.

Table 1. Zest of IoT history

Year	Description
1997	"The Internet of Things" is the seventh in the series of ITU internet reports originally launched in 1997 under the title " Challenges to the Networks"
1999	Auto-ID center founded in MIT Keven Ashton
2003	EPC Global founded in MIT
2005	In this year four important technologies of the internet of things was proposed in the WSIS conference
2008	In 2008, the first international conference was held at Zurich.
From 1990s to 2010s	Around this period, it is said to be the Internet of People because till this date about 5 million devices are connected.
From 2010s to 2020s	It will be predicted that trillions of devices are connected around that period and it is termed as the Internet of Things

Figure 3.

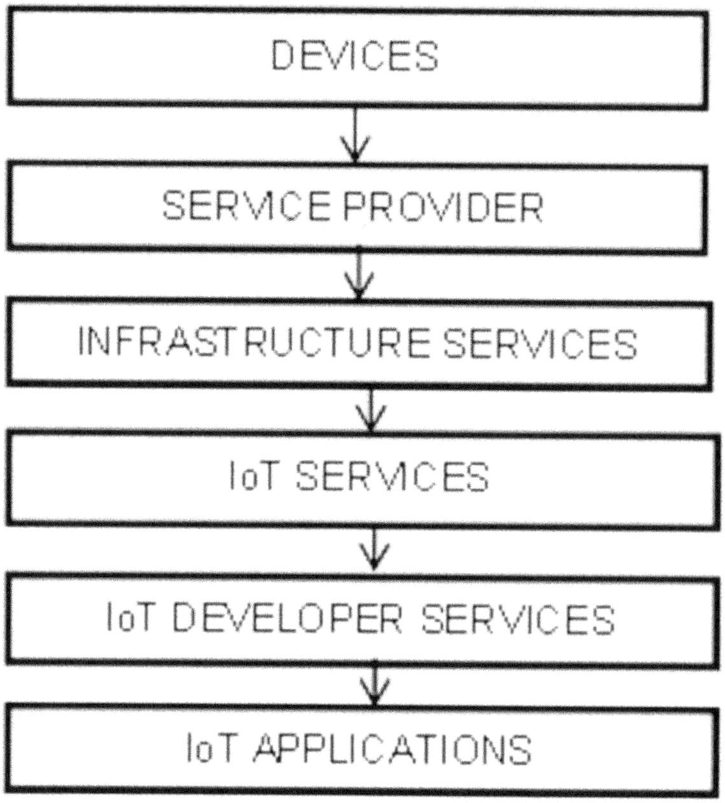

Service Provider Layer

This is the second layer of IoT. Device layer interacts with this layer. This layer consists of business and operation support, and network and security management. This layer is responsible for communication. Basically, the layer provides services to the device layer. The IoT gateway aggregates data from numerous sensing devices and relays it to the cloud. The various IoT devices need to make connectivity to the internet via a powerful computing device that is gateway which primarily acts as a networking device. The IoT gateway is responsible for aggregating data from numerous sensing devices and transmitted or relays it to the cloud.

Infrastructure Services

This is the third layer. It is a very important layer of IoT. It acts as the glue between the physical world of sensors and virtual enterprise infrastructure. This layer provides scalability and reliability to millions of devices when they start communicating with each other. An IoT gateway transmitted or relays all the sensor data to the cloud-hosted servers, where all the data stored. Further, these data are accepted by servers, store and processing are done for the decision making process. This layer is also responsible

Table 2. Characteristics, features, advantages, and disadvantages of IoT

Characteristics	Features	Advantages	Disadvantages
Intelligence	Device Virtualization	Integration of Devices	Security
Connectivity	High-Speed	Reliable, Secure Bi-Directional Communication	Privacy
Sensing	End Point Management	Enhanced Data Collection	Complexity
Expressing	Small Devices	Reduce Waste	Flexibility
Energy	Steam Processing	Real-Time Analysis Improved Engagement	Compliance
Safety	Data Enrichment	Automation and Technology Optimization	Tier Management

for the creation of a live dashboard where decision makers monitor and take the proactive data driven-decision. Nowadays, almost every cloud computing companies offer custom services for IoT solutions.

IoT Services

This is the fourth layer of IoT. This layer is completely independent of the underlying devices, communication protocols, and connectivity semantics. This layer includes event processing, integration, security, identity management, user interface, and big data analytics. This layer contains an important set of services which is responsible for building IoT applications (Ding et al. 2014).

IoT Developer Services

This is the fifth layer of IoT. It includes developer portal, developer API management, 3rd party APIs and service provider APIs.

This layer enables the developer to develop IoT based applications by using IoT services and development kits.

IoT Applications

This is the topmost layer of IoT. In the application layer custom application built using things data. It includes applications like smart home, wearables, connected cars, smart cities, etc (Borgia, 2014).

In table 2, characteristics, features, advantages, and disadvantages of IoT are listed.

TECHNOLOGIES OF IoT

IoT is based on various technologies like RFID, WiFi, Barcode, Zigbee, Sensors, and Smartphones. Each of them is discussed in detail below.

Table 3. The IoT value chain

Product	Description
Radios	It is responsible for providing connectivity
Sensors	It is a chip which is responsible for measuring various environmental/ electrical variables
Microcontrollers	Processors/ storage that allows low-cost intelligence on a chip
Modules	It combines radios, sensors, and microcontrollers in a single package
Platform Software	Platform software are those software's which is responsible for activation, monitoring, and analyses of device networks
Application Software	It is responsible for presenting information in a usable/analyzable format for an end user.
Device	It is responsible for the integration of modules with application software into a more usable form factor
Airtime	It is used licensed and unlicensed spectrum for communication
Service	It is responsible for the deployment, managing, and supporting of IoT solution

RFID

It stands for Radio Frequency Identification. It is an IoT based technology. It uses an electromagnetic field to automatically identify and track tags attached to it. This technology is widely used in transport and logistic. It brings revolutionary digital transformation in the area of transport and logistics. In figure 4, RFID is shown.

RFID is a backbone of supply chain management. There are two tags used in this technology that is active tag and passive tags. Active tags have a local power source like battery power and it may operate at hundreds of RFID reader. Passive tags collect energy from nearby RFID readers. Tags are generally attached to products during production processes. RFID is more advantageous over the barcode or manual system because tags are read even if it is covered by an object. It can be read inside a case, carton, box or other containers.

Figure 4.

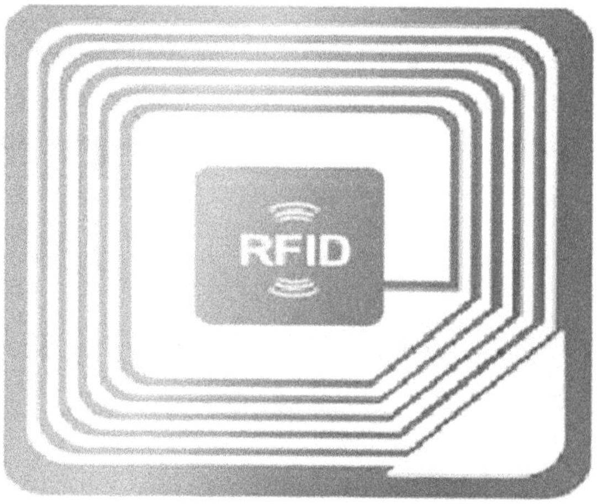

Uses of RFID

- Access Management
- Tracking Goods, person, and animals
- Toll collection and contactless payments
- Smart dust
- Airport baggage tracking logistics

RFID works on different frequency bands which are shown in table 4.

BARCODE AND QR CODE

Barcode and QR Code is the technology on which IoT is based. Barcode is responsible for optical representation of data. It is machine readable. Initially, barcodes are scanned by special optical scanners called barcode readers. After that application software's are available for devices that could read images, such as smartphones with cameras. The QR code is a quick response code. A QR code uses four standardized encoding modes that are numeric, alphanumeric, binary, and kanji to store data. It provides the facility of fast readability and more storage capacity.

Barcodes are More Successful Because

- Barcodes labels are cheap.
- More Accuracy.
- User-friendly technology.
- Standard symbology is developed.

There are **two** types of barcode

Table 4. Different RFID frequency bands

Band	Range	Data speed	Application	Approx. Cost
120-150 KHz (low frequency)	10 cm	Low	Animal identification, factory data collection	$ 1
13.56 MHz (High Frequency)	10 cm- 1m	Low to moderate	Smart Cards	$ 0.50 to $5
433 MHz (UHF)	1-100 m	Moderate	Defense application with active cards	$ 5
865-868 MHz (Europe) ISM band	1-12 m	Moderate to high	EAN, various standards, used by Railroads	$0.15 (passive tags)
2450-5800 MHz	1-2 m	High	WAN, Bluetooth Standards	$ 25 (active tags)
3.1-10 GHz Ultra-Wide Band	Up to 200 m	High	Not defined Semi-active or active tags	-

- **Linear Barcode**

it consists of black bars and white spaces.
Example:- CODE 11, UPC, CODE128, CODE 39.

- **2D Barcode**

It is consists of a grid of light and dark modules.
example:- data matrix, Aztec code, OR- code
In figure 5 and 6, barcodes and OR code are shown.

Advantages

- It represents the unique identity of the product.
- It is less expensive and error free.
- Bidirectional Symbology.

Disadvantage

- System failure may cost more delay.
- Scratched or crumbled barcodes.

Application Of Barcode

- Departmental stores
- libraries
- health agency
- document tracking

Figure 5.

Figure 6.

WIFI

It stands for wireless fidelity. It is one of the important technologies which are used by the IoT and this technology makes IoT more successful in remote areas. It is used in both environments whether it is indoor and outdoor.

Cost of this technology is very low. In figure 7, WiFi is shown.

Advantages of Wifi

- Less expensive.
- No cabling required.
- Quick and easy to set up.
- Scalable
- It provides the facility of mobile access.

Disadvantages of Wifi

- Speed is low.
- The range is affected by various medium
- Less Secure.

Figure 7.

ZigBee

ZigBee is a new wireless technology. It created by the ZigBee Alliance. It is a low-cost technology and it has long battery life. This technology is easily deployable and it has a large number of nodes that is 64770. This technology is globally used. It is highly securable. It is ideal for WPAN and mesh networks and also supports multiple network topologies. In figure 8, ZigBee technology is shown.

This technology specially designed for low power consumption so its batteries last for month or year

Characteristics of ZigBee Technology

- Low power consumption with battery life ranging from a month to year.
- Less expensive.

Figure 8.

- Easy implementation.
- low data rate
- Small packet devices.

Application of ZigBee

- Home Networking
- Industrial control and management

SENSORS AND SMARTPHONES

Sensors are one of the most important technology on which IoT is based. Without the sensor, IoT is not workable. Sensors help in sensing changes in the nearby object or things. There are various sensors used in IoT that is temperature, humidity, proximity, etc. which makes IoT successful, workable, and adaptable.

Characteristics of Sensor

- High Accuracy
- High Resolution
- Easy to integrate
- Energy efficient
- High Reliability

Uses of Sensor

- Detection of light, motion, temperature, and pressure.
- The speed of vehicles etc.

A smartphone is that kind of mobile, which is able to perform various functions of a computer like internet access, having touch screen interface and OS, able to download apps. Smartphones have a lot of facilities and sensors which can be used in IoT. Smartphone comprised of various advanced features. It is used for multipurpose. The smartphone is one of the key elements of IoT technology only by this people are able to control whole devices connected to the internet and messages can also be displayed or received on the smartphone. In the upcoming future, most of the people will have smartphones. In figure 9, a smartphone is shown.

In Table 5, 6 Comparison between different technologies and the growth of IoT is shown as follows:-

This table depicts a comparison between different technologies of IoT on the basis of frequency, data rate, range, power usage, and cost.

This table depicts the growth of IoT, in the year 1990 about 0.3 million devices were connected but in 2000 about 90 million devices were connected. Every year there is rapid growth in IoT technology. In the year 2010, it became 5 billion in number and recently in a survey of 2015 it was found that about 9 billion devices are connected to IoT technology. And it is expected that about 1 trillion devices will be connected to IoT technology in upcoming future.

Figure 9.

Table 5. Comparison of IoT technologies

Technology	Frequency	Data Rate	Range	Power Usage	Cost
ZigBee	2.4 GHz	250 kbps	~ 300	Low	Low
Wi-Fi	Sub GHz, 2.4 GHz, 5GHz	0, 1-54 Mbps	< 300 feet	Medium	Low
Bluetooth	2.4 GHz	1,2,3 Mbps	~ 300 feet	Low	Medium
Z-Wave	Sub GHz	40 kbps	~ 100 feet	Low	Medium
RFID	3.1- 10 GHz	High	Up to 200 m	High	High

Table 6. Growth of internet of the things

Year	Number of Connected Devices
1990	0.3 Million
2000	90 Million
2010	5 Billion
2015	9 Billion
2025	1 Trillion (EXPECTED)

Table 7, depicts a list of hardware, middleware, and software. IoT required different development tool like Arduino, Raspberry-Pi, IBM Watson, Azure, AWS, INTEL JOULE, Netduino. It also needs middleware and software like OpenIoT, OpenRemote, OpenHUB, Kaa, Oracle Fusion, Eclipse IoT, Google Brillo, IBM IoT Foundation, Azure IoT Suit, Cloud Sensor, Ninja Sensor, Control Any (Amaral et al. 2015). In table 8, a comparison is made between the top platform of IoT as following:-

Table 7. Development tools of IoT

Hardware	Middleware	Software
Arduino Nano Pro Mini	OpenIoT	Eclipse IoT
Raspberry-Pi	OpenRemote	Google Brillo
IBM Watson	OpenHUB	IBM IoT Foundation
Azure	Kaa	Azure IoT Suit
AWS	Oracle-Fusion	Cloud Sensor
INTEL JOULE		Ninja Sphere
Netduino		Control Any

Table 8. Comparison between Top Platforms of IoT

Platform Name	Predix	IoT Hub	AWS IoT	IBM Watson
CRM/ERP Integration	Manual	Manual	Manual	Manual
Machine Learning	√	√	√	×
BigData-Hadoop	√	√	√	×
Device security	√	X.509, TLS	X.509	TLS
Device Protocols	√	AMQP,MQTT,HTTP, Web sockets	MQTT, HTTP, Web sockets	MQTT, HTTP
Object Storage	√	√	√	√
Visualization	√	√	√	√

ISSUES OF IoT

There are various challenges faced by human beings after the application of this technology. Major challenges of IoT are Security, Privacy, and Connectivity Challenges

Security Issues

The word security means the state of being free from danger or harm. Due to advancements and growth in IoT, our security is also at risk and it draws the attention of the technology firms and government agencies all over the world.

The hacking of baby monitors, smart fridges, Barbie dolls, drug infusion pumps, cameras, and even assault rifles are portending a security nightmare being caused by the future of IoT.

IoT is risky because the future of the human is more and more connected to IoT. Our lives, health, infrastructure may become a target of this attack. So it's very necessary to incorporate security in such a way in things, human life and its assets free from risk.

Privacy Issues

The word privacy means personal or private to a particular individual or group. Privacy is also defined in another way like the ability to seclude themselves, or the information about themselves and by doing this they selectively express themselves. It is also a state of the personalhood. Some of the data collected by the devices are very sensitive and personal to an individual. But its security is at risk because of this user's personal information exposed to everyone (Yang et al. 2017). In the coming future, everything will be available on the cloud so government's confidential information might be exposed to terrorists or intruders which can disturb or destroy our whole country so it requires the high level of security.

Connectivity Issues

Maintaining connectivity of millions of device is one of the biggest issues of the future of IoT. Every day there is a new addition of devices to IoT, it also leads to a change in the model. At present centralized model is used mostly, but in the near future, it will require the decentralized model for maintaining the whole system. In the upcoming future, the centralized model will not be sufficient for the IoT system. Connectivity of IoT devices also depends upon the weather conditions if the weather condition is not good various devices will be unable to connect with the internet. So connectivity between devices will be disturbed and the whole system turned into a bottleneck.

APPLICATION OF IoT

There are various applications of the Internet of Things and in coming future more application will be discovered. Few applications are listed as follows in detail:

Smart Energy

It is one of the most important applications of IoT. Smart energy concept basically deals with minimum wastage of energy. Generally, people waste a lot of energy in day to day life by forgetting to switch off the light, fan, AC, etc. leading to wastage of electricity. Both residential and Office buildings waste a huge amount of electricity by virtue of overusing appliances due to different reasons like Ignorance, low maintenance or lack of awareness. Most of this wastage can be prevented by using home and office appliances smartly. Since time is of the essence for anyone such mistakes and inefficiency is only natural. Thus, the best solution is to automate its use. Figure 10, shows the different percentage of electricity used in the office building.

Solution and Application

- To build sensor modules which can identify in which direction an object passed and whether it had legs or not.
- A pair of infrared beam breakers can be used to identify the direction of motion while another set of the beam in coordination with the above-mentioned sensor can identify if the body moved on legs by gap due to the count of two legs per object.

Figure 10.

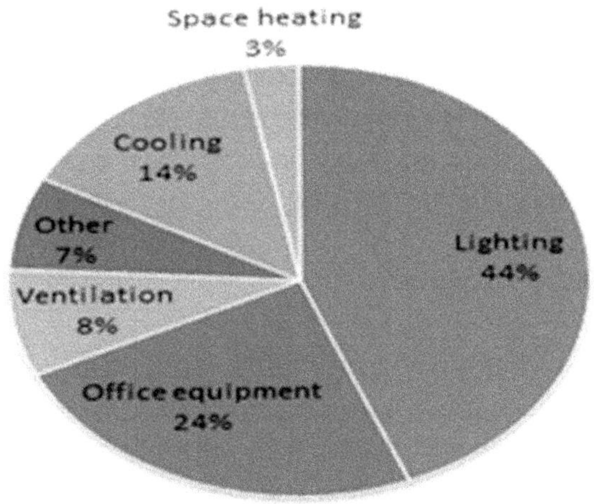

- These Gate modules will be working in collaboration with each other to compute the desired output for this they will communicate to a Central Unit which will, in turn, take required actions.
- This Central Unit is the Raspberry Pi which makes decisions based on predetermined factors and user made policies.

In figure 11, the flow of a smart energy system is shown.

The directional sensor module, beam breaker and Arduino Nano interacts with the gate module where sensor modules help in identifying in which direction an object passed and whether it had legs or not. Directional Sensor used to find the direction, in which any passing object moved. It is based on the principle of IR beam break detection. Obstacle sensor used to find which part of the gate was occupied at any instance of time. It also helps in recognizing whether the object is moving on legs or not. Beam

Figure 11.

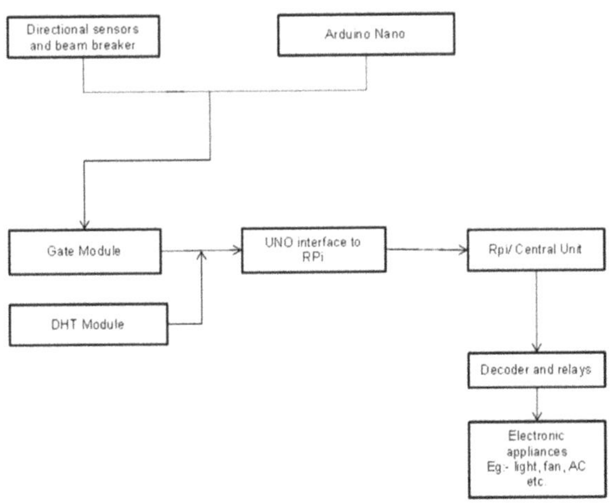

breakers can be used to identify the direction of motion. Arduino Nano used for taking data from each gate module (NANO) and forwarding it to RPi. Gate modules used to compute the desired output for this system. For this, they communicate to a central unit which will, in turn, take required actions. Then gate module and DHT module interacts with UNO interface to RPi, then it goes to a decoder. RPi is the central unit which makes a decision based on predetermined factors and user made policies. A decoder is used to convert incoming binary input into the index of a relay which is to be switched on or off.

In figure 12, 13, 14 and 15 codes for smart energy sensor is shown.

Code for Direction Sensor Module

The system can detect entry and exit as well as the number of people with high accuracy in the premises and take actions like switching appliances on, off or set them at a given rate. Profiles according to residents can be created. It is also capable of detecting intrusion and alarming the correct authorities. Nearly up to 70% of the wasted electricity is saved in a medium-sized setup.

Figure 12.

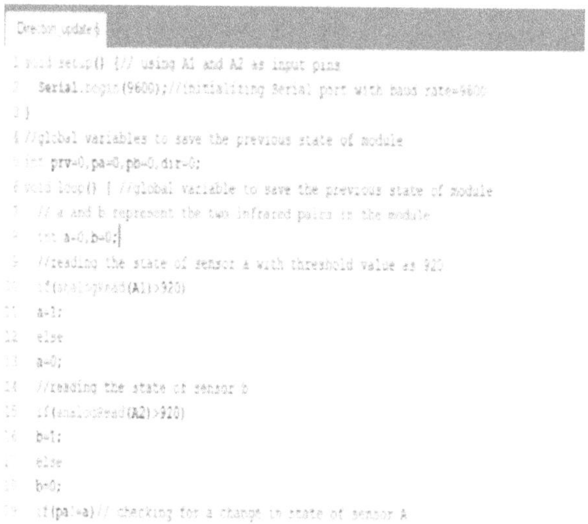

```
void setup() {// using A1 and A2 as input pins
  Serial.begin(9600);//initializing Serial port with baud rate=9600
}
//global variables to save the previous state of module
int prv=0,pa=0,pb=0,dir=0;
void loop() { //global variable to save the previous state of module
  // a and b represent the two infrared pairs in the module
  int a=0,b=0;
  //reading the state of sensor a with threshold value as 920
  if(analogRead(A1)>920)
  a=1;
  else
  a=0;
  //reading the state of sensor b
  if(analogRead(A2)>920)
  b=1;
  else
  b=0;
  if(pa!=a)// checking for a change in state of sensor A
```

Figure 13.

Figure 14. Code for DHT sensor module

Figure 15. Code for interfacing gates to RPi

Smart Home

It is one of the best applications of IoT. IoT technology adds smartness to their home. For example, if you forget to lock your garage then the sensor will sense that there is no activity in the garage will automatically lock the garage and send a message to your smartphone. By doing this it will make human life easier and also provide comfort and security.

Assistants like Siri, Google assistant, and Alexa make it easier to control smart home product by simply speaking commands. It helps in controlling the light, fan, heat, etc.

With our virtual assistant, we can set up your home a number of ways to make it smarter (Dempsey, (2017).

Wearables

Wearables devices like rings, watches etc. those devices are connected with the internet and human beings can wear all this very easily, this wearables carry out information related to bank, health, etc. if any fraud or issue related to this is found a message is sent to the user's smartphone and by this the user becomes more informed. In wearables devices, sensors and software are installed so it can collect data and information about the user.

Connected Cars

It is one of the best applications in the automobile. It brings a revolution in automobiles industry. Giant automobile companies are working on the connected car. Top Brands like BMW, Apple, and Google are also working on the connected car and are trying to bring a revolution in the automobile industry. Connected car optimizes its operation and maintenance and also provide solace to passengers with the help of sensors and internet connectivity.

Smart car brings revolutionary change for us to drive and stay in touch with the world around at the same time.

Kaa offers a stack of plug and plays IoT components. It provides streamline the development of connected car applications by times and ensures smooth integration between modules of the connected car within a secure cloud environment.

Kaa is highly scalable and with this, it is easy to enable new services over the air.

With the help of Kaa software followings things are easily done like:-

- Connected car system monitoring
- Smart driving assistance
- Cloud-based infotainment solutions
- Automated emergency management
- Real-time fleet management
- Driver safety systems

Smart Cities

It is one of powerful application of IoT, it brings a big revolution in cities and population of cites. "Smart city includes smart surveillance, automated transportation, smarter energy management system, water distribution, urban security and environmental monitoring all are examples of IoT applications for smart cities" (Zanella et al. 2014, p 22-32). Smart cities application is a combination of IoT applications. IoT solves various problems of cities like pollution, scarcity, traffic, etc. Smart cities also include WiFi enabled cities and the availability of internet connectivity 24*7.

IoT in Agriculture

IoT in agriculture brings tremendous change in agriculture, it increases its productivity and also provides comfort to farmers by reducing their manual work. As everyone knows there is huge growth in population and the demand for food also increases. With the help of IoT technology, productivity, and availability of food increases. Smart farming is one of the fastest growing fields of IoT. "It senses soil moisture and nutrients, controlling water usage for plant growth and determining custom fertilizer are some simple uses of IoT in agriculture".

From the last decades, there is a tremendous transformation in farming, it has become more industrialized and technology driven. By using various smart agriculture gadgets, productivity increases, and efficiency increases.

IoT can improve agriculture because

- Millions of Data collected by smart agriculture sensors.
- Better control over the internal processes and, as a result, lower production risk.
- Cost management and waste reduction
- Increased business efficiency through process automation
- Enhanced product quality and volumes

There are many sensors used in IoT applications and in agriculture.

Monitoring of Climate Conditions

Most popular agricultural gadgets are weather stations; it consists of various smart farming sensors. It monitors the weather conditions and also helps in choosing appropriate crops for farming according to weather condition, so it helps in surplus productivity.

Some examples of such agricultural device are **allMETEO, Pycno,** etc.

Greenhouse Automation

Weather stations automatically adjust the weather condition to match the parameter. It works on the same principle.

Example: GreenIQ product also uses smart agricultural sensors.

Crop Management

IoT sensors also help in managing the crop just like weather stations, its sensor is also placed in the field for collecting information about the crop and helping to manage the crop from a different perspective.

Example: Arable and Semios

Cattle Monitoring and Management

Just like crop monitoring, there are IoT agriculture sensors that can be attached to the animals on a farm to monitor their health and log performances.

Example: SCR by Allflex and Cowlar use smart agriculture sensor.

These sensors provide information about temperature, health, activity, and nutrition insights on each individual cow as well as collect information about the bird.

IoT in Healthcare

If IoT comes in healthcare it brings revolution in medical science. Research shows that in the coming year there will be a huge growth in healthcare and it is only because of IoT. For example, if a person meets with a heart attack then its medical condition is sent to the nearby connected hospital, then an ambulance is sent to the patient and the patient is bought to the hospital. Another example is that if a person meets with an accident then report/ notification will be sent to a nearby police station and after that, the person is sent to the hospital. That time we only use a few % of IoT, here are a lot more uses and growth in this technology. The main aim of IoT in healthcare is to empower people by providing a healthier and long life by wearing connected devices. The collected data will help in the personalized analysis of an individual's health and provide tailor-made strategies to combat illness.

Smart Airports

IoT applications are seen in each and every area. It provides ease and comfort in the airport area too. Using IoT technologies at the airport adds smartness to the airport area that is it is named as the smart airport.

IoT technologies have the potential of making air travel experience much easier and comfortable. A recent report of Deloitte shows that IoT could improve passengers experience while simultaneously increasing revenue for the operators.

Passengers basically face two types of process stage while using airport service that is pre-departure and posts arrival. Using air service passenger have to reach airport almost two hours before the departure of their respective flight and during this time passenger gone through numerous checking processes. So by implementing or using IoT technologies in airport area changes will be done for the passenger's comfort and ease and successful in providing the much better experience.

By using advanced biometric systems at the airport, the person can be easily authenticated and gets the entry in the airport. By using digital IoT beacons at airport navigation become easier, by strategically placing it throughout the airport. The digital beacon uses Bluetooth technology, which is responsible for providing or getting real-time information on parking availability, airport shuttle schedules, restaurant table availability, and shopping stores.

Top of Form

IoT-enabled airports also assist passengers in check-in procedures, take-off and more with digital beacon-linked maps for easier routing and wait-time updates through SMS notifications. The smart airport navigation experience frees up the passengers' time for them to relax and encourages them to spend more time eating, shopping and doing other leisure activities, thereby generating greater revenue for the airport.

In today era, nearly all commercial airports provide free Wi-Fi facility for passengers. By applying IoT technology at the airport provide a new level of experience to the passengers. By implementing it at airport transforms airports to intelligent eco-systems by offering location-based services to the travelers. Lots of expenses made on lost luggage that causes domestic flights up to $3,300 as expenses and up to $1,742 for international flights. With the help of RFID which is used in IoT, tracking of the product becomes easier and less expensive. With the help of IoT technologies, passengers are able to get real-time location information about their luggage. At each point of the journey, passengers can track their luggage.

The massive expense of flight delays can also be addressed through IoT implementation, with schemes for weather-related delays available 24/7. Airports also can survey/monitor conditions for take-offs, approach and landings by connecting simple weather monitor beacons across runways to the airport's larger network. Connected surveillance also provides precise and timely information for passenger-centric apps as well, helping passengers understand the reasons for delays or stoppages delays in their journey better.

Implementation of IoT technology at airline makes airport smart and responsible for creating comfort to passengers, smooth working of airlines and it also increases the overall performance of airport authority. "Linking each device around an airport through a robust IoT cellular network allows the airport to conduct in-depth data collection and performance tracking for improved operations and provides better statistical summaries for airport management".

In table 9 and 10 vulnerabilities of IoT application and funding details of various applications are shown as follows:-

Table 11 shows a comparison between Machine 2 Machine (M2M) and IoT. These two technologies are very close to each other so comparisons are established on the basis of the different domain like technology, connectivity, target market, value chain, business model, and delivery model (Cao et al. 2016).

Table 9. Vulnerabilities of IoT applications

Applications of IoT	Problems (Weak Points)
Smart Home	• limited AAA services • security in web-based interfaces • lack of cryptographic support
Smart City	• limited privacy • insecure cloud connectivity • insecure device connectivity • lack of cryptographic support
Smart Health	• insecure mobile connectivity • limited availability, limited privacy • insecure cloud connectivity • insecure device connectivity

**Note: AAA stands for authentication, authorization, and accounting.

Table 10. Funding Details of various application

Applications	Total Funding	Startup	Funded Startup	Status
Agriculture	$ 960 M	61	23	More mature
Automotive	$ 728 M	49	28	More Mature
City and Buildings	$ 1.3 B	64	40	More mature
Corporate	$ 2.3 B	26	18	More mature
Industrial	$ 536 M	24	16	Newer
Retail	$ 320 M	30	18	More mature
Security	$ 338 M	17	10	Newer

Table 11. Comparisons between M2M and internet of things

DOMAIN	M2M	INTERNET OF THINGS
Terminology	More frequently used by telcos	More frequently used by the broader technology industry
Connectivity	Applications are oriented to telco networks (fixed, mobile or satellite)	Greater variety of hybrid connectivity approaches, combining wide- area with short-range networking technologies (Bluetooth, WiFi, ZigBee, etc.)
Target Market	Primarily for B2B applications such as asset tracking, fleet management, point of sale terminals, etc.	Broader market applicability that includes consumer applications (B2B, B2B2C)
Value Chain	Traditional, linear value chain	Multi-sided value chain that accommodates several users of IoT application data
Business Model/ Value Proposition	Supplier-driven ("push") models focusing on operational efficiency	Extended and multi-sided business models focusing on innovation, Emergence of consumer self-serve (data pull) business models
Delivery Model	Custom designs, reliant on systems integration expertise	Increasing tendency towards pre-configured or plug and play components to build IoT applications.

RESEARCH TOPICS IN INTERNET OF THINGS

There are various areas for research in the Internet of Things which is the very emerging and wide scope of studies like Wireless Sensor Networks, Cloud Computing, Big Data, Distributed Systems, Mobile Computing, Artificial Intelligence, Machine Learning, Security, Semantic Web, RFID, and Smart Cities.

Wireless Sensor Networks

"Wireless sensor networks consisting of the spatially distributed autonomous device using sensors to monitor physical or environmental conditions". It is a wide area for research. There is so many research topic or area in WSN in Circuits (low power, on- chip sensor MBMS, Miniaturized size, low cost imaging, energy scavenging), Wireless (Multi–hop routing, energy efficient, very low duty cycle, efficient MAC, cooperative communication), Networking (self – configuration, scalability, ad-hoc, hybrid networks, distributed routing and scheduling), and Computation (Processing power, embedded software, collaborative processing methods, onboard processing) (Bellavista et al. 2013). There are various tools used in carrying out simulation in wireless sensor networks. For example SensorSim, TOSSIM, GloMoSim, TOSSF, Qualnet, OPNET, EmStar, SENS, J-Sim, Diego, NS-2 and NS-3, Shawn, GTSNets, CNET, etc.

Cloud Computing

Cloud computing is an emerging area for research. It is one of the most important areas for research which is related to IoT. It means storing and accessing data over the internet instead of our personal computer storage device. Cloud computing basically represents infrastructure. There are various challenges in cloud computing like cost, service provider reliability, downtime, password security, data privacy, vendor lock-in, etc. These challenges provide the area for research for resolving these issues. There are various tools available for carrying out simulation are KVM, Delta cloud, Eucalyptus, Open Nebula, etc.

Big Data

It is a broad area for research work where IoT plays a vital role. This term basically used to denote large scale dataset or complex dataset where the traditional method inadequate to handle that kind of large dataset. Big data basically represents vast content which is difficult to handle (Sezer et al. 2018).

Datasets grow very fastly because of so many reasons like scientific research, educational field, medical field, etc. There are so many challenges in big data like capturing data, data storage, data analysis, search sharing, transfer, visualization, querying, updating and information privacy, etc. Big data is a part of the data mining field. There are so many tools of big data like Hadoop. **Hadoop** is a framework used for the storage and processing of data. Others tools are HDFS, MapReduce, SAS, SPSS, RapidMiner, KNIME, ORANGE, MySQL, etc.

Some of the companies of big data where research work on big data is carried out:

- Alpine data labs, an analytics interface working with Apache Hadoop and Big Data.
- CBIG Consulting, a consulting group that specializes in business intelligence (BI).

It has application in various fields like government, media, healthcare, etc.

Distributed Systems

It is also a research area which is related to IoT. A lot of research is done in a distributed system and a lot more is running. A distributed system is a collection of computers that are spatially separated and do not share a common memory. The processes executing on this computer communicate with one another by exchanging messages over communications channels. There are various tools available for carrying out simulation these are Zabbix, Nagios, Ganglia, Hyperic, ManageEngine AppManager, IBM SmartCloud Monitoring, HP Operations Manager, AppDynamics, Datadog, Graphite, New Relic, Prometheus, Riemann, Sensu, and TICK by InfluxData.

Mobile Computing

It is a popular area for research work. IoT is also concerned with this. Mobile Computing defined as the computing environment over physical mobility. It allows a user to perform a task from anywhere. There are vast areas for research in mobile computing. There are various challenges in mobile computing like diversity in portable device types the users are operating (hardware and software variation), Wireless causes network variations, and mobility causes network connection or disconnection. It has various applications in vehicles, emergencies, business, replacement of wired networks, infotainment, and location dependent services, etc. There are various tools available for a simulation like MATLAB, NS-2, and NS-3, etc.

Artificial Intelligence

It is an emerging area for research. It is the most important area for research which is concerned with IoT. "Artificial is defined as the science of making computers, do things that require intelligence like humans". It is a branch of computer science that deals with the automation of intelligent behaviors. There are various open source tools available for simulation in artificial intelligent these are TensorFlow, Keras, Scikit learns, Microsoft Cognitive Toolkit, Theano, Caffe.

Machine Learning

It is emerging and wide is an area for research. It is an application of artificial intelligence. It provides facility and the ability to automatically learn and improve from experience without being explicitly programmed. It basically focuses on the development of computer programs that can access data and use it to learn from themselves. Tools required for carrying out simulation are Weka, Rapid Miner, KNIME, ORANGE, SAS, SPSS, and NIME, etc.

Security

Security is one of the major challenges of an IoT. So it provides a lot more area for research. The word security means the state of being safe from dangerous happening or act. There are basically two types of attack that is an active attack and passive attack. The passive attack consists of the release of message content and traffic analysis. An active attack consists of masquerade, modification, and fabrication. There are various tools are used for carrying simulation these are Wireshark, Metasploit, Nessus, Aircrack, Snort, Cain, and Abel, BackTrack, Netcat, etc.

Semantic Web

Nowadays, the semantic web is emerging and wide area for research work. The Semantic Web is just an extension of the World Wide Web. It is the branch of linguistics and logic concerned with meaning. It is the vision for the future of the web. The semantic web is of two types that is logical semantics and lexical semantics. Lot more research is running in this sector. It is also called the web of data. It works on structured data. Its major technologies are RFID, URI, RQL, XML and many more. Main goals of the semantic web are to spend less time searching, looking at things which is unnecessary, and less time in explaining what need from the computer (Ding et al. 2014). Major challenges of the semantic web are vastness, vagueness, uncertainty, inconsistent, and deceit. Its applications are seen in the digital library, the legal sector, and telecommunication sector, etc. There are various tools available for carrying out simulation in the semantic web are as following that is Pellet, RDFBroker, PlatypusWiki, SEAGENT Platform, GFacet, SWOOKI, Paggr, HyperGraphDB, EasyRDF, CEL, DBin, MashQL, KOMMA, PoolParty, AceRules, YARS.

RFID

It is one of the major areas for research which is concerned with IoT. It stands for radio frequency identification. It is used in transport and logistic. It is a backbone of supply chain management. It is one of the major technologies of IoT. It is a vast area for research work. Its major challenges related to implementation and standardization, expensive tags, cost of installation, new technology not available to all suppliers, the threat of security from competitors and many more. Its simulation is carried out in MATLAB, it has various toolbox which is concerned with RFID.

Smart Cities

It is also a trending area for research. It is a major project for research work. The government also involved in this. The Smart Cities means all facility available in the cities 24*7 wifi facility, pollution free city, water facility, traffic-free city, Availability of lighting in the city, facilities of metros and many more(Dohler et al. (2011). The smart cities face lots of challenges like technology challenges with coverage and capacity, digital security, legislation and policies, Lack of confidence or reluctance in the citizen, Existing infrastructure for energy, water, and transportations (Walravens & Ballon, 2013). These challenges provide the scope of research. Some tools used in the smart city are NS-2 and NS-3, MATLAB, etc.

FUTURE SCOPE

This technology is highly acceptable by a human. It is sure to make tremendous growth in the coming future. It can also be termed as the internet of everything. IoT is growing rapidly, in near future IoT will be seen everywhere. There are a lot of applications developed based on IoT technology like agriculture and farming, daily life, shopping and tracking, and health, etc.

Future of IoT is seen in daily life, traffic issues, production, logistics, resource and power control and retailing. Basically, there are three major areas of the IoT that is enterprise, home and, government. EIOT being the largest among them, by 2019 the EIOT sector is estimated to account for nearly 40% or 9.1 billion devices.

CONCLUSION

In this chapter, the Internet of Things is reviewed critically. It is found that IoT is a highly accepted technology. It is also found that in coming future about 1 trillion devices will be connected to the internet. It is observed that IoT delivers smart, connected and secure customer experiences. The conclusion is also made that in coming future IoT will be seen everywhere.

REFERENCES

Amaral, L., Tiburski, R., Matos, E., & Hessel, F. (2015). Cooperative middleware platform as a service for Internet of Things applications. *Proceedings of the 30th Annual ACM Symposium on Applied Computing*. 10.1145/2695664.2695799

Bellavista, P., Cardone, G., Corradi, A., & Foschini, L. (2013). Convergence of MANET and WSN in IoT urban scenarios. *IEEE Sensors Journal*, *13*(10), 3558–3567. doi:10.1109/JSEN.2013.2272099

Borgia, E. (2014). The internet of things vision: Key features, applications and open issues. *Computer Communications*, *54*, 1–31. doi:10.1016/j.comcom.2014.09.008

Cao, Y., Jiang, T., & Han, Z. (2016). A Survey of Emerging M2M Systems: Context, Task, and Objective. *IEEE Internet of Things Journal, 3*(6), 1246–1258. doi:10.1109/JIOT.2016.2582540

Cuff, D., Hansen, M., & Kang, J. (2008). Urban sensing: Out of the woods. *Communications of the ACM, 51*(3), 24–33. doi:10.1145/1325555.1325562

Da Cruz, M. A. A., Rodrigues, J. J. P. C., Al-Muhtadi, J., Korotaev, V. V., & de Albuquerque, V. H. C. (2018). A Reference Model for Internet of Things Middleware. *IEEE Internet of Things Journal, 5*(2), 871–883. doi:10.1109/JIOT.2018.2796561

Dempsey, P. (2017). The Teardown: Google Home personal assistant. *Engineering Technology, 12*(3), 80–81. doi:10.1049/et.2017.0330

Dohler, M., Vilajosana, I., Vilajosana, X., & Llosa, J. (2011). Smart Cities: An action plan. *Proceedings Barcelona Smart Cities Congress*, 1–6.

Firner, B., Moore, R. S., Howard, R., Martin, R. P., & Zhang, Y. (2011). Poster: Smart buildings, sensor networks, and the internet of things. *Proceedings of the 9th ACM Conference on Embedded Networked Sensor Systems*, 337–338.

Sezer, O. B., Dogdu, E., & Ozbayoglu, A. M. (2018). Context-Aware Computing, Learning, and Big Data in Internet of Things: A Survey. *IEEE Internet of Things Journal, 5*(1), 1–27. doi:10.1109/JIOT.2017.2773600

Somayya, M., Ramaswamy, R., & Siddharth, T. (2015). Internet of Things (IoT): A Literature review. *Journal of Computer and Communication, 3*(05), 164–173. doi:10.4236/jcc.2015.35021

Walravens, N., & Ballon, P. (2013). Platform business models for smart cities: From control and value to governance and public value. *IEEE Communications Magazine, 51*(6), 72–79. doi:10.1109/MCOM.2013.6525598

Wu, Q., Ding, G., Xu, Y., Feng, S., Du, Z., Wang, J., & Long, K. (2014). Cognitive Internet of Things: A New Paradigm beyond Connection. *IEEE Internet of Things Journal, 1*(2), 129–143. doi:10.1109/JIOT.2014.2311513

Wu, S., Rendall, J. B., Smith, M. J., Zhu, S., Xu, J., Wang, H., & Qin, P. (2017). Survey on Prediction Algorithms in Smart Homes. *IEEE Internet of Things Journal, 4*(3), 636–644. doi:10.1109/JIOT.2017.2668061

Yang, Y., Wu, L., Yin, G., Li, L., & Zhao, H. (2017). A Survey on Security and Privacy Issues in Internet-of-Things. *IEEE Internet of Things Journal, 4*(5), 1250–1258. doi:10.1109/JIOT.2017.2694844

Zanella, A., Bui, N., Castellani, A., Vangelista, L., & Zorzi, M. (2014). Internet of Things for Smart Cities. *IEEE Internet of Things Journal, 1*(1), 22–32. doi:10.1109/JIOT.2014.2306328

Chapter 13

IoT-Based Smart and Secure Health Monitoring System

Parul Verma
Amity University, Lucknow, India

Brijesh Khandelwal
Amity University, Raipur, India

ABSTRACT

IoT is not a new keyword. The latest gradations in the technology are now leaving no stone unturned to make IoT a part of every domain. Be it education, agriculture, transportation, business, or healthcare, every domain is now ready to exploit the benefits of IoT. By end-to-end connectivity and simultaneous monitoring and reporting, it is improving the efficiency of the healthcare systems and thus improving the health of patients. The chapter focuses on how IoT can be integrated with healthcare systems and draw maximum benefits from its ubiquitous presence. The chapter also covers various security concerns of an IoT-based healthcare system and their suggested solutions to overcome those concerns.

INTRODUCTION

According to Kevin Ashton, as published in RFID journal in year 2009, almost a decade after his introduction to phrase IoT, the IoT can be better understood in its reality in a way where, without taking human assistance, if all possible data can be gathered with the help of computers, and making it possible to know all about everything by tracking, recording, and sensing etc. Such mechanism would certainly reduce the cost, loss, and waste too. By counting, tracking, sensing things would be in a condition at all time to know about their good time for replacement, repairing, and recalling etc. S. Sinha (2018)

It is high time; we need to understand computers can be empowered to be capable of getting things done as per the requirement generated by gathering information. The computers or devices would be able to sense in possible ways viz., hear, see, smell, touch, etc. and confirms the actions' initiation accordingly.

DOI: 10.4018/978-1-5225-9574-8.ch013

If we conclude Ashton words, it means that making our devices and the users smart by simply sharing each and every whereabouts by the help of connecting devices and make information available for processing. This is possible only by connecting all such devices by some means in order to share data and information.

The IoT is a huge pool of interconnected devices, sensors, vehicles, embedded software and many other things. Such things are connected seamlessly using wired/wire-less connection media. Each such device/sensor/thing is treated as a node which can be configured by itself. This globally connected network of things/nodes works on infrastructure that is dynamic in nature and well supported by the communication technologies like - GSM– Wi-Fi, RFID, GPRS etc. Verma P. et. al. (2017)

IoT is not limited to a particular domain. It is vast and its usage is also versatile. Its usage can be justified in various domains like HealthCare systems, Transportation, Education, Energy Management, Agriculture and Business to name a few. IoT has been used in versatile domain and had justified its role in each domain. The use of IoT is no more restricted to only few domains. Nowadays if we turn around we can see use of IoT in every domain in some or other way. The chapter will discuss about the usage of IoT in Health Care Systems and its security concerns.

IoT and Health Care Systems in Current Scenario

In Health Care Systems earlier there was AI which was supposed to help doctors and patients to communicate in a better way. Doctors and Patients are connected through a device which helps in better diagnosis by doctors. AI enabled Chatbots communicate to the patients and suggest primary medical help to them.

IoT is one step ahead to all what AI enabled bots had offered. Healthcare IoT can be understood as a coupled mechanism of software applications and medical instruments capable to communicating with numerous healthcare/ medical IT systems. The objective of IoT is to ascertain a better health in a record time in the competitive environment, IoT in healthcare unites patients and healthcare support providers via software and technologies. The Internet of Things for healthcare industry is an example of technology that enables physicians/ medical practitioners to experience healthier relationships with patients and eventually it nurture patients' involvements and establishing their trust in this mechanism.

Various researchers had worked and showed that remote health monitoring is not a dream now. IoT has shown promising results in many other fields by remote monitoring and data collection and processing. Similarly IoT can be used for the similar purpose of health monitoring of individuals and reporting their health status to the caretaker parties like doctors, health care centers or emergency services.

Such remote monitoring systems are quite helpful for non-critical patients where doctors or health centers can guide them by getting updated status of their health based on various parameters. These systems can be a boon to the elderly people where people living in the rural areas have critical issue of reachability to the doctors or hospitals.

The vision of Smart Health Care System is to cater the solutions for following two needful areas-

1. Improvised disease management systems which may result for better experience of patients
2. Reducing the healthcare associated costs so that it's affordable for wider demographics of patients

As per the recent literature published the global growth of Internet of Things (IoT) in the world's healthcare market is expected to be $158.07 billion in year 2022 as compared to $41.22 billion in year 2017 with a Compound Annual Growth Rate of 30.80%. This projected phenomenal growth is significant and perhaps that is why pharmaceutical products are now being meticulously planned, manufactured in good volume and quality and also been used, stored and consumed by the good variety of demographics. S Sivagami & D Revathy (2016)

Remote monitoring and controlling of IoT enabled healthcare devices/ machines have been emerged as most demanding technologies in the health care sector. IoT based system is considered as one of the sought for technology which has empowered physicians not only to deliver superlative care to their patients but also guarantees safe and healthy patients. Eventually, it has resulted in reinforcement of patients' satisfaction. Now, patients are finding it more convenient to get convenient engagement with physicians and to have easy interactions with doctors/ practioners. IoT enabled healthcare systems have made things more efficient. In addition to this, overstay in hospitals and re-admissions like bottlenecks of prevailing systems have also been checked and reduced. IoT enabled healthcare devices and systems as a whole resulted into reducing healthcare costs extensively and improving treatment results.

IoT is no doubt benefitting the way the whole network of Health and Care industry works. The benefits are manifolds and there are various application areas where IoT is directly or indirectly impacting the way traditional health care systems works. Monitoring systems helps doctors in many ways by early symptoms predictions and also sometimes emergency situations can also be predicted. IoT devices are playing quite promising role in patient care and offers lot of opportunities for the present health care systems. These devices are vulnerable to various security threats and the risk factor for such devices is comparably high in health care zone than consumer zone. Patients share their health related data through multiple devices to the centralized health care system. The data is being used by hospital management at various levels, hence the security of such data is crucial and since the IoT devices are quite vulnerable the chances of data breach is also higher.

IoT based Health Care Monitoring Systems and Devices

Besides other industries healthcare is the one which has adopted IoT and its benefits at the first step. Integration of IoT with various health devices is giving promising results. The effectiveness these devices bring high value of health care especially in elderly patients suffering from some chronic disease and require regular attention. An estimation says that IoT healthcare solutions are a new buzz and by 2025 the total spend amount will reach to $1 trillion.

KAA

Kaa is one of the leading enterprises based on IoT platform. It allows Original Equipment Management (OEM) and healthcare systems to implement cross device communication connectivity and also implement latest and smart features in medical devices and health care system. Kaa functionality can be integrated with the hardware and software devices produced by Healthcare companies in order to meet their IoT objectives in very less expenses.

Kaa had made a strong impact on scalability and flexibility of IoT based health care solutions. The flexibility feature lets you expand the size of Kaa cluster, so that as many number of devices can be supported with the assurance of optimal performance. The use risk of downtime can be minimized by available deployment of Kaa. Security is of the prime importance in IoT health care devices. The best part of Kaa is that it makes use of modern security protocols like TLS supports it in secure data exchange between devices.

OpenAPS

Dana Lewis and her husband Scott Leibrand had hacked two health devices based on IoT . The devices hacked by them were Insulin pump and Continuous Glucose Monitor (CGM). They developed their own software which utilizes the data from CGM and modified the quantity of insulin delivered by Dana Lewis insulin pump. In Austin, during OSCon it was observed that more than 50 people are using Open Source Software for the modifications in their own health device. Western PA (2018)

Cancer Treatment through Activity Trackers

This IoT device is of great importance in treatment of cancer patients. The Memorial Sloan Kettering Cancer Center (MSK) and cloud research firm Medidata are collecting data from task trackers in order to collect information regarding the living style of patients that are being gone under treatment for multiple myeloma. Patients are supposed to wear this activity tracker prior to their treatment for one week and later on for months till the treatment carry on. Digital Intelligence (2016)

The tracker logged down all information regarding the activities of patients like activity level, exhaustion and hunger etc, all such information is saved to Medidata's Patient Cloud ePRO app on their own mobile phones. The analysis of data collected through these activity trackers and wearables are being analyzed by various techniques and on the basis of results treatment can be upgraded in many circumstances. The device help in planning strategically treatment of the patients where after effects of a particular therapy can be monitored for further planning of their treatment.

Monitoring Inhalers

The IoT technology is gone far away in health care and they are used not only in spotting disease, besides that it also helps doctor in keeping the precise record of whether patients are following their specified treatment plan or not. The other side of this technology is that patients get reminders through mobile apps attached to the devices to check their devotion level towards treatment.

The pharmaceutical company Novartis along with Propeller Health and Qualcomm are trying to create connected inhaler. This can be used successfully for the patients of protracted disruptive pulmonic disease. Propeller's technology has been linked through sensors to its digital platform which records and conveys useful information.

1.2.5 Proteous Discover

Ingestible sensors are an example of digital medicine. It has been introduced by Proteus Digital Health. A study by WHO in 2003 shows that almost 50% of the prescribed medicines are not eaten as they should be. Proteus system is one of the solutions to decrease this percentage.

Proteous discover comprises of ingestible sensors, a small wearable sensor patch, an application on a mobile device and a provider portal. The application is claimed to give an ever seen in-sight of the patient's health parameters. It also reports about the medication effectiveness status which leads to take meaningful decisions regarding treatment of patient.

Google Smart Lens

Google has developed smart lens technology which is being used successfully by Alcon which is a part of Novartis. Smart lens comprises of in-built non-intrusive device. The glucose level of the patients is measured through their tears and the lens will further transfer the data to a mobile device.

AutoBed

It is quite painful to wait outside the emergency rooms. Sometimes it takes hours to complete. Mt. Sinai Medical Center in New York City used IoT devices to reduce the wait times for their emergency patients by 50%. They used AutoBed program which tries to solve the problem of hospital by managing the available beds.

AutoBed is an algorithm which utilizes the recommendation for the nurses who admit patients and the data produced in real time about the availability of hospital beds to find out the best possible match as per the requirements of the patient to be admitted. The nurses keep data in the form of electronic medical record which is vast that included data on gender too. The real time data is collected using devices like radio-frequency identification tags, infrared to name a few. The AutoBed is capable of monitoring 1200 beds on the basis of fifteen various parameters and can process eighty requests for bed.

Telehealth

Telehealth is a new wave in health care domain. Patients living in a remote area or elderly patients need not to visit hospital at a preliminary stage of illness. Telehealth is a solution for such patients. It is one of the most popular applications of IoT based health care systems. Telehealth facilitates long distance patient and clinician in many ways like contacting and care, advising patient, reminders to the patient and doctors too, educating patients, monitoring health of patient and remote admissions as well.

It helps in reducing number of visits to the hospital and also improves patient's quality of life by avoiding travel for them. Some patients have limited mobility and depend largely on public transportation; Telehealth can be one of the solutions for them.

Philips-e-Alert

The hospitals now require next-gen hardware and software. There are many devices in hospital whose improper working can be a huge risk for the patient's life. One of them is MRI system. Such equipment's are susceptible to various risks be it power breakdown or system failure- each one of them can be matter of life or death. A new IoT based device called e-Alert developed by Philips aims to resolve such issues.

e-Alert is a smart device which does not wait for causality to occur rather than it follows preemptive approach by monitoring medical hardware and it also alerts hospital staff members in prior if any such device reports problem. Recently Philips revealed this new product with the collaborative effort with OpenMarket.

Drug Monitoring

The advancement of technology is all set to touch new benchmarks in healthcare. It seems science fiction that upcoming pills will be sensitized by microscopic sensors which can transmit information to the external device connected to them in order to ensure prescribed dosage of a medicine.

The information passed to the external device can be useful in ensuring that patients are taking prescribed medication and in proper dosage and also suggesting new medications. Patients can also access that data through app and have a check on their performance and improvements.

Health Net Connect

The basic aim of HealthNet connect is to improve the accessibility of healthcare and also decrease cost for healthcare. The health net connect Telehealth resource provide a combined package of workflows, use-cases, process, education and technology to various sectors like schools, enterprises, hospitals, health care centers to name a few, whoever are keen in providing distant health care.

IoT is been a big help for many revolutionary treatments of chronic diseases. There is a combination of devices and technology that are helping in curing chronic diseases. Utilities like Fitbit use the IoT to monitor personal health — such information can be shared with a doctor to help solving chronic issues. Very recently Health Net Connect started a program for diabetes management to improve the clinical treatment and reducing cost of treatment for patients. Their program got success and they produced exciting results.

LITERATURE REVIEW

Recent advancements in IoT technology has opened up various opportunities for health care systems. Latest Biosensors and other IoT based medical devices are capable of monitoring various health related parameters and are setting new benchmarks in this field. This section will discuss work of various contributors in this field.

Bui N. and Zorzi M. (2011) mentioned that IoT can be the key enabler for health care applications working in a distributed manner. A. A. N. Shirehjini, A. Yassine et. al. 2012 also used passive RFID for locating equipment in hospitals. Li et. al. 2012 proposed a secured data access framework for patients to access their PHR (Patient Health Record) which are stored on servers.

A. Redondi,M. Chirico et. al. 2013 proposed the successful usage of WSN for various healthcare applications in nursing institutes. Caro De N. and Colitti W. (2013) They focused on CoAP and MQTT protocols which are light weight in nature and compared their quantitative and qualitative performance.

C. Occhiuzzi, C. Vallese et. al. (2014) developed RFID based system called NIGHT-Care. The system is supposed to monitor health status of disabled and elderly people during night.

D. De Donno, L. Catarinucci in 2014 proposed REMA which is used for monitoring and tracking of patients. In the 6LoWPAN standard and smart mobile communication techniques are combined to monitor the health condition of patients and provide several effective healthcare services. More in detail, the proposed solution makes use of WSN devices to measure photo plethysmo gram (PPG) signals and deliver them to a server through the Internet.

Donno De D., Catarinucci L., and. Tarricone L (2014) combined the standard 6LoWPAN and mobile communication technique to provide efficient healthcare services for the health monitoring of the patients.

Bezawada et al. 2015 proposed PASS (Pattern Aware Secure Search) which used the concept of symmetric key based approach for secure encrypted searching of health data in cloud environment. Khoi Manh N. and Ahlund C. (2015) discussed various issues related to network like low band width consumption, remote health monitoring and many more. They suggested IReHMo architecture which is capable of combining many automation sensors with IoT healthcare devices at the sensing layer level.

Li et. al. 2016 also proposed WBAN (Wireless Body Area Network) which uses secure key management and authentication scheme. All health data items generated from WBAN will be encrypted before transmission. Lounis et. al. 2016 proposed an architecture for wireless network for cloud based health care systems. They introduced a secured access control architecture which depends on CP-ABE (Ciphertext-Attribute Based Encrytpion).

Liao Y. (2016) proposed a precise I2O (in to out) human body path loss model which rely on a safe heterogeneous human body model. Mio et. al. 2016 developed multi keyword search based on various attributes. Their approach is used to search encrypted data of patient's health record.

Hu et. al. 2017 proposed a secure IoT based health monitoring system for elderly people which uses cloud computing. The system utilizes digital certificate, envelope, time-stamp and asymmetric encryption technology for the transmission of related health data in the system.

Salunke P. and Nerkar R. (2017) proposed a system which is valuable for elder people who have limitations for arriving to healthcare facilities and also for doctors who are puzzled because of patient load. Lu and Li 2018 introduced k-anonymity algorithm to maintain the security of the wearable health devices.

Various researchers have worked in the area of secure health care systems. Many of them have provided secure solutions for the processing and accessing of electronic health data in health care systems.

ARCHITECTURE OF IOT BASED HEALTH CARE SYSTEM

The complete architecture of Health Care System based on IoT can be divided into various layers. The overall architecture can be viewed as a four layered system where each and every layer had a different role to support the functioning of this system.

Innermost Layer (Wearable and Monitoring Devices)

The innermost layer of the architecture is wearable or monitoring devices. The whole Health Care System initiates its working from this layer itself. These devices are an integral part of the system and are used to monitor any significant changes in a regular pattern and issue an alert message in case of any imperfection in a pattern. There are various wearable and monitoring devices nowadays present in the market. All such devices are helping the end user in some or other way. Devices like ***Wearables, Activity Tracking, Weight Loss, Consultation with Virtual Doctor, Hydration Tool, BioSensors, Sensiotec Virtual Medical Assistant, Freestyle Libre Flash, Health Patch*** *are quite handy to the end users.*

Machine to Machine Data Transfer Layer

Next to innermost layer is a Data Transfer Layer. This layer requires a device named Multi Service IoT gateway. The purpose of this device is to amalgamate protocols for networking, and to help in managing storage and edging analytics on the data, and also smooth the progress of data flow between edging devices and the cloud in a secure manner.

IoT enabled systems are quiet complex with versatile devices and consumers of data as well. There is a need to unify various dissimilar devices over the network environment such as sensors, actuators, measuring devices, smart machines, automated human/machine interfaces, and the legacy systems for making data communications amongst consumers easy and comfortable.

There are various businesses which are looking forward to adopt IoT and its robustness but they don't have that technical expertise to implement it. Implementing an IoT infrastructure needs skilled professionals on protocols that can write TCP/IP stacks, operating system kernels or many other algorithms required. Besides that integration of the devices with the back-office business information system also requires expert professionals. In short, they need help to create a functional IoT system.

The Multi-Service Gateway Approach

Enterprises require such expertise to solve the complexity of the IoT system. Complexity of IoT systems can be solved by using a multi-service gateway approach. This is done by the integration of disparate which will ensure claimed better decision making by having access to essential and valuable data.

In a nutshell, with the multi-service gateway approach:

- Several related business relevant jobs are tackled and consolidated technically.
- Data delivery to destination point uses an open protocol efficiently segregating the data consumers and the data providers
- IT centric IoT enabled device application development by making use of a programmed framework to implement application layer business logic in smart edge devices having different protocols using gateways.

The Multi-service gateway is used for the complex systems and is used as an integration device for two different environments. It simplifies the working of the complex systems by using cloud environment for the integration purpose. These gateways can tackle bi-directional communication. The customers are capable of connecting their devices, actuators and sensors to the various enterprises with the help of Multi-Service Gateway. The gateway supports real time management of the data and the customers have their control on devices that are part of the field.

Third Layer (M2M Integration)

Next layer is of M2M integration. In such system sensors or devices monitor themselves and keep a track of each and every communication taken place. Hence removing the human intervention is avoided. Patient's vital signs drop can also be taken as good example where a device connected to the devices' cloud (private or public) may control vital oxygen and cater care additionally till healthcare experts reach on the spot of concern. With M2M based technology, the patient or consumers don't even have to be in the healthcare units to get any such care. Instead, all this can be delivered effectively to the patient at his/ her home. M2M integration is a new paradigm which supports the IoT systems whether it is a data transfer through wearable. In broader way one can say that M2M is a technology which integrates disparate networked devices and help them in communication and performing action without human intervention.

Following claimed actions/ tasks must be performed by an effective and efficient M2M integrated platform-

- Firstly, it should work like an operating system for the IoT, capable of ensuring transfer of device's data irrespective of programming language and operating system
- Secondly, it should offer the possible means to carry out effective thorough management of the devices in the concerned field
- And thirdly, it should flawlessly put together with the organizational computational world, using best practices of information technology approaches at the same time implementing optimal M2M based technologies. Andres R. (2013)

The basic aim of M2M technology is to assemble data received from sensors, devices, actuators and then transmit it to a public networks using public access methods like mobile network of Ethernet. M2M is quite different from SCADA(Supervisory Control and Data Acquisition is a control system architecture that uses computers, networked data communications and graphical user interfaces for high-level process supervisory management) or other remote monitoring tools as they use cost effective means of data transmission. The major constituents of M2M based systems incorporate sensors, RFID systems, cellular communications connection along with a software program which performs automatic computing which helps the particular device to understand data and statistics to go with decisions.

The popular M2M communication is telemetry in which there is an automated communication process to collect data from some remote location and transmit it to receiving equipment for monitoring. In early days of telemetry telephone line were used and later on it started using radio waves for the transmission of performance dimensions congregated from sensing gadgets from remote areas. The revolution in wireless and radio technology has increased and extended the working domain of telemetry with engineering technologies and science to electric meters and various connected devices through IoT. There are several ways the M2M is enhancing telemedicine, helping providers offer a quality of care that's potentially better than many in-person treatment modalities. Rouse M.(2018)

The perfect M2M based integration platform is more or less middleware that behaves like a governing platform or operating system for the Internet of Things. It is an in-between system of disseminated units and the software applications that make use of the data arriving from these IoT enabled devices. Any established communication amongst devices should be a two-way communication in nature, and should allow these kinds of software applications to manage the devices where ever required. The M2M based system must ensure the relocation of device's data irrespective of language, and governing operating system. It must finally provide accommodation for the complicated nature of M2M based projects.

Uppermost Layer

The uppermost layer can be considered as output layer or response layer. In IoT based health care system output may be in the form of some alert message or some reports. The uppermost layer will receive data through M2M integration and will be analyzed and processed based on some previous history or patterns. On the basis of such processing, alerts or reports must be delivered to the end users.

For example in a health care system patient's will be notified on a particular day by their wearable devices for their appointment with the doctor. The device will also notify for the medicine schedule of the patient. Old age patients are highly benefitted as they most of the time they do not remember to take their medicines on time and also their appointments with their doctors. In such cases these health care systems are really a boon for them. M2M integration provides range of benefits like social connectivity, health and fitness monitoring and tracking to name a few. Figure 1 shows the architecture of IoT based healthcare system.

SECURITY ASPECTS OF IOT BASED HEALTH CARE SYSTEMS

Security of IoT data is a critical issue from the collection point at the device level to the destination point. IT fraternity has a great concern as there are no such defined security standards for IoT based health care systems hence high risk factors are associated to IoT device and possibility of data breaches is high. The diversity of the various endpoints and absence of proper tools to manage them is a big challenge for the security team. The devices from different vendors require different methods for updation which puts a significant burden on IT teams who deals with security.

FDA took a step in defining common security standards for all medical devices FDA released a draft of "Postmarket Management of Cybersecurity in Medical Devices". Draft addresses various guidelines which manufacturers need to follow to address security risks associated to the IoT devices.

The more and more devices are getting connected in various hospitals and health care systems. Hospitals and health care centers must pay attention and should be very cautious about the security of these devices. The hospital management should be proactive and should implement full proof plan to secure the devices and protect them from any threat.

"There are so many benefits that come with these new connected devices," mentioned by Mike Nelson, vice president of a healthcare solutions at DigiCert, a security certification company in Lehi, Utah. "But they also present some new risks and vulnerabilities that as an industry we haven't, firmly dealt with

Figure 1. Architecture of IoT based healthcare systems

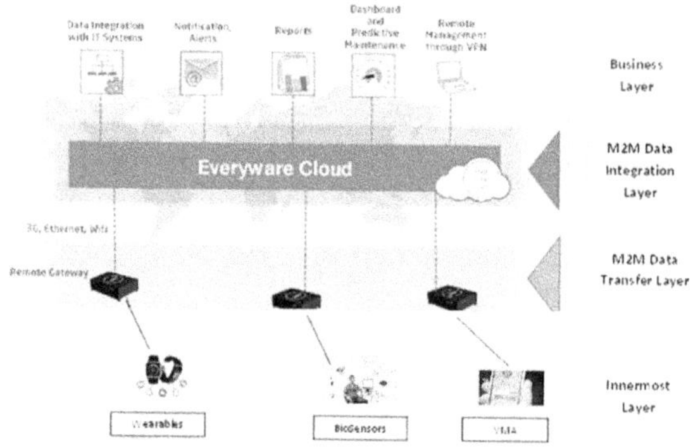

to this point". Those risks include possible harm to the patient's safety and health, loss of PHI (Patient Health Information) and unauthorized access to devices, Nelson said. Lee K. (2016)

Security Vulnerabilities in Health Care Devices and Systems

IoT health applications might be seen in hospitals or medical centres, in the home, or while on the move (tying into some aspect of a person's location through GPS tracking). Following are the diversified applications of IoT in Health Care Systems and their security vulnerabilities –

Wearables

There are numerous such type of devices available in the market and are used to monitor daily activities and their effects on human being. These devices store the metrics of various data like heartbeat, blood pressure, glucose level of the person. Such devices have their own specific features and way of collecting data. In general all such devices are used to track calorie burn and route used while running exercises through GPS. There is no proper security check on such data which is generated through these devices. GPS reveals the physical location of the person using wearable devices. Besides that other health related data can also be spoofed. For example suppose a fitness tracker display's a person's current heart rate and urges them to move quicker in order to increase their rate and burn more calories – an attacker able to spoof the heart rate (to be lower than the actual value) might be able to put a victim into an exercise state that increases their heart rate to dangerous or lethal levels.

Health Care Monitoring and Communication

Remote patient monitoring is one of the biggest advantage of IoT based health care systems. The people who live in remote areas or not physically fit to visit health care stations often can be monitored by the health care systems staff from remote locations. They can monitor their sugar levels, heart rate etc. and can give primary medication and suggestions. Inside hospitals also patients are allowed to roam here and there with the sensors connected to it. These sensors keep on updating health related data to the central location. These systems are in general support Bi-directional communications which allows to direct patients in the hospitals to report back at their times of medication also calling remote patients and advising them in adverse situations. However there is serious threat of unavailability of the concerned. The attackers can disrupt this bi-directional communication and this will pose a serious situation for the patients at remote areas as well as in the hospitals.

Drug Supply System

There are some IoT based health care system equipment's which works autonomously. These systems used to assess health status of a patient in a real time and after analysis disperse medication. For example – a remote, immobile patient might suffer from diabetes and require periodic insulin depending on blood sugar levels. Some connected devices might automatically monitor blood sugar levels and automatically inject insulin when necessary. However there can be very serious implications, if such systems are being attacked and can be a threat for the health of the patients and insufficient doses or overdoses of medication can be really deadly for the patients.

Health Implant Devices

A pacemaker which is a kind of implant device is quite common in health monitoring devices. With the improvement in technology the device is also upgraded its Wi-Fi enabled version. Such wireless version of pacemakers allows monitoring of health related data which can be used for diagnosis. The serious security threat associated to such devices. Any person who wishes to harm any patient can send dangerous commands and can stops or malfunction the working of such devices and can put patient in a serious situation.

e-Calling

In April 2018 e-calling is made mandatory for all the new cars in Europe. It is basically an innovative initiative which is meant to provide quick assistance to the cars involved in accident or crashing. When any eCall equipped vehicle crashes and automatic call goes to the nearest health centre in order to provide them first hand medical assistance to the victims of vehicle crash. In case passengers are seriously injured and are not able to speak then too minimum set of data is sent to the nearest emergency health centre. However any attacker can spoof the data sent to the emergency health centre and this will mislead the team to some other location where there is no causality.

Health Drones

Health Drones are also called as Air Ambulance. People living in remote areas who require urgent medical aid can be served using these Health Drones. It's already more than one year since the first trials of AirMule which can airlift two people and autonomously man oeuvre itself in locations where helicopters would not be able to manoeuvre and land. Attackers can make these drones unavailable or can hijack them and can make its illegal use.

General Security Vulnerabilities In Iot Devices

The IoT devices are quite vulnerable and hence chances of data spoofing, breaching and threat for the people is higher. There are numerous reasons behind the vulnerability of IoT devices. The reasons behind their vulnerabilities are quite many –

- IoT devices are the products which are commonly released and do not undergo regressive security testing phase.
- Some IoT health devices are quite small and also have limited capabilities. Like small chipsets of memory are not able to support larger upper limit of encryption. Hence it makes the device vulnerable as it gets quite easier to exploit the smaller upper limits of encryption.
- Due to small upper limits of small devices they are more vulnerable for attacks. In devices like implanted pacemaker or an insulin pump the installation of updates is not an easy task because the responsibility of update installation cannot be identified among manufacturers, vendors or hospital/medical centers.
- Communication is the core aspect of IoT health devices and in case communication is disrupted the patient's health is at the risk. The small size of these devices also hinders implementation of

any backup feature. In any case if device goes out of the network channel by no means it can report it to the healthcare system that it is out of action.

RECOMMENDATIONS FOR SECURE IOT BASED HEALTH CARE SYSTEMS

Recommended Consumer/Patient Actions

To minimize the risk, consumers/patients should take following actions:

- Consumers should not buy devices from unknown or less popular vendors. The devices should be bought from the reputable manufacturers which are globally known.
- Consumers should take care while selecting passwords so that it cannot be hacked. Consumers should read the license terms and conditions related to security properly in order to avoid any kind of legal issues in future.
- In case the health device is issued by the hospital or health care center, consumers must ensure about the security of device.
- Consumers should be careful while using IoT health devices online through mobile or computers.

Recommended Manufacturer Actions

The manufacturer of IoT devices should also take measures to improve the safety and security of devices. Following are the recommended actions for IoT health device manufacturers:

- A **Privacy Impact Assessment** (**PIA**) is a technique of identification and taking actions against the prospective security and privacy risks in new projects. PIA helps manufacturers to design effective and efficient process in order to reduce security risks. Many devices in spite of processing data transfer it to some Apps. In that case the data security of the particular app should be assessed.
- The implementation of Secure Software Development Lifecycle by manufacturers. This means implementing security at the time of designing of products is mandatory. Besides that not only implementing security but the testing and validation of security is also obligatory throughout the lifecycle of a product.
- The post product release phase of an S-SDLC is of great importance. The device should be engaged in rigorous security testing in this phase.
- Manufactures should try to implement standard security measure in order to maintain consistency.

Recommendations for IoT Devices Infrastructure

At infrastructure level following are the recommendations for security of health care systems-

Encryption of Data

Patient's health related data is of high importance hence it is advisable to use encryption techniques to encrypt useful data. Sensors and other biomedical devices used in IoT based health care systems have limited processing capabilities. The complicated encryption algorithm which demands for high computational power cannot be a solution. Hence the demand is to have a balanced security protocols to manage between data and the computational energy.

Authorized Access

Another way of making healthcare systems secure is by authorized access of various devices involved in the system. Every authorized user should have a certified key. Strong key management is an important aspect of cryptosystem. ABE (Attribute Based Encryption) is considered as the scalable option for such cryptosystems for flexible access management. Various technologies are being used at Access Control level- like ABE and CP-ABE (Ciphertext-Policy Attribute Based Encryption)

Trusted Third Party Authorization

The data stored on cloud servers by health care systems can be compromised by modification and deletion of data by unauthorized users. For security purpose trusted third party is introduced in the system to provide unbiased auditing results.

Searching Secure Data

In health care systems sensitive data of patients should be encrypted and data search should be avoided in a plain text manner. Hence the need of encrypted cloud search is raised. Few popularly used encryption techniques for data search are SSE (Searchable Symmetric Encryption) and PEKS (Public-key encryption) with keyword search. Following are the secure data search technology – Symmetric Key, APKS and CP-ABE

Anonymity of Data

Patient's data can be divided into explicit identifiers (personal details like – id, name, phone no.), quasi-identifiers (details lie – Age, date of birth etc) and privacy attributes (patient's health related data). Nowadays k-anonymity and l-diversity are the techniques used to maintain anonymity of the data in health care systems W Sun et. al. 2018

Authentication Layer

The authentication layer is required to be added for the correct identification of these connected medical devices. The data encryption is also required so that chances of data breaches should be minimized. The device that is sending data also needs to be authenticated. The authenticated users should only be able to send commands to the device.

Digital Certificate

Another way of securing these health devices is to provide authentication and authorization by issuing them a digital certificate. There should be one control and monitoring software to manage all such devices in a particular health care system. The system can then be configured to accept signals only from the authentic and authorized devices which are able to produce authentication certificate. This way a protected and secure network channel is established between the server and medical device. Some metadata information can be added to these digital certificates with the specification of the people allowed to handle certain device.

SOLUTIONS FOR SECURE IOT BASED HEALTH CARE SYSTEMS

HIMSS (Healthcare Information and Management Systems Society) focuses on the security of new technologies especially IoT devices security, related to health care system. The machine data is harnessed in order to ensure the security of the various critical systems in health care centers. The success of healthcare systems depends largely on their ability of harnessing machine data. The more intelligently systems harness this data it will result in better focus on care delivery systems.

Conventionally organizations assets and processes should be HIPAA-complaint, but the current scenario is that the healthcare industry is exploded with variety of IoT medical devices and you have patients roaming wearing health care devices hence it is getting difficult to manage all those devices and make them HIPAA-complaint.

The latest HITECH Act issues notification to the individuals to ensure that their health device is following security norms and information is transmitted in encrypted form. This helped in forming the updated HIPAA Breach Notification Rule to send notifications to the individuals who are in significant risk position in terms of finance, health or any other risk. The rule mentioned that notification should not be delayed and in any case it should not be late than 60 days after breach is discovered. (Hipa Journal)

Researchers mentioned that basic authentication of health care devices is must. This will limit the device access and firmware of the device is verified and proper scrutiny of communication among devices is being done. Secured booting of a device also ensures that device configuration is not tampered.

Hence the need of innovative security measures is required for such systems. CodeBlue is a prevalent healthcare project which has been developed at Harvard Sensor Network Lab. It has been deployed in hospital emergency care where various medical sensors are placed on various parts of patient's body. However the authors of CodeBlue admit lack of expected security and also explained its necessity. D. Malan et al (2004)

Following are the solutions suggested for the IoT based health care systems –

Secure REST(Representational State Transfer) API and OAuth

API's have significant roles in implementing secure health care systems. These API allow you to connect customers and application in a secure manner. Since APIs connect various devices in the IoT framework in the health care system hence it is necessary to deploy flexible and scalable API management. REST is basically a concept which is being used most popularly in the modern web as it uses JSON and XML for data transfer over HTTP.

REST is a suggested model for IoT based applications as each and every device can make its state available and can use standard methods to create, read and update that data. To fulfill the goal of IoT framework we need REST API for every device involved in a system. REST allows flowing data over internet protocols and also managing authorization.

Security integration with REST API makes it robust. REST API can be secured in many ways by using basic auth or OAuth but it should be stateless in nature. The authorization and authentication should be done for each API request on the server by authentication credentials rather than using cookies or sessions.

OAuth 2.0 is an open standard for authorization and authentication which does not provide any authentication protocol it provides a framework for authenticating decisions and mechanisms. Following are the major components of OAuth –

Resource Owner – It controls the data which is being exposed by the API.
Authorization Server – It issues, controls and revokes tokens in the OAuth system. It is also called as Security Token Service (STS)
Client – It is the application or a website or any other system that is requesting for data.
Resource Server – It exposes and stores/sends the data. It is basically an API.

OAuth follows implicit flow by delegating access to the resources. Client sends its request to the Authorization Server which request for authentic user-id and password which is further passed to the Authentication Server for its approval or denial. After its approval now client can request to access resource from Resource Server.

OAuth uses various security tokens like JWT tokens, CWT tokens, WS- Security tokens, legacy tokens to name a few. JWT stands for JSON web token which is a security token used for the secure deployment of various standard applications. The contents of JWT are secured by making use of JWS (JSON web signature) and JWE (JSON web encryption). However JWT is considered inefficient for IoT based systems which basically use low power radio technologies. To remove these anomalies CWT is being introduced known as Consent Web Token. The basic role of CWT is transmission of secure claims between two different parties. CWT uses a cryptographic key for signing the consents given by the users. The key guarantees the origin and validity of the consent. CWT grants security by using signatures to assure integrity. But it does not create a secure token themselves.

OAuth allows the secure communication and data exchange in client-server applications. The basic purpose of OAuth framework is to enable secure authorization on top of the transport layer.

MQTT (Message Queue Telemetry Transport)

It is popularly being used by IoT based applications. It is basically a light weighted message protocol which uses PUB/SUB operations for message transfers between client and servers. This protocol is considered ideal for IoT based health applications because it is smaller in size, uses low power, packets used in message passing are smaller in size and quite easy to implement. MQTT allows multiple clients to connect with broker at the same time for two basic purposes –

- Subscribe for a particular topic and then receiving updates on it.
- Publish messages related to a particular topic.

MQTT exchanges all information in the form of plain text which can be accessed and readable easily. Hence it is the utmost requirement to make MQTT secure and this can happen by following these steps-

- Using a private key (CA key)

- Using this private key produce a certificate (CA cert)
- Produce a certificate for Mosquitto MQTT server using private key
- Finally configuring Mosquitto MQTT so that it can use those certificates.

MQTT protocol is preferred over HTTP protocol because of fewer payloads of data; publish/subscribe architecture which uses small message size hence making it ideal for small size memory devices.

LoraWAN

It is Long range, low power Wide Area Network designed specifically for the IoT setup. It emphasizes on smart, secure and seamless bidirectional communication between various things connected. It uses varied frequency channels and data rates for end to end communication between devices. The data rates ranges from .3 to 50 kbps, hence various communication channels do not interfere with each other. It creates virtual channels to increase the capacity of the gateway. It uses Adaptive Data Rate (ADR) plan to improve the capacity of network and battery life of the devices. Adaptive data-rate lets LoRa Server control the data-rate and power of the device, so that it uses less airtime and less power to transmit the data.

LoRa is a secure way of connecting to the devices end to end. It addresses the security issue by providing encryption at various levels like network, application and device by making use of unique network key (EUI64), application key(EUI64) and a unique device specific key (EUI128). Each device is authenticated by a 128 bit AES key and a global unique identifier EUI-64. The security of LoRa is also strengthening by the use of AES cryptographic primitive along with CMAC2 for integrity and CTR3 for encryption.

It could be exploited well by the medical devices. The devices that record and send signals or reports to the health care system like hospital bed, blood glucose sensor, scale that reports your weight to name a few can utilize LoRaWAN as it utilizes wide spectrum of data rates and it is a complete solution for sending noiseless and secure data without any obstructions.

Symphony Link

Link Labs (2016,2017). It is an alternative synchronous protocol which uses LoRa at its physical layer. All messages send through Symphony Link are acknowledged and all messages can be of longer length. Though it is a secure and robust protocol but It is not energy efficient as lot of energy is consumed in synchronizing nodes frequently with the network. The increased power consumption and lack of deployment are the main limiting factors in using Symphony Link for healthcare purposes.

DTLS (Datagram Transport Security Layer)

E. Rescorla and N. Modadugu have suggested that end-to-end security protocols are one of the solutions for implementing security in health care systems. Datagram Transport Layer Security (DTLS) is considered as most relevant protocols in this regard. It is based on Transport layer security protocol and provides security which is equivalent to it. It is basically designed for the security of data between two communication ends. E. Rescorla and N. Modadugu (2012)

Figure 2. Architecture of IoT Base Health Care System Using Smart E-Gateway

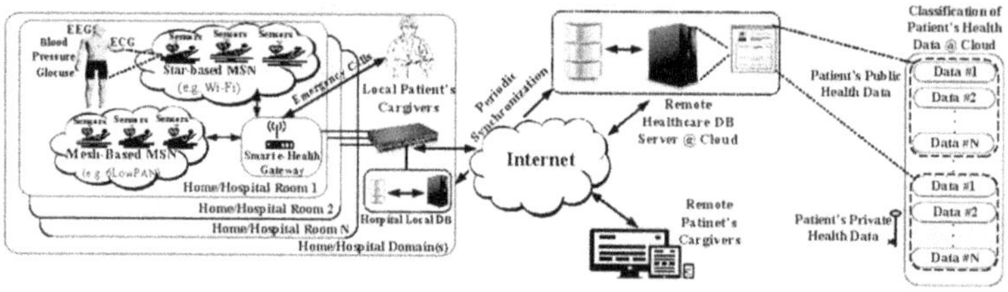

E-Health Gateway

It is proposed by R. Mueller et al (2007) which was called SwissGate used for the optimization of sensor networks. They employed their proposed gateway on home automation systems. In the year 2015, Rahmani et.al. (2015) proposed a smart e-health gateway called UT-GATE which provides intelligence to the ubiquitous healthcare systems. The proposed gateways work as a temporary storage of users and sensors information (like local database). This database is further utilized for various data interpretation techniques and facilitates intelligence to the system hence making it smart gateway.

These intelligent gateways are supposed to provide preliminary results and hence reduce the repeated communication between the cloud servers for performing data aggregation and compression techniques. In this way, remote cloud computers will just provide premium services which are often computationally intensive and require to access to the central database.

SEA – Smart Gateway

Sanaz Rahimi Moosavi et. al. (2015) suggested the secured architecture for healthcare systems which includes following major components –

- **Medical Sensor Network (MSN):** The network is supposed to collect contextual signal from the medical sensors on the body of the patients or may be by biomedical devices kept in the rooms of the patients. The network is capable of ubiquitous identification, intelligent sensing and also communicating the signals received. These signals are further transmitted to the smart gateways via wireless or wired communication protocols such as Wi-Fi, SPI, Serial or Bluetooth.
- **Smart e-Health Gateway:** The purpose of this gateway is to work as a bridge between MSN and the Internet or local switch. It supports various communication protocols. The main function of this gateway is to receive data from various sub-networks and perform versatile activities like data filtering, aggregation, protocol conversion and dimensionality reduction.
- **Back-End System:** Rest of the part makes back-end system which comprises of a switch, cloud platform for broadcasting, datawarehouse, big analytic server and hospital database. The role of back-end system is to synchronize with the remote healthcare server.
- **Web Client:** The GUI's at client are meant for data visualization. Lot of data is collected regarding health of patients and also contextual information. The collected health data can be used by analyst and researches for drawing various inferences.

BSN (Body Sensor Network)

Gope P. et. al. (2016) proposed BSN –Care which is Body Sensor Network based secure health care system. They implement security measures at two places-Network Security and Data Security. They used Lightweight Anonymous Authentication Protocol for the security of network and for Data Security is managed by using OCB (Offset Codeback) authenticated encryption mode. Figure 3 shows their Secure BSN network.

ECC(Eliptic Curve Cryptography) Model

Amjad Ali Almar et. al. (2016) proposed an identification authentication protocol which is based on eliptic curve cryptography (ECC) to avoid the vulnerabilities of RFID based systems. Most of the IoT based system deployment is based on RFID as it consists of most robust features in comparison of other peer technologies. However the technology is prone to various security issues due to wireless communication media. Almar along with ECC also used ECDH (elliptic curve Diffie–Hellman) key agreement protocol to implement security measures for IoT deployment.

The researchers have tried to overcome the hurdle of security in IoT based healthcare systems. Many of them as discussed above have given solutions for implementation of secure IoT based health care system and got success also. However there is still a long road to go ahead.

CONCLUSION

Health care systems are most promising area of IoT applications. IoT based health care systems deal with personal data of patients hence security is the major aspect. Most of the communication in IoT based health care systems is wireless in nature hence it demands high level of security. The researchers have worked in this area and shown great concern. Many solutions have been provided as already discussed in the chapter. The security of data, authorization and access control are major concerns in health IoT.

Figure 3. BSN-Care secured framework

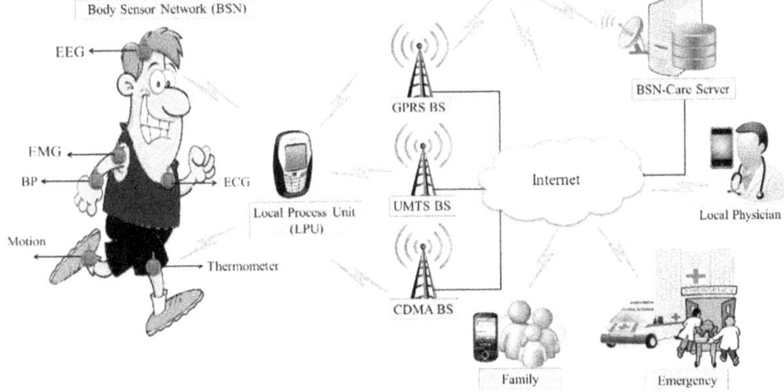

Various parties are being involved in IoT based health care systems like patients, doctors, health care staffs, manufacturers etc. hence the need of secure and robust data communication is of high importance.

REFERENCES

A Comprehensive Look at Low Power, Wide Area Networks for 'Internet of Things' Engineers and Decision Makers. (2016). Annapolis, MD: Link Labs.

Alamr, Kausar, Kim, & Seo. (2016). A secure ECC-based RFID mutual authentication protocol for internet of things. *Journal of Super Computing, 1*.

Andres, R. (2013). *M2M integration platforms enable complex IoT systems*. Received December 1, 2013, from http://www.embedded-computing.com/embedded-computing-design/m2m-integration-platforms-enable-complex-iot-systems

Bezawada, B., Liu, A. X., Jayaraman, B., Wang, A. L., & Li, R. (2015). Privacy Preserving String Matching for Cloud Computing. *Proceedings of the 35th IEEE International Conference onDistributed Computing Systems, ICDCS '15*, 609–618. 10.1109/ICDCS.2015.68

Bui, N., & Zorzi, M. (2011). Health Care Applications: A Solution Based on The Internet of Things. Department of Information Engineering, University of Padova Via Gradenigo. doi:10.1145/2093698.2093829

Caro De, N., & Colitti, W. (2013). *Comparison of two lightweight protocols for smartphone -based sensing*. Department of ETRO-IRIS Vrije University Brussels. doi:10.1109/SCVT.2013.6735994

De Donno, D., Catarinucci, L., & Tarricone, L. (2014, July). RAMSES: RFID augmented module for smart environmental sensing. *IEEE Transactions on Instrumentation and Measurement, 63*(7), 1701–1708. doi:10.1109/TIM.2014.2298692

Digital Intelligence Blog. (2016). *Wearable's impact on cancer patient's quality of life to be tested.* Received July, 26 2016 From http://www.pmlive.com/blogs/digital_intelligence/archive/2016/july/wearables_impact_on_cancer_patients_quality_of_life_to_be_tested

Donno De, D., Catarinucci, L., & Tarricone, L. (2014, April). A battery-assisted sensor-enhanced RFID tag enabling heterogeneous wireless sensor networks. *IEEE Sensors Journal, 14*(4), 1048–1055. doi:10.1109/JSEN.2013.2293177

Gope, P., & Hwang, T. (2016). BSN-Care: A Secure IoT-Based Modern Healthcare System Using Body Sensor Network. *IEEE Sensors Journal, 16*(5), 1368–1376. doi:10.1109/JSEN.2015.2502401

Hipa Journal. (n.d.). *Healthcare Data Security*. Retrieved from https://www.hipaajournal.com/category/healthcare-data-security/page/6/

Hu, Chen, Fan, & Wang. (2017). An intelligent and secure health monitoring scheme using IoT sensor based on cloud computing. *Journal of Sensors*.

Khoi, M. N., & Ahlund, C. (2015). *IReHMo: An Efficient IoT Based Remote Health Monitoring System for Smart Regions*. Department of Computer Science, Electrical and Space Engineering Lulea University of Technology Skelleftea.

Lee, K. (2016). *Healthcare IoT security issues: Risks and what to do about them*. Retrieved from https://internetofthingsagenda.techtarget.com/

Li, C. (n.d.). *Introduction and overview of Biosensors and Electrochemistry*. Retrieved from http://nanohub.org/resources/2261/download/nanobiotechnology_and_biosensors.pdf

Li, C.-T., Lee, C.-C., & Weng, C.-Y. (2016). A secure cloud-assisted wireless body area network in mobile emergency medical care system. Journal of Medical Systems, 40(5), 1–15. doi:10.100710916-016-0474-9

Li, M., Yu, S., Zheng, Y., Ren, K., & Lou, W. (2012). Scalable and secure sharing of personal health records in cloud computing using attributebased encryption. *IEEE Transactions on Parallel and Distributed Systems*, *24*(1), 131–143. doi:10.1109/TPDS.2012.97

Liao, Y., & Leeson, S. M. (2016). *Analysis of In -to-Out Wireless Body Area Network Systems: Towards QoS-Aware Health Internet of Things Applications*. School of Engineering, University of Warwick.

Liu, F., & Li, T. (2018). A clustering k-anonymity privacy-preserving method for wearable IoT devices. *Security and Communication Networks*, *2018*, 1–8. doi:10.1155/2018/1635081

Lounis, A., Hadjidj, A., Bouabdallah, A., & Challal, Y. (2016). Healing on the cloud: Secure cloud architecture for medical wireless sensor networks. *Future Generation Computer Systems*, *55*, 266–277. doi:10.1016/j.future.2015.01.009

Malan, D. (2004). CodeBlue: An Ad hoc sensor Network Infrastructure for Emergency Medical Care. WIBSN'04.

Mathews, K. (2018). *6 Exciting IoT Use Cases in Healthcare*. Received May 03, 2018 From https://www.iotforall.com/exciting-iot-use-cases-in-healthcare/

Miao, Y., Ma, J., Liu, X., Wei, F., Liu, Z., & Wang, X. A. (2016). m2-ABKS: Attribute-based multi-keyword search over encrypted personal health records in multi-owner setting. *Journal of Medical Systems*, *40*(11), 246. doi:10.100710916-016-0617-z PMID:27696175

Moosavi, Gia, Rahmani, Nigussie, Virtanen, Isoaho, & Tenhunen. (2015). SEA: A Secure and Efficient Authentication and Authorization Architecture for IoT-Based Healthcare Using Smart Gateways. *Proceedings of 6th International Conference on Ambient Systems, Networks and Technologies (ANT 2015)*.

Mueller, R. (2007). Demo: A Generic Platform for Sensor Network Applications. MASS'07, 1–3. doi:10.1109/MOBHOC.2007.4428693

Occhiuzzi, C., Vallese, C., Amendola, S., Manzari, S., & Marrocco, G. (2014). NIGHT-care: A passive RFID system for remote monitoring and control of overnight living environment. *Procedia Computer Science*, *32*, 190–197. doi:10.1016/j.procs.2014.05.414

Rahmani, A. (2015). Smart e-Health Gateway: Bringing Intelligence to IoT-Based Ubiquitous Healthcare Systems. CCNC'15.

Redondi, A., Chirico, M., Borsani, L., Cesana, M., & Tagliasacchi, M. (2013). An integrated system based on wireless sensor networks for patient monitoring, localization, and tracking. *Ad Hoc Networks*, *11*(1), 39–53. doi:10.1016/j.adhoc.2012.04.006

Rescorla, E., & Modadugu, N. (2012). Datagram Transport Layer Security (DTLS) Version 1.2. RFC 5238.

Rouse, M. (2018). *IoT analytics guide: Understanding Internet of Things Data*. Received May 2018, from https://internetofthingsagenda.techtarget.com/definition/machine-to-machine-M2M

Salunke, P., & Nerkar, R. (2017). IoT Driven Healthcare System for Remote Monitoring of Patients. Department of E&TC, SITRC.

Sebastian, S. (2018). *How to protect against IoT security threats in healthcare*. Received December 13 2018 from https://www.cabotsolutions.com/how-to-protect-against-iot-security-threats-in-healthcare www.cabotsolutions.com

Shen, W. (2011). Smart Border Routers for eHealthCare Wireless Sensor Networks. WiCOM'11, 1–4. doi:10.1109/wicom.2011.6040606

Shirehjini, A. A. N., Yassine, A., & Shirmohammadi, S. (2012, November). Equipment location in hospitals using RFID-based positioning system. *IEEE Transactions on Information Technology in Biomedicine*, *16*(6), 1058–1069. doi:10.1109/TITB.2012.2204896 PMID:24218700

Sinha, S. (2018). *Introduction to Internet of Things: IoT Tutorial with IoT Application*. Retrieved December 24, 2018, from https://www.edureka.co/blog/iot-tutorial/

Sivagami, S., & Revathy, D. (2016). Smart Healhcare System implemented using IoT. *International Journal of Contemporary Research in Computer Science and Technology*, *2*(3), 641–646.

Sun, W., Cai, Z., Li, Y., Liu, F., Fang, S., & Wang, G. (2018). Security and Privacy in the Medical Internet of Things: A Review. *Security and Communication Networks*, *2018*, 1–9.

Symphony Link vs. LoRaWAN. (2017). Link Labs. Available: https://www.link-labs.com/whitepaper-symphony-link-vs-lorawan

Verma, P., Rajnish, R., & Fatima, S. (2017). Challenges: Wearable Computing for Internet of Things. *International Journal of Science and Research, 6*(4), 1306-1311.

Western PA Healthcare News Team. (2018). *The Market vs. technologists: Treating type one Diabetes*. Received November 21, 2018, from https://www.wphealthcarenews.com/?s=OpenAps

Chapter 14
IoT-Based Smart Water Treatment Plant of GIFT City

Parth Vinayak Brahmbhatt
Sankalchand Patel University, India

Rajan G. Patel
Gandhinagar Institute of Technology, India

Nimisha Patel
Sankalchand Patel College of Engineering, India

ABSTRACT

Water is a vibrant source for life, and its management and treatment are a crucial issue at the present time. A smart solution for city-level water treatment plant is gaining importance with the development in communication technology. We can avoid the issues pertaining to water management and treatment, such as distribution, quality, monitoring of overflows, pH (pouvoir hydrogène) value of water, etc., with help of information communication and technology. Based on this standard, the authors proposed a smart water treatment plant for the entire city using IoT with business processes direction and decision support systems. The suggested IoT architecture not only provides real-time monitoring for supply of pure water being transferred to the citizens but maintains the quality and purity of water supplied along with automation of motor operation, automatic valve controlling, and health monitoring of equipment. The system also delivers an alert to a remote user when there is a deviation of water quality parameters from the pre-defined set of standard values.

INTRODUCTION

Every living organism needs water to survive. Water treatment plant helps in controlling the quality of water and thereby reduces toxins in wastewater to a level by which nature can handle. Water Treatment Plant removes all sediment, bacteria, and other impurities before arriving at the water tap.

DOI: 10.4018/978-1-5225-9574-8.ch014

Water is one of the most valuable ordinary resources. With hasty development, the resource is becoming rare swiftly. For organizations and systems that functions in the water industry, should have Supervisory control and Data Acquisition (SCADA) systems to monitor every parts of the water distributions systems but yet the real-creation limit of its installation points has restricted its practice. As stated in IoT and Hospitality, "The number of connected devices grow by two billion objects in the year 2006 to a projected 200 billion by the year 2020" (Evans et al, 2011). With the growing number of connected devices, the water industry is captivating gain of IoT sensors to monitor water levels, chemical leaks, and even control the water flows. Internet of Things (IoT) in water treatment uses the idea of smart sensors connected at numerous points in the water system. These sensors gather data and send it back to the monitoring systems. This statistic could contain, water quality, temperature changes, pressure changes, water leak detection, and chemical leakage detection. Our goal is to explain how we can use IoT technology more effectivity for water treatment.

The proposal is divided in the following sections: first section refers the quality of water to protect public health, support the economy and maintain a rich ecosystem. Second section refers the implementation of existing smart technologies for high quality service and strict control of operating costs for water treatment process. Third section is about water treatments (as a word originally means the act or) process of making water more useful by purifying, clarifying, softening or deodorizing it. Here, it defines about the combination of various process involved in making the water hygiene and make it more acceptable for specific end use. In this section, also illustrated the plant level technologies to meet its objectives of supplying safe drinking water used for smart city concept and it extended to a diagram with devices been used in the Water Treatment Plant (WTP). The IoT driven water treatment solutions helps in more productive and end up making well informed decisions for better operational strategies. Fourth section refers to the permission been granted for completing the project report on Water Treatment Plant at GIFT. The main aim is to build the most community focused large city in the world with a planned working in the process of Water Treatment Plant. Section 5 refers to the support and assistance provided by reviewers. Section 6 refers to the List of References for this proposal.

SECTION 1: IMPORTANCE OF WATER TREATMENT PLANT

Smart cities consist of six essential sectors which require to work in unison to accomplish a common aim i.e. making a city more liveable, sustainable and economical for its residents. These segments are smart energy, smart integration, smart public services, smart mobility, smart buildings, and smart water. One of a city's key parts of critical infrastructure is its water system. With growing populations in cities, it is certain that water consumption will increase as well. The term "smart water treatment plant" points to water infrastructure that ensures this valuable resource - and the energy used to transport it - is managed effectively. A smart water system is proposed to collect significant and actionable data about the flow, pressure and distribution of a city's water. Further, it is critical that the consumption and calculating of water use is precise. An effective smart water treatment solution can measure and keep a track on flow of the water running in an entire plant. The smart water treatment solution stands strong on three pillars namely; Sensors, Communication medium and IoT platform. A smart water treatment solution offers other benefits to the treatment plant such as;

- Custom apps as per the need of business

- Perform composite calculations simply
- Collect data from various points
- Allows reactive and predictive maintenance for the treatment plant
- Create actionable insights

SECTION 2: EXISTING SOLUTIONS

Almost 80 percent of water supplied for domestic use, comes out as wastewater. In most of the cases wastewater is kept untreated and sinks into the ground as a pollutant of ground water or is discharged into the natural drainage system which further cause pollution in downstream areas. So in this case there is an urgent need in treating waste water by modern technology and thereby recover much usable water as possible. But in fact, the majority of towns and cities in India have no sewerage treatment facilities are inadequate. IoT enabled smart water treatment solution helps water treatment plants in maintaining the consistency and purity of the water, thereby cutting the costs involved and increasing the efficiency of the plant as well. It helps in smart water treatment solution, measuring pH, oxgen, carbon and chlorine level of the water with the help of different sensors. With the use of these smart IoT technologies in hand can further analyze the data and manage the plant more effectively. The smart leak detector enables all the users to be notified about issues in real time. The focus should be on adaptive IoT modern technologies that not only treats waste water but can also easily be added in the existing infrastructure and thereby having lower cost implication.

SECTION 3: WATER TREATMENT PLANT OVERVIEW

GIFT Information Communication & Technology vision is to create leading-edge infrastructure, services and platforms and offer financial services enterprises a significant competitive advantage to operate regionally and globally. GIFT City which is India 's first operational Smart City, is expected to proactively monitor and manage City's Infrastructure using Information and communications technology advances for ensuring better services to the citizens. The infrastructure which needs to be monitored with the help of IoT consists of multiple utilities which are water treatment plant, electrical infrastructure, automatic waste collection and recycling system, district cooling system, city users' safety, security and surveillance, Traffic management and so on.

GIFT City consists of four section in Water treatment plant i.e. Raw water section, filter section, dosing section and Reverse Osmosis (RO) Section.

Raw water section: The water is collected from Nabhoi (Place of station) Pumping station to RAW Tank which is placed at our utility at GIFT WTP Plant. The stored water undergoes a filtering process by media filter being installed for filtering process.

Filter section: There are three DMF filter tanks available at filtering section which improves the quality of water during the process. In addition, the backwash tank stores some water which help during the maintenance of these dual media filters. Further the water is transferred to next section which is dosing section.

Dosing section: In this process some required chemicals like SMBS (Sodium metabisulfite), antiscalant, acid which improves the quality of water i.e. pH, chlorine is added. Further this water is trans-

Table 1.

Ref.No	Purpose	Platform	Technique/ Methodology	Findings
(Robles et al, 2013)	Smart Water Management	MEGA Architecture	Object Linking and Embedding for Process Control (OPC)	Defined a reference architecture for water management based on integrating IoT capabilities to achieve a scalable and feasible industrial system.
(Kamienski et al, 2018)	IoT-Based Precision Irrigation for Agriculture	Personalised hardware and software platform	Semantic Web Abstract Messaging Protocol (SWAMP)	Developed IoT-based methods for smart water management in precision irrigation, and pilots them in Italy, Spain, and Brazil.
(Goel et al, 2017)	Smart Water Management	Novel semantic based context-aware ramework	Multimedia Ontology (MOWL)	Presented a novel semantic based context-aware framework which can dynamically acquire a range of water-specific information in an IoT environment.
(Pavana et al, 2016)	Real Time Monitoring of Water Quality Parameters	Cloud Architecture & Linux OS	Raspberry pi B+	Monitor the different parameters in regard to water quality using cloud server
(Patawala et al, 2017)	Water Management System for Smart City	Internet-based approach	IoT Platform	Explained that how to be measuring water quality and usage on a real-time basis. A flow sensor for measuring of quantity supplied, eliminating the drawbacks of traditional water metering systems.
(Pradeep et al, 2017)	Sustainable Water Management	IoT & MongoDB Database	IoT Platform with Mongo DB database	Developed the system which reduce human intervention in water management system.
(Lade et al, 2018)	Water Management System	IoT Architecture (Sensor Layer)	Wireless Sensor Network	This concept has helped to improve the basic outline of water management techniques by keeping the user up to date by storing the information regularly, obtained using certain variables with the help of network sensors
(Barnaghi et al, 2012)	Sharing of Information in IoT	Nanotechnology and use of Intelligent Device	Nanotechnology	Expressed the future technology based on the Internet which involves sharing of information. also use nanotechnology and miniaturization for placing intelligence in various devices. These devices with intelligence are called smart devices, which plays a very important role in the world of IoT.
(Endler et al, 2014)	The massive amount of data is generated from various sensors by (Data Management)	Cloud base Architecture	Data Management Technique with Cloud	Explanation on data generated which is further used in analysing and providing solutions were ever needed. The cloud and data management services of which can be addressed with an appropriate architecture.
(Zhou et al, 2011)	Key Considerations and Challenges in Interoperability and Security	Proposed New Architecture	IPV6 & SDN	Study architecture of Internet was developed in 70's and was found that the existing architecture is not capable to handle increased traffic for which they proposed existing architecture which suggested changes like use IPV6 and SDN
(Padmadevi t al, 2016)	To measure the water level of the sand	Arduino Microcontroller	GPRS, GSM	Define technique for integration of Water level sensor, pH sensor, Turbidity sensor, Moisture sensor
(Kulkarni et al, 2016)	To measure the water level of water in sump or tank with automatic on off control	Raspberry pi, Arduino	GSM, Relays as actuators	Explained mythology for monitoring the water tank related parameters
(Peruma et al, 2015)	To alert based on the water level	Ultrasonic sensor	Bespoke cloud server	Defined automated escalation matrix for water controlling in smart system.
(Ravichandran et al, 2016)	Leakage detection	Load sensor	Zigbee	Defined integrated system for water leakage detection in water treatment plant.
(Vidya et al, 2016)	Water level monitoring	Water level sensor	LED, GSM, DAS, WSN	Derive architecture which used for monitor water level inside tank with the use of DAS & WSN technology.
(Malche et al, 20170	Water level monitoring for smart village	Liquid levels sensor	CarrIoTs Cloud base server	Done improvisation on sensor layer and storage layer for gathering data related to water level monitoring
(Deepiga et al, 20115)	Smart water monitoring in home/office	Water level sensor, Led, Light sensor, Turbidity sensor, Temperature sensor, pH sensor	Force sensing register (FSR)	Study on integration of various sensors for monitoring water quality and quantity related parameters of water
(Viswanath et al, 2015)	To create low power wireless liquid monitoring system	Ultrasonic sensor	GSM, Dc power analyzer	Done inventiveness on sensor layer and us Edge layer to create low power wireless of liquid monitoring system
(Jagtap eta l, 2016)	Monitor the water level remotely through website/ app	Node.js Server	MQTT	Done study on data storage layer and edge layer for improving the response time in application used for water level monitoring
(Gupta et al, 2016)	Create a mobile app that monitors the water level	Ultrasonic sensor & Arduino	BLYNK server	Developed application for monitoring water level by use of BLYNK server and ultrasonic sensor
(Pappu et al, 2017)	Intelligent IoT based Water Quality Monitoring system	Machine to Machine (M2M)	Cloud base IoT architecture	Express M2M concept for water quality monitoring by which water quality can be monitored automatically with no human involvement.

ferred to next filtering section which is Micro Cartage Filter (MCF) and finally it moves to RO section with high pressure pump.

RO section: RO section includes 15 nos. of membrane at both the sides i.e. of 30 nos. During the process only, one-sided membrane is working. So, if any problem exists with the existing one the other can be kept in use. Waste water generated during the process is collected in suck back tank which is further used for gardening purposes. GIFT City is carrying out city infrastructure monitoring and maintenance by integration of all utilities like water treatment plant, raw water pumping station, Electrical substation on a single platform which is termed "City Command and Control center (C-4) by using Internet of things.

Being a Smart City, GIFT City authorities are expected to proactively monitor and manage City's Infrastructure using Information and communications technology advances for ensuring better services to the citizens. The infrastructure which needs to be monitored consists of multiple utilities which are water management system, electrical infrastructure, automatic waste collection and recycling system, district cooling system, city occupier's safety, security and surveillance, traffic management etc. Along with other utilities, GIFT authorities' approach is for the smart and the best use of Information and communication technology in managing Water Infrastructure for the entire city. At GIFT the water treatment plant is managed and maintained in semi automation mode. It is difficult to keep on monitoring on start/stop operation, level monitoring and other day to day operations at Raw Water pumping station (RWPS) & Water Treatment Plant (WTP) remotely from City Command and Control Centre. Besides this process of manual monitoring on parameters like Quality/pH/Turbidity/ Conductivity etc., which further leads in keeping record of water consumed by the citizens. GIFT by using IoT platform further leads water treatment plant for operating smartly and fully automated which helps to achieve below mentioned aspects.

1. Start and stop operations, level monitoring and other day to day operations of Raw Water pumping station (RWPS) & Water Treatment Plant (WTP) remotely from City Command and Control Centre (C-4). This also removes the requirement of having physical presence of the pump operator at RWPS & WTP.
2. Monitoring of water supply distribution system by using flow sensors and leak detection system and automation build at various places across the water distribution system, these sensors and automation are monitored from City Command and Control center (C-4). The connected system helps GIFT City authorities to minimize the losses in water distribution network and will help to become a "Zero Discharge City".
3. GIFT City is India's first city which supplies potable water to each tap in the city. Various parameters like Quality / pH / Turbidity / Conductivity etc. are checked at source level and monitored from City Command & Control Centre (C-4). This help to proactively address any issue in the quality of water. In due course of time, GIFT City also plans to display water quality parameters over Public Display panels spread across the city and GIFT City Mobile App, where citizen can find out the quality of water being supplied on their mobile phone itself.
4. Effective monitoring and management of waste collection system consisting of Sewage Treatment Plant (STP) & Sewage Conveyance System will help GIFT to be a "Water neutral" city where waste water collected will be filtered and used at various places like irrigation of gardens, construction of buildings and District cooling system to provide centralized air conditioning services to the Citizens of the City. Various ICT initiatives like water accounting automation and Analytics are going to play to a major role in achieving this objective.

Plant Level Technology

Considering the requirements of make water treatment plant fully automated and connected via city command and control center, we have designed an architecture to efficiently manage and maintain the water treatment plant without any kind of human intervention. The recommended system contains of four different layers as shown in Figure 1 which are discussed in detail below:

Things Layer

This layer consists of monitoring and management of modules, its hardware and interactions with the digital world. Sensors enable the users to measure the water flow, energy consumed, pH, salinity, chlorine, and temperature. Sending data and alerts continuously to an IoT platform again allows the users to move from preventative to predictive maintenance. The data gathered from this connection are stored in the server and monitored on a continuous basis via city command and control center.

Networking Layer

The communication layer also plays a very crucial role in sending and receiving the data from physical objects to computer systems over a wireless and wired medium. These mediums of communication can provide technicians, engineers and other facility management staffs to have full access to the required data from sensors to city command and control center.

Middleware Layer

The middleware layer has multiple functionalities which are: to interface protocols to support semantic and syntactic interoperability, i.e. to shield the heterogeneity of the system; providing central control of the managing and monitoring modules associated with resource, distribution and storage; to identify and process context information that will have an impact on the basic computation; data management

Figure 1. Water treatment plant high level architecture with device detail

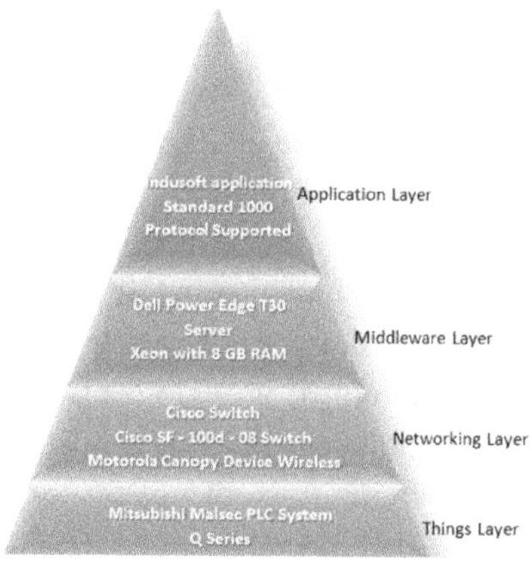

to support the sensor data generated and logging control actions, data analytics and learning systems to support the decision module .

Application Layer

The application layer hosts an alert system to the local community regarding the water level and its availability. The city command and control center operator will have access to a web-based dashboard consisting of real time visualizations for usage, availability and health of the system. The intended audience of the data generated from these systems are primarily plants level operators, city command and control center operator & maintenance team from water department.

Diagram and Device Detail Use at Plant Level

The main concept behind this project can be described in a simple block diagram shown in Figure 2 Shows to make water tenement plant fully automated and do integration with city command and control centre.

There are four stages in the water treatment process that can be monitored and managed with IoT solutions: Raw water section, filter section, dosing section and Reverse Osmosis(RO) Section.

1. Though the sensing requirements, some of the factors monitored by the sensors include water temperature, conductivity, pH, turbidity and dissolved oxygen content, as well as atmospheric conditions such as pressure, humidity and solar radiation.

Figure 2. Water treatment plant - GIFTCITY

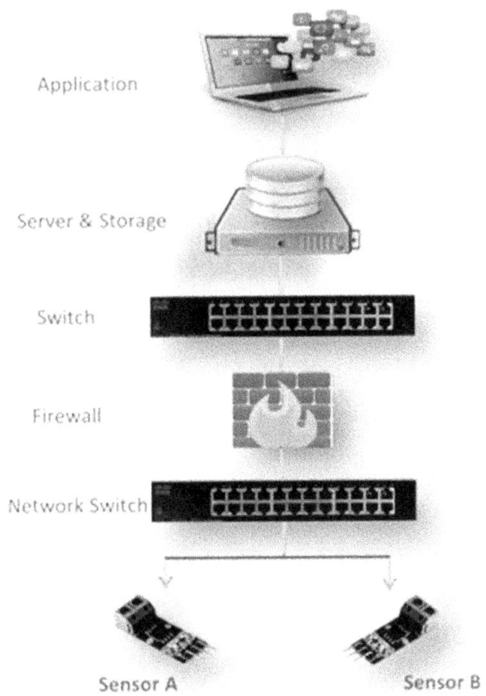

Table 2.

Sr.no.	Layers	Devices	Device Configuration
1	Sensors/Hardware	Mitsubishi malsec PLC System	Q Serise
2	Edge Layer	Cisco Switch	Cisco SF-100D-08 Switch
3	Enge Layer (Wireless Connectivity between plant to C4)	Motorola Canopy Device Wireless	Point to Point Connectivity
4	Data Processing Layer	Dell Power Edge T30 Server	Xeon with 8GB RAM & 1TB Hard Disk
5	Application Layer	Indusoft Application	Standard 1000 protocol supported

2. After collecting the relevant data, the sensors communicate with IoT gateway via the edge layer Next, the gateway uploads the sensor data to the storage system. From there, users can access and visualize the data in the application dashboard, based on the data integration and business analytics platform.

Below are the devices used for making Water treatment Plant IoT enable.

1. PLC System: The Programmable logic Controller (PLC) is a purpose-built machine control computer designed to read digital and analog inputs from various sensors. We have to designed WTP controller with all required facilities and then need to install and start only.
   a. In electrical control panel auto manual switches & its associated wiring can be minimized as we have auto-manual mode in our controller in built-saves cost and space. In manual & auto mode in our controller each tank level /pressure/flow/pump status etc. Can be monitored on LCD screen.
2. SCADA: supervisory control and data acquisition is a system operating with coded signals over communication channels so as to provide control of remote equipment.
   a. In control room single engineer can see status of valves, pumps, flow-meter etc. & operate manually also.
   b. The data can be display in anyway the user requires.
   c. The computer can record and stored a very large amount of data.

3. INDUSOFT Application: The design of cluster distribution mains has been done based using INDUSOFT software.

   a. It performs extended-period simulation of hydraulic and water-quality behavior within pressurized pipe networks.
   b. INDUSOFT provides an integrated environment for editing network input data, running hydraulic and water quality simulations, and viewing the results in a variety of formats.
   c. It allows users to analyze the movement of a reactive or non-reactive tracer material which spreads through the network over time.

Figure 3. Mitsubishi PLC (Q Series)

By using the INDUSOFT application at presentation layer we can control and monitor water treatment plant related from city command and control centre like,

1. Raw water Section**:** Monitoring the tank level, status for valve and quantity of the water stored can be done after treatment in RO water tank.

2. Filter Section:

During filtering we can monitor water quality related parameters and air compressor status. For this filtering tank clearing process can be done in a single click

3. Dosing Section:

In dosing section, we can add some minerals in water such as SMBS, ACID and antiscalent in water. We can control and monitor water pH value in dosing section.

Figure 4. Mitsubishi PLC (Q Series)

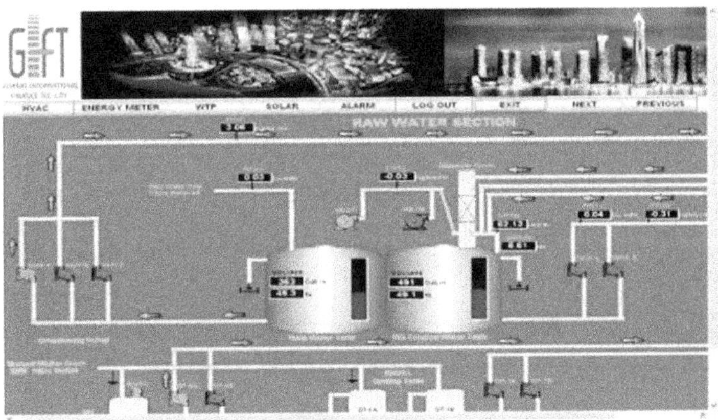

Figure 5. Mitsubishi PLC (Q Series)

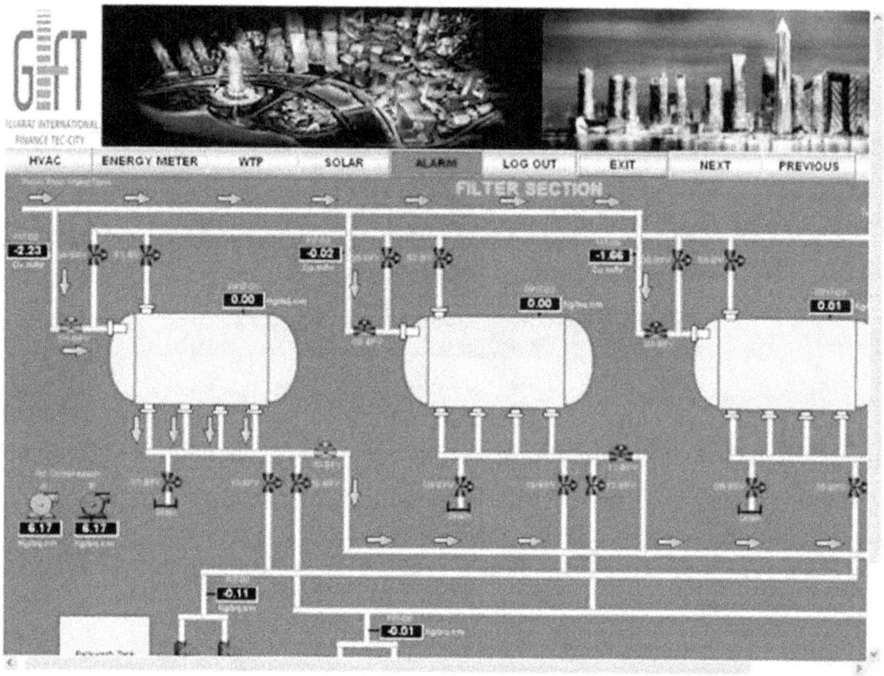

Figure 6. Mitsubishi PLC (Q Series)

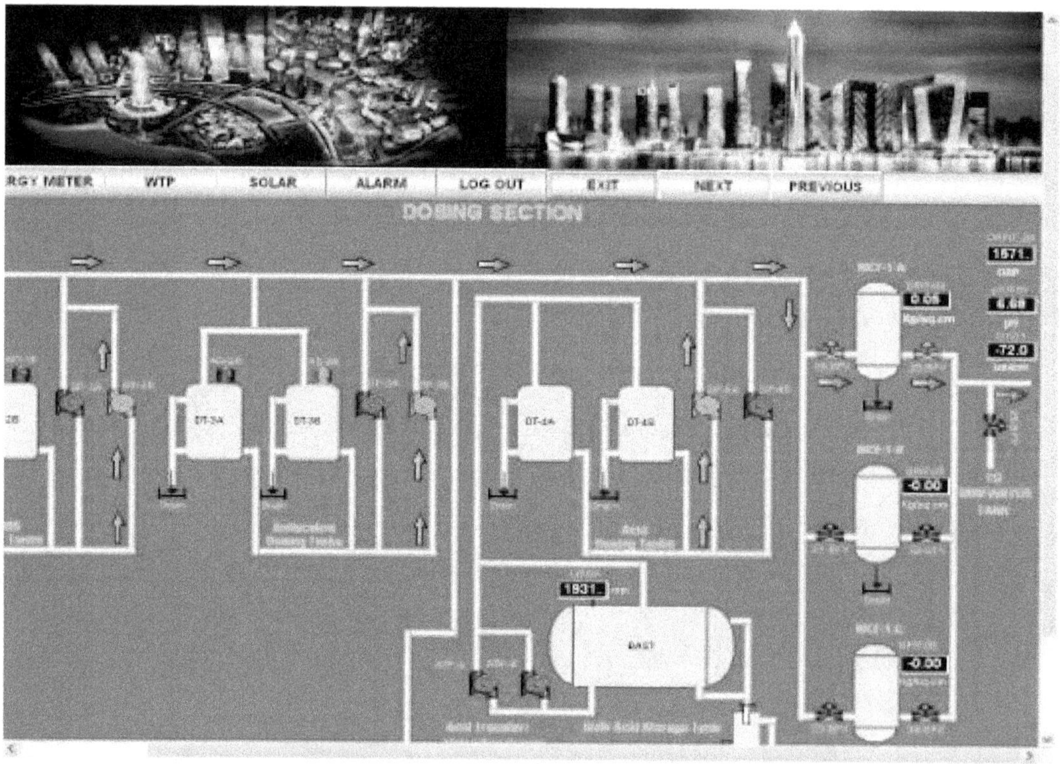

Figure 7. Mitsubishi PLC (Q Series)

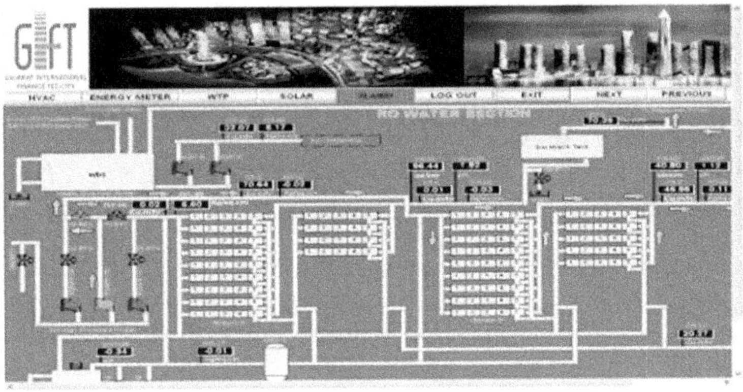

4. RO section:

In RO section we can monitor quantity of the waste water after treatment, health of the membrane and will be able to monitor tank level of waste disposal system.

CONCLUSION

We have evaluated to be the best comprehensive solution for water treatment plant based on several criteria: high reliability and service, low downtime due to faults or failure, short delivery time, slashed Total Cost of Ownership, and potential to make available a fully automated monitoring and control system.

We have used the equipment and system solutions which have an impressive record of performance all over the world. For this huge project, we have planned and deployed a pre-engineered "iQ Plant Suite" solution featuring fully automated monitoring and control systems for water processing. The length and complexity of the several processes including the collection and purification of river water for the supply of safe drinking water, the treatment of ground water for industrial water consumption, and the treatment of routine wastewater for its environmentally safe return to the river, submitted many challenges for the development of an automation control system.

ACKNOWLEDGMENT

We would also like to show our gratitude to the Gujarat International Finance Tec-City for sharing their knowledge during the research and helped for comments that greatly improved the manuscript. Any errors are our own and should not tarnish the reputations of this esteemed organization.

REFERENCES

Amatulla, P., Navnath, B., Bhoug, Y. P., & Zadbuke, A. S. (2017). IOT based water management system for smart city. *Journal of International Journal of Advance Research*, *3*(2).

Barnaghi, P., Wang, W., Henson, C., & Taylor, K. (2012). Semantics for the IOT: Early progress and back to the future. *Journal of Centre of Communication and System Research. International Journal on Semantic Web and Information Systems*, *8*(1), 1–21. doi:10.4018/jswis.2012010101

Deepiga, T., & Sivasankari, A. (2015). Smart water monitoring system using wireless sensor network at home/office. *International Research Journal of Engineering and Technology*, *2*(4), 1305–1314.

Endler, M. (2014). Microsoft Azure Intelligent Systems: 4 Facts. *Information Week, 906*, 174.

Evans, D. (2011). The internet of things how the next evolution of the internet is changing everything. *Advances in Internet of Things*, *6*(4).

Goel, D., Chaudhury, S., & Ghosh, H. (2017). Smart water management - an ontology-driven context aware IOT application. *Journal of Pattern Recognition and Machine Intelligence*, 639-646.

Jagtap, A. M., Sikandar, B. S., Shivaji, S. S., Pramod, K. V., & Sarangdhar, N. K. (2016). Aquarius-smart IOT technology for water level monitoring system. *International Journal of Computational Science*, *5*(8).

Kamienski, C., Soininen, J.P., Taumberger, M., Dantas, R., Toscano, A., Cinotti, T.S., … Neto, A.T. (2018). Smart water management platform: IOT-based precision irrigation for agriculture. *Journal of MDPI Sensors, 19*(276).

Kulkarni, P. V., & Joshi, M. S. (2016). An IOT based water supply monitoring and controlling system with theft identification. *Journal of Innovation Research in Science*, *5*(9), 16153–16157.

Lade, P. S., Vyas, P., Walavalkar, V., Wankar, B., & Yadav, P. (2018). Water management system using IOT with WSN. *Journal of International Research of Engineering and Technology*, *5*(3), 3079.

Malche, I., & Maheshwary, P. (2017). Internet of things IOT based water level monitoring system for smart village. *Journal of INT International Conference on Communication and Networks*, *508*(12), 1–6.

Padmadevi, S., & Deepigha, V. (2016). Smart e-agriculture monitoring using internet of things. *International Conference on Emerging trends in Engineering and Technology*, 47-50.

Pappu, S., Vudatha, P., & Niharika, A. V. (2017). Intelligent IOT based water quality monitoring system. *International Journal of Applied Engineering and Research*, *12*(16), 5447–5454.

Pavana, N. R., & Padma, M. C. (2016). Design of low cost system for real time monitoring of water quality parameters in IOT environment. *International Journal of Advanced Research in Computer Science and Management Studies*, *4*(5), 12–15.

Peruma, T., Sulaiman, M. N., & Leong, C. Y. (2015). Internet of things enabled water monitoring system. *IEEE 4th Global Conference on Consumer Electronics,* (8), 42-43.

Pradeep, P., Narendran, S., & Ramesh, M. V. (2017). An internet of things (iot) based sustainable water management. *Journal of IEEE Global Humanitarian Technology Conference*, *6*(2), 68–72.

Ravichandran, S. (2016). Liquid level monitoring system using IOT. *International Science Press*, *9*(31), 215–218.

Robles, T., Alcarria, R., Mart'ın, D., Navarro, M., Calero, R., Iglesias, S., & Lopez, M. (2013). An IOT based reference architecture for smart water management processes. *Journal of Wireless Networks*, *6*(1), 4–23.

Sukriti, G. S., & Indumathy, K. (2016). IOT based smart irrigation and tank monitoring system. *International Journal of Innovative Research in Computer and Communication Engineering*, *4*(9), 15718–15722.

Vidya, B., Poonam, K., Priyanka, G., Gaurav, D., & Chandgude, A. S. (2016). Water level monitoring system in real time mode using WSN. *International Journal of Emerging Technology and Advanced Engineering*, *6*(9), 212–214.

Viswanath, S., Belcastro, M., Barton, J., O'Flynn, B., Holmes, N., & Dixon, P. (2015). Low-power wireless liquid monitoring system using ultrasonic sensors. *International Journal on Smart Sensing and Intelligent Systems*, *8*(1), 26–44. doi:10.21307/ijssis-2017-747

Zhou, L., & Chao, H. C. (2011). Multimedia traffic security architecture for the IOT. *Journal of IEEE Networks*, *25*(3), 35–40. doi:10.1109/MNET.2011.5772059

Chapter 15
IoT in Education:
A Future of Sustainable Learning

Jyoti Batra Arora
Institute of Information Technology and Management, New Delhi, India

Suruchi Kaushik
Institute of Information Technology and Management, Delhi, India

ABSTRACT

Internet of things (IoT) is an emerging technology that is looked upon as a technology that can bring positive changes in development of the society. Using IoT, humans and things embedded with sensors, actuators, and other hardware can communicate in an intelligent manner, thus creating a sea of innovative ideas which can make life easy, simple, and smart. IoT has touched upon different areas such as health-care, home security, and agriculture. In this chapter, the authors give an insight about the technology and its applications in various fields. The major focus is on emerging trends in education and potential applications of IoT in the field of education. IoT can bring a major shift in traditional educational practices and make the teaching-learning process more effective by actively engaging learners. Use of IoT devices can also help teachers in deciding their future plan of action and giving more attention to the slow learners. This chapter also highlights the challenges that this technology faces in the education sector.

INTRODUCTION

Internet of Things (IoT) is considered to be the most trending topic in the world of Information Technology. Since its inception, Internet of Things has been an area of interest for people in various fields being it developers and innovators in the IT world or academicians and researchers and still its full potential has not yet been realized. IoT brought a complete change in the usual accepted ways of connecting devices through the internet. It has touched every area for development- being it is medical, horticulture, biotechnology, science and education. The basic framework of IoT technology is that almost everything around us can be connected anytime to the internet using easily available sensors which can be embedded into any object through which there can be communication between the things and the environment in an

DOI: 10.4018/978-1-5225-9574-8.ch015

intelligent manner. This ability of connecting things to things and to other devices and finally connecting things to humans has revolutionized the way wired or wireless communication can be set up among various non-living and living systems in a much more affordable and effective way.

IoT finds its applications in various sectors like agriculture, manufacturing, healthcare and military and can help in developing smart buildings, smart agriculture and smart environment etc. A smart home is the place where one can access and control over all devices even from his office or may be from other city. It makes life easier and also provides better security.

Developing smart cities require smart planning taking into consideration different aspects such as infrastructure, manufacturing, transportation, health of the citizens with smart education being one of the most important concerns. IoT technologies consist of people, sensor and networking device to make connection. These technologies are different from other previous technologies as these technologies follow ubiquitous computing and provide more intelligent and autonomous solutions. It has developed the ability to impose or embed sensors into any object and use machine to machine communications to develop connection among billion of people and devices using internet as infrastructure. According to Cisco (2016) IoT technologies are used to make a network around the world by making connectivity of different objects for anything and for anyone, and any where using any network services. It fulfills the goal of tracking, identifying intelligent and managing the things. The communication between human to things and things to things has been expanded to a large extend. There are huge developments in the areas where a communication between human and machine is required. Like in Agriculture, farmers can put sensors in their fields that may help them in determining the temperature and humidity condition of soil. It may help them to decide the water requirement of their field. The sensors can also be put to determine the requirement of specific fertilizers for the crop in field. Moreover, it is not just knowing about the requirement, IoT technology may also remind farmers about each and every requirement by sending an alert on their mobile phones using connecting devices. Similarly, in medical, information about patient, maintaining and sharing the history of cure of patient with his family's member has become easy.

Many research organizations described the future of IoT and its impact on the development of communication and connection among various devices. The devices are trillion in numbers that may have digital sensors and may be connected by leveraging any network. According to a research conducted by Juniper in 2015,the number of connected devices is 13.4 billion which can be raised to triple i.e. 38.5 billion by 2020.

Still, there is one field that is lacking by the touch of IoT is Education. Though there are huge researches that talk about the development of smart campuses, attendance system and other development in the area of academics. But all these researches talked about the connection between different devices. As mentioned earlier, these researches are on the basis of machine to machine and human to machine interaction. Education symbolizes the interaction between human and human. Education is not only the communication between humans, rather it is learning process, where interaction, understanding, gaining of facts, interest, and accumulation of knowledge are various parameters. IoT technologies may manifest the education field in many ways, whereas in others it may not. This chapter will present the implication of IoT in education. It will present the implication of these connected devices in academics and also present that how it could emphasize the learning for future generation. IoT can provide a modern education system by converting traditional classrooms into smart classrooms. This chapter comprises of four sections. The first section will explain about the introduction to topic, architectureof IoT. The second section will explain the applications of IoT in general. The third section will explain emerging trends of education. It will also present the implication of IoT in education sector that may be helpful in achieving the goal of sustainable learning. This section also has a proposed model TISAO to describe the usage of IoT in education. The fourth section will explain the future of IoT in education along with few of challenges regarding the IoT in education.

Here, the education refers to education in middle and higher level. As the researchers believe that education is the way of making people better informed and wiser through acquiring knowledge and skills, therefore the collaboration of people and technology can be achieved by modernizing the education sector with emerging technologies of IoT.

IoT Architecture

IoT is a combination of two words, i.e. '**internet**' and '**things**'(Madakam et al., 2015).Internet is network of networks which connects billions of users across the globe. It is a network through which people working in various domains are connected anytime from anywhere in the world. Billions of private, government, business and public networks are linked through the internet by use of various media such as electronic, wireless or optical fiber technologies (Nunberg, 2012). According to Cisco forecast, globally, from 2016 to 2021 Internet traffic will grow 3.2 times with growth rate of 26% annually. A large number of world's population will be using the internet in forthcoming years.

The second word 'things' represents any object which can be identified by the real world (Kosmatos et al.,2011). Every day we come across a variety of electronic devices which are technologically advanced, programmed and designed according to the needs of the users. These appliances can range from a simple electronic digital toaster to a Smart Voice Assistant using advance technologies of machine learning and artificial intelligence. 'Things' in IoT can be the objects around us which we do not consider as electronic at all, such as the chair on which we sit the entire day, or a pen which we use to sign the official documents. Here, things symbolizes both living things like people, animals and plants and non-living things like furniture, home appliances, industrial apparatus, buildings, monuments etc. So, all the objects in the real world can be considered as things in IoT.

IoT provides a unique identity to every object and these things can be connected to each other in an intelligent manner giving rise to useful information for communication between various things. Today, IoT gained a special place as a technology in the world of information technology attracting innovators, academicians, researchers, developers and people from the corporate world. Kevin Ashton, a new age technology expert, the co-founder of the Auto-ID Centre at the Massachusetts Institute of Technology in 1999 has used the term "Internet of Things". He stated that like internet, IoT has also huge potential to change the world.

Majority of the data available on the Internet has been created by human beings; it was always the people communicating with each other through internet. Now the data can be created by things (Madakam et al., 2015) which can communicate with each other in an intelligent manner by use of **IoT** technology. By using ubiquitous sensors and other hardware, things can communicate with not only humans but also with things or machines. And this will make IoT more pervasive revolution affecting the ways we think making it possible to remain connected anytime from anyplace with anyone. Thus IoT can be described as a network which allows communication between humans, between things and also between humans and things and allows them to stay connected all the time. By staying connected we do not necessarily only mean the electronic devices such as smart phones or computers but in internet of things this also means the objects in which sensors and actuators are embedded and linked through wired or wireless networks. These physical objects or 'things' can communicate with the environment and sense the

Figure 1. Five layered structure of IOT (Fuqahaetal., 2015).

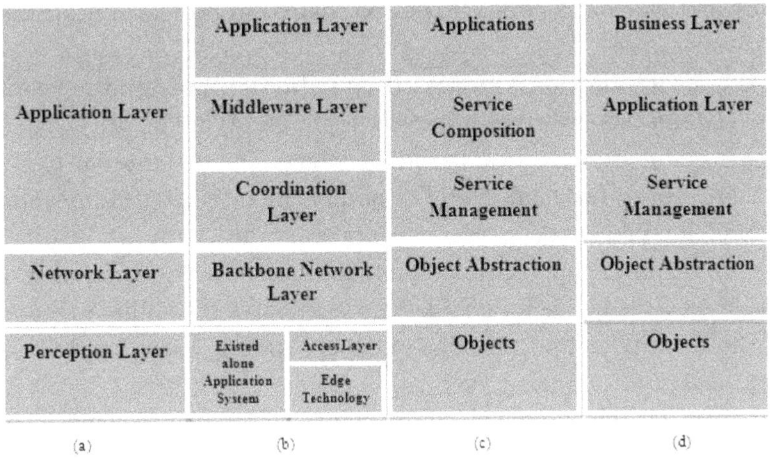

changes and thus become tools for understanding the stored information in an intelligent manner and more importantly analyze the data without human intervention.

IoT architecture is three layered architecture comprises of Perception layer, network layer and application layer. Kumar and Patel (2014) also described the three layered architecture. Fuguha(2015) has expanded these three layers and made it to five layer architecture.

The perception or objects layer includes sensors and actuators to process and collects the information (Sonar & Upadhyay,2014). It is also called recognition layer as it gathers information from physical world. Farooq (2015) poised that it works on motion, location, weight, temperature identification and acceleration etc. The information is converted into digital signals and then these signals are passed to network layer.

Network layer is also called "Central Nervous System" that is used to perform the global services in IoT. Object abstraction layer works on cloud computing and data management processes (Fuqahaetal., 2015). The transferring of data can be done through Wifi, Bluetooth, 3G and Zigbee etc. Service Management or middle ware layer processes the received data to make decision and delivers the required services (Granjaletal., 2015). The information received from sensor embedded devices is processed by using some intelligent processing equipment's (Farooq etal.2015).

Application layer is used for large scale development of IoT. This top most layer carries formulas, business logic and UI to user end (Upadhyay, 2015). It provides the services requested by customer such as temperature and humidity measurement. Business layer receives the data from application layer and manages overall system services and activities using graphs flowchart and models etc. This layer manages and monitors all the four underlying layers. It enhances the services by comparing the output of each layer with expected outcomes (Faquhaetal., 2015, Farooq etal.,2015)

The challenges and preventive measure related to each layer are described as following:-

1. **Perception layer:** This layer includes object properties, environmental conditions, all types of sensors and RFID reader. As this layer has sensor data associated with it therefore, security is more concerned about authenticity, confidentiality and integrity. It has small computer power and storage capacity therefore; public key encryption and frequency hopping communication algorithm

cannot be used here. Ding (2011) poised that authentication as basic requirement to protect illegal access of nodes and data encryption. It is also required to protect confidentiality of transmitted information between nodes.

2. **Network Layer:** The information is processed, classified and polymerized in this layer which leads to congestion. Though this layer has relatively more ability of security yet two types of attacks still exists here ie counterfeit and man in middle attack. The major security measures are identity authentication, encryption mechanism and communication security for prevention of illegal nodes and attacks from vulnerable node.
3. **Support Layer:** This layer supports intelligent decision of network behavior and mass data processing so, the biggest challenge for this layer is to improve the ability to recognize the malicious information. The security architecture like cloud computing, secure multiparty computation and antivirus are needed in this layer.
4. **Application Layer:** The layer is involved in sharing of data that creates issues like privacy, disclosure and access control of information. It also requires security, education and management for information security and password management.

IoT is emerging area which is actually a combination of different streams of computer science. A huge amount of money is spent on research of IoT enabled technologies while much more is expected to come in coming years. The embedded sensors, light communications and ubiquitous computing enhances the significance of IoT however, it also imposes a lot of challenges. There is also a requirement for development of communication protocols and specialized standards. As aforementioned in the layer architecture of IoT, the layers of IoT is built on above each other and each layer required different security and communication protocols. For routing among sensors, these specialized protocols have been designed.

POTENTIAL APPLICATIONS OF IOT

Since IoT is used in different sectors or areas depending upon the requirement and usage of IoT. IoT is a network of intelligent devices connected together which communicate and interact with other smart devices, machines and the environment (Sharma and Tiwari,2016). According to Cisco, IoT links objects to the internet and it predicted in 2016 that by 2030 the total number of devices around the world that will remain connected through internet would be 500 billion. An enormous amount of data will be generated which can be stored and monitored to interact and control the 'things' around us to make life easier and safer. Things as aforementioned refer to a large variety of objects which can be identified uniquely. These objects can be embedded with sensors and actuators and connected through a network which enables them to collect information and exchange it with other things (Tiwaryetal., 2018). The applications of IoT in science and technology and everyday life range from smart homes, intelligent healthcare systems to education, agriculture, smart governance and human rights in special perspective of education.

The potential applications of IoT technology in various domains are discussed here in following section:

Smart Homes and Buildings

In today's busy lives we need more control over home appliances so that it makes life easier and safe. Kousalya et al.(2018) presented IoT based home automation and home security method. The home appliances such as fans and lights can be controlled using Wi-Fi. Home automation provides the comfort, ease, security and efficiency. In case of a person trying to get inside the home, an image of the person will be captured and owner will be alerted by sending the captured image through email or message. If there is an unexpected guest the owner can let him in the home. The owner can lock all the entry and exit point of home while sitting in his office. All appliances in home can be controlled via long distance by use of internet. The use of IoT can be extended to make smart buildings which can further include smart lightning, smart air quality control, smart waste management, better surveillance and safety of individuals. The future of IoT can provide more intelligent means of automation, security and energy management in homes and buildings (Mohammed & Ahmed, 2017).

Smart Cities

The different aspects of smart cities will include smart homes and buildings, smart healthcare, smart manufacturing and retail, smart waste management, smart transport and traffic management, smart agriculture, smart education, smart energy management and smart security (Mohammed & Ahmed,2017). To meet the requirements of ever growing population and to cater to the enhanced facilities of urban lifestyle the concept of smart cities is looked at as a feasible alternative in various cities of the world. The smart city concept requires IoT devices which can provide real time information and help in smart living. Smart cities require collaboration of government and citizens to enable smooth implementation of the various projects.

According to IEEE (IEEE International conference on smart cities, 2015) technology, government and society are integrated together in a smart city to enable the following characteristics: A smart environment, smart people, smart living, smart economy, smart mobility and Smart governance. Smart cities over the globe like Singapore, Barcelona and San Jose are providing smart services to their citizens. IoT enabled services will play a major role in development of smart cities and providing smart services to all the citizens in different spheres of life.

Smart cities are also coming up with innovative solutions for people with disabilities thus becoming inclusive. For example, Singapore city which has a 'Smart Nation' vision, hasdeveloped a prototype of a four-wheel personal mobility device which is self driven.This device will help in providing mobility to disabled persons and complementthe existing transportation system (Smart Nation and Digital Government Office, 2016). Cisco began experimenting with turning Barcelona into a smart city in 2011. The three key visions for digital transformation in Barcelona are:1. Technology for better government2. Urban technology3. City data commons (Barcelona Digital City, 2018)

Smart Health Care

IoT applications in healthcare showed a great potential and it is an emerging field with a lot of new areas to be explored. Using various sensors, vital health parameters can be monitored; the data can be collected and analyzed without any human intervention. This is less costly and differs from traditional methods of care where one has to be always present to monitor the patient's health condition. With the help of IoT technology they can be brought into the mainstream and also their health can be monitored and further steps can be taken for improvements.

Smart Environment

A healthy environment is required to build a healthy body and mind. Today, it is a matter of concern to make efficient use of the available resources with minimum wastage and also take steps to protect the environment.Using IoT new technologies can be developed to monitor as well manage the environment. Air quality can be monitored by collecting data from various points in a city and it can be analyzed to take further actions for improving air quality (Mohammed and Ahmed, 2018). IoT can provide real time monitoring and management systems to take informed decisions to reduce waste and hence improve the quality of environment. The impact of natural and manmade disasters can be avoided or reduced by using different types of sensors in IoT devices and by linking these with various disaster relief agencies to make announcements on time and take precautionary measures.

Smart Agriculture

IoT will also help in agricultural growth and eventually contribute to the economic growth of a country. Sensors such as soil and moisture sensor and temperature sensors can be used to monitor the growth of crops and ensure proper management of water resources for timely irrigation. Real time data from IoT devices can help in reducing damage to crops by giving accurate information about weather, soil conditions and control of pests. IoT based devices can turn around the face of agriculture by not only enhancing the quality and production but also by making it cost effective and reducing wastage.

Smart Factories and Smart Manufacturing

IoT can improve the efficiency and lower the costs of manufacturing. Also there is a huge amount of data that is being generated for which edge computing is implemented to manage the data. Connecting machines to intelligent systems by means of IoT, Artificial Intelligence and machine learning can change the way factories work (Benioff,2015). Smart factories will change the ways of manufacturing and shipping and make better utilization of resources in a more productive way. It will also improve the working environment and make it more safer, secure and smarter (Hartmann and Kuschel, 2015).

Smart Transportation

The challenge is to work in an innovative and organized way taking into account all parameters to improve the safety and ease of use of different transport. Use of IoT in transportation requires different architectures for exchanging information between vehicles and also between vehicles and other infrastructure (Goel, 2017). For example if a bus is arriving at a bus stop then the information has to shared with the fleet of buses as well as the main control so that same bus number does not arrives at the same time at the same bus stop. Internet of things will make the services in transportation sector faster thus reducing wastage of time, increase the accuracy and safer.

Smart Education

IoT will make all phases of learning more efficient. All educational disciplines have special place for technology. Porter (2013) poised that IoT is unifying the virtual world and physical world, many new training practices and cross cutting areas will be generated in future. The cost of IT technologies continues to grow, and educational institute do not have strategy for sharing cost and identifying total cost of ownership for an IoT infrastructure. Academic institutes need to come up with new ideas for the same. IoT and stake holder collaboration is the symbiotic approach that possesses the potential to impact educational practice in improvement of learning issues and also improving teacher's role and skills.

The benefits of IoT in education sector is not just connected to enhance learning but also increases safety security improves access to different sources of information and ensures emergence of techniques to monitor resources utilization and consumption contributing to energy and saving and cost saving (Brandt,2016).

IoT enabled devices get connected to each other and communicate without human intervention. Utilization of IoT devices and technologies and collaboration among people and devices process and data will bring numerous benefits in all aspect of the education sector.

Emerging Trends in Education and implications of IOT in Education Sector

As Nelson Mandela said, "Education is the most powerful weapon which can be used to change the world". Education forms the backbone of the society on which overall development of a nation depends. It is very important to build a strong foundation in education if an individual has to grow not only intellectually but also as a sensible citizen of nation. There is a strong relationship between education and technology. Today we live in digital world where information can be accessed anytime and anywhere. Traditional educational practices are no more relevant in today's education system. The learner's needs are changing with time so technology now plays an important role in dissemination of knowledge. Today's learners are smart using smart phone and different types of electronic gadgets from a very young age. Thus apart from delivering the content in an effective way the teachers have to engage them to retain their attention. There has been a rapid technological growth resulting in widespread use of electronic media in education sector. Use of Information Communication and Technology(ICT) in education has emerged as an important factor in achieving the learning objectives. ICT provides a better access to different sources of information thus enhancing skills and knowledge and helping in better understanding and retention of information. Advancements in technology have impacted learning in many ways. IoT in the context of ICT can be used to enhance the learning outcomes by making learning process simpler and smarter. IoT has great potential for educational institutions and universities where it can be implemented at all levels of teaching and administrative work. Keeping at pace with today's developments in technologies there is a huge demand for having digital classrooms and smart campuses. Entry of students and staff can be authenticated using RFID tags making the campus safer and advance. New training methods can be explored for teaching and progress of students can be monitored in real time. Education is the most effective medium of making the word better, safer and more sustainable place to live what makes it the best investment(UNESCO,1997). IoT brings dramatic transformation in educational technology.

Blewitt (2010) stated that the education for sustainable development and integrating aspect of sustainability cannot be realized without using techniques that encourage participatory learning and use higher order thinking skills. IoT can reform the educational practices and lead to changes in educational technology, and completely change the teaching learning environment. With the development of IoT it will be easier for students to understand. IoT has the ability to make learning process much more effective because it can help in referring to the real time insights into the topic of study.

IoT brings enormous challenges in the field of higher education. This chapter presents the outlook and challenges of IoT in education. Linand (2010) poised that not only in classroom but outside that connected devices can be used to monitor staff, student and teacher and equipment at a reduced operating cost. A full engagement and collaboration of all stakeholders is required to completely redesigned education sector as technology supported, enhanced and economically, ecologically and socio-culturally sustainable.

Zhiqiang, and Junming (2011) stated the three major aspects of IoT in education- student progressive evaluation, integration of current teaching platform and development of educational middle ware. The connected devices and technology allows the academician to focus on actual learning than performing routine task. It enhances the learning experience of student by providing real time experience and useful insight.

Gupta and Fisher (2012) in their paper poised that technology used in education makes academic activities are more student focused and issue focused. Johnson etal. (2015) stated the seven different strategies and technologies that has revolutionize the education sector- consumer technologies, digital strategies, enabling technologies, learning technologies, visualization technologies, social media technologies and internet technologies. IoT being the part of Internet technologies connected all devices and overcome the limitations of traditional classrooms by internet and use it as one of major communication tools.

Ghazal etal (2015) stated that IoT can help the professors to collect the data about student's performance and the collected data can be used to determine which student needs more care and attention. This data analysis helps the teacher to plan methodology for future classes. The asset intelligence enables the institute to take informed decision which added value to operational proficiency, learning experience of student and security of institute. Moksimovic (2017 a.) stated that usage of IoT in educational institute is not only to raise the quality of education but also to realize economically socially and environmentally sustainable education sector.

Following are the parameters to describe the impact of IoT in Education

Digital Campus

This is the system that helps the students to get all types of information. The digitization of contents and academic activities are highly in demand as it allows researcher and academician to work effectually in digital environment. Porter(2013) poised that a digital campus has the technology that enables teaching, learning and empower collaborative research. IoT help to reduce the operational cost, improve security and offer a platform to support different professional and academic activities at campus. Cisco in its report stated that conventional network supports user and user data whereas digitization network of campus with IoT supports sensor and sensor data. IoT reform and change the teaching and learning in the campus. The campus will become ubiquitous learning and training. IoT in educational institute commence to develop and leverage solution to analyze and manage big data. Students can have access to information anytime from anywhere in the campus on their devices and thus learning experience is more engaging.

Educational institutes and universities can use connected devices to monitor the activities in the entire campus thus keeping a check on any untoward incident. By using IoT technology it is easier to monitor and maintain the resources and equipment's of the campus at a low cost and with less requirement of manpower. Furthermore, IoT enables universities to improve the security of the campus and design their campus in safer ways keeping the safety of students as well as staff in mind. Cata (2015) described about the smart lightening and smart tracking of students, good and equipment using RFID technology. Simic etal.(2015) poised that the smart education institute has smart classrooms, smart corridors with info boards and data centers for processing all types of data.

Student Life

All the connected devices along with networks make the classroom a dynamic one. The students can have wearable devices that may track ECG pattern. It enables the logging of attendance. To monitor cognitive activities of students in the class, ECG sensors can be used. A card can be applied for vision impairment student so that they can view the text with enlargement. The mobile app can be helpful to the students that have reading or writing related disability and cannot understand the material. The security and safety can be achieved through utilization of digitized wristband and identity card that help to track education establishment of student and staff (Ralhan,2017).

Class Room

IoT can support classroom in many ways like improving learning methods, upgrade learning resources, improvement in techniques of learning and saving of management cost. Nowadays students look for various electronic media to learn new concepts than text books. The advanced technologies give students advantage of learning at their own pace and learning what they want to learn thus making a difference in traditional classroom teaching approach where all students have to learn at the same pace. Moreover, through IoT technology data about students' performance can be collected and slow learners can be identified who need more attention and time. A careful analysis of data collected through sensors and other IoT devices can be used to make a plan of action for future classes. Various devices can be used to keep a check on students' attendance and also their attention span in the class by studying their ECG patterns and their cognitive activities can be further monitored by using EEG sensors. As a result educational institutions can improve the quality of education they are providing to students and improve the learning outcomes. Rytivaara (2012) stated that smart devices can also check when the teacher need to speak louder, when students are losing interest or their concentration level is decreasing.

Chang(2011) stated that smart classroom help students to understand the real purpose of using technology which also makes the learning process easier. A study proposed an efficient smart classrooms roll caller system(SCRCS) using iot architecture to collect or record student attendance after every period accurately and timely(Alghamdi and Shetty,2016). Chew(2015)stated that iot based classroom can provide real time feedback on lecture to improve the quality of lecture . It help to monitor and observe student's reaction to a lecture after sensing and monitoring technology.

Edge Computing

Institute which have on corporated with internet of things, they can be benefitted from edge or fog computing. Edge computing is a possible remedy to data privacy concerns. Students across all stages of education have right to access modern technology that would help them to prepare well with digital focused for the competitive market. Edge computing is opposite of cloud computing. It brings high performance computing as close to the action as possible. It can make human communication smoother and make the augmented reality more realistic. Edge computing could help academician / researcher to take full advantage of IoT devices. It also help the schools or institutions which districts lacking bandwidth to send sensor information to cloud process that data could power learning system designed to enhance performance with real time feedback to locally.

Edge computing is also a risk to security, as linking of many devices could unleash new vulnerabilities to institution.

Fog Computing

It is emerging structure for computing which is used for computing, storing and securing information beneath cloud layer. It spread these actions to entire network not only on central node. Fog computing can be used more in higher education because of its powerful support in AI, Mixed reality devices, high performance computing and 5G. It helps the institutions to install smart building infrastructure and in secure entry system etc. by using real time security cameras it also help to monitor traffic through and around the institution. A research done by duke university state that fog computing could potentially lead to responsive augmented reality hologram on campus.

Edge and fog computing along with IoT provides aggregation and data processing from sensor network and individual devices. Increasing of productivity of boundary network nodes can significantly reduce data access delay traffic. And bandwidth of internet access channels.

A research done by Lund etal.(2014) poised that IoT connected and communicated devices would be 28.1 units in2020. There is no such model of using IoT in education but it can be used as course and as an opportunity to learn different courses (Byrne o'sullivan & Sullivan (2017), Krotov (2017).

Gaggal (2011) poised two approaches on teaching IoT concept and using IoT in subject. The use of IoT in academics brought new opportunities and possibilities for improvement of both the learning process and institution's infrastructure. IoT has an important impact on education field. IoT is used as a tool to enhance academic infrastructure and as a subject to teach the concept of computer science. Everyone being it a student or teacher can be benefitted with IoT in education. Many advanced and innovative concept are being proposed or introduced in education like introducing IoT technology with crowd sourcing in e-education can be useful for improving learning and teaching process.

Recent Implication of IoT in Education

Barakat (2016) in his paper has given an example to show the impact of IoT on education. He described the product super mechanical's twin 7 product which is a small box to connect the user to almost any physical object in local area. Twine integrates sensors with cloud based service. Sensors are immediately recognized by webapp which reflect what the sensors see in real time. It will help the people with no knowledge of software coding to receive text and emails updates on whatever items or environments the box is sensing. Marquez (2016) stated *"integrating IOT as a new actor in educational environment can facilitate the integration of peopleand physical objects in the academic environment"*. UK has realized the importance of IoT and introduced a new subject **My Digital Life** for undergraduate students (Chin and Callaghan, 2013). It is based on IOT to teach English language. Cheng and Liao (2012) described about the use of IoT technologies in University of Padova. The services are used in the information engineering department of university. The paper examined the use of cloud computing and IoT in incorporating the structure of education resources and provides an integration model. Castellani etal. (2010) poised the impact of four different technologies including IoT, cloud computing, data mining and triple play on new distance education. Chen and Dong (2013) discussed the application of IoT and cloud computing in education and also differentiate smart campus with digital campus.

Proposed System for Smart Education

Education industry is undergoing revolutionary changes as the technology is many researchers have given the innovative practices are being used in teaching processes and educational institutes. IoT as a technology can be used to form innovative models in Education which can be adopted to transform teaching learning process, improve the efficiency and make campus more secure and safe. There is no such model defined or prescribed by different researcher. Many research papers proposed some important parameters that are required to make smart campus. But smart education involves smart campus and innovation and digitization in teaching and learning process. Therefore, the proposed model **TOIAS** defined certain parameters where IoT can be efficiently used at different level of education. The proposed model **TOIAS** covers different spheres from considering smart classrooms to digital and secure campus.

1. **T**eaching learning process,
2. **O**utside learning,
3. **I**nclusive learning,
4. **A**dministrative work
5. **S**afety and security
    1. **Teaching Learning Process:** Internet of things technology can convert traditional classrooms into smart classrooms. Teachers can make use of IoT devices in marking attendance, grading students and delivering lectures. This can help teachers in designing curriculum, planning lessons and taking assessments. Learning process can also be improved by the use of IoT devices (Pandharkar and Suryawanshi, 2018). Use of tablets, smart phones makes learning process more interesting and effective. Students can use real time information and make informed decisions. Use of IoT technology makes teaching learning process smoother, interesting and effective.
    2. **Outside Learning:** Use of IoT devices can help learners to stay connected anytime from anywhere. Nowadays, Traditional methods of using notebooks and taking notes being physically present in the classroom are not required. Digital information can be shared with students outside the classroom also. IoT devices can connect the learners and researchers having same areas of interests across the globe.
    3. **Inclusive Learning:** IoT can be a boon for special children. IoT devices can be helpful in making the learning process easier for such students. Slow learners and learners with other impairment like dyslexia can learn with the help of smart devices using advanced applications converting the concepts of study into meaningful information for such students, thus catering to all types of learners according to their needs. Use of this technology can make classrooms more inclusive and thus giving special people their rightful place in the society.
    4. **Administrative Work:** IoT can help educational institutes in making the administrative system smooth and will reduce the wastage. From admission process to examinations the process can become less time consuming, cost effective as it reduces usage of paper thus more efficient. IoT sensors can track and monitor consumption of power in the premises thus it can make the campus smarter and energy efficient. The movement of staff, students and faculties can also

be monitored and used to keep a check on their attendance. Smart technology can be used to take feedback from teachers and learners thus again improving the quality of teaching learning practices.

5. **Safety and Security**: Smart campus requires real time notifications and alerts in case of security breach and in case of any untoward incident disciplinary action if required can be taken accordingly. In today's world parents want to have information about their children. Use of IoT devices can make it possible to track the location of the child all the time. With the help of IoT based devices such as sensors, cameras, RFID etc. Real time monitoring can be done in the entire campus to keep a track of activities and making thus making the campus safe and secure.

CHALLENGES FACED BY IOT IN EDUCATION

IoT is helping in developing a new society with a new digital culture. Big data analytics along with ubiquitous computing and cloud computing improves the core value of teaching and quality of research. It is dramatic shift in the traditional instructional paradigm. Following are the few IoT Challenges in educational sector

Cloud Computing: Educational institute depend on hybrid cloud infrastructure with computing platform on private clouds whereas enterprise applications have moved to public cloud. The enterprise infrastructure of educational institute is needed to reduce its latency time because of huge demand and increase in audio video instructions.

Mobility of Applications and Technologies: IoT applications are used to integrate mobile learning applications for grading and assessment system (JIN 2102). Academician can use these technologies to teach highly specialized concept, complex physical, scientific simulations and other social topic.

Security and Privacy: As education sector produces future challengers therefore, this sector requires embrace IoT platform along with challenges of IoT Financing, digital educational pedagogy and interdisciplinary research. It is required to develop new strategies and policies that consider individual privacy choices and expectation.

Cost: The setup of an IoT based educational institute can be expensive. So, the cost of devices and equipments are other challenges that are faced to use IoT in academics.

Research Computing: The reduction in the cost of hardware has given momentum to interdisciplinary research. It uses unmanned aerial vehicle, high performance computing and microcontrollers. Social science researches along with big data are constantly using distributed computing platform such as HPC, GPU clusters and hadoop clusters to improve IoT research. IoToffers unique opportunities to raise the standard and content of digital education. It also offers challenges to maintain quality of lectures and evaluation of student. In educational sector it requires to perform these challenges with ethics.

The inclusion of IoT helps teachers to enhance efficiency of learning methods, creating high quality lessons and moving more quickly through material. Stosic (2015) poised that the aid of technology in teaching the teacher interaction with students is significantly improved. Thompson (2014), Pappas (2014) stated that the numerous collaboration software solutions used from both learners and educators for communications. Ronfeldt etal. (2015) stated that promoting high quality collaboration across all levels is a better way to improve student achievement and enhance whole education sector than promoting collaboration at any single level.

FUTURE OF IOT IN EDUCATION

IoT will give better operational efficiency in all phases of learning. All educational disciplines have special place for technology. Porter(2013) poised that IoT is achieving the unity of the virtual world and physical world, many new training methods and cross cutting areas will be generated in future. The cost of IT technologies continues to grow, and educational institute donot have strategy for sharing cost and identifying total cost of ownership for an IoT infrastructure. Academic institute must come with new ideas for the same. IoT and stake holder collaboration is the symbiotic approach that possesses the potential to impact educational practice in improvement of learning issues and also improving teacher's role and skills.

The benefits of IoT in education sector is not just connected to enhance learning but also increases safety security enhances access to a variety of information and ensures the ability to monitor resources utilization and consumption contributing to energy and saving and cost saving (Brandt,2016).

IoTenabled devices connect to each other and communicate without human intervention. The utilization of IoT devices and technologies and collaboration among people and devices process and data will bring numerous benefits in all aspect of the education sector.

CONCLUSION

IoT has given a new definition to growth and connectivity. Earlier, only human are connected with human, which is followed by connection between human and machine. Now machine are considered as object and connected to each other. This paper has described the architecture of IoT layers that explains briefly about the security challenges and protocols used in different layers. IoT is used in every sector to enhance its connectivity with users. Medical and healthcare are very well developed but education is the area which is lacking behind. This paper explains the feature of IoT that can be used to enhance the growth of educational sector which would be beneficial to both students and teachers. It not only enhances the learning but also improvises the security and information sharing.

REFERENCES

Abhik. (2019). *Internet of Things, for things, and by things*. CRC Press.

Aditya, Manish, Mayank, & Mohit. (2018). Internet of Things (IoT): Research, Architectures and Applications. *International Journal on Future Revolution in Computer Science and Communication Engineering, 4*(3).

Aijaz, A., & Aghvami, A. (2015). Cognitive machine-to-machine communications for internet-of-things: A protocol stack perspective. *IEEE Internet of Things Journal, 2*(2), 103-112. Available at http://ieeexplore.ieee.org/xpl/articleDetails.jsp?tp=&arnumber=7006643

Ala Al, Guizani, Mohammadi, Aledhari, & Ayyash. (2015). Internet of Things: A Survey on Enabling Technologies, Protocols, and Applications. *IEEE Communication Surveys & Tutorials, 17*(4).

Aldowah, H., Ul Rehman, S., Ghazal, S., & Naufal Umar, I. (2017). Internet of things in higher education: A study on future learning. *Journal of Physics: Conference Series*, *892*, 012017. doi:10.1088/1742-6596/892/1/012017

Alghamdi, A., & Shetty, S. (2016). Survey toward a smart campus using the internet of things. *Proceedings - 2016 IEEE 4th International Conference on Future Internet of Things and Cloud*, 235–239.

Ashton. (2009). Internet of things thing. *RFID Journal*. Available at Http://www.rfidjournal.com/article/print4986

Barakat, S. (2016). Education and the internet of everything. *International Business Management*, *10*(18), 4301–4303.

Barcelona Digital City. (n.d.). *Digital transformation, 2018*. Retrieved from http://ajuntament.barcelona.cat/digital/en/digital-transformation.`

Brandt, C. (2016). *The Internet of Things and Its Impact on Education*. University Herald.

Byrne, J., O'Sullivan, K., & Sullivan, K. (2017). An IoT and Wearable Technology Hackathon for Promoting Careers in Computer Science. *IEEE Transactions on Education*, *60*(1), 50–58. doi:10.1109/TE.2016.2626252

Castellani, A. P., Bui, N., Casari, P., Rossi, M., Shelby, Z., & Zorzi, M. (2010). *Architecture and Protocols for the Internet of Things: A Case Study*. PERCOM.

Cata, M. (2015). Smart university, a new concept in the Internet of Things. *2015 14th RoEduNet International Conference - Networking in Education and Research (RoEduNet NER)*, 195–197.

Chang, C. H. (2011). Smart classroom roll caller system with IOT architecture. *Proc. - 2011 2nd Int. Conf. Innov. Bio-Inspired Comput. Appl. IBICA 2011*, 356–360. 10.1109/IBICA.2011.94

Chen, Y., & Dong, X. (2013). The Development and Prospect of New Technology in Modern distance education. *International Conference Information Science Computer Application*, 40–44. 10.2991/isca-13.2013.7

Cheng, H., & Liao, W. (2012). *Establishing an lifelong learning environment using IOT and learning analytics*. Advanced Communication Technology.

Chew, C.B. (2015). Sensors-Enabled Smart Attendance Systems Using Nfc and Rfid Technologies. *Int. J. New Comput. Archit. their Appl.*, *5*(1), 19–28.

Chin, J., & Callaghan, V. (2013). Educational living labs: A novel internet-of-things based approach to teaching and research. *Proc. - 9th Int. Conf. Intell. Environ. IE 2013*, 92–99. 10.1109/IE.2013.48

Decuir, J. (2010). *Bluetooth 4.0: Low Energy, Presentation slides*. Retrieved from http://chapters.comsoc.org/vancouver/BTLER3.pdf

Ding, C., Yang, L. J., & Wu, M. (2011). Security architecture and key technologies for IoT/CPS. *ZTE Technology Journal*, *17*(1).

Farooq, M. U. (2015). A Review on Internet of Things (IoT). *International Journal of Computers and Applications*, *113*(1), 1–7. doi:10.5120/19787-1571

Gaggl, L. (2011). *Education and the Internet of Things*. Retrieved from: http://www.brightcookie.com/wp-content/uploads/ 2011/11/IoT_in_Education-3.pdf

Gomez, C., Oller, J., & Paradells, J. (2012). Overview and evaluation of Bluetooth low energy: An emerging low-power wireless technology. *Sensors (Basel)*, *12*(9), 11734–11753. doi:10.3390120911734

Granjal, J., Monteiro, E., & Sa Silva, J. (2015). Security for the internet of things: A survey of existing protocols and open research issues. *IEEE Communications Surveys and Tutorials*, *17*(3), 1294–1312. doi:10.1109/COMST.2015.2388550

IETF. (n.d.). *IPv6 over Networks of Resource-constrained Nodes (6lo)*. Retrieved from https://datatracker.ietf.org/wg/6lo/documents/

Karagiannis, V., Chatzimisios, P., Vazquez-Gallego, F., & Alonso-Zarate, J. (2015). Asurvey on application layer protocols for the internet of things. *Transaction on IoT and Cloud Computing, 3*(1), 11-17. Available at https://jesusalonsozarate.files.wordpress.com/2015/01/2015-transaction-on-iot-and-cloudcomputing.pdf

Kim, A., Hekland, F., Petersen, S., & Doyle, P. (2008). When HART goes wireless: Understanding and implementing the Wireless HART standard. *IEEE International Conference on Emerging Technologies and Factory Automation (ETFA 2008),* 899-907. Available at https://library.e.abb.com/public/eb20fe80a-391ca8485257bc600667573/When%20HART%20Goes%20Wireless%20Understanding%20and%20 Implementing%20the%20WirelessHART%20Standard.pdf

Kosmatos, E.A., Tselikas, N.D., & Boucouvalas, A.C. (2011). Integrating RFIDs and Smart Objects into a Unified Internet of Things Architecture. *Advances in Internet of Things*: *Scientific Research*, *1*, 5-12. Available at doi:10.4236/ait.2011.11002

Krotov, V. (2017). The Internet of Things and new business opportunities. *Business Horizons*, *60*(6), 831–841. doi:10.1016/j.bushor.2017.07.009

Kumar, J. (2014, March). A survey on Internet of Things: Security and Privacy Issues. *International Journal of Computers and Applications*, *90*(11).

Lund, D., Turner, V., MacGillivray, C., & Morales, M. (2014). Worldwide and Regional Internet of Things (IoT) 2014-2020 Forecast: A Virtuous Circle of Proven Value and Demand. Framingham: International Data Corporation (IDC).

Marc, B. (2015). *Industrial Internet of Things: Unleashing the Potential of Connected Products and Services*. Retrieved from http://www3.weforum.org/docs/WEFUSA_IndustrialInternet_Report2015.pdf

Marquez, J., Villanueva, J., Solarte, Z., & Garcia, A. (2016). *IoT in Education: Integration of Objects with Virtual Academic Communities. In New Advances in Information Systems and Technologies, no. 115* (pp. 201–212). Springer International Publishing.

Mirjana. (2017). IOT concept application in educational sector using collaboration, series of Teaching. *Learning and Teacher Education*, *1*(2), 137–150.

Mohammeda & Ahmedb. (2017). *Internet of Things Applications, Challenges and Related Future Technologies*. Available at www.worldscientificnews.com/EISSN

Norbert. (2015). *Partialiternal models-techniques for integration into the standard formula*. Available at: https://www.munichre.com/site/corporate/get/documents_E352744188/mr/assetpool.shared/Documents/0_Corporate%20Website/1_The%20Group/Focus/Emerging%20Risks/CROF-ERI-2015-The-Smart-Factory.pdf

Nunberg, G. (2012). *The Advent of the Internet: 12th April*. Courses.

Object Management Group. (2015). *Data Distribution Service V1.4*. Retrieved from http://www.omg.org/spec/DDS/1.4

Park, M. (2015). IEEE 802.11ah: sub-1-GHz license-exempt operation for the internet of things. *IEEE Communications Magazine, 53*(9), 145-151. Available at: http://ieeexplore.ieee.org/xpl/articleDetails.jsp?arnumber=7263359

Prachi. (2017). *IOT Fundamentals- networking technologies, Protocols and use cases for the Internet of Things*. ciscopress.com.

Raza, S., & Voigt, T. (2010). Interconnecting WirelessHART and legacy HART networks. *6th IEEE International Conference on Distributed Computing in Sensor Systems Workshops (DCOSSW)*, 1-8. Available at: http://ieeexplore.ieee.org/xpl/articleDetails.jsp?arnumber=5593285

Rytivaara, A. (2012). Collaborative classroom management in a co-taught primary school classroom. Academic Press.

Saint-Andre, P. (2011). *Extensible messaging and presence protocol (XMPP): Core*. IETFRFC 6120. Available at: https://tools.ietf.org/html/rfc6120

Simic, K., Despotovic-Zrakic, M., Đuric, I., Milic, A., & Bogdanovic, N. (2015). A Model of Smart Environment for E-Learning Based on Crowdsourcing, RUO. *Rev. za Univerzalno Odlicnost, 4*(1), A1–A10.

Singh, M., Rajan, M., Shivraj, V., & Balamuralidhar, P. (2015). Secure MQTT for Internet ofThings (IoT). *Fifth International Conference on Communication Systems and NetworkTechnologies (CSNT 2015)*, 746-751. Available at http://ieeexplore.ieee.org/xpl/articleDetails.jsp?arnumber=7280018

Smart Nation and Digital Government Office. (2016). *Self-Driving Vehicles (SDVs): Future of Mobility in Singapore*. Retrieved from https://www.smartnation.sg/initiatives/Mobility /self-driving-vehicles-sdvs--future-of-mobility-in-singapore

Smart Nation and Digital Government Office. (n.d.). *Self-Driving Vehicles (SDVs): Future of Mobility in Singapore, 2016*. Retrieved from https://www.smartnation.sg/initiatives/Mobility /self-driving-vehicles-sdvs--future-of-mobility-in-singapore

Somayya, Ramaswamy, & Siddharth. (2015). Internet of things (IOT): A literature review. *Journal of Computer and Communications, 3*, 164-173.

Sonar, K., & Upadhyay, H. (2016). *An Approach to Secure Internet of Things Against DDoS*. Singapore: Springer Singapore.

Stankovic, J. (2014). Research directions for the internet of things. *Internet of Things Journal, IEEE, 1*(1), 3–9. doi:10.1109/JIOT.2014.2312291

Sudha, Reddi, Vasanthi, & Venkatesh. (2018). IOT Based Smart Security and Smart Home Automation. *International Journal of Engineering Research and Technology*, *7*(4).

Vandna & Ravi. (2016). A Review paper on "IOT" & it's Smart Applications. *International Journal of Science, Emerging and Technology Research, 5*(2).

Zeinab, Aldein, & Ali. (2017). Internet of Things Applications, Challenges and Related Future Technolofies. *World Scientific News*, *67*(2), 126–148.

Zhiqiang & Junming. (2011). The Application of Internet of Things in Education and Its Trend of Development. *Modern Distance Education Research, 2*, 19.

Zhou, Z., Yao, B., Xing, R., Shu, L., & Bu, S. (2015). E-CARP: An energy efficient routing protocol for UWSNs in the internet of underwater things. *IEEE Sensors Journal*. Available at: http://ieeexplore.ieee.org/xpl/articleDetails.jsp?arnumber=7113774

ZigBee Standards Organization. (2008). *ZigBee Specification*. Document 053474r17. Available at http://home.deib.polimi.it/cesana/teaching/IoT/papers/ZigBee/ZigBeeSpec.pdf

Chapter 16
Smart Plant Monitoring System Using IoT Technology

Ankur Kohli
University of Petroleum and Energy Studies, India

Rohit Kohli
Parkland Fuel Corporation, Canada

Bhupendra Singh
Schematics Microelectronics, India

Jasjit Singh
https://orcid.org/0000-0001-9574-0324
University of Petroleum and Energy Studies, India

ABSTRACT

Plants play a vital role in maintaining the ecological cycle, and thus, to maintain the plant's proper growth and health, adequate monitoring is required. Hence, the aim of the chapter is to create a smart plant monitoring system using automation and internet of things (IOT) technology. This topic highlights various features such as smart decision making based on soil moisture real-time data. For this purpose, sensors like soil moisture sensor, DHT11 sensor, level sensor, etc. are used. The soil moisture sensor measures the level of moisture (i.e., water content of different plants). The signal will be sent to Arduino board when the moisture level drops below the marginal value, which triggers the pumping of water into the plant by the pump. When the moisture level reaches absolute value, the pump is halted. The other condition for this process is level sensor. Level sensor senses the water level in the tank and sends the information of water level value to Arduino board and Arduino board to cloud. The whole data about the plant monitoring will be sent to the cloud server.

DOI: 10.4018/978-1-5225-9574-8.ch016

OVERVIEW

This chapter outline an IOT based Plant Parameters Monitoring System. Wireless Network is prepared by using NodeMCU board, Arduino UNO/Atmega interfacing board. This network helps with obtained data through Internet of Things. The planning of system is designed to locate the numerous needs of the plant parameters monitoring. The whole system consists of nodes which are situated at different part of the board. The data or the value from the sensors is collected by Arduino UNO/Atmega interfacing board and uploaded to cloud i.e. ThingSpeak Server through NodeMCU board.

The Other Operations of Smart Monitoring of Plants and Crops are as Follows

- Soil moisture monitoring
- Crop Monitoring
- Temperature and Humidity Monitoring
- Water level monitoring

Smart Plant Monitoring System using IOT Technology: Internet of Things technology is suitable for the smart plant monitoring system. This is also suitable to check the soil conditions as well, which may help to start a better growth of a plant. It may support to control water usage and wastage of water.

INTRODUCTION

Plant plays a vital role in maintaining the ecological cycle and forms the foundation of a food chain pyramid and thus to maintain the plant's proper growth and health adequate monitoring is required. Hence the aim at making plant monitoring system smart is using automation and Internet of Things (IOT) technology. This topic highlights various features such as smart decision making based on soil moisture real time data.

The computerized water system framework with IOT is practically and financially sufficient for planning water resources for plantation (group of a plant). Adopting the automatic water system framework we can demonstrate that the utilization of water can be decreased for various plantations (group of plants) usages. The system framework has an appropriated microwaves (wireless) chain of moisture content in the soil through soil moisture sensor, humidity and temperature sensor set in the root zone of the plants and level of water (ultrasonic) sensor is set in tank for checking the water level in tank. The data will gather from the sensors and send to the web server (cloud).

The background of chapter highlights the study of IOT in the field of agriculture. This shows how we can implement the IOT technology to make our planting smart and reliable with the real time updated data. This chapter also helps the beginners to implement the IOT technology and learn the basics of this technology.

REVIEW OF LITERATURE

Anita,G discuss et al Hands on approach on Application of Internet of Things in Various Domain of Engineering studied about the various applications of Internet of Things (IOT) in the domain of engineering such as in environmental monitoring, healthcare, home automation, fire and safety engineering etc (Anita et al, 2018). Sushabhan,C discuss et al Internet of Things [IoT] Enabled Automation in Agriculture explain about the automation in the domain of agriculture (Sushabhan et al, 2018). Kumar,N Arduino (2017) discuss based Automatic Plant Watering System with Internet of Things (IOT) this paper presents that making agriculture smart using automation and IOT technology which focus on smart agriculture with smart control and making perfect decision based on real time data. Michal,S. (2016) discuss Internet of Things (IoT) in Agriculture - Selected Aspects. 83-88. 10.7160/aol.2016.080108 this article focuses on a use of agriculture where IOT technology ac easily be implemented in which the Czech Republic environmental condition also take into consideration while writing this article .J. H. Gultom (2017) discuss Smart IoT Water Sprinkle and Monitoring System for chili plant this paper presents the sprinkle the water automatically and monitoring the system which is based on IOT platform. Thamaraimanalan,T. (2018) discuss et al Smart Garden Monitoring System using IOT to maintain the parameters of the plants by observing regularly which tends to improve the life of plant i.e. proper growth of a plant and human beings life as well because plant plays a vital role in ecological cycle and maintain the carbon dioxide and oxygen level in the atmosphere. Athani,S. (2017) discuss et al Soil moisyure monitoring using IoT enabled arduino sensors with neural networks for improving soil management for farmers and predict seasonal rainfall for planning future harvest in North Karnataka-Indiato install a computerized or automatic technique to measure the moisture content in the soil and monitoring the parameters using different technology and constructing the system which is economical to the end users. Sadasivan,S. (2015) discus et al Smart Plant Monitoring System about the execution of new devices to develop a smart plant monitoring system with a use of ambient intelligent concept. Elangovan,R. (2018) discuss et al A Plant Monitoring and Gardening System using IoT about the automation in a field of planting and reducing the manual work and improving the parameters needed for the proper plant growth without any interfere of human being. Kaur,S.(2017) discuss et al A Survey on Automatic Irrigation System Using Wireless Senor Network about the use of Wireless Sensor Network (WSN) technology to monitor and control the soil parameters in the field which improves the new plant growth, gathered the data. Keerthana,B.(2018) discuss et al Literature Review on IOT Based Smart Security and Monitoring Devices for Agriculture about the smart way of automatic farming which helps in the proper growth of the plants and fulfilling all the parameters. Siva Kumar,P. (2015) discuss et al Design and Development of Indore Plant Monitoring System about the implementation of sensors and arduino processor which helps in saving wastage of water and minimizing the labour cost and a manual work as well. Siddagangaiah.S. (2016) discuss et al A Novel Approach to IoT Based Plant Health Monitoring System about plant health monitoring system which will examine the environmental parameters such as humidity, temperature, etc. and all this information transmit from arduino uno board to the IOT cloud platform. Anushka,K. (2018) discuss et al Automatic Plant Monitoring and Watering System using Raspberry Pi to developed the Embedded System for plants monitoring and watering system based on IOT technology by implementing different sensors which analyze environment current situations and connects these sensor through internet and monitor the real time data. Jayaraman,P.P. (2016) discus et al Internet of Things Platform for Smart Farming: Experiences and Lessons Learnt about to developed the smart farming which is based on IOT platform and automate plant parameters data.

RELATED WORK

A related research work was done in a tomato nursery on the south of Italy by Mancuso et all. The envirnment temperature, humidity and content of moisture are estimated utilizing Sensicast device with remote sensor arrange or wireless sensor network (WSN). Online plant parameters monitoring application has additionally been created. The Greenhouse client can gather the parameters data over Internet .ie. fetched on cloud server. If the data changes or not upto the requirement, at that point a alert will be sent to the holderr's cell phone by SMS. In one moment interim the extension hub assembles information from other sensor hubs and transmits the estimations of temperature and relative humidity.

Teemu Ahonen completed research on Smart Plant Monitoring. He did the exploration in the Martens Greenhouse Research focus in Narpio town in Western Finland. Four natural key factors in nursery control were estimated by incorporating three business sensors with Sensinode's sensor.

PROBLEM IDENTIFIED

While surveying, the problems identified that the plant parameters are still fulfilling by humans, who leads to increase in improper growth of plants and decreasing the ecological cycle. It leads to an irregular plant health which results into unreliable new plant yield in a particular pot or a small land which is not beneficial for future scope.

A country like India, which is a main origin of agriculture and planting in a garden. Where citizens use an old method for planting which leads to improper growth of plants in a garden. These old methods need to be changed. To vanquish these problems this "smart plant monitoring system using IOT technology" is introduced in this modern technology, which is economical and beneficial for proper growth of plants.

Proposed Solution

Solution proposed for this chapter is to discuss the growth of plant or a group of plants which is being monitored by the sensor nodes and through Internet of Things. This system consists of sections such sensor nodes, local network (*LCD*), main server. Sensor nodes transmit the data to micro-controller (Arduino UNO) and the micro-controller transmits the data to NodeMCU board and NodeMCU board send data to cloud with the help of an ESP8266 WiFi module. Fig 1.1 shows the simplified form of the system framework.

Keynote: Sensor Nodes and Pots can be 1, 2, 3, 4, 5, and so on. These sensor nodes and pots are added or removed as per demand or requirement. Sensors can be joined or detached as per demand or requirement.

Block Diagram and System Description

Figure 2 is the simplified form of the sensor nodes. This figure shows sensor nodes, which consists of Micro-Controller (Arduino UNO), battery, water pump, soil moisture sensor, temperature/humidity Sensor (DHT11), and ultrasonic sensor.

Figure 3 simplified form of the local network and main server (cloud, mobile phone and computer). This figure shows Local network consists of LCD, Arduino UNO, and battery. The main server consists

Figure 1. Simplified form of the system framework

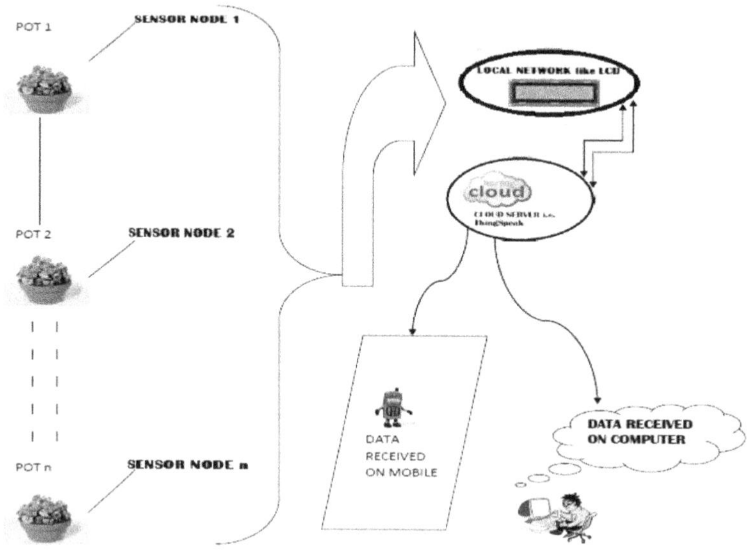

of WiFi Modem, cloud server, Computer, and Mobile Phone. Table 1 below shows the components list and the quantity of components need to establish the sensor node. In table1 abbreviate like M-M, M-F, F-F where,

M-M = Male to Male
M-F = Male to Female

Figure 2. simplified form of sensor node

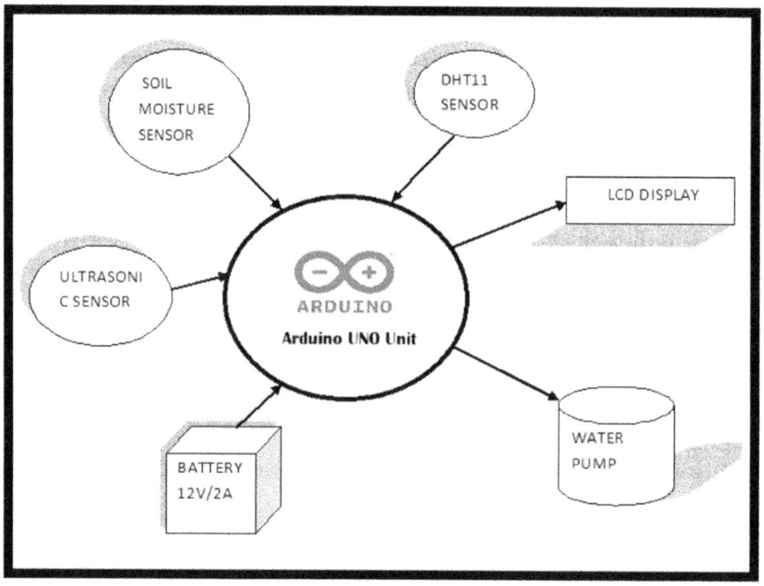

Figure 3. Simplified form of the local network and main server

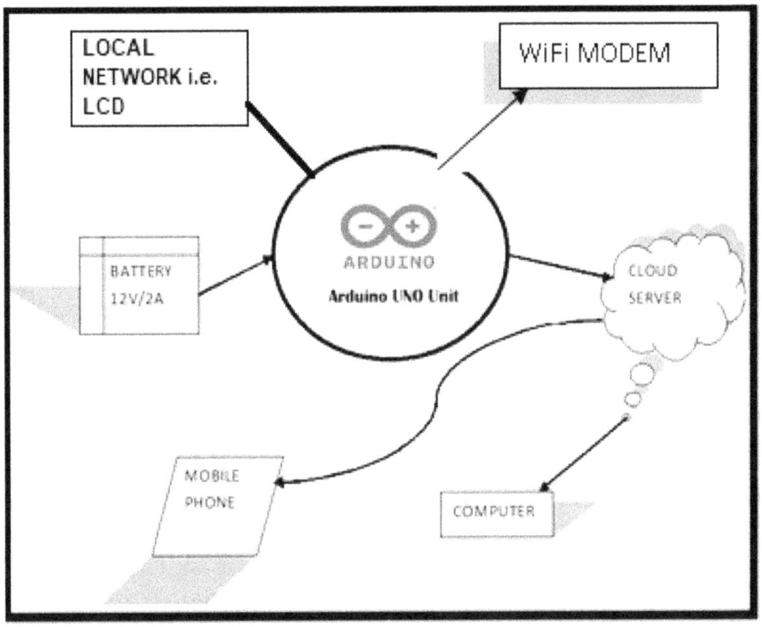

F-F = Female to Female

These abbreviate M-M, M-F, F-F are jumper wires used for the connections of sensors, water pump/ motor etc.

Table 1. Components list for sensor nodes

COMPONENTS REQUIRED	QUANTITY
Battery: Power supply of 12V/2A	1
Arduino UNO/Atmega interface board	1
Jumper wire M-M	20
Jumper wire M-F	20
Jumper wire F-F	20
Power supply extension to get +5V and GND	1
Soil moisture sensor	1
DHT11 sensor	1
Ultrasonic sensor	1
Water pump/ Water Motor	1
Relay	1
Water reservoir(bucket or water tank)	1
A thin water tube(water pipe)	1

Table 2 shows the components list and the quantity of components need to establish the local network and main server. In table 2 abbreviate like M-M, M-F, F-F where,

M-M = Male to Male
M-F = Male to Female
F-F = Female to Female

These abbreviate M-M, M-F, F-F are jumper wires used for the connections of LCD patch, GSM, NodeMCU patch etc.

Methodology Of The System

In this section we design a simple architecture of the system or we can simply say methodology of the system as shown in figure 4. All the sensors like DHT11, soil moisture, and ultrasonic are linked with internet which help us to send the data on IOT cloud server i.e. ThingSpeak. The fetched data help us to improve the proper plant growth. Ultrasonic sensor helps to check the level of water present in the tank and send the message according to the water level presents as well as the motor turn on or off. This simple architecture help us to demonstrate the main goal of this chapter.

DETAILED STUDY AND HARDWARE IMPLEMENTATION

In this section we will study in detailed about the, basics of Internet of Things (IOT) modules like NodeMCU, ESP8266, GSM and its interfacing with the devices, features of the modules as well as specifications of the modules. We will also discuss about the specifications of sensors, Pin Diagram of sensors and NodeMCU board, ESP8266 WiFi module, Arduino UNO/Atmega board as well.

Table 2. Components list for local network and main server

COMPONENTS REQUIRED	QUANTITY
Battery: Power supply of 12V/2A	1
Arduino UNO/Atmega interface board	1
Jumper wire M-M	20
Jumper wire M-F	20
Jumper wire F-F	20
Power supply extension to get +5V and GND	1
LCD 16*2/LCD 20*4	1
GSM	1
NodeMCU	1
LCD patch	1
NodeMCU patch	1

Note: All components are available at www.nuttyengineer.com

Figure 4. Architecture of the system

Key Note: *Sensors Plants can be 1, 2, 3, 4, 5, and so on. These sensors and plant are added or removed as per demand or requirement. Sensors can be joined or detached as per demand or requirement.*

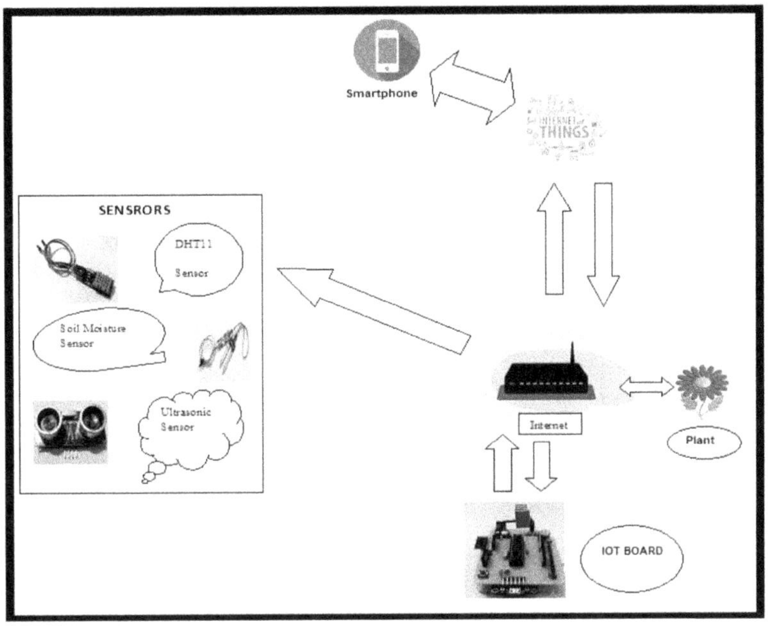

IOT stands for ***"Internet of Things"***. Internet of Things is divided into two words **Internet** and **Things**. The word **internet** means "a worldwide computer network/system or connectivity, giving a collection of data, information and transmission means, comprising of interconnected systems/networks utilizing normalized transmission conventions and backbone of connectivity" and word **things** means "a device or an object".

The Internet of Things (IOT) indicates to the usage of wisely linked devices/objects and system frameworks to use information or a data collected by installed sensors and actuators in machines and other objects/devices (things). As it were, the IOT (Internet of Things) can be called to any of the objects/devices linked with network and send data to the main server or a cloud.

The things (objects or devices) which is connected to the internet or installed can be focused into different classes such as:

- Devices or objects those are connected to the internet which gathered information and send it to the main server.
- Devices or objects that accept information and then take a step on it.
- Devices or objects that do both the actions like gathered information and send it to the main server and accept the information, then take a step on it.

Figure 5 is the schematic form of IOT (Internet of Things) that shows how the things those are connected to the internet which collect or gathered information and send it to the main server.

Let us take a general example to explain the concepts behind the IOT (Internet of Things) or the working of IOT (Internet of Things).

Figure 5. Schematic form of IOT (Internet of Things)

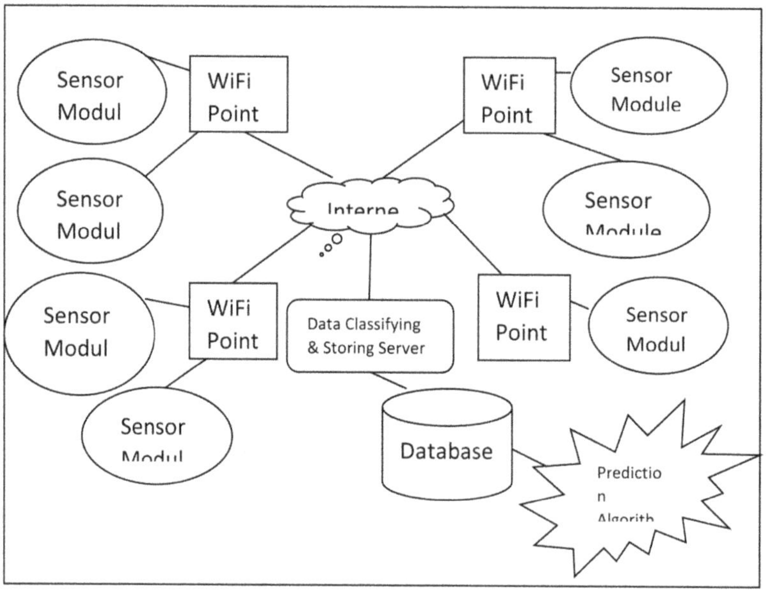

Principal of a school is connected to Vice Principal, vice principal is connected to all the class teachers of different classes and each class teacher is connected to his/her assign class students. Now, vice principal collect or gathered information from the principal then vice principal send information to all class teachers and each class teacher spread information among class students. Like this way IOT works. Where Principal of school acts like a sensor module, Vice Principal like a WiFi Module, Class teacher like a Cloud where data or information is collected/gathered and Class students like a Personal Computer or a Mobile Phone where the collected information is displayed.

DETAILED STUDY OF COMPONENTS LIST

ESP8266 WiFi Module: ESP8266 is a majestic, low-cost WiFi module relevant for adding WiFi purpose to a current microcontroller venture by means of a UART sequential network. The module can even be reinvented to go about as an independent WiFi network object/device.The exciting factor of ESP8266 is 802.11 b/g/n protocol Wi-Fi Direct (P2P), delicate-AP Integrated TCP/IP protocol stacks. ESP8266 is a WiFi microchip which was manufactured by Chinese organization name Espressif Systems in Shanghai. It is a processor with L106 32-bit RISC microchip center depending on the Tensilica Xtensa Diamond Standard 106Micro running at 80 MHz with memmory 32 KiB guidance RAM, 32 KiB guidance reserve RAM, 80 KiB user-information RAM and KiB ETS system-information RAM. Fig 1.6 shows the view of ESP8266 WiFi Module.

After ESP8266 WiFi Module was created, then NodeMCU created.

NodeMCU: NodeMCU is a wide origin IOT phase. It integrates with an unchangeable program which keeps running on the ESP8266 WiFi Module SOC (System on Chip) from convey if system framework and equipment or hardware which depends on the ESP-12 module. The expression "NodeMCU" as a matter of course specify to the firmware as opposed to the gadget or device packs. The firmware uti-

Figure 6. View of ESP8266 WiFi Module

Figure 7. View of NodeMCU

Figure 8. Arduino IDE windows

lizes the Lua scripting dialect Advanced API for equipment IO, which can significantly decrease the repetitive work of arranging and controlling equipment or hardware. NodeMCU Greatly accelerates our IOT application creating process. The principle favorable position of NodeMCU is less cost; size of the board is shrinking, less power utilization. NodeMCU is broadly utilized in IoT applications. It depends on C++ firmware.

Figure 9. File menu dialog box

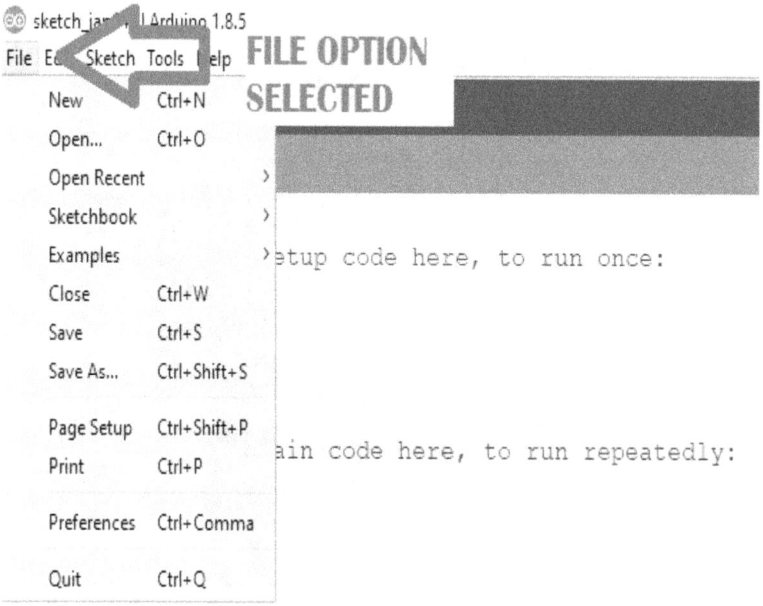

Figure 10. Preference Option Selected

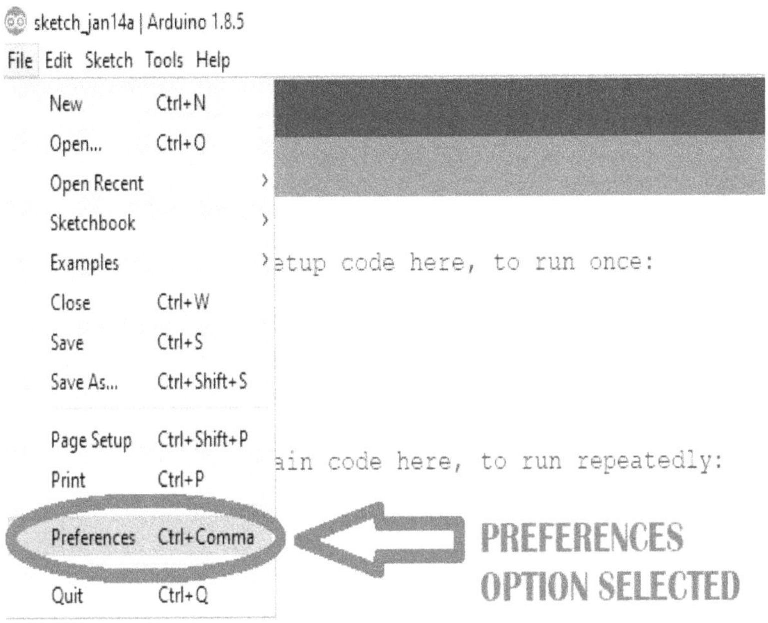

Figure 11. Preference Windows

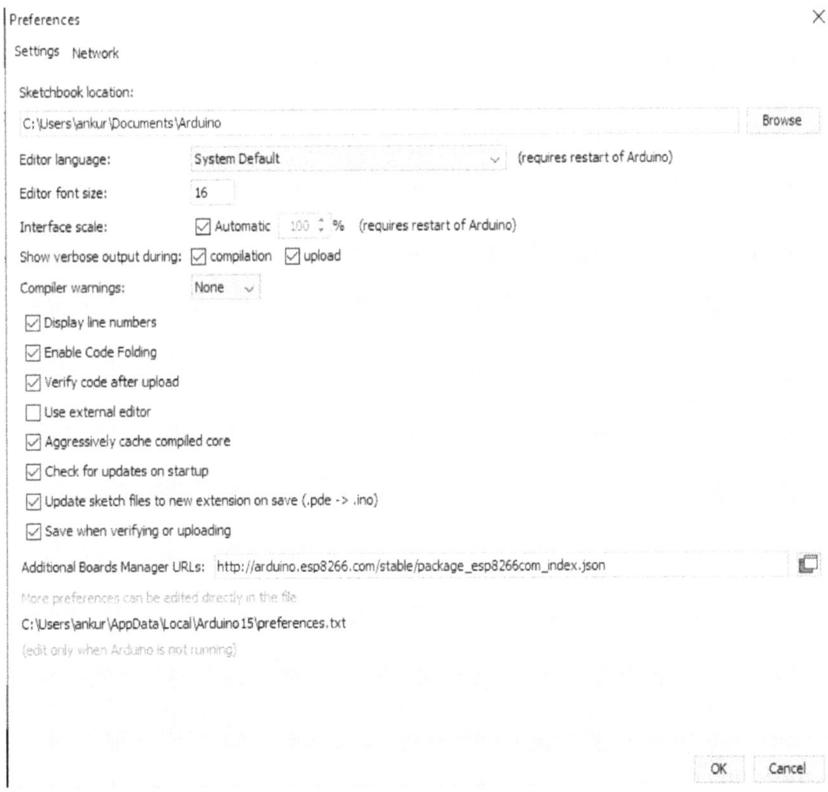

Figure 12. Additional Board Manager URLs Field Windows

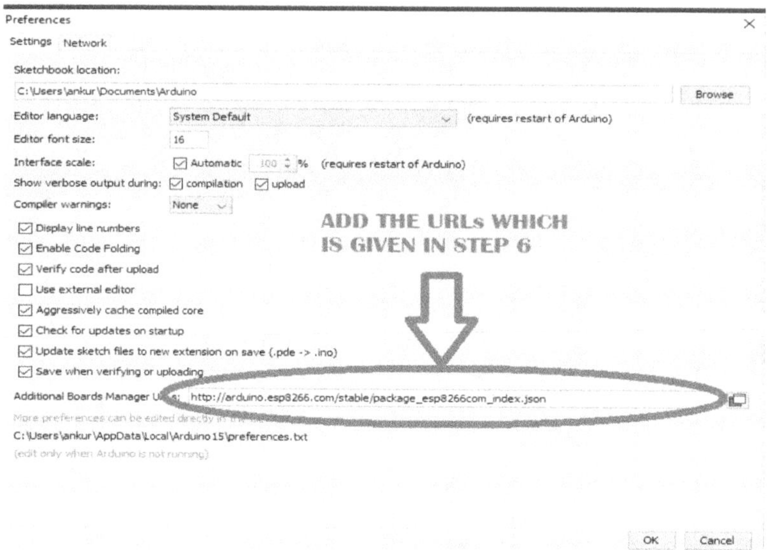

How to Install NodeMCU Board on Arduino IDE?

Step 1: Install Arduino IDE software in your PC (personnel computer). The latest version of Arduino IDE is at the given link https://www.arduino.cc/en/main/software, download it from this link. If already installed Arduino IDE then ignore the first step.

Step 2: Start Arduino and window of Arduino will open as shown in fig 1.8.

Step 3: Click on File menu dialog box as shown in figure 9

Step 4: Select Preference option as shown in figure 10

Step 5: Open Preference Window as shown in figure 11.

Step 6: Enter the given url http://arduino.esp8266.com/stable/package_esp8266com_index.json_into Additional Board Manager URLs field to add or download additional board like NodeMCU 1.0(ESP-12E Module), NUTTYFI IoT(ESP-12x Series), etc. as shown in fig 12.

Step 7: NodeMCU board is added to your Arduino IDE.

How to Make Connection?

Figure 13 shows how the connection of a circuit is to be made.

Interfacing of NodeMCU with Arduino UNO/Atmega Board

This part demonstrates the interfacing of NodeMCU and Arduino UNO/Atmega Board with the support of block diagram, and pictures.

Figure14 displays the block diagram of the NodeMCU its interfacing with Arduino UNO/Atmega Board. It consists of +12V/2A power supply DC jack, NodeMCU, LCD (acts like a display device), and Arduino UNO/Atmega. The main goal is to show how the interfacing of NodeMCU with Arduino UNO/Atmega is done and this interfacing forms a board which is known as **IOT Board**. The status or data of sensors which linked with this board is displayed on the LCD (Liquid Crystal Display). Table 3 displays the components list used to design the board.

Figure 13. Connection of a circuit

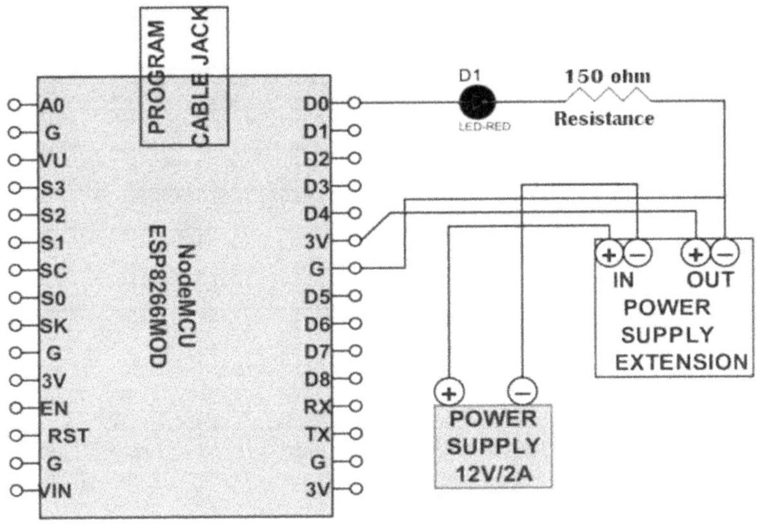

Figure 14. Block diagram of the interfacing

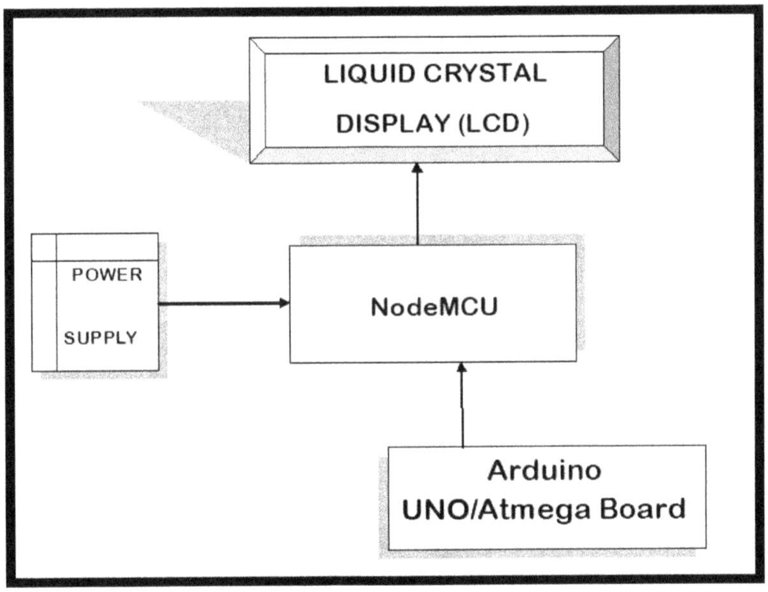

How to Interface NodeMCU with Arduino UNO/Atmega?

There are some simple steps to interface NodeMCU and Arduino UNO/Atmega. The steps are as follows:

Step 1: First take Arduino UNO/Atmega board as shown in figure 15.

Step 2: Now take NodeMCU board which is shown in figure 16.

Step 3: Assemble Arduino UNO/Atmega board and NodeMCU board as shown in fig 17. Assembling of board is known as **IOT Board**.

Table 3. Components list used for interfacing

COMPONENTS REQUIRED	QUANTITY
Power supply of 12V/2A	1
NodeMCU	1
Jumper wire M-M	20
Jumper wire M-F	20
Jumper wire F-F	20
Power supply extension (to get more +5V and GND)	1
LCD 16*2/LCD 20*4	1
LCD patch/LCD breakout board	1
Arduino UNO/Atemga	1
Heat synk	1

Note: All components are available at www.nuttyengineer.com

Figure 15. Arduino UNO/ Atmega board

Figure 16. NodeMCU board

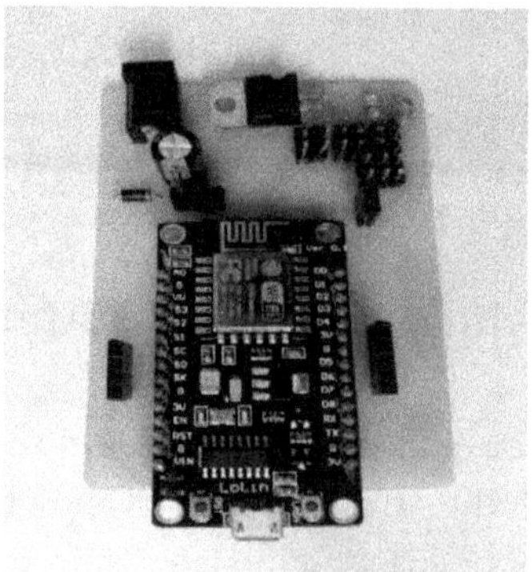

Figure 17. IOT Board

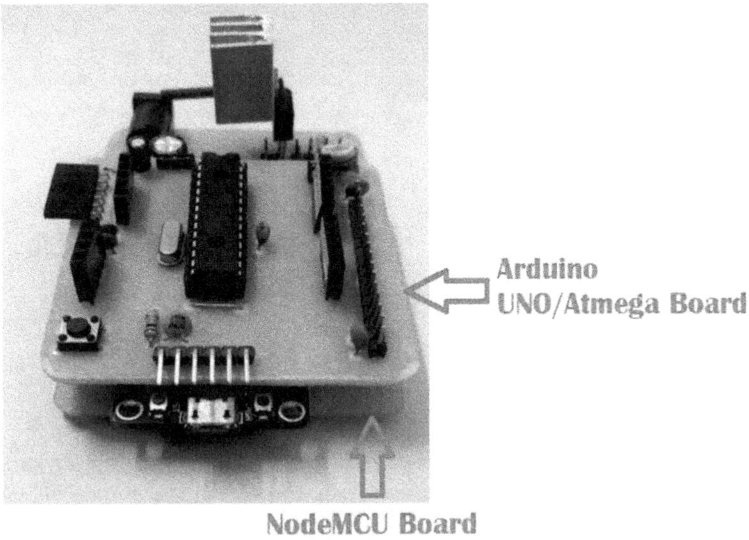

Note: IOT Board is available at www.nuttyengineer.com

Interfacing Of IoT Board With Sensors

The sensor is a present-day gadget or a device that is generally used to detect and react to electrical signals (or) optical signals. For the most part, the yield of the sensor will be as an electrical signals or optical signals. In our everyday life, we utilize various kinds of sensors generally used in different applications like the Infrared sensor is utilized to work the TV remote, PIR sensor (Passive Infrared Sensor) is utilized in opening and shutting the door, LDR sensor is utilized in to detect light-intensity, and so on.

Classifications of Sensor: There are two different classifications of sensor are analog sensor and digital sensor.

Digital Sensor: Electronic sensors or electrochemical sensors in which data change and data transmission appears digitally i.e. HIGH or LOW are called as advanced sensors. For example, PIR sensor, IR sensor, etc.

PIR Sensor (Passive Infrared Sensor): This sensor gives output signal digitally (High or Low) and it is used as a motion detector sensor. Figure 18 shows the view of PIR Sensor.

IR Sensor (Infrared Sensor): This sensor also gives output signal digitally (High or Low) and it is used to work the TV remote. Fig 1.19 shows the view of IR sensor.

Analog Sensors: Analog Sensors generates a constant output signal. They measures signal of physical quantities, for example, Temperature, Speed, Pressure, Displacement, and so on. Some examples of analog sensors are Lm35 sensor, LDR sensor, Soil moisture sensor, DHT11 sensor, DHT22 sensor etc.

Lm35 Sensor: Lm35 is also known as Temperature Sensor. This sensor gives the analog signal as an output signal and it is used to detect air temperature. Figure 20 shows the view of Lm35 sensor.

LDR (Light Dependent Resistance) Sensor: This sensor gives the analog signal as an output signal and it is used to detect light intensity. This is sensor also for a smart city light project. Figure 21 shows the view of LDR sensor.

Figure 18. View of PIR sensor

Figure 19. View of IR Sensor

Note: All sensors are available at www.nuttyengineer.com

Figure 20. View of Lm35 (temperature) sensor

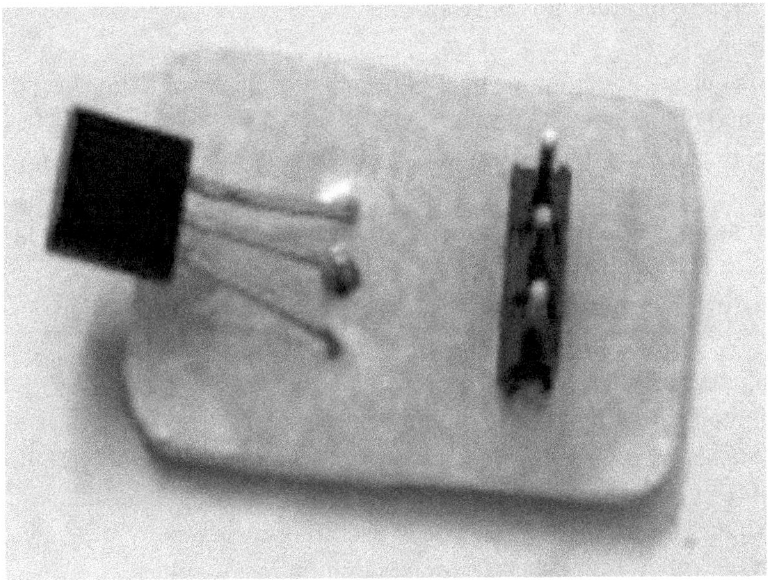

Figure 21. View of LDR sensor

Sensors and Components used in this Chapter for Smart Plant Monitoring System Using IOT Technology are as Follows

1. **Soil Moisture Sensor:** Soil Moisture Sensor is utilized to determine the proportion of water in the soil. This creates it perfect for fulfilling tests in area, for example, soil science, farming (or

agriculture), ecological science, plant science, and natural science. Fig 22 displays the view of soil moisture sensor. Use of Soil Moisture Sensor:

a. It is used to check or determine the fall of moisture after sometime because of disappearance of water and plant take-up.
b. It is used to calculate optimal value soil moisture substance or content for different types of plants.
c. It is used to supervise soil moisture substance to control water system in nurseries.

Pinout of Soil Moisture Sensor: The various pinout of soil moisture sensor are as follows:

1. **A0:** Analog pin to get analog value or analog data.
2. **D0:** Digital pin to get digital value or digital data.
3. **GND:** GND stands for ground. GND pin is used to connect with GND pin of microcontroller (Arduino UNO).
4. **VCC:** VCC stands for voltage. VCC pin is used to connect with VCC pin of microcontroller (Arduino UNO). Soil moisture sensor requires minimum 5V voltage.

Figure 23 displays the block diagram of the interfacing of soil moisture sensor with IOT board. It consists of 12V/2A power supply, IOT board, Liquid Crystal Display (LCD), and Soil Moisture Sensor. In this chapter soil moisture sensor is used to check the level of moisture present in the soil.

How to Make a Connection of Soil Moisture Sensor With IOT Board?

Figure 24 shows how to make a connection of Soil Moisture Sensor with IOT Board to check the working of sensor and displaying the data obtained by sensor on Liquid Crystal Display (LCD).

Figure 22. View of soil moisture sensor

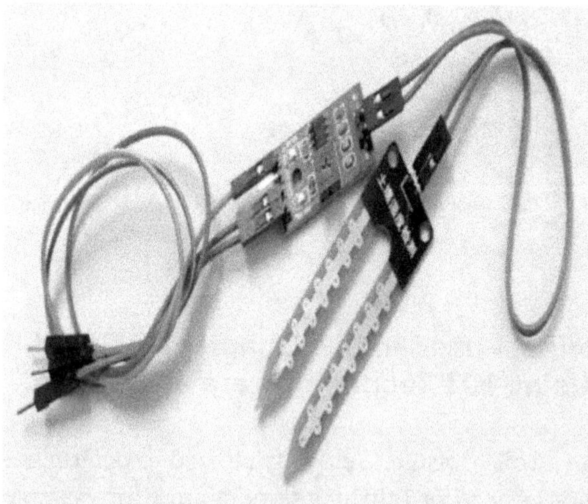

Figure 23. Block diagram of the interfacing of soil moisture sensor

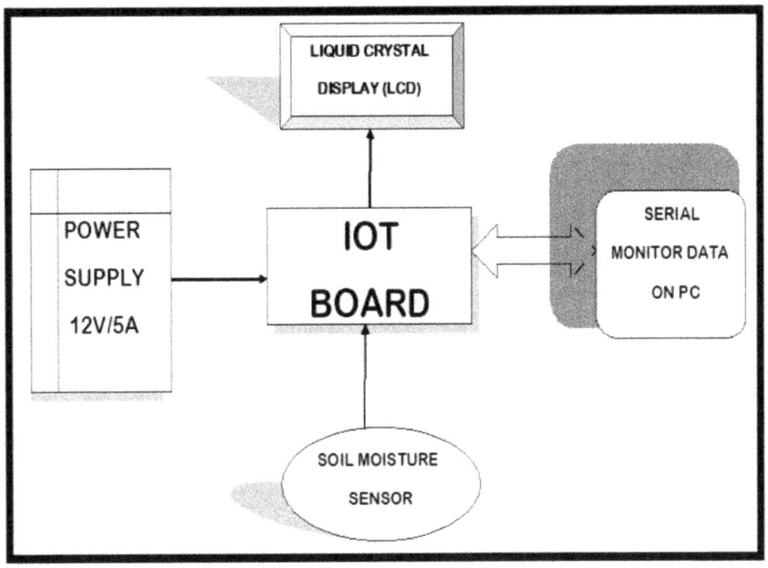

Figure 24. Connection of soil moisture sensor with LCD

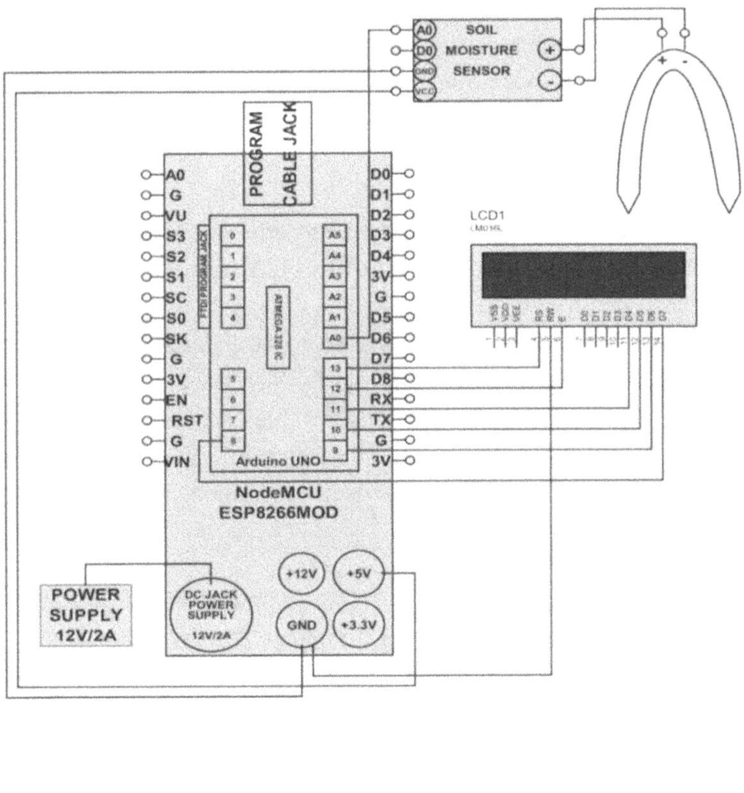

Figure 25. View of DHT11 Sensor

2. **DHT11 Sensor:** DHT11 is a needed, in-expensive computerized temperature and humidity sensor. It employs with an electrical device humidity sensor and a thermistor to calculate or measure the environment air, and discharge a digital signal on the data pin (no analog pins are required). Figure 25 shows the view of DHT11 Sensor.

Note: DHT11 sensor available at www.nuttyengineer.com
Pinouts of DHT11 Sensor: The various pinout of DHT11 sensor are as follows:

1. **VCC:** VCC stands for voltage. VCC pin is used to connect with VCC pin of microcontroller (Arduino UNO). DHT11 sensor requires minimum 5V voltage.
2. **DATA:** Data pin which is used to read the sensor value and forward data to the micro-controller (Arduino UNO) and micro-controller display the data on LCD or on Serial monitor.
3. **GND:** GND stands for ground. GND pin is used to connect with GND pin of microcontroller (Arduino UNO).

Figure 26 displays the block diagram of the interfacing of DHT11 sensor with IOT board. It consists of 12V/2A power supply, IOT board, Liquid Crystal Display (LCD), and DHT11 sensor. In this chapter DHT11 sensor is used to check the environment humidity and temperature and shows whether it is beneficial for plant proper growth or harmful for plant growth.

Figure 26. a Block diagram of the interfacing of DHT11 sensor

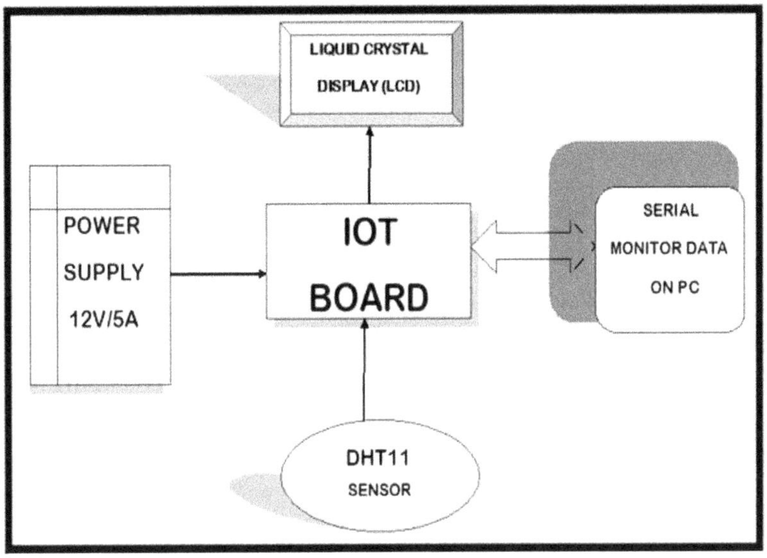

Figure 27. Connection of DHT11 sensor with LCD

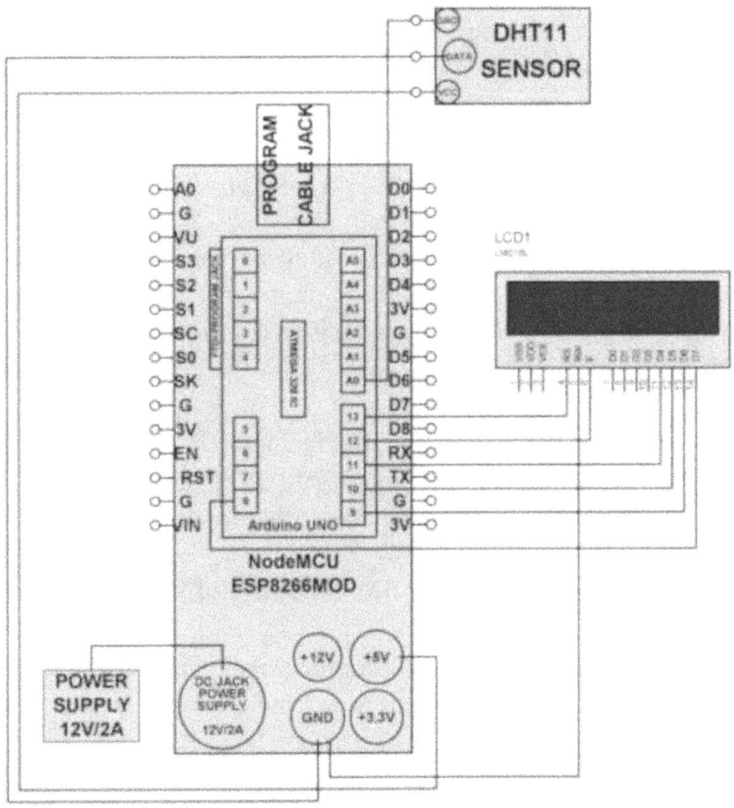

How to Make a Connection of DHT11 with IOT Board?

Figure 27 shows how to make a connection of DHT11 with IOT Board to check the working of sensor and displaying the data obtained by sensor on Liquid Crystal Display (LCD).

How to Add DHT11 Library to Arduino IDE?

Step 1: Download the library from the given url https://www.brainy-bits.com/dht11-tutorial/ .

Step 2: Downloaded file is a zip file, select the zip file >> right click >> then select extract here option and extract the file and copy the extracted file.

Step 3: Now open C drive or local disk C: >> then open Program Files (x86) >> open Arduino folder >> now open library folder and paste the extracted which was copied earlier (see step 3).

Step 4: Restart Arduino IDE. Now you can write your DHT11 program and execute it and check data.

3. **Ultrasonic Sensor:** As the name suggests, ultrasonic sensors estimate separation between ultrasonic sensor and object by utilizing ultrasonic waves. The sensor imparts ultrasonic wave and gets the wave reflected yet again from the goal or a target. Ultrasonic Sensors estimates the partition to the goal or a target by assessing the time between the release or discharge of ultrasonic sensor waves and encounter (object). Figure 28 shows the view of ultrasonic sensor.

Note: DHT11 sensor available at www.nuttyengineer.com

Pinouts of Ultrasonic Sensor- The various pinout of Ultreasonic sensor are as follows:

1. **VCC:** VCC stands for voltage. VCC pin is used to connect with VCC pin of microcontroller (Arduino UNO). DHT11 sensor requires minimum 5V voltage.
2. **Trigger:** Trigger pin acts as an Output pin. This pin must be kept high for 10us to boot (start) estimation by sending US (ultra sonic) wave.
3. **Echo:** Echo pin acts as an Input pin. This pin goes high for duration of time which will be equivalent to the time taken for the US (ultra sonic) wave to return back to the sensor.
4. **GND:** GND stands for ground. GND pin is used to connect with GND pin of microcontroller (Arduino UNO).

Figure 29 displayss the block diagram of the interfacing of Ultrasonic sensor with IOT board. It consists of 12V/2A power supply, IOT board, Liquid Crystal Display (LCD), and Ultrasonic sensor. In this chapter ultrasonic sensor is used to show the water level present in a water tank.

How to Make a Connection of Ultrasonic Sensor with IOT Board?

Figure 30 shows how to make a connection of Ultrasonic sensor with IOT Board to check the working of sensor and displaying the data obtained by sensor on Liquid Crystal Display (LCD).

4. **Water Pump:** A pump is an apparatus that maneuver fluids (liquids or gases), or every so often slurries, by mechanical activity. Pump can be arranged into three classes according to the method

Figure 28. View of Ultrasonic Sensor

Figure 29. Block diagram of the interfacing of Ultrasonic sensor

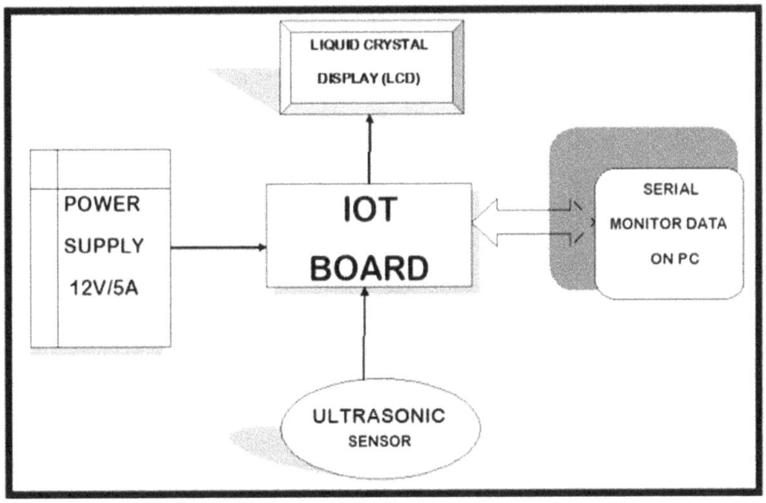

they use to move the fluid: clear lift, evacuation, and gravity pump. Pump work by some instrument (usually reacting), and it consume essentialness to perform mechanical work by maneuvering the fluid. Pumps work by methods for various essentialnesses's including manual assignment power, engines, or wind control, come in various sizes, from insignificant for use in medicinal activities

Figure 30. Connections of ultrasonic sensor

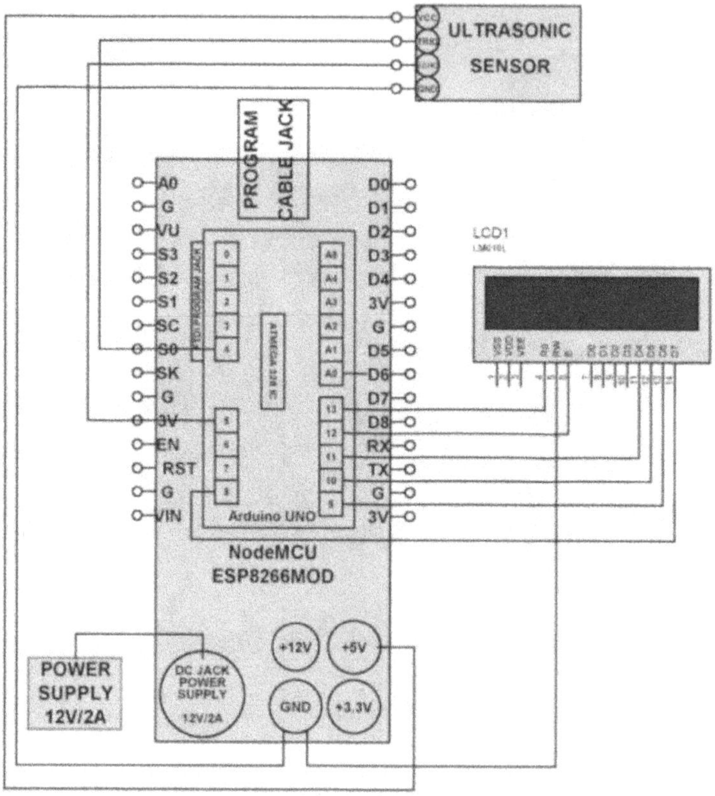

Figure 31. View of DC water pum with LCD

Figure 32. View of Relay

Figure 33. Block diagram of the interfacing DC water pump and relay.
Note: *Relay available at www.nuttyengineer.com*

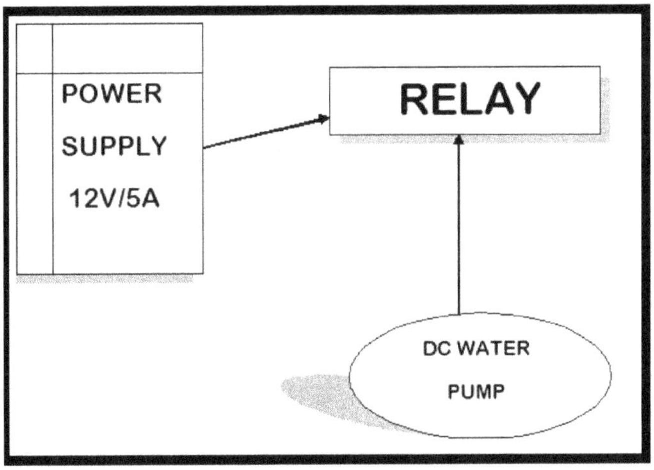

to significant mechanical pumps. Fig 1.31 shows the view of dc water pump. Dc water pump operates with the help of relay driver. Relay helps in regulating the current. This draws low current and converts it into high current. Fig 1.32 shows the view of relay driver. In this chapter water pump is used for watering the plant and pumping the water in a water-tank.

Fig 1.33 displays the block diagram of the interfacing of Relay with DC water pump. It consists of 12V/2A power supply, DC water pump, and Relay. In this chapter relay is used to turn on motor or turn off motor.

How to Make a Connection of DC Water Pump with Relay?

In this we will check the connections of relay and dc water pump with micro-controller (Arduino UNO). There is a relay board with three headers name as +12V, GND, INPUT on one side. The+12V pin on relay board is the voltage supply and it is connected with +12V pin of micro-controller board, GND pin is on relay board is the ground and it is connected with GND of micro-controller and IN pin is the INPUT pin of relay on relay board and it is connected with IN pin of micro-controller. The other four pin name as AC, OUT, AC, IN on other side of relay board. AC pin is the AC source pin and OUT pin is the OUTPUT pin. These two pin is the ac source pin which will insert in the ac source such as socket on electricity switch board. These two pin AC and OUT will insert in socket like a plug insert. The other pins such as AC pin is the ac pin and one wire of dc water pump is connected with this pin and other wire of dc water pump should be connected with IN is pin port. Figure 34 shows how to make a connection of DC water pump with relay to check the working.

How to Program Relay which helps to Start a DC Water Pump?

5. **GSM:b** GSM is a mobile to mobile conversation modem; it is means global system for mobile communication (GSM). The concept of GSM was produced at Bell Laboratories in 1970. It is broadly utilized mobile conversation (communication) on the planet. GSM is an open and advanced cell

Figure 34. Connection of relay with dc water pump

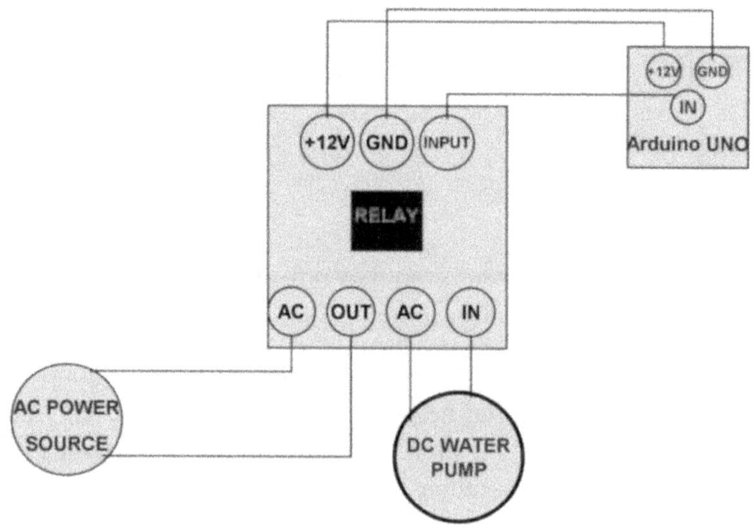

Figure 35. Vie of GSM modem
Note: *Relay available at www.nuttyengineer.com*

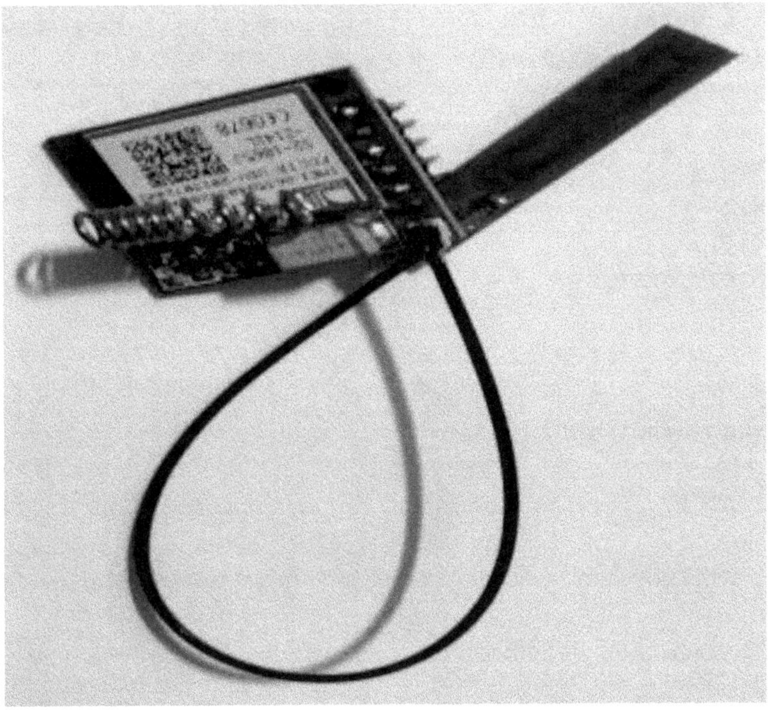

innovation utilized for transmitting mobile voice and information supply works at the 850MHz [GSM], 900MHz [EGSM], 1800MHz [DCS] and 1900MHz [PCS] frequency bands. Figure 35 displays the view of GSM modem. In this chapter use of GSM is to send the sms(short message service) when the level of water in the tank is less or full and whether the motor turns on of off.

AT Commands to Send SMS are give below-

1. Send AT+CMGF=1 using Serial.println command in Arduino IDE to initialize GSM module in text form.
2. Send AT+CMGMS=\”mobile number with country code\”\r using Serial.println command in Arduino IDE to send the message to given number. For example, Serial.println(“AT+CMGS=\”+ 91XXXXXXXXXX\”\r”).
3. Send (char)26; using Serial.println command in Arduino IDE which is ASCII of cntrl+Z to stop the process.

Implementing the above commands in program-

```
#include <LiquidCrystal.h> //LCD header file (external LCD library)
LiquidCrystal lcd(13,12,11,10,9,8);
void setup()
{
Serial.begin(9600);
lcd.begin(16, 2);
lcd.setCursor(0,0);
lcd.print("GSM INITIALSING "); // to initialize GSM
delay(10000);
lcd.clear();
delay(100);
lcd.print("Message Sending");
delay(1000);
Serial.println("AT");     //Thus command is used to send SMS in Text Mode
delay(2000);
Serial.println("AT+CMGF=1");     //This command is used to send SMS in Text Mode
delay(2000);
Serial.println("AT+CMGS=\"+9196XXXXXXXX\"\r"); //change to the phone number
according to your utiliy
delay(2000);
Serial.println("Welcome to SME Dehradun ");
delay(500);
Serial.println((char)26); //This command is used to stopping character
delay(2000);
lcd.clear();
lcd.print("message send                ");
delay(2000);
} // end of setup()
void loop()
{
}  // end of loop()
```

Main Server: The main server is on ThingSpeak.com. Follow the some simple steps to create account on ThingSpeak.

Steps to sign up on ThingSpeak server

Step 1: Visit on the given URL https://thingspeak.com/. After visiting the given URL ThingSpeak server windows will appear as shown in fig 1.36.

Step 2: Now check the sign up option at right-hand corner of the thingspeak server window and create the MathWorks account and filled the necessary details which is shown in fig 1.37.

Step 3: Enter your current Email ID for example; an*****@gmail.com / an*****@hotmail.com / an*****@yahoo.com etc. then >> enter your location for example; INDIA, CANADA, UNITED STATES, AUSTRALIA, AUSTRIA etc. then >> enter your first name for example; Jhon, Bhupi, Raj, Wilson, Akon etc. then >> your last name for example; Singh, Kohli, Kumar, Starc, Stoinis etc.

Step 4: After filling the details click on continue option on the same window which is shown on figure 37.

Step 5: Then select option use this account (the account you entered) for my MathWork account and after this selection verified Email is sent to your mail ID, verify your account and you can sign in ThingSpeak server with your account.

Steps to Create Channel on ThingSpeak Server

Step 1: First sign in to ThingSpeak after creating a MathWork account.

Step 2: Click on New channels option in figure 38

Figure 36. ThingSpeak server window

Figure 37. Create MathWork account window

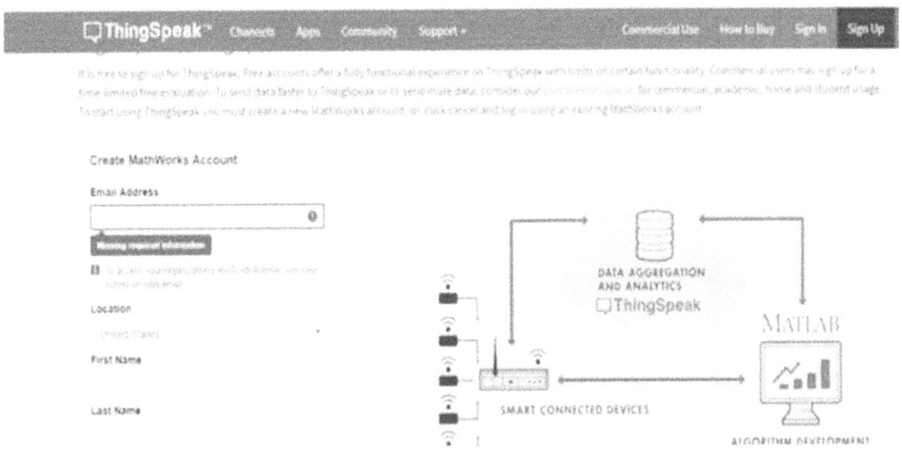

Figure 38. Window for ThingSpeak

Figure 39. New channel window

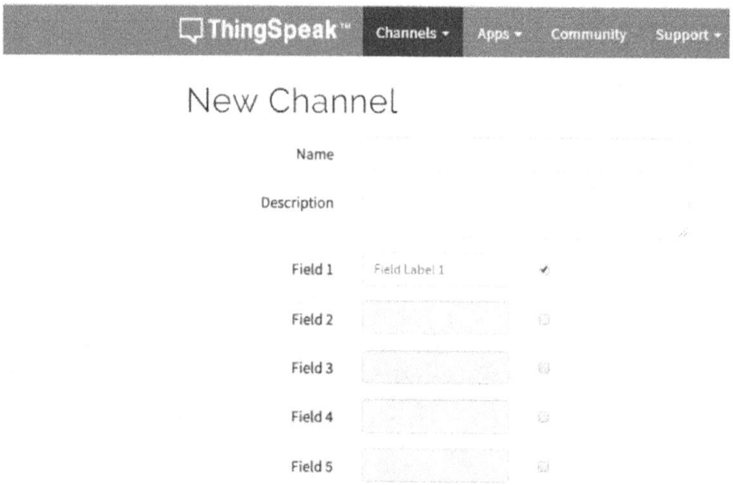

Figure 40. Sharing option window

Figure 41. API keys window

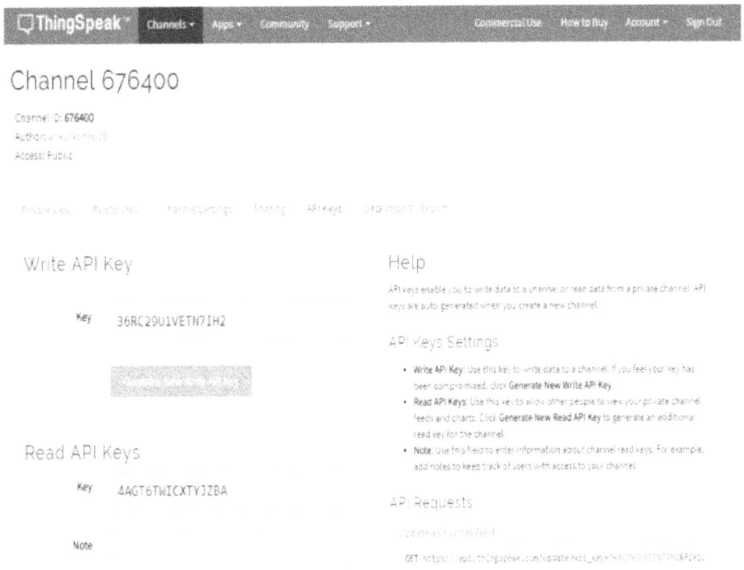

Step 3: New channel window will appear as shown in figure 39. Write your channel name for example Plant Monitoring System and write field name in the boxes, first selecting the field and save the channel at the bottom of the window. Enter the value like:

Field 1: Soil Moisture
Field 2: Humidity
Field 3: Temperature

Step 4: Now click on sharing option and sharing option window will appear as shown in below and click on option "Share channel view with everyone" as shown in figure 40

Step 5: Click on API Keys option >> API Keys option window will appear as shown in figure 41 and create your own API Key. "Write API Key will used in a program".

IOT Based Water Tank Monitoring and Plant Parameters Monitoring

This part will discuss about water tank monitoring system and plant parameters monitoring system, fetch data on cloud server with the help of IOT board (Arduino UNO board + NodeMCU board). Sensors are connected on Arduino UNO board. Then packets of data which is collected on Arduino UNO board is transmits to NodeMCU board through serial communication. The process of collecting data on Arduino board and transmit data to NodeMCU board and then fetched on cloud server i.e. ThingSpeak is called IOT monitoring.

Figure 42 is the block diagram of the monitoring system. This consists of IOT board (i.e. comprises of Arduino board and NodeMCU board), Liquid Crystal Display (LCD), Soil Moisture Sensor, Water Level Sensor (i.e. Ultrasonic Sensor), DHT11 Sensor, 2 Relay Board, 2 DC Water Pump.

Figure 42. Block diagram of the system

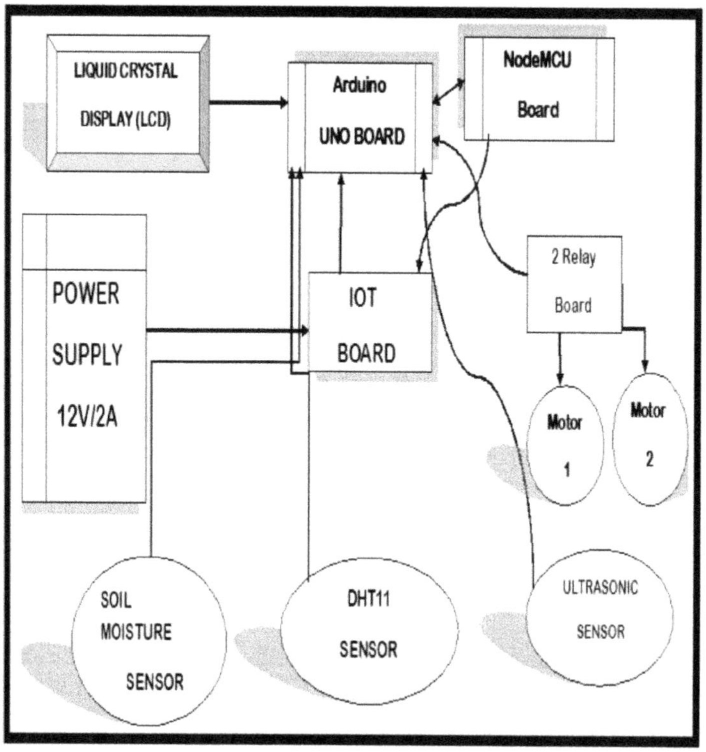

Figure 43. Connection of the devices
Note: *All components available at www.nuttyengineer.com.*

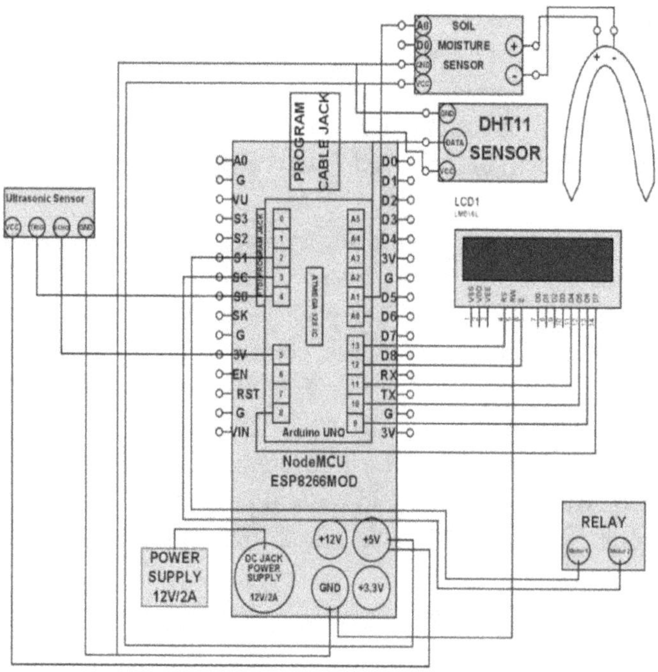

Connections of Water Tank Monitoring System and Plant Parameters Monitoring System

First interface Arduino Board and NodeMCU board as discuss earlier in this chapter and this interfacing of board is known as IOT (Internet of Thing) Board. Connect all the sensor nodes such as Soil Moisture Sensor, DHT11 Sensor, and Ultrasonic Sensor on Arduino Board. Then connect Relay input pin with Arduino UNO. Now connect Liquid Crystal Display (LCD) on Arduino board. Connections of system is done and this shown in figure 43.

Program: This part will demonstrate the program of Arduino UNO and NodeMCU to run the connected devices with it. This is an IOT code.

Arduino UNO Code

```
#include "dht.h" //to include dht library
#define dht_apin A0 // Analog Pin sensor is connected to Aalog pins
#include<LiquidCrystal.h> //LCD header file
int rs=13, en=12, d4=11, d5=10, d6=9, d7=8;     // initializing LCD pin
LiquidCrystal lcd(rs, en, d4, d5, d6, d7); //lcd pins
const int trigPin = 4;          // acts as input pin
const int echoPin = 5;      // acts as output pin
// defines variables
long duration;     // to store duration
int level;      // to calculate distance
const int depth=74;      // assuming depth of water tank
int Y=0;
float Z=0;
const int relay_pin1 = 2;  //initializing relay pin for motor 1
const int relay_pin2 = 3;  //initializing relay pin for motor 2
dht DHT;
float H, T;        // H for humidity & T for Temperature
#include<SoftwareSerial.h>
SoftwareSerial mySeries(6, 7);     // RX=6, TX=7
int soil_sensor_pin=A1;
int moisture=0; //to store moisture value
int level=; //to store level of water in a tank
void setup()
{      //setup() started
mySeries.begin(9600);
pinMode(trigPin, OUTPUT);              // Setting the trigPin as an Output
pinMode(echoPin, INPUT);                // Setting the echoPin as an Input
```

```
Serial.begin(9600);   // This command is used to begin the serial communica-
tion
lcd.begin(16,2);    //to beging LCD and this is 16*2 LCD; for 20*4 replace
only (16,2) with (20,4) like lcd.begin(20,4)
lcd.clear();    // to clear lcd
lcd.setCursor(0,0);    // This command is for to set the cursor to print a
string(sentence)
lcd.print("WELCOME TO SME");    // welcome message is displayed on LCD to check
the working of LCD
lcd.setCursor(3,1); //to set the cursor where the message has to display on
LCD
lcd.print("DEHRADUN"); //print command to print the message
delay(1000);     // dipalying welcome message for 1000 milliseconds
lcd.clear();   // clear lcd
Serial.println("DHT11 Humidity & temperature Sensor\n\n"); //printing the
message serially
Serial.println("Reading value from sensor:"); //booting up the soil moisture
sensor
lcd.setCursor(0,0);    // to set cursor
lcd.print("Reading Sensors:");
delay(500);
lcd.clear(); //clear command to clear the lcd message
Serial.println("Depth of tank: 74%");  //assuming the initial level of water
tank
lcd.setCursor(0,0);    // to set cursor
lcd.print("Tank Depth: 74%");
delay(1000);
lcd.clear();
}       // end of setup()
void loop()
{       //loop()started
soil_moisture_sensor(); //calling soil moisture sensor()
dht11_sensor(); //calling DHT11 sensor()
if (moisture<=10) //comparing the moisture content in the soil
{
  relay_water_motor_on();   //calling dc water motor funtion to start motor
  gsm_watering_plant();    //calling GSM function for watering a plant
} // end of if condition
else if (moisture<=50) //comparing the moisture content in the soil
{
  relay_water_motor_on();   //calling dc water motor funtion to start motor
  gsm_watering_plant();     //calling GSM function for watering a plant
```

```
} //end of else if condition
else if(moisture=90) //comparing the moisture content in the soil
{
  relay_water_motor_off();    //calling dc water motor funtion to stop motor
  gsm_watering_plant_stop();  //calling GSM function to stop watering a plant
} //end pf else if condition
else
{
  relay_water_motor_off();     //calling dc water motor funtion to stop motor
  gsm_watering_plant_stop();   //calling GSM function to stop watering a plant
} end of else condition
universal_ultrasonic(); //calling ultrasonic()
depth_tank(); //caaling depth_tank()
mySeries.print('\r');
mySeries.print(moisture);
mySeries.print('|');
mySeries.print(H);
mySeries.print('|');
mySeries.print(T);
mySeries.print('|');
mySeries.print(level);
mySeries.println('\n');
Serial.print('\r');
Serial.print(moisture);
Serial.print('|');
Serial.print(H);
Serial.print('|');
Serial.print(T);
Serial.print('|');
Serial.print(level);
Serial.println('\n');
delay(2000);
 } //end of loop()
void soil_moisture_sensor()
{
moisture=analogRead(soil_sensor_pin); //reading soil moisture sensor value
moisture=map(moisture,1023,10,0,100); // map function used to mapping values
}   // soil moisture sensor program
void dht11_sensor()
{
DHT.read11(dht_apin);
H=(DHT.humidity);
```

```
T=(DHT.temperature);
delay(2000);      //Wait 5 seconds before accessing sensor again.
}  //dht11 sensor program
void universal_ultrasonic()
{
// Clears the trigPin
digitalWrite(trigPin, LOW);
delayMicroseconds(2);
// Setthing the trigPin on HIGH state for 10 micro seconds
digitalWrite(trigPin, HIGH);
delayMicroseconds(10);
digitalWrite(trigPin, LOW);
// Reading the echoPin, returns the sound wave travel time in microseconds
duration = pulseIn(echoPin, HIGH);
// Calculating the distance
level= duration*0.034/2;
// Printing the distance on the Serial Monitor
Serial.print("Level of tank: ");
Serial.println(level);
delay(500);
lcd.setCursor(0,0);
lcd.print("Tank level: ");
lcd.print(level);
delay(500);
} // ultrasonic sensor program function
void gsm_less_water()
{
Serial.print("GSM INITIALSING ");
delay(10000);
Serial.println("AT");    //This command is used to send SMS in Text Mode
delay(2000);
Serial.println("AT+CMGF=1");    //This command is used to send SMS in Text
Mode
delay(2000);
Serial.println("AT+CMGS=\"+9196xxxxxxxx\"\r"); // change to the phone number
according to your requirement
delay(2000);
Serial.println("Water is less in a tank and Motor switched ON");// to check
message serially
delay(500);
Serial.println((char)26);    //This command is used to stopping the character
Serial.println("message send                    ");
```

```
delay(2000);
} // GSM modem program function for less water in a tank
void gsm_full_tank()
{
Serial.print("GSM INITIALSING ");
delay(10000);
Serial.println("AT");    //This command is used to send SMS in Text Mode
delay(2000);
Serial.println("AT+CMGF=1");    //This command is used to send SMS in Text
Mode
delay(2000);
Serial.println("AT+CMGS=\"+9196xxxxxxxx\"\r"); // change to the phone number
according to your requiremnt
delay(2000);
Serial.println("Water is full in a tank and Motor switched OFF");  // to check
message serially
delay(500);
Serial.println((char)26);  //the stopping character
Serial.println("message send                ");
delay(2000);
}   // GSM modem program function for full tank function
void gsm_empty_tank()
{
Serial.print("GSM INITIALSING ");
delay(10000);
Serial.println("AT");    //This command is used to send SMS in Text Mode
delay(2000);
Serial.println("AT+CMGF=1");    //This command is used to send SMS in Text
Mode
delay(2000);
Serial.println("AT+CMGS=\"+9196xxxxxxxx\"\r"); // change to the phone number
accordinf=g to your requirement
delay(2000);
Serial.println("Water is full in a tank and Motor switched ON"); // to check
message serially
delay(500);
Serial.println((char)26);  //This command is used to stopping the character
Serial.println("message send                ");
delay(2000);
}   // GSM modem program function for empty tank
void relay_water_pump_on()
{
```

```
digitalWrite(relay_pin2, LOW); // The relay is active low, so it turns on when we set pin to low
Serial.println("Pump ON");
lcd.setCursor(0,0);
lcd.print("Motor ON");
delay(1000);
lcd.clear();
}    // dc water pum function to start motor
void relay_water_pump_off()
{
digitalWrite(relay_pin2, HIGH); // The relay is active high, so it turns off when we set pin to high
Serial.println("Pump OFF");
lcd.setCursor(0,0);
lcd.print("Motor OFF");
delay(1000);
lcd.clear();
}    // dc water pump function to stop motor
void depth_tank()
{
  if(level>=74)
  {
    level=74;
  }
  Y=depth-level;
  Z=Y*1.3513; // divide 100 by depth of tank i.e. 74
  if(Z>=95)
  {
    Z=95;
  }
  Serial.print("TANK ");

  Serial.print(Z);
  Serial.print(" %");
  if(Z==95)
  {
    Serial.println(" TANK OVERFLOW ");
    relay_water_pump_off();    // calling dc water pump funtion to stop motor
    gsm_full_tank();    // calling GSM full water in a tank function
  }
  else if(Z>=50)
  {
```

```
    Serial.println(" TANK HALF FILLED ");
    relay_water_pump_on();    // calling dc water pump funtion to activate
motor
    gsm_less_water();    // calling GSM less water in a tank function
  }
  else
  {
    Serial.println(" TANK EMPTY ");
    relay_water_pump_on();     // calling dc water pump funtion to activate
motor
    gsm_empty_tank();     // calling GSM empty tank function
  }
    delay(1000);
}
void gsm_watering_plant()
{
Serial.print("GSM INITIALSING ");
delay(10000);
Serial.println("AT");     //To send SMS in Text Mode
delay(2000);
Serial.println("AT+CMGF=1");     //To send SMS in Text Mode
delay(2000);
Serial.println("AT+CMGS=\"+9196xxxxxxxx\"\r"); // change to the phone number
you using
//    Serial.println("AT+CMGS=\"+91xxxxxxxxxx\"\r"); // change to the phone
number you using
delay(2000);
Serial.println("Watering to Plant because soil is dry)");    // to check mes-
sage serially
delay(500);
Serial.println((char)26);       //the stopping character
Serial.println("message send                 ");
delay(2000);
}    // GSM modem program function to start watering to plant
void gsm_watering_plant_stop()
{
Serial.print("GSM INITIALSING ");
delay(10000);
Serial.println("AT");     //To send SMS in Text Mode
delay(2000);
Serial.println("AT+CMGF=1");     //To send SMS in Text Mode
delay(2000);
```

```
Serial.println("AT+CMGS=\"+9196xxxxxxxx\"\r"); // change to the phone number
you using
//    Serial.println("AT+CMGS=\"+91xxxxxxxxxx\"\r"); // change to the phone
number you using
delay(2000);
Serial.println("Watering to Plant is stop because soil is wet");   // to check
message serially
delay(500);
Serial.println((char)26);     //the stopping character
Serial.println("message send                    ");
delay(2000);
}    // GSM modem program function to stop watering to plant
void relay_water_motor_on()
{
digitalWrite(relay_pin1, LOW); // The relay is active low, so it turns on when
we set pin to low
Serial.println("Motor ON");
lcd.setCursor(0,0);
lcd.print("Motor ON");
delay(1000);
lcd.clear();
}    // dc water pum function to start motor
void relay_water_motor_off()
{
digitalWrite(relay_pin1, HIGH); // The relay is active high, so it turns off
when we set pin to high
Serial.println("Motor OFF");
lcd.setCursor(0,0);
lcd.print("Motor OFF");
delay(1000);
lcd.clear();
}    // dc water pump function to stop motor
NodeMCU
#include<SoftwareSerial.h> //Software Serial header file
SoftwareSerial mySeries(D6,D7, false, 256); // RX=D6, TX=D7
#include <ESP8266WiFi.h> //ESP8266 WiFi module header file
int moisture=0;    // to store content of moisture in a soil
float Humidity;    // to store humidity
float Temperature;  // to store temperature
int X_Time=0;
String apiKey = "VGE8XXNTI4FTBTPC";  //generate your own api key from Thing-
Speak server
```

```
const char* ssid = "Schematics Microelectronics Dehradun";  // wifi  name;
write your own wifi name
const char* password = "XXXXXXXXXXX";   // write wifi password
const char* server="api.thingspeak.com";
WiFiClient client;
void setup() //start of setup()
{
  Serial.begin(9600); //initializing Serial board rate
  mySeries.begin(9600);
  delay(10);
  WiFi.begin(ssid,password); //initializing WiFi
  Serial.println();
  Serial.println();
  Serial.print("Connecting TO");
  Serial.println(ssid);
  WiFi.begin(ssid,password);
  while(WiFi.status() != WL_CONNECTED)
{
  delay(500);
  Serial.print(".");
}
Serial.print("");
Serial.print("WiFi Connected");
delay(2000);
} // end of setup()
void loop()  //start of loop()
{
   X_Time++;
  if(client.connect(server,80)) //connected to server
  {
    seriesEvent_NodeMCU_BOARD(); //calling of SeriesEvent_NodeMCU_BOARD()
    if(X_Time>=10);
    {
      send_data_to_thingspeak();
    X_Time==0;
    }
  }
  client.stop();
  Serial.println("WAITING");
  delay(2000);
}  // end of loop()
void send_data_to_thingspeak() //start of send_data_to_thingspeak()
{
```

```
  String str=apiKey; //storing apiKey
  str +="&field1="; //creating field1
  str += String(moisture); //storing moisture value in field 1
  str +="&field2="; //creating field2
  str += String(Humidity); //storing Humidity value in field 2
  str +="&field3="; //creating field3
  str += String(Temperature); //storing Temperature value in field3
  str +="\r\n\r\n";
  client.print("POST /update HTTP/1.1\n");
  client.print("Host: api.thingspeak.com\n");
  client.print("Connection: close\n");
  client.print("X-THINGSPEAKAPIKEY: "+apiKey+"\n");
  client.print("Content-Type: application/x-www-form-urlencoded\n");
  client.print("Content-Length: ");
  client.print(str.length());
  client.print("\n\n");
  client.print(str);
  Serial.print(moisture);
  Serial.print('|');
  Serial.print(Humidity);
  Serial.print('|');
  Serial.print(Temperature);
  Serial.print('|');
  Serial.println("DATA SEND TO SERVER");
}    //end of send_data_to_thingspeak()
void seriesEvent_NodeMCU_BOARD() //start of seriesEvent_NodeMCU_BOARD()
{
  while(mySeries.available()>0) //start of while loop
  {
    if(mySeries.available()<1)
    return;
    char data=mySeries.read();
    if(data != '\r')
    return;
    moisture=mySeries.parseInt();
    Humidity=mySeries.parseFloat();
    Temperature=mySeries.parseFloat();
    delay(600);
  }//end of while loop
}     //end of seriesEvent_NodeMCU_BOARD()
```

Note: The map function used for mapping the soil moisture values which was fetched by soil moisture sensor.

Figure 44. Arduino board

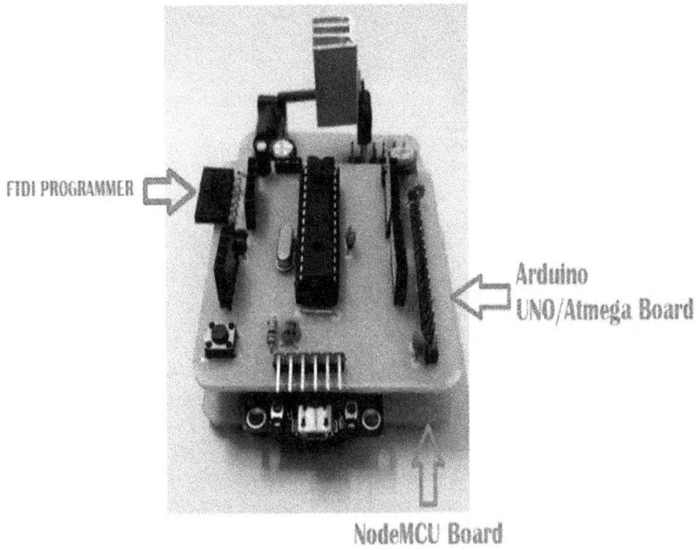

Figure 45. FTDI programmer and cable

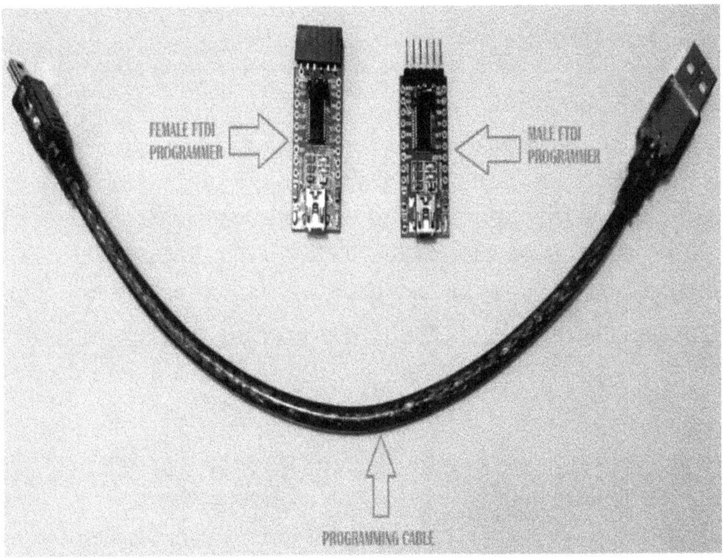

Map Function

Function is used to re-map the values like from one span to another i.e. a Low value (from_Low value) will get generalized to maximum range (to_Low value) and a High value (fromHigh value) will get generalized to minimum range (toHigh value) and the process of re-generalizinging the values is known as MAP FUNCTION. Let's taken an example of the soil moisture sensor sensor value

Figure 46. NodeMCU board and android phone data cable

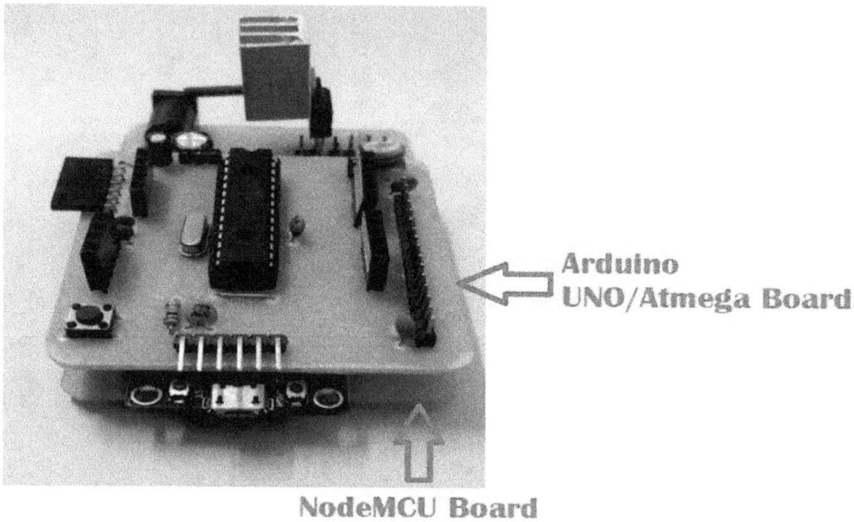


```
                    output_val=map(output_val,1023,10,0,100); ----------------
-----(i)
Syntax of equation (i) is
                    map(value, fromLow, fromHigh, toLow, toHigh) -----------
(ii)
Variables used in equation (ii) are:
              value: the number or value to map
              fromLow: the lower bound of the value's current range
              fromHigh: the upper bound of the value's current range
              toLow: the lower bound of the value's target (maximum) range
              toHigh: the upper bound of the value's target (minimum) range
```

<u>Example Code to implement Map Function:</u>

```
   void setup() {}
   void loop()
{
int output = analogRead(sensor);
output = map(val, 0, 1023, 0, 100);
}
```

How to Upload a Program on IOT Board?

Step 1: First upload Arduino Code on IOT board i.e. Arduino UNO Board as shown in figure 1.44.

Step 2: Upload a program on Arduino Board with the help of FTDI (Future Technology Devices International) programmer as shown in Figure 45

Figure 47. Charts from the field data received on cloud server (ThingSpeak)

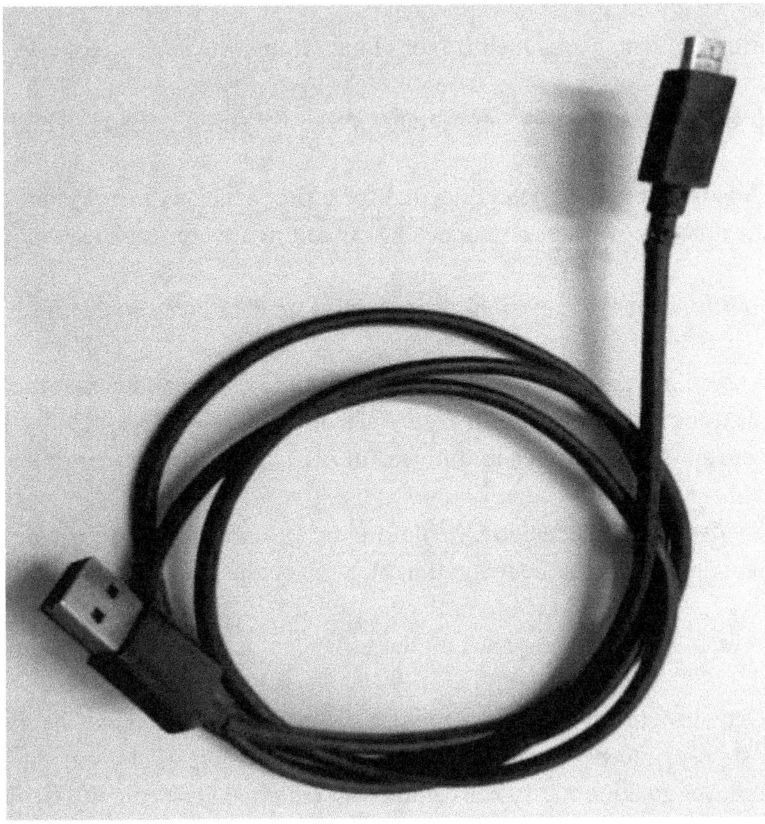

Table 4.

Graph 1	Graph 2	Graph 3
This graph demonstrates the content of moisture in the soil, fetched by soil moisture sensor. This data help us to check the moisture level present in the soil and fulfill the water level in the soil which helps the plant proper growth.	This graph demonstrates the level of humidity presents in the atmosphere, fetched DHT11 sensor. This data helps to check the humidity and clarify that the humidity level is good for plant health or not and also help us take precautions against this fetched data.	This graph demonstrates the environment temperature, fetched by DHT11 sensor. This data also helps to check the temperature present in the environment as humidity did and clarify that the temperature level is good for plant health or not and also help us take precautions against this fetched data.

Step 3: Now NodeMCU Code on NodeMCU board with the help of android phone data cable as shown in fig 1.46 with the help of Android Phone cable.

Step 4: Connect all the sensors, motor as described earlier in Connection of a devices part.

Note- All components are available at www.nuttyengineer.com

PROS AND CONS

Pros: There are various advantageous method for plant automation such as follows

- **Time Saving:** Once the system is setup, many systems can be integrated and controlled without wasting time.
- **Parameters Absorption:** Parameters needed for a proper growth of a plant can be fulfilled automatically like amount of water content in a soil and in a water-tank as well, moisture level in a soil, etc.
- **Real Time Monitoring:** We can monitor the real time data and surveys for the better growth of the plants.
- **Economical:** The main advantage of IOT technology is cheap and economical. IoT in a general sense turns out to be extremely useful to people in their daily schedules by influencing the apparatuses to convey to one another in a powerful way monitoring strength and cost and makes a perfect solution.
- **No Labour Involvement:** By making the automatic system for plants growth there is no involvement of labour and this reduces the manual work which helps the plants to fulfill all the parameters.

Cons: There are various disadvantageous method for plant automation such as follows

- **Equipment Deterioration:** There is a biggest problem for equipment deterioration because of lack of power supply, internet connectivity, garbage values fetched by sensors etc.
- **Unemployment:** By making the system automatic which leads to unemployment.
- **Installation Cost:** Whether the system is for one or two or more plants the basic cost for installation is expensive. Considering the complete system we need to install all the equipments for the automation of plants to make a smart plant monitoring system which is too expensive for single plant automation.
- **Regular Monitoring of the System-** There is always require regular monitoring. This is because there can be equipment failure, internet connectivity etc. which needs to be regular monitoring.

RESULTS AND DISCUSSIONS

Figure 47 shows the charts from the field data i.e. Soil Moisture, Humidity, and Temperature showing the data fetched on cloud server.

In figure 47 first graph displays the data fetched by the soil moisture sensor, second graph displays the humidity data fetched by DHT11 sensor and the last graph displays the temperature data fetched by DHT11 sensor. These graph data fetched at cloud server i.e. ThingSpeak server. Soil moisture sensor graph explains the content of the moisture present in the soil which was mapped from the scale of 0 to 100. Humidity sensor graph explains the humidity level in the atmosphere and Temperature sensor graph explains the temperature level in the atmosphere which demonstrates that humidity and temperature level is beneficial for plant growth or not. These graphs fetched the real time data from which we can monitor the parameters of the plants.

CONCLUSION

This chapter concludes successful implementation of smart plant monitoring system using IOT technology by merging following tasks -

- Automatic turn on of the motor if the level of tank decreases below 20% and "empty tank" message will be send to the user.
- Automatic turn off the motor if the level of tank reaches t its maximum value.
- The pump will water the plant if the moisture content decreases below 30%.
- The pump will stop watering if moisture content increases above 90%.
- The real time data value of temperature and the humidity of the surrounding atmosphere will be uploaded on the cloud.

This process helps to develop a smart way of gardening which is based on IOT technology using ThingSpeak Server.

FUTURE SCOPE

Smart Plant Monitoring System using IOT Technology may prevent plant from detorating under harsh weather condition and help to promote sophisticated farming methods by growing different species which need extreme care.

This may help to increase the production of different types f plants and also help to improve the plant growth which may have different medicinal and economical benefits. Thus is may help to preserve nature and reduces harmful toxic substances which have imbibed our nature since existence of mankind.

REFERENCES

Anita, G., Rajesh, S., Mamta, M., Bhupendra, S., Ravindra, S., & Vikasgarg, G. (2018). *Hands on approach on Application of Internet of Things in Various Domain of Engineering*. Delhi: International Research Publication House.

Anushka, K., & Mahadevasawamy, U.B. (2018). Automatic Plant Monitoring and Watering System using Raspberry Pi. *I.J. Engineering and Manufacturing,* (6), 55-67. Doi:10.5815/ijem.2018.06.05

Athani, S., Tejeshwar, C.H., Patil, M., Patil, P., & Kulkarni, R. (2017). *Soil moisture monitoring using IoT enabled Arduino sensors with neural networks for improving soil management for farmers and predict seasonal rainfall for planning future harvest in North Karnataka-India.* Academic Press. doi:10.1109/I-SMAC.2017.8058385

Deepali. (2017). A Survey on Automatic Irrigation System Using Wireless Senor Network. *International Journal of Current Engineering And Scientific Research, 4*(8), 69-73.

Elangovan, R., Banu, A., Rozario, R., & Santhanakrishan, N. (2018). *Tome: A Plant Monitoring and Gardening System using IoT*. Academic Press.

Gultom, J. H., Harsono, M., Khameswara, T. D., & Santoso, H. (2017). Smart IoT Water Sprinkle and Monitoring System for chili plant. *2017 International Conference on Electrical Engineering and Computer Science (ICECOS)*, 212-216. 10.1109/ICECOS.2017.8167136

Jayaraman, P.P., Yavari, A., Morshed, A., Zaslavsky, A., & Geograkopolous, D. (2016). *Internet of Things Platform for Smart Farming: Experiences and Lessons Learnt*. Academic Press. Doi:10.339016111884

Keerthana, B., Nivetha, P., Boomika, M., & Mathivatahni, M. (2018). Literature Review on IOT Based Smart Security and Monitoring Devices for Agriculture. *International Journal on Future Revolution in Computer Science & Communication Engineering, 4*(2), 342-345.

Kumar, N., Kumar, P., Archana, B., & Nishant, G. (2017). Arduino based Automatic Plant Watering System with Internet of Things (IOT). *International Research Journal of Advanced Research in Electrical, Electronics and Instrumentation Engineering, 6*(3), 1-8. Doi:10.15662/IJAREEIE.2017.0603144

Michal, S., Jiri, V., Jan, M., & Palvik, J. (2016). *Internet of Things (IoT) in Agriculture - Selected Aspects*. Academic Press. . doi:10.7160/aol.2016.080108

Sadasivan, S., Vadhri, V., & Ramesh, S. (2015). *Smart Plant Monitoring System*. Academic Press. Doi:10.19140/RG.2.1.2915.6563/1

Siddagangaiah, S. (2016). A Novel Approach to IoT Based Plant Health Monitoring System. *International Research Journal of Engineering and Technology, 3*(11), 880-884.

Siva Kumar, P., Jilani, S.A.K., & Venkata Raman, B. (2015). Design and Development of Indore Plant Monitoring System. *International Journal of Advanced Research in Electronics and Communication Engineering, 4*(10), 2531-2535.

Sushabhan, C., Bhupendra, S., Rajesh, S., & Anita, G. (2018). *Internet of Things [IoT] Enabled Automation in Agriculture*. NIPA Publication House.

Thamaraimanalan, T., Vivekk, S. P., Satheeshkumar, G., & Saravanan, P. (2018). Smart Garden Monitoring System using IOT. *Asian Journal of Applied Science and Technology, 2*(2), 186-192.

Chapter 17
Temperature and Humidity Sensors With Arduino and Android

Kavita Srivastava
Institute of Information Technology and Management, Delhi, India

Sudhir Kumar Sharma
Institute of Information Technology and Management, Delhi, India

ABSTRACT

Future internet will not only comprise computers and computing things but also everything else. IoT is one of the emerging technologies. This chapter explores different kinds of temperature and humidity sensors, their specifications, usefulness, and prospective applications. Arduino boards are also explained in detail. This chapter explores the working principle and applications of different types of humidity sensors. The chapter contains the specification of common temperature and humidity sensors, pinout diagram, and schematic of the interconnection of sensors with Arduino Uno microcontroller. Working and interfacing of the ESP8266 wifi module are also explained. The chapter also provides the architecture and code examples of Android app for sensor data view and manipulation.

INTRODUCTION

Origin of IoT

In the year 2000 we have seen a new era of ubiquitous computing. It means connectivity every time and everywhere. Number of humans on the earth will become minority and number of connected things will outnumber. The challenge is that all the connected things generate data all the time. This data need to be handled and analysed.

Internet of Thing will become more complex network with much wider scope. It has many more complexities as compared to the current network. Internet, which is considered as a global network

DOI: 10.4018/978-1-5225-9574-8.ch017

connecting computers will now connect not only computing devices but it will interconnect anything and everything.

Therefore, objects such as lights, fans, heating devices, gas stoves, microwaves, wheel chairs, refrigerators, air conditioners, locks, doors, tables, chairs and anything that you can imagine can be connected and form the part of Internet.

IoT has lot to offer to the society. Imagine that a blind person is crossing a road with IoT enabled stick in his hand which is guiding him about directions and warning for obstacles. Consider another case where IoT enabled wheel chair is operated remotely by patient's wards. This is the whole vision of IoT. IoT has potential to make everything smart whether it is cities, homes, farms, schools, colleges, laboratories or gardens.

Physical devices are of different types. Several technologies like low powered embedded systems, cloud computing, data science, computer vision, big data, machine learning, deep learning and technologies from Electrical engineering and mechanical engineering are required.

Over 9 billion things have been interconnected till now and it is projected that in near future over 20 billion things will be interconnected. IoT will form a big Internet.

The term "Internet of Things (IoT)" was coined by Kevin Ashton in a presentation given by him at Proctor & Gamble in 1999. Kevin Ashton conceptualized Internet of Things as

"If we had computers that knew everything there was to know about things – using data they gathered without any help from us – we would be able to track and count everything, and greatly reduce waste, loss and cost"– **Kevin Ashton** *(1999)*[1].

According to CISCO-IBSG (CISCO Internet Business Solution Group), IoT refers to exactly the time when connected "things or objects" grow more in number as compared to human being.

However, the Internet of Things is considered to be known in the year 2008 – 2009 according to a report published by CISCO IBSG[2]. The total number of interconnected device will reach to 25 billions by the year 2015 and this number will be doubled in next 5 years. That is, it is estimated that by the year 2020 number of interconnected devices will become 50 billions. Figure 1 shows the increase in connected devices with increase in world population.

Figure 2 shows the growth in usage of connected devices by people.

Internet of Things evolved gradually in several stages. The inventor of Berkeley Unix and the founder of Sun Microsystems, Bill Joy, in his speech at the World Economic Forum in Davos in 1999, described the concept of "Six Webs"[3]. First, "The Near Web", that is, the Internet accessible from a Laptop or Desktop. Second, "The Here Web", that is the Internet always available to you in Mobile Phones. Third, "The Far Web", that is, the Internet available on Television. Fourth, "The Weird Web", that is, the Internet that you use to interact through voice. Fifth, "B2B", where machines interact with each other and the Sixth one is "D2D", that is the Internet of Sensors.

In 2000, LG launched first Internet Refrigerator that was capable of sensing the types and quality of products being stored. In 2001, David L. Broke at MIT proposed Electronic Product Code (EPC)[4] that uniquely identifies products and thus enabling the tracking of products throughout their life cycle. The objective was to enable coordination among physical things that can be controlled remotely. In the year 2003, Bernard Traversat and co-authors presented a research paper "Project JXTA-C: Enabling Web of Things"[5] in 36th Hawaii Conference on System Sciences – 2003" that describes implementation of JXTA protocol stack that can be embedded at system level.

Figure 1. Growth in Connected Devices

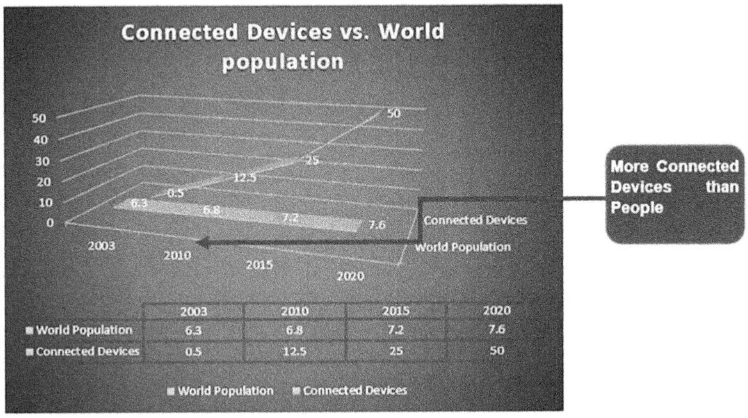

In the same year Interaction Design Institute Ivrea (IDII) in Ivrea, Italy started Arduino project. In 2005, single board microcontroller was developed at IDII that was used in various interactive projects developed by students. In 2008, IPSO Alliance was formed to promote Internet Protocol (IP) to enable communication between smart objects in the fields of energy, consumer, healthcare, and industry applications. Table 1 shows the events resulting in emergence of Internet of Things.

In the next section the literature survey in the direction of automation with emphasis on temperature and humidity monitoring and controlling is briefly described.

LITERATURE SURVEY

Temperature and Humidity monitoring have been used in industrial applications for long in various thermal processes such as boilers, furnaces, ovens, vats etc. (Roots et al., 1969). Bernstein et al. (1969)

Figure 2. Each person will have more number of connected devices in coming years

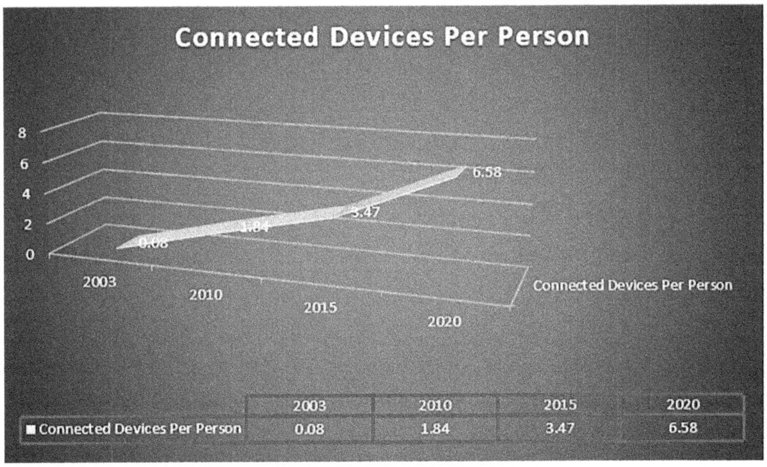

Table 1. Important events related to the evolution of Internet of Things

Year	Event
Year 1999	Bill Joy created the term Device to Device Communication (D2D).
Year 2000	The Internet Refrigerator (LG Internet DIOS – RS73CT) was launched by LG.
Year 2001	David L. Broke at MIT proposed Electronic Product Code (EPC) for tracking physical things.
Year 2003	Bernard Traversat and co-authors presented a research paper "Project JXTA-C: Enabling Web of Things"
Year 2005	Single board microcontroller was developed at IDII.
Year 2008	IPSO Alliance was formed to promote the use of IP for communication among smart things.
Year 2016	We have seen lots of applications of IoT in connecting homes, coordinating vehicles, smart agriculture, manufacturing and many more fields.

demonstrated the use of Resistance Wire Temperature Sensor for measuring the environmental temperature. Lardear, 1991 proposed a method for controlling temperature without conventional sensors. In this approach, cable resistance was used to determine the temperature. The control system utilized 24 individual controllers.

Beale et al. (1997) presented the design of temperature controller for controlling the temperature of ceramic samples. The required temperature range is 1200°C to 1500°C. At this range Thermocouple type temperature sensors can be used which operate from -200°C to 1750°C.

Folea et al., 2012 proposed a Smart home Automation System using Tag4M system with TC1046VNB Temperature Sensor and HIH-5030 Humidity Sensor.

Online Temperature Control System proposed by Ikhlef et al., 2014 gives the LabView simulation. The proposed system was intended to be used for educational purpose. In another application of temperature control, PID controller using fuzzy logic was proposed. (Mungisha et al. 2015). However, its implementation was not discussed.

Singh et al., 2015 proposed Hybrid Particle Swarm Optimization (HPSO) algorithm. They have used ATMega8 and ATMega16 microcontrollers and LM35 Temperature Sensor. For wireless communication Zigbee WiFi module was used. Behera et al., 2015 emphasised the use of thermostats for temperature controlling. Chen et al., 2015 proposed Bluetooth Low Energy (BLE) Based system for In-Home Data Acquisition. In this system, Bluetooth communication has been used to transfer the data to connected PC. Newberry et al., 2015 proposed Wireless Sensor Based Home Automation using ATMega328 MCU.

Temperature Control Solution with PLC was proposed by Rata et al., 2016. In this experiment, a PLC XC-CPU-101 was used which connects the analog I/O module. The temperature sensor used in this experiment was LM35 temperature sensor. For communication CANopen protocol was used.

Reddy et al., 2016 proposed IoT based Home Automation System using Arduino Board, LM35 Temperature Sensor and HLK-RM04 WiFi Module.

Wahab, 2016 proposed an automation system for people with disabilities using Arduino Uno board and Cytron BlueBee wireless communication. Wang et al., 2016 proposed a temperature controller using thermostats.

Kumar, 2017 proposed Smart Home Control using a microcontroller interfaced to PC through RS-232 serial communication. Al-Kuwari et al., 2018 proposed Smart-Home Automation using IoT-based Sensing and Monitoring Platform. In this system authors have proposed the use of Arduino, NodeMCU or Raspberri Pi for data processing. The sensors used in proposed system are LM35 Temperature Sensor and HIH4000 Humidity Sensor.

Next section discusses the problem that is being addressed later in the chapter.

Problem Identified

Temperature and Humidity monitoring has been used in research field for several decades. Most of the solutions published are either theoretical in nature or require costly experimental setup. There is a need of a simple and cost effective solution which can speed up the automation process and leverage the research. It is also required for the solution to be feasible in home environment so that anybody can build innovative automation applications. Arduino based IoT applications have potential for addressing the challenges in automation. A comparison of this approach with existing solutions is given next.

Comparative Study of Existing Solutions

Advanced Virtual RISC (AVR) microcontrollers, Programmable Logic Controller (PLC) and Programmable Interface Controllers (PIC) have been used largely in industrial automation applications. However, AVRs and PICs are bare chips. In order to build applications one has to build the supporting circuitry as well. Hence the use of these chips in application makes it very complex as well as expensive. On the other hand, Arduino has supporting software library and it can be programmed over USB.

PLC is bulkier and complex and consumes more power as compared to Arduino. Most PLCs are proprietary. It restricts them to be used in industrial automation. Arduino is suitable for creating a home robot and other automation applications. In the next section a solution is proposed using Arduino.

PROPOSED SOLUTION

Temperature and Humidity Sensor

Temperature and Humidity sensors offer a variety of IoT applications. Applications of these sensors can be found in various industrial, biomedical and environmental applications for measurement of temperature and relative humidity. *Humidity* is measured as the amount of water vapours present in surrounding air.

Humidity effects the health of human beings in several ways. Humidity shows the relationship with temperature. At very high temperature, if the air is dry or humidity is low we feel comfortable. However if humidity is also high along with temperature it poses several health risks such as dehydration. If temperature is low and air is also dry that means low humidity, then also there are several health issues. Hence both Temperature and Humidity are significant in human well-being. The relationship between these two factors must be measured, monitored and analysed. Temperature and Humidity sensors can be used in variety of Personal Comfort Systems. Other domestic applications of Temperature and Humidity monitoring is in the area of cooking and indoor gardens.

In medical field Temperature and Humidity sensors have applications in ventilators, sterilisers and incubators.

In this chapter, the use of Temperature and Humidity Sensor in an IoT application is discussed. The three components – Arduino board, Sensors and WiFi module are the main focus of this chapter.

Table 2. Applications of Temperature and Humidity Sensors

Sensor	Application
Analog LM35 Temperature Sensor (Arduino Compatible)	The functional range of this sensor is 0 – 100°C. Therefore it can be used in industrial systems, aeronautics and astronautics systems.
SHT1x Humidity and Temperature Sensor	It also a high range temperature sensor. Therefore this sensor has applications in HVAC, medical equipment, laboratory experiments and automobiles. This sensor also measures humidity, hence it is also used in humidity regulators.
DHT22 Temperature and Humidity Sensor	This sensor is widely used in weather monitoring applications. It is also used in indoor temperature and humidity monitoring. It has potential applications in air purifiers, humidifiers, exhaust fans and dimmable lights, also used in HVAC, automobiles, medical equipment etc.
DHT11 Temperature and Humidity Sensor	This sensor is used in applications which operate in conditions where the temperature is not very high. It is mostly used in weather monitoring, medical equipment and humidity regulators.
I2C BME280 Environmental Sensor	This sensor is used in weather forecasting, room change detection, indoor and outdoor navigation and Internet of Things applications.
I2C BME680 Environmental Sensor	Since this sensor uses four parameters – Temperature, Humidity, Pressure and Gas present in the atmosphere, it is widely used in Air Quality Monitoring applications. It is also used in home automation systems, wearable devices and other Internet of Things Applications.
Waterproof DS18B20 Sensor	This sensor has found its applications in Industrial Control Systems. It is also used in consumer electronics and telecommunications. Other applications of this sensor are refrigerator and tanks.

Developing Temperature and Humidity Sensor Application

Temperature and Humidity sensors have a variety of applications everywhere. Temperature sensors are widely used in thermocouples and semiconductor chips. However in recent years these sensors are tremendously used in Internet of Things applications. A variety of Temperature and Humidity sensors are available in market. Availability of these sensors has given rise to their usage in various innovative IoT projects with their applications ranging from simple room weather monitoring to navigation, forecasting and industrial systems. Table 2 shows the applications of these sensors. Temperature and Humidity sensors have much wider applicability as compared to other sensors. These sensors have been used in weather station, weather monitoring, indoor air quality monitoring, navigation control, house hold devices, consumer electronics, healthcare, automobiles, industrial and military applications.

Sensor Applications

Temperature and humidity sensing application has two main components – Embedded System and Sensor. Sensor is used to read environmental attributes. Embedded system is used to derive the application. Embedded system comprises of a microcontroller unit (MCU) which runs the program for the application. There is another component required to upload the data read from the sensor permanently. It is a WiFi module. WiFi module establishes the TCP/IP connection with the web server and sends the data bytes.

Figure 3 shows the schematic of our application. The embedded system used here is Arduino Uno board. Sensor is DHT11 Temperature and Humidity Sensor. The WiFi module used is ESP8266.

As shown in the figure, the WiFi module ESP8266 is connected to the Arduino Uno board. The connections are made as follows. Vcc and CH_PD (Chip Enable) pin of ESP8266 are connected together and further connected to 3.3V pin of Arduino Uno board. Ground pin of ESP8266 is connected to one

Figure 3. Schematic of Application

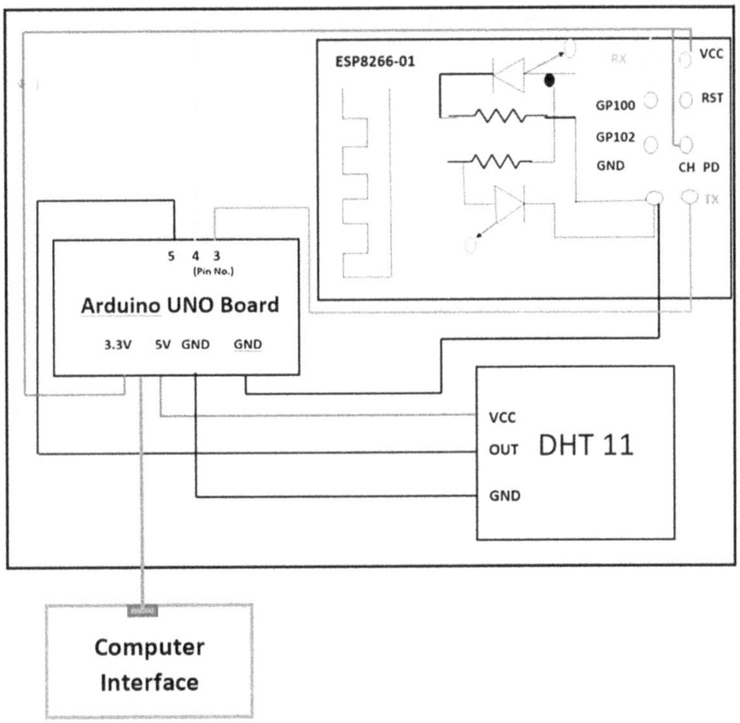

of the Ground Pin of Arduino Uno board. Transmission (Tx) pin of ESP8266 is connected to pin 3 of Arduino. Receiver (Rx) pin of ESP8266 is connected to pin 4 of Arduino Uno board.

DHT11 Sensor is connected to the Arduino Uno board as follows. Vcc pin of DHT11 sensor is connected to 5V pin of Arduino Uno board. Output pin of DHT11 is connected to pin 5 of Arduino Uno. Ground pin of DHT11 is connected to Ground pin of Arduino Uno board.

The Arduino Uno board is powered up by connecting to the Laptop by USB connector. The board can also be powered up by connecting to 9V battery. Figure 3 shows connection schematic.

In this chapter, these three components namely Arduino, ESP8266 WiFi module and Temperature and Humidity Sensor are described in detail.

Arduino

Arduino project began in 2003 by Massimo Banzi, David Cuartielles, Tom Igoe, Gianluca Martino, and David Mellis at Interaction Design Institute Ivrea (IDII) in Ivrea, Italy. Arduino is both a hardware and software platform that allows building low-cost projects by people who are having limited or no knowledge of electronics engineering. Arduino board is physically programmable. Multitudes of logic circuits can be built using Arduino board.

Arduino is complete solution for building a variety of projects for day-to-day activities as well as intelligent scientific applications.

Arduino was based on Wiring Platform. Wiring platform was made up of ATmega168 microcontroller on Printed Circuit Board (PCB) [6]. Later, ATmega128 microcontroller was replaced by ATmega8 microcontroller (much cheaper as compared to ATmega168 used in Wiring Platform). Hence the first prototype of Arduino was created in the year 2005.

Arduino is basically an open source hardware and software platform. It is a combination of hardware board with Atmel Mega series MCU such as Atmega328, Arduino programming language and the Arduino software for development and compilation.

In the beginning, Arduino was just an embedded environment targeted to help students build projects with low cost. Now a number of Arduino variations exist ranging from simple boards for building students' projects to more complex ones to be used in IoT applications.

Another advantage of Arduino is that it is open source platform. That is why, source code, PCB files, and hardware design schematics is freely available. This gives users the flexibility to adapt and develop the design to make innovative projects. Arduino is a combination of hardware board with MCU, kits, motor shields, accessories and also an Integrated Development Environment to develop Arduino programs.

Arduino boards come in a number of varieties with different specifications and different target applications. Table 3 gives an overview of the applications supported by specific Arduino board.

Arduino Uno is the most widely used microcontroller board. People with limited technical knowledge can use this board easily.

Arduino Nano is a smaller board as compared to Arduino Uno. Arduino Lilypad and derivatives of it are primarily used for wearable devices. Arduino motor shield are stand-alone boards to enhance the functionality of the board. Table 4 shows the specifications of some of the commonly used Arduino boards.

As Arduino is an open source platform, there are other alternatives available designed by other vendors who are inspired from actual Arduino board design. Table 5 shows three commonly used boards.

Arduino board mostly used Atmel Mega Series of AVR Microcontroller such as Atmega328. These are 8-bit microcontrollers with On-Chip Flash memory to accommodate programs.

- ATmega8
- ATmega168
- ATmega328

These microcontrollers are primarily used in Arduino board.

Table 3. Applications of Arduino

Arduino Board	Applications
Arduino Uno (R3)	A variety of applications such as weather monitoring system, gas sensor, smart irrigation system, obstacle detection etc.
Arduino Nano (R3)	Traffic light controller, LED cube, temperature control device etc.
ArduinoLilypad	e-textiles, wearable products
Arduino Motor Shield	Controlling DC motor, robotics etc.
Arduino Due	Industrial Automation, Home and Security Systems, Virtual Reality Applications, Android Applications, GSM Based Projects, Embedded System
Arduino Mega	Developing 3D printer, Controlling and handling more than one motors, Interfacing of number of sensors, Sensing and detecting temperature, Water level detection projects, Home automation and security systems, Embedded Systems, IoT applications, Parallel programming and Multitasking
Arduino Leonardo	Pick and place machine, GPS, Bluetooth application, reading RFID tag

Table 4. Arduino Board Specifications

Arduino Board	Specifications
Arduino Uno (R3)	ATmega328P Microcontroller, Operating voltage: 5V, Input Voltahe:7-12V, 14 Digital I/O pins out of which 6 are PWM (Pulse Width Modulation), 6 Analog Input pins, 32 KB Flash Memory, 2KB SRAM, 1KB E^2PROM
Arduino Nano (R3)	ATmega328 Microcontroller AVR Architecture, Operating Voltage: 5V, 22 digital I/O pins out of which 6 are PWM, 8 Analog Input pins, 32 KB Flash Memory, 2KB SRAM, 1KB E^2PROM
ArduinoLilypad Simple	ATmega328 Microcontroller, Operating Voltage: 2.7-5.5V, 9 Digital I/O pins, 5 PWM channels, 4 Analog Input pins, 32 KB Flash Memory, 2KB SRAM, 1KB E^2PROM
ArduinoLilypad Main Board	ATmega168 or ATmega328V Microcontroller, Operating Voltage: 2.7-5.5V, 14 Digital I/O pins, 6 PWM channels, 6 Analog Input pins, 16 KB Flash Memory, 1KB SRAM, 512 Bytes E^2PROM
Arduino Motor Shield	L298P Motor Controller capable of Driving 2 DC motors or 1 stepper motor, Operating Voltage: 5V to 12V, Current Sensing: 165V/A
Arduino Due	AT91SAM3X8E Microcontroller, Operating voltage: 3.3V, Input Voltage:7-12V, 54 Digital I/O pins out of which 12 are PWM (Pulse Width Modulation), 12 Analog Input pins, 2 Analog Output pins(DAC), 512 KB Flash Memory, 96 KB SRAM
Arduino Mega 2560	ATmega2560 Microcontroller, Operating voltage: 5V, Input Voltage:7-12V, 54 Digital I/O pins out of which 15 are PWM (Pulse Width Modulation), 16 Analog Input pins, 256 KB Flash Memory, 8KB SRAM, 4 KB E^2PROM
Arduino Leonardo	ATmega32u4 Microcontroller AVR Architecture, Operating voltage: 5V, Input Voltage:7-12V, 20 Digital I/O, 7 PWM pins, 12AnalogI/O pins, 32 KB Flash Memory, 2.5 KB SRAM, 1 KB E^2PROM

Microcontroller is basically a computer which is smaller in size and built on an integrated circuit which performs a specific task based on an uploaded program.

Major components of microcontroller are Arithmetic Logic Unit (ALU), Control Unit (CU), Memory and Input-Output Ports.

CPU is the most significant part of a microcontroller. Major components of CPU include

- ALU
- CU
- General Purpose Registers
- Interrupts
- I/O Ports

Table 5. Alternatives of Arduino Board

Board	Description
Teensy2.0	This board is similar to the Arduino compatible ATmega32U4 chip and is a complete USB based microcontroller. It has 32 KB Flash memory, 2.5KB RAM, 1 KB E^2PROM.
Nanode	Nanode is another alternative to Arduino, but it is built specifically for web based projects. It has ATmega328 microcontroller and can be programmed using Arduino IDE.
STM32 Discovery	It is another alternative of Arduino from ST Microelectronics. It uses STM32F100RBT6B microcontroller. It has 128 KB Flash memory and 1 KB RAM. It can be programmed with GCC based IDE.

It consists of the ALU, general purpose registers, Interrupts and Instruction Control Unit (CU). ALU performs arithmetic and logic operations. General purpose registers store operands and address of instructions temporarily.

Interrupts form an efficient system to make the microcontroller respond to external changes, compared to repeatedly checking the status of devices. The Instruction Control unit fetches instruction from memory, decodes them and executes them.

The result of an instruction execution is also stored back in one of the general purpose register before transferring in the memory.

The memory in any microcontroller is either a volatile memory or a non-volatile memory. The volatile memory is cleared when power is turned off. The static RAM (SRAM) is volatile memory.

The non-volatile memory can retrieve information even after it is powered on following a switch off. E^2PROM – Electrically Erasable Programmable Read Only Memory is a type of non-volatile memory. This memory stores that data which need to be retrieved when power is turned off. E^2PROM is used to store device configuration. The data in E^2PROM can be retrieved, erased and re-written. The large size in bytes E^2PROM is called Flash memory. Flash memory is also a non-volatile memory. Flash memory holds the program to be executed. Flash memory retrieves data faster compared to the E^2PROM.

A microcontroller also consists of Timers. Timers and Counters are most important to a microcontroller.

Timers set the frequency at which the instruction will be executed, interrupts occur, and also the frequency of Pulse Width Modulation signals. Timers provide clocks to the microcontroller.

The I/O ports communicate with the external peripheral devices. Each bit corresponds to one I/O pin of the 8-bit port. Each pin can be referred one at a time and the whole port can be addressed as well. The Analog ports can handle analog data such as Analog to Digital Conversion (ADC). ADC converts an analog signal to digital value in binary form, that can be used in specific applications like Pulse Width Modulation.

Figure 4 shows the pin out diagram of ATmega328 microcontroller. The pin out diagram is significant for interfacing the MCU with other components such as sensors.

Figure 5 describes the I/O Ports available in this microcontroller. Figure 6 shows the components placed on Arduino Uno board.

Arduino Uno has ATMega328 Microcontroller.

Arduino IDE – The Development Environment of the Arduino

Arduino IDE allows us to write programs using Arduino library. Name of the Arduino file is called as sketch. The IDE gives main menu and several buttons. File menu provides the options to create a new file, open an existing file, save file and upload the program on Arduino board. There is also an Examples option which provides several examples to help learning Arduino programming. Sample programs are also available in the IDE. These programs help in controlling specific devices that may be connected to the Arduino board, such as an LCD, a Stepper Motor, Servo motors and so on. The Edit menu has the option to Edit the existing code.

The Sketch menu has option to verify the code for error checking and compile the code to machine level before it is uploaded to the Arduino board.

The Tools menu allows the selection of a specific Arduino board that is being used. It also allows to choose the kind of programming cable being used. The Help menu contains the reference to several documents that come with Arduino software such as Getting Started guides, References etc.

Figure 4. ATmega328 Pinout Diagram

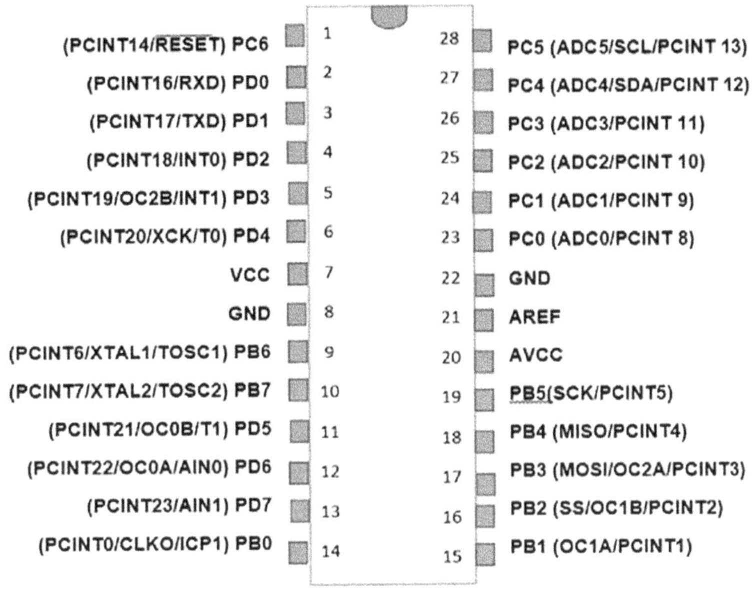

Figure 5. ATmega328 I/O Ports and other pins

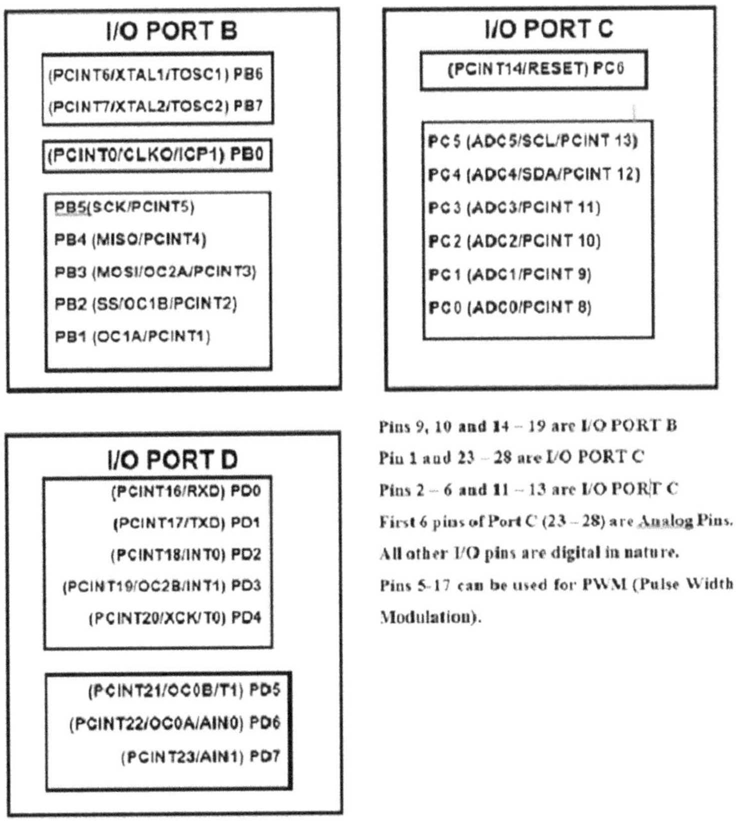

Figure 6. Arduino UNO Components

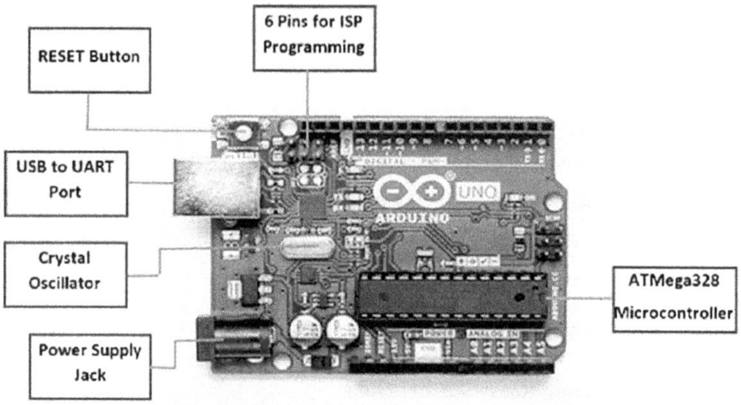

Arduino IDE also contains a toolbar. The first button in the toolbar is Verify/Compile. It is used for error checking and compilation. Next button is Upload button. It is used to upload the compiled program to the configured board. The New button creates a new Arduino sketch file.

The Open button provides a list of all the sketches in the sketch book. The Save button saves a sketch file. Figure 7 shows the components of Arduino IDE.

Structure of an Arduino Program

Arduino programs are called sketches. Sketches are written in C++ programming language. Every sketch is made up of two void type functions – setup() and loop().

The statements specified in the setup() function are executed first by the Arduino. The statements specified in the loop() function run forever as long as the Arduino board gets power.

Figure 7. Arduino IDE

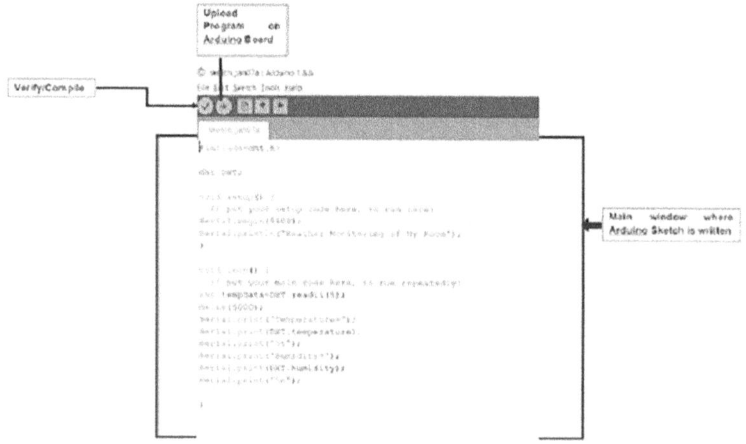

Any initialization code can be added in the setup function. The function loop() should contain statements that should be executed over and over again. Headers are the components of Arduino board that make connecting components easier onto the Arduino board. The location at which the headers connect on the board are called pins. All the pins of Arduino have a number.

In order to program the Arduino board, it is essential to know the pins to be used for input, output, Vcc and Ground.

For instance, suppose we want to control the blinking of LED that is present on the Arduino board. Pin 13 connects the LED.

Example. Switching off LED always

```
int pinnumber=13;
void setup()
{
pinMode(pinnumber, OUTPUT);
}
void loop()
{
digitalWrite(pinnumber, LOW);
}
```

Now, verify and compile the code. Finally, upload the compiled code onto the Arduino board.

Example. Blinking On Board LED

```
Int pin_no=13;
void setup()
{
pinMode(pin_no, OUTPUT);
}
void loop()
{
digitalWrite(pin_no, HIGH);
delay(5000);
digitalWrite(pin_no, LOW);
delay(5000);
}
```

Save the program. Verify and compile. Connect the Arduino board to the USB port of your system. Finally upload the compiled program onto the Arduino board.

ESP8266 WiFi Module

The ESP8266 is a major low-cost platform for IoT that provides connectivity. The importance of ESP8266 is mainly due to its dual functionality.

ESP8266 can work either as a self-contained module or as a wifi adapter. This functionality of ESP8266 can be used in an IoT application that is used to monitor Temperature and Humidity values and transfer those values on a web server.

ESP8266 has a number of GPIO pins which can directly connect sensors to the MCU. The ESP8266 contains a 32-bit microcontroller, which provides communication, power mode, and GPIOs on the board. It works in the power range of 3.0 – 3.6 Volts. Therefore, 3.3V pin of Arduino can be used to power up the ESP8266.

ESP8266 is basically a System on Chip (SoC) which has functionalities of 2.4 GHz WiFi (802.11 b/g/n IEEE standards). It also has TCP/IP communication protocol stack on it. SPI and UART serial communications are provided in ESP8266.

They can be used to send command from the ESP to Arduino for performing different functions.

There have been multiple variance of ESP8266 platform such as ESP-01, ESP-02, ESP-03 and ESP-06.

These versions differ mainly in number of GPIOs available. In advance versions more GPIOs are available for interfacing parallel devices and the development of stand- alone IDEs and libraries for enabling self-contained functionalities of ESP8266.

Although ESP8266 – 01 is a simple low cost WiFi module, it exhibits a number of features that are sufficient to develop a variety of IoT projects. The specifications of ESP8266 – 01 are shown in Table 6.

Pin Configuration of ESP8266 – 01

ESP8266 contains only 8 pins. Pin 1 is Ground. Pin 2 is Tx (Transmission). Its alternative name is GPIO-1 since it can be used for General Purpose I/O when not being used as Tx pin. Pin 3 is GPIO-2, a General Purpose I/O pin. Pin 4 is CH_EN (or CH_PD) or Chip Enable. It is Active High. Pin 5 is GPIO-0, a General Purpose I/O pin. Its alternative name is Flash. It enables Serial Programming when set to low at start up. Pin 6 is Reset pin. Pin 7 is Rx (Receiver) Pin. Its alternative name is GPIO-3. When it is not used as Rx, it can be used as a GPIO pin. Pin 8 is Vcc.

Applications of ESP8266 range from Home Automation, security and surveillance to Industrial Wireless Operations. However, in order to design IP Camera and Sensor Networks, additional microcontroller units are required. They can be interfaced with Internet using ESP8266.

AT Commands of ESP8266

AT commands are also known as Hayes commands. Hayes commands or AT commands is the command language used to control the communication modems. AT commands control communication modems

Table 6. Specifications of ESP8266 – 01

Input Power	+3.3 V
Current	100mA
I/O Voltage	3.6V
I/O Source Current	12mA (maximum)
Flash Memory	512 KB
Programmed with	Arduino IDE and AT commands

like GSM, Bluetooth or ESP8266. These commands are made up of a series of short text messages sent using serial communication between MCU and the device which is communicating.

AT denotes Attention. It is used to start any command. It informs the modem that whatever appears next, it is the actual command that needs to be executed.

A number of uses of AT Commands is given below which are significant in an IoT project.

- Connecting with a WiFi network.
- Allow other devices to connect with WiFi network.
- Connect microcontroller unit to WiFi module and send data over internet.
- Open TCP connection without requiring to have TCP/IP stack running on the connected microcontroller unit.

AT Commands are categorised as given below.

- Test – The Test AT commands are used to query whether the commands or any parameter used is supported by the ESP8266.
- Read – The Read commands retrieve the current settings or parameters used by ESP8266 module for any specific function.
- Set – values of specific parameters can be set using these commands. These commands can also be used to change the settings using the Set operations.
- Execution – The Execution AT commands perform predefined functions.

The syntax of each of the command is shown below

Test Commands: AT + <cmd>=?

Read Commands: AT + <cmd>=?

Set Commands: AT + <cmd>=<….>

Execution Commands: AT + <cmd>

Table 7 shows basic AT commands. Table 8 shows WiFi AT commands whereas Table 9 shows the TCP/IP AT commands.

List of AT Commands[8]

Temperature and Humidity Sensor

Any IoT application is basically made up of a group of sensors, embedded systems, communication system and necessary protocols and some mechanism of analysing and representing the data.

Sensors acquire the data from real world whereas embedded systems are used to drive entire physical system of IoT application. Figure 8 shows the flowchart of sensor application.

Sensors provide an interface between the environment and application components. Data captured through sensors is real time data. MEMS (Micro-Electro-Mechanical-Systems) technology has really enabled widespread use of sensors because of their low cost and small size. For instance, a smart phone can have multiple sensors embedded in it like accelerometer, barometer, GPS, gyroscope, and proximity sensor.

Table 7. Basic AT Commands

Command	Description
AT	This command is used to test the status of AT command when ESP8266 is powered up.
AT+RST	This command is used to restart the ESP8266 module.
AT+GMR	This command provides the version information of the device.
AT+RESTORE	This command is used to restore factory settings of the module.
AT+UART	This command is used to configure UART.
AT+UART_DEF	This command is used to save the default UART configuration in Flash memory.
AT+SYSRAM	This command provides the information about the available RAM size.
AT+SYSIOGETCFG	This command provides the information about configuration of I/O pins.

Table 8. WiFi AT Commands

Command	Description
AT+CWMODE	This command sets the WiFi mode. Wifi mode can be • Station • Access Point (AP) • Station + Access Point
AT+CWMODE_CUR	This command is same as above. However, the configuration is not saved on the Flash memory.
AT+CWMODE_DEF	This command also performs the same function as the above mentioned command. Additionally, it saves the configuration on the Flash memory.
AT+CWJAP	This command is used to connect with the Access Point.
AT+CWJAP_CUR	This command is also connects with AP. However, configuration is not saved in the Flash memory.
AT+CWJAP_DEF	This command is used to connect with the Access Point and also saves the configuration on the Flash memory.
AT+CWLAP	This command displays a list of available Access Points.
AT+CWQAP	This command is used to disconnect from an Access Point.
AT+CWHOSTNAME	This command is used to set the hostname of the ESP8266 Station.

Table 9. TCP/IP Related AT Commands

Command	Description
AT+CIPSTATUS	This command retrieves the connection status.
AT+CIPSTART	This command can be used to establish a TCP connection, a UDP connection, or an SSL connection.
AT+CIPSEND	This command is used to send data on an established connection.
AT+CIPSENDBUF	This command is used to write data into the TCP-send-buffer.
AT+CIPBUFSTATUS	This command provides the status information about TCP-send-buffer.
AT+CIFSR	This command is used to retrieve the local IP address.
AT+CIPMUX	This command is used to configure multiple connection modes.
AT+CIPMODE	This command is used to configure the transmission mode.
AT+CIUPDATE	This command is used to upgrade the software through network.

Figure 8. Flowchart of a Sensor Based IoT Application

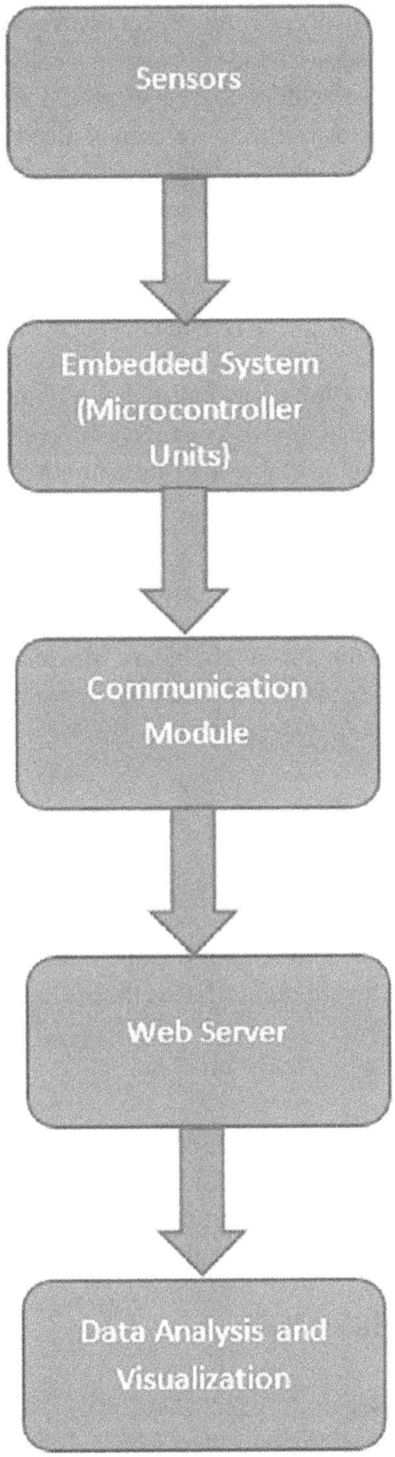

Working Principle of Sensors

Sensors are based on transducers. Transducer converts energy from one form to another.-

Transducers can be used as both sensors and actuators. Usually the sensors convert some form of energy into electrical energy whereas actuators convert electrical energy into another form of energy.

Electric motor is an actuator. It performs the conversion of electrical energy into mechanical energy. The Air Flow sensor is used to convert kinetic energy into electrical energy.

Sensors can be categorised on the basis of their functionality. Different kinds of sensors are discussed below.

Thermal Sensors

Thermal sensors enable devices to measure temperature of environment and objects.

These are the simple thermocouples and resistive devices as well as more sophisticated non-contact based on Infrared and Laser Technology. Temperature measurement and control is significant to an IoT application.

Motion & Position Sensors

Motion and Position sensors measure movement of objects. Movement is either linear or angular. These sensors are also capable of detecting acceleration, vibration and tilt of objects. These sensors find their applications in aircraft and drone design.

Optical Sensors

Presence and Absence of objects can be detected using Optical Sensors. They work along with light transmitters and photo-electric receivers. These sensors are extensively used in digital cameras. Digital cameras, in turn, play significant role in a number of IoT applications. Apart from these, digital cameras are also used extensively in industrial applications of surveillance, mining operations and chemical plants.

Pressure Sensors

Pressure Sensors measure Gas and liquid pressure. Pressure sensors convert physical energy into electronic signals on the basis of pressure imposed. Additionally, other parameters measured using pressure sensors include speed of the fluid, rate of flow and altitude. Pressure sensors are also used in weather monitoring application. These sensors also have applications in automobile industry.

Usage of Sensors in IoT Applications

An IoT device may combine several sensors together. Collective usage of multiple sensors in an IoT application is termed as ***Sensor Fusion***. Microcontrollers used in IoT application fetch data from disparate source, where each source corresponds to a different sensor, and then visualize this data to form more reliable estimate of the environment conditions. Sensor fusion is more than just a collection of different sensors in that it represents a unified representation of information collected separately.

Sensor Fusion Configurations

Sensor fusion can be ***Competitive Sensor Fusion*** where two or more sensors are used in same application to provide more reliability and accuracy of the environment conditions. For example, in order to get more accurate weather data, an IoT application can employ two different temperature sensors. This is essentially a redundant configuration.

When sensors are independent of each other but data gathered from them is used in an application to compute a specific single value, the configuration is called ***as Complementary Sensor Configuration***. For example, a home security system may employ multiple cameras to obtain the complete view of house.

When a single sensor is not capable of providing complete information required by an IoT application, application may use more sensors. These sensors measure different properties. This configuration is called ***Cooperative Sensor Fusion***. For example, for air quality monitoring application, gas sensor, chemical sensor and smoke sensors can be used together.

Architecture of Sensor Node

A sensor node is made up of MCU, transceiver, storage memory, power source, and specific sensors. Figure 9 shows the components of a sensor node.

Microcontroller in the sensor node is used to perform a specific predefined task. It performs data processing and control functions. Typically a microcontroller is used for this to optimize the node's power consumption. Microprocessors as well as personal computers can be used which don't have restriction on power consumption.

Figure 9. Architecture of Sensor Node

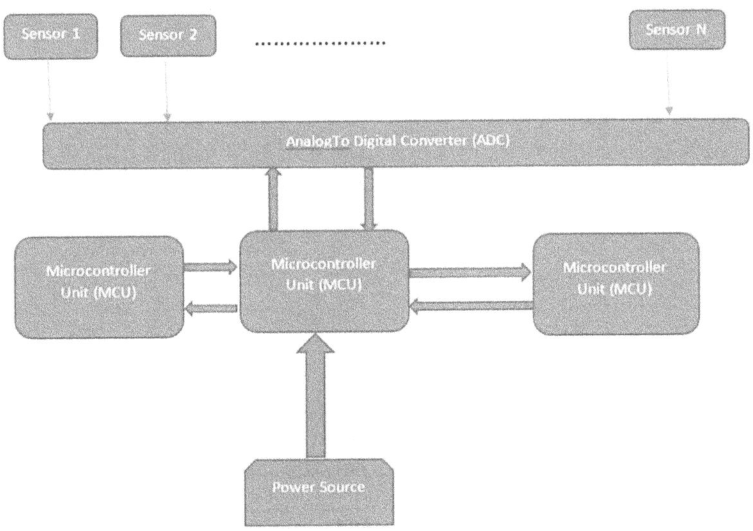

Wireless Sensor Networks

Wireless sensor networks work wirelessly as the name suggests. Each sensor node contains the transceiver with the transmitter and receiver to send and receive data. Radio frequency or infrared communication can be used as wireless technology. The data is logged on to the external storage source in addition so that backup can be there in case of transmission failure. Sensor nodes should be employed in places which are difficult to reach in order to ensure the security of system. Therefore ensuring adequate supply of energy is vital. The battery on sensor nodes cannot be recharged frequently. Hence the use of other kind of power sources is required such as solar energy or wind energy.

The sensor node performs the task of sensing data for a particular property such as temperature and humidity. It also communicates with other elements of IoT application. The sensor node can also possess the capability of processing the data. There must be adequate provision for providing power supply in order to support the functionality of sensor nodes.

Design of Wireless Sensor Network has to address following challenges:-

- Most critical design consideration is power consumption and implementing the sensor network. This challenge can be addressed by employing Solar, Thermal, Wind, RF energy and Kinetic energy sources.
- Another challenge is wireless communication network range.
- Node Failure. Redundant nodes must be employed in order to address this issue.
- Scalability is another challenge. How easily we replicate the nodes and employ in broader geographic area is significant to the success of Wireless Sensor Networks.

DHT 11 Sensor

DHT 11 sensor uses the thermistor to sense the changes in temperature. Thermistors are resisters whose resistance changes in accordance to the change in temperature.

This sensor operates in the range of 3V to 5V. Specification of DHT11 is shown in Table 10.

Sensing the Temperature by DHT 11 Sensor

Temperature is the degree of heat present in a device or thing or in atmosphere. Humidity is the quantity of water vapour present in the atmosphere.

It means quantity of water vapour in its gaseous state. DHT 11 sensor uses the thermistor to sense the changes in temperature. Thermistors are resisters whose resistance changes in accordance to the change in temperature. Thermistors are of two types

Table 10. Functional Specification of DHT 11 Sensor

Sensor	Temperature Range	Temperature Accuracy	Humidity Range	Humidity Accuracy
DHT 11	0 – 50°C	±1°C	20 – 90% RH	±5%RH

- NTC – Negative Temperature Coefficient
- PTC – Positive Temperature Coefficient

Thermistor in DHT 11 is NTC type. NTC type thermistors are semiconductors based. There are two types of bands of electronic energy in every solid – the Conduction Band and the Valance Band.

A material's electrical conductivity depends upon the number of electrons present in the energy range of conduction band. Valance band has electrons present at 0K. Metals can conduct at 0K since both valance and conduction bands overlap. However, semiconductors have a band gap also called as the forbidden band that separates valance band and conduction band. As soon as the air temperature increases electrons in valance band start jumping to the conduction band. Electrons in conduction band are energised by the thermal energy that leads the increase in conductivity of the semiconductor.

Sensing the Humidity by DHT 11 Sensor

Humidity Sensors can be either capacitive or resistive. Resistive type humidity sensor has been used in DHT 11 Sensor. Resistive type humidity sensors have non-metallic conductors. Their conductivity has dependence on humidity present. Humidity sensor has three layers which are described next. The layer that has the electrodes with metal contacts has the ends of this electrode dipped in another layer of a material which is sensitive to the moisture present. The insulating substrate layers contains the sensing layer. In sensing layer, the number of ions present in it increase with increase in humidity since in an aqueous medium, water molecules dissociate into ions. That is why conductivity of the medium increases. As the humidity value changes, the resistivity of the material also gets changed. The sensor has resistance varying from 1KΩ to 1MΩ.

Table 11 shows pin structure of DHT11 sensor.

DHT 11 Sensor has four pins. Pin 1 is connected to 5V supply. Pin 2 transfers the data. Pin 3 is not required, therefore it is not connected. Pin 4 is the Ground pin. Figure 10 shows interfacing of DHT11 with microcontroller.

The communication between Microcontroller Unit (MCU) and DHT11 Sensor occurs in three stages.

1. Request from MCU
2. Response by DHT11
3. Data Transfer by DHT 11

Table 11. Pin Structure of DHT 11 Sensor

Pin	Description
1	Vcc
2	Data
3	NC (Not Connected)
4	Ground

Figure 10. Interfacing of DHT11 with MCU

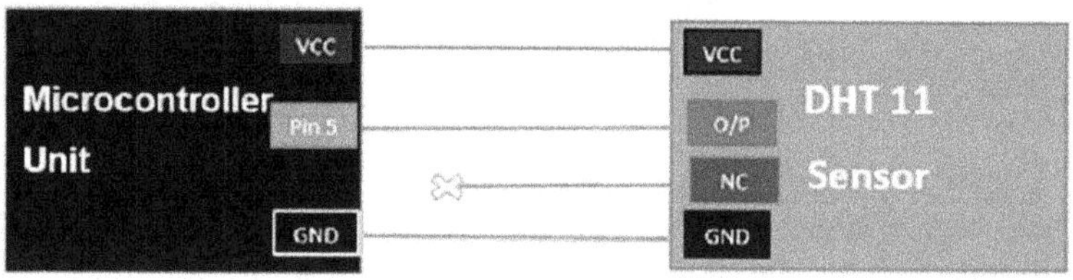

Request

The request is 80 ms low signal followed by 20 μs to 80 μs high signal.

Response

When DHT11 sensor receives the request signal, it sends 54 – 80μs low signal followed by 80 μs high signal as response.

Data Transfer

Total 40 bits are transferred by DHT11 sensor as data. It sends the data in 5 segments of 8-bit each. The first 8 bit of data denote integer bits of Relative Humidity (RH). Next 8 bits denote fractional value of RH. Next 8 bits denote integer bits of temperature. Next 8 bits denote fractional value of temperature reading. The last 8 bits denote the check sum bits. Figure 11 shows the waveform diagram.

Sensor Specifications

Table 12 shows the specifications of various kind of Temperature and Humidity sensors. These sensors support specific applications in different temperature ranges.

Interfacing of Arduino with Android

Integration of various technologies is the key to improve quality of life. The complete automation system requires the use of different technologies like Internet of Things, Android, Data Science, Machine Learning and Network technologies.

Smartphones have gained popularity in recent decade with first android app launched by Google in 2008. Smartphones have been used in controlling the home automation system.

A smart home system employs different kind of sensors for lighting, video capturing and motion detection. The data captured by these sensors must be available on smartphone for easy tracking by people. Also the use of smartphone in automation system allows for the authentication of people using the system. Smartphone based automation system allows the user to take control of the system. User can receive updates and send the commands to control the system. Use of smart phones makes fast adoption of IoT applications by the common people. It allows everybody to be involved in the process of generating and collecting sensor data.

Figure 11. Waveform Diagram of DHT 11 Sensor

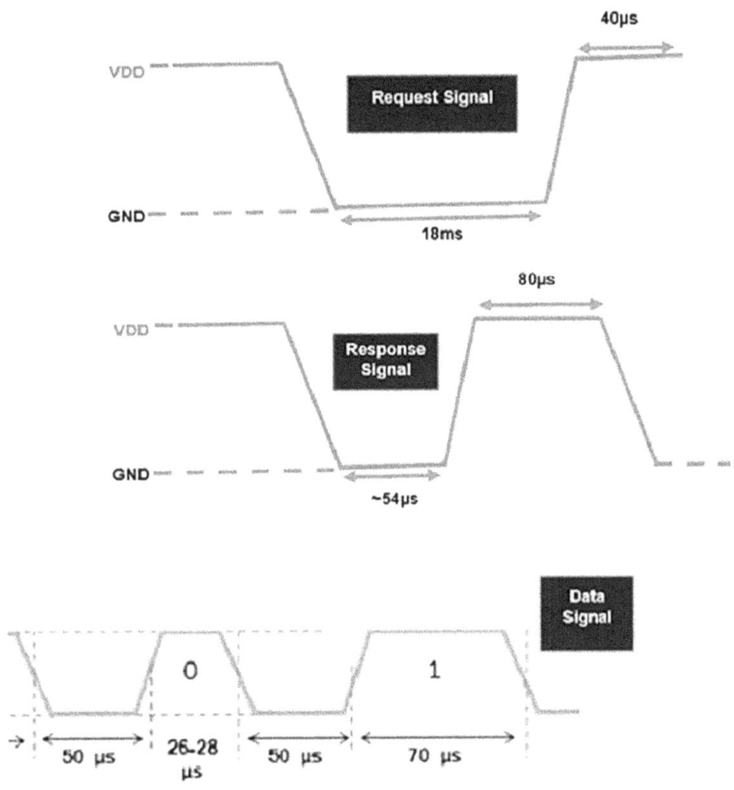

Use of smartphone in IoT applications has additional advantage. Users can remotely monitor and control the status of the system. The use of smartphone in controlling an automation system also provides for a centralized control. Nearly each and every house hold device comprising of embedded systems and sensors can be controlled remotely.

The control can be provided by a number of methods. These methods could be a Short Message Service (SMS), a Bluetooth connection, or through an Android application using Rest API.

The automation systems involving the use of Temperature and Humidity Sensors can be significantly useful in both indoor and outdoor environments. A remote operated fire alert system can alert the possibility of a fire outbreak if the temperature increases abruptly and humidity is lowered to a significant level. Another application can be found in monitoring of transportation of inflammable goods.

A patient monitoring system can employ a number of sensors for measuring temperature, heart-beat, blood pressure and other parameters. The data obtained by these sensors can be stored on a web server and further can be fetched on a smartphone for monitoring the health of patient.

Smartphone operated IoT systems have tremendous potential to improve the health and well-being of people and can reduce the cost of healthcare dramatically. These type of systems can transform our daily lives.

Table 12. Temperature and Humidity Sensors Specifications

Sensor	Specification
Analog LM35 Temperature Sensor (Arduino Compatible)	Detects: Ambient Air Temperature Functional Range: 0 – 100°C Operating Voltage: 3.3V/5V Operating Current: 2.7μA Temperature Deviation: ±0.5°C
SHT1x Humidity and Temperature Sensor	Detects: Both Temperature and Humidity Functional Range: 40°C – 128°C Operating Voltage: 3.3V/5V Operating Current: 2.0μA Temperature Deviation: ±0.4°C
DHT22 Temperature and Humidity Sensor	Detects: Both Temperature and Humidity Functional Range: 40°C – 80°C Operating Voltage: +5V Operating Current: 1.5mA Temperature Deviation: ±0.5°C
DHT11 Temperature and Humidity Sensor	Detects: Both Temperature and Humidity Functional Range: 0°C – 50°C Operating Voltage: 3.3V~5V Operating Current: 0.5mA~2.5mA Temperature Deviation: ±0.1°C
I2C BME280 Environmental Sensor	Detects: Temperature, Humidity and Barometer Functional Range for Temperature: 40°C – 85°C Humidity Range: 0 – 100%RH Humidity Resolution: 0.1%RH Pressure Measurement Range: 300~1100Pa Operating Voltage: 3.3V~5V Operating Current: 2mA Temperature Deviation: ±0.5°C
I2C BME680 Environmental Sensor	Detects: Temperature, Humidity, Pressure and Gas Functional Range: 40°C – 85°C Operating Voltage: 3.3V~5V Operating Current: 5mA Humidity Range: 0-100%RH Humidity Precision: ±3%RH Pressure Range: 300-1100hPa Pressure Measurement Precision: ±0.6hPa Temperature Deviation: ±0.1°C
Waterproof DS18B20 Sensor	Detects: Temperature Functional Range: -55°C to 125°C Operating Voltage: 3.3V~5V Operating Current: <3μA Temperature Deviation: ±0.5°C

Architecture of Application

Architecture of application involving Android and Arduino is given in Figure 12.

Application Components

The application consists of both hardware and software components. These components are described below:

Figure 12. Block diagram of Android application with Arduino and DHT11 Sensor

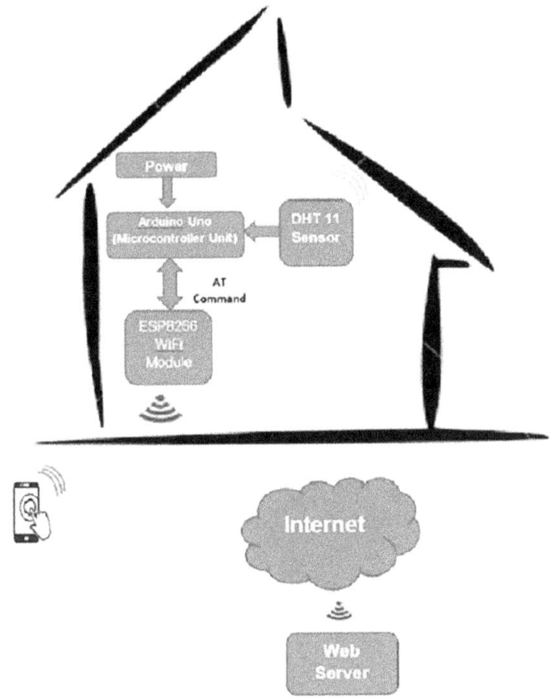

Hardware Components

1. Android Device

Android device in the application is Android smartphone that runs the app. The app retrieves the data from web server using REST API.

2. Arduino UNO Board

Arduino UNO board consists of ATmega328 microcontroller unit (MCU). It has several on-board features such as Flash memory to store programs. Arduino UNO board runs the sketch uploaded on it from the Arduino IDE. The board can be connected to laptop using the USB cable. It can also be powered through AC or DC adapter or simply from a battery.

3. ESP8266 WiFi module

This is a WiFi module that provides Internet connectivity to the Arduino. ESP8266 runs AT commands to perform serial communication. The module is responsible for establishing TCP/IP connection with the web server. The WiFi module sends the data to web server and receives the status information from the API.

4. DHT11 Sensor

DHT11 Sensor senses the Temperature and Humidity data. The data can be read from Output pin of the sensor in digital format. This sensor uses Negative Temperature Coefficient for sensing temperature. Humidity sensing is based on resistive technique.

5. Web Server

Web server can be a remote server running TCP/IP protocol suite. It stores and maintains the data obtained from the WiFi module of application. Request and response is performed using REST API.

Software Components

1. Arduino IDE

Arduino Integrated Development Environment (IDE) is used to develop Arduino programs called sketches. An Arduino sketch is written C++ language. The IDE provides the facility to compile and debug programs. The compiled program can be uploaded on the Arduino board.

2. AT Commands

AT commands or Hayes commands are used by ESP8266 to communicate using Serial Monitor available on Arduino IDE. These commands are used in Arduino sketch to begin communication, connect with a WiFi Access Point, establish TCP/IP communication with web server and send the data.

3. Java Development Kit (JDK)

The android application uses Java language. While developing the application, Android libraries and Java libraries are used.

4. Android Studio

Android Studio is a software that provides an Integrated Development Environment (IDE) to develop Android application. The IDE provides facility to organize files, build application and run the application on a target device such as an emulator or an Android phone. Android Studio also allows the creation of signed APK in order to install and run the application on smartphone.

The android application for the sensor data is shown in figure 13.

The sensor data has been uploaded by ESP8266 on a ThingSpeak channel. The ThingSpeak API fetches data to be fed to the Android application. Figure 14 shows JSON feed from web server.

Android code to fetch the data.

```
class DownloadWeatherInfo extends AsyncTask<String, Void, String> {
    @Override
```

Figure 13. Android Application

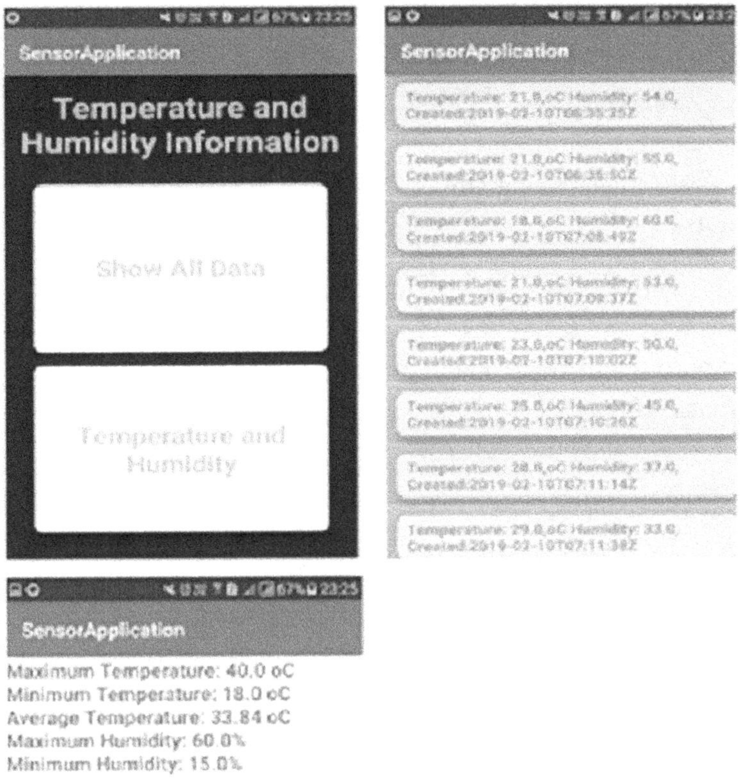


```
protected void onPreExecute() {
super.onPreExecute();
```

Figure 14. Sensor data uploaded on Web Server in JSON Format

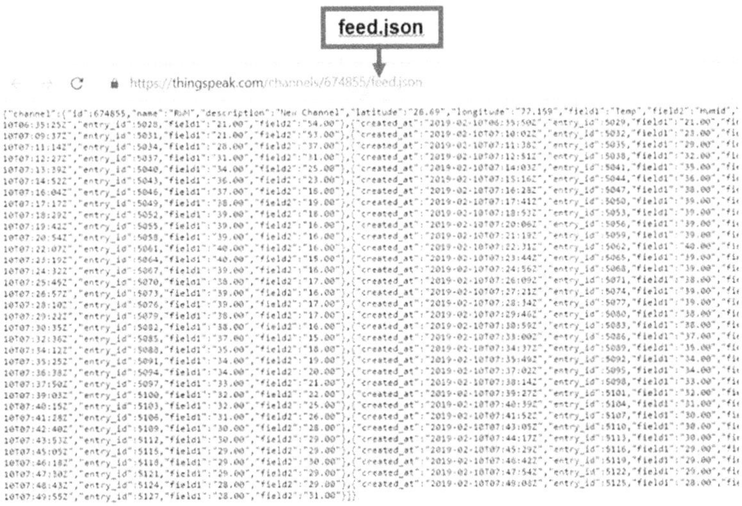

```
}
protected String doInBackground(String... args) {
String xml = Function.excuteGet("https://thingspeak.com/channels/674855/feed.json");
return xml;
}
@Override
protected void onPostExecute(String xml) {
Log.i("RWM", xml);
data1=xml;
Intent in1;
if(type.equals("1")) {
in1 = new Intent(MainActivity.this, ShowAll.class);
in1.putExtra("mydata", data1);
startActivity(in1);
}
if(type.equals("2")) {
in1 = new Intent(MainActivity.this, ViewSummary.class);
in1.putExtra("mydata", data1);
startActivity(in1);
}
}
}
```

Handling JSON Data

```
Bundle bundle=getIntent().getExtras();
    String str=bundle.getString("mydata");

    String str1="";
    try{
    JSONObject jsonobj = new JSONObject(str);
    JSONArray jsonArray=(JSONArray)jsonobj.getJSONArray("feeds");

    for (int i=0; i<jsonArray.length(); i++){
    JSONObject jsonobject=(JSONObject)jsonArray.getJSONObject(i);

    String s1;
    String s2, s3;
    double d1, d2;
    s1=jsonobject.getString("created_at");
    s2=jsonobject.getString("field1");
    s3=jsonobject.getString("field2");
    d1=Double.parseDouble(s2);
    d2=Double.parseDouble(s3);
```

```
Log.i("Result", s1);
TempData s=new TempData(s1, d1, d2);
all.add(s); }
}
catch (JSONException e){
e.printStackTrace();
}
```

Working Principle

The microcontroller unit (MCU) on the Arduino boards obtains the Temperature and humidity data sensed by the DHT11 sensor using the WiFi module. The MCU on the Arduino board, then sends the data by running AT commands through the ESP8266 WiFi module. The data is sent to a web server or cloud server. The server stores the data and provides REST API for reading and updating data. Android application retrieves the data from server using API and displays the data to user in desired format. The data presentation can vary according to specific user requirements.

Different Approaches

1. The internet connection via WiFi and 3G or 4G network

This approach requires an HTTP server to be running in order to upload the data retrieved in a database. The web server may be a XAMP server running PHP with MySQL database. The web server allows request and response using REST API.

2. The Bluetooth connection

This approach involves the use of a Bluetooth module. Bluetooth technology allows the communication between two devices within a range of 10 meters.

Pros and Cons of Proposed Solution

Pros

The most prominent advantage of the proposed solution is that it uses Arduino board which has open source hardware and software and has huge community. It makes the applications development easier and less costly enabling more and more people to work with it. It makes automation not only confined to industrial environment rather people can build applications in home environment.

Cons

Arduino is not meant to work in large scale industrial applications which operate in harsh environment. It has very few I/O ports which are not sufficient for industrial applications.

Arduino is not meant to work in harsh environment. Another drawback of Arduino is that it can be used for a single application at a time.

CONCLUSION

This chapter reviews the some of the very significant aspects of Internet of Things (IoT) applications. The chapter starts with brief introduction of IoT. The origin and history of IoT is discussed in the chapter.

This chapter provides insight into some of the most important building blocks of an IoT application. These components include the Arduino board, ESP8266 WiFi module and Temperature and Humidity sensors in particular.

First section of the chapter provides the information on how the Arduino board is emerged from Wiring platform. A detailed discussion on Arduino board and its architecture is discussed in this section. Variations and alternatives of Arduino board are also discussed.

The third section describes the components for communication with Arduino, in particular, the ESP8266 WiFi module. AT commands are also discussed.

The fourth section provides the details of Temperature and Humidity Sensors. This section discusses the working principle of Temperature and Humidity sensors and explores different types of Temperature and Humidity Sensors.

Finally the schematic of IoT application for weather monitoring is provided and the architecture and interfacing of Android application with IoT application is also provided.

Future Scope

Arduino offers a solution to bring robotics and automation in helping the daily tasks for everybody. However it has one limitation that it can run one application at a time. In order to use Arduino in another application the program needs to be erased and uploaded again. In future a Raspberri Pi board can be used instead of Arduino board since it can run multiple applications simultaneously enabling the development of more powerful automation applications.

REFERENCES

Al-Kuwari, M., Ramadan, A., Ismael, Y., Al-Sughair, L., Gastli, A., & Benammar, M. (2018, April). Smart-home automation using IoT-based sensing and monitoring platform. In *2018 IEEE 12th International Conference on Compatibility, Power Electronics and Power Engineering (CPE-POWERENG 2018)* (pp. 1-6). IEEE. 10.1109/CPE.2018.8372548

Beale, G. O., & Li, M. (1997). Robust temperature control for microwave heating of ceramics. *IEEE Transactions on Industrial Electronics*, *44*(1), 124–131. doi:10.1109/41.557507

Behera, A. R., Devi, J., & Mishra, D. S. (2015, October). A comparative study and implementation of real time home automation system. In *2015 International Conference on Energy Systems and Applications* (pp. 28-33). IEEE. 10.1109/ICESA.2015.7503308

Bernstein, T., & Chang, P. P. (1969). Calculation of Self-Heating Current for a Resistance Wire Temperature Sensor. *IEEE Transactions on Industrial Electronics and Control Instrumentation*, (1), 99-102.

Chen, A., Gil-de-Castro, A., Palacios-García, E. J., Flores-Arias, J. M., & Bellido-Outeiriño, F. J. (2015, June). In-home data acquisition and control system based on BLE. In *2015 International Symposium on Consumer Electronics (ISCE)*(pp. 1-2). IEEE. 10.1109/ISCE.2015.7177780

Folea, S., Bordencea, D., Hotea, C., & Valean, H. (2012, May). Smart home automation system using Wi-Fi low power devices. In *Proceedings of 2012 IEEE International Conference on Automation, Quality and Testing, Robotics* (pp. 569-574). IEEE. 10.1109/AQTR.2012.6237775

Ikhlef, A., Kihel, M., Boukhezzar, B., Guerrouj, A., & Mansouri, N. (2014, November). Online temperature control system. In *2014 International Conference on Interactive Mobile Communication Technologies and Learning (IMCL2014)* (pp. 75-78). IEEE. 10.1109/IMCTL.2014.7011108

Kumar, P. (2017, February). Design and implementation of Smart Home control using LabVIEW. In *2017 Third International Conference on Advances in Electrical, Electronics, Information, Communication and Bio-Informatics (AEEICB)* (pp. 10-12). IEEE. 10.1109/AEEICB.2017.7972317

Lardear, J. L. (1990). Control of self-regulating heating cable for use in pipeline heating applications. In *37th Annual Conference on Petroleum and Chemical Industry*. IEEE. 10.1109/PCICON.1990.218089

Mugisha, J. C., Munyazikwiye, B., & Karimi, H. R. (2015, November). Design of temperature control system using conventional PID and Intelligent Fuzzy Logic controller. In *2015 International Conference on Fuzzy Theory and Its Applications (iFUZZY)* (pp. 50-55). IEEE. 10.1109/iFUZZY.2015.7391893

Newberry, S., & Gupta, G. S. (2015, April). Wireless sensor based home automation as an educational springboard. In *2015 IEEE Sensors Applications Symposium (SAS)* (pp. 1-6). IEEE. 10.1109/SAS.2015.7133610

Rata, G., & Rata, M. (2016, October). Temperature control solution with PLC. In *2016 International Conference and Exposition on Electrical and Power Engineering (EPE)* (pp. 571-575). IEEE. 10.1109/ICEPE.2016.7781405

Reddy, P. S. N., Reddy, K. T. K., Reddy, P. A. K., Ramaiah, G. K., & Kishor, S. N. (2016, October). An IoT based home automation using android application. In *2016 International Conference on Signal Processing, Communication, Power and Embedded System (SCOPES)* (pp. 285-290). IEEE. 10.1109/SCOPES.2016.7955836

Roots, W. K., & Gonenc, G. (1969). Temperature control in industrial processes. *IEEE Transactions on Industrial Electronics and Control Instrumentation*, (2), 147-156.

Singh, R., Kuchhal, P., Choudhury, S., & Gehlot, A. (2015). Wireless controlled intelligent heating system using hpso. *Procedia Computer Science*, *48*, 600–605. doi:10.1016/j.procs.2015.04.141

Wahab, M. H. A. (2016, September). IoT-based home automation system for people with disabilities. In *2016 5th International Conference on Reliability, Infocom Technologies and Optimization (Trends and Future Directions) (ICRITO)* (pp. 51-51). IEEE. 10.1109/ICRITO.2016.7784923

Wang, Y., & Dasgupta, P. (2016, April). Intelligent and adaptive temperature control for large-scale buildings and homes. In *2016 IEEE 13th International Conference on Networking, Sensing, and Control (ICNSC)* (pp. 1-6). IEEE. 10.1109/ICNSC.2016.7478971

Wu, C., Gu, W., Jiang, P., Li, Z., Cai, H., & Li, B. (2018). Combined economic dispatch considering the time-delay of district heating network and multi-regional indoor temperature control. *IEEE Transactions on Sustainable Energy*, *9*(1), 118–127. doi:10.1109/TSTE.2017.2718031

Yue, Y., Xu, Q., Luo, A., Guo, P., He, Z., & Li, Y. (2018). Analysis and control of tundish induction heating power supply using modular multilevel converter. *IET Generation, Transmission & Distribution*, *12*(14), 3452–3460. doi:10.1049/iet-gtd.2017.1274

ENDNOTES

1. https://www.embitel.com/blog/embedded-blog/unraveling-the-story-of-evolution-of-iot-and-its-rapid-adoption
2. https://www.cisco.com/c/dam/en_us/about/ac79/docs/innov/IoT_IBSG_0411FINAL.pdf
3. https://www.technologyreview.com/s/404694/etc-bill-joys-six-webs/
4. http://cocoa.ethz.ch/downloads/2014/06/None_MIT-AUTOID-WH-002.pdf
5. http://citeseerx.ist.psu.edu/viewdoc/download?doi=10.1.1.10.1455&rep=rep1&type=pdf
6. https://hackaday.com/2016/03/04/wiring-was-arduino-before-arduino/
7. https://www.arduino.cc/
8. https://ieee-sensors.org/wp-content/uploads/2018/05/4a-esp8266_at_instruction_set_en.pdf
9. https://www.dfrobot.com/category-64.html

Compilation of References

"The state of food and agriculture: Climate change agriculture and food security," Food and Agriculture Organization of the United Nations (FAO), Rome, Tech. Rep., 2016.

A Comprehensive Look at Low Power, Wide Area Networks for 'Internet of Things' Engineers and Decision Makers. (2016). Annapolis, MD: Link Labs.

A. Abdullah, S. A. Enazi, and I. Damaj, "Agrisys: A smart and ubiquitous controlled-environment agriculture system," pp. 1–6, March 2016.

Abhik. (2019). *Internet of Things, for things, and by things*. CRC Press.

ABI Research. (n.d.). *Internet of robotic things*. Retrieved from https://www.abiresearch.com/market-research/product/1019712-the-internet-of-robotic-things

Ackerman, E. (2019). *ElliQ, A Social Home Robot for Older Adults, Now Available for Pre-Order*. Retrieved from https://spectrum.ieee.org/automaton/robotics/home-robots/elliq-a-social-home-robot-for-older-adults-now-available-for-preorder

Aditya, Manish, Mayank, & Mohit. (2018). Internet of Things (IoT): Research, Architectures and Applications. *International Journal on Future Revolution in Computer Science and Communication Engineering, 4*(3).

Aditya, Nagateja, Vishvakarma, Yellappa, Cho, & Vaddi. (2018). Tunneling Field Effect Transistors for Enhancing Energy Efficiency and Hardware Security of IoT Platforms: Challenges and Opportunities. *IEEE International Symposium on Circuits and Systems (ISCAS).*

Agarwal, A., Mehandiratta, E., Sanket, R., Samkaria, R., Gupta, T., Singh, R., & Gehlot, A. (2016). Smart Door Lock System for Elderly, Handicapped People Living Alone. *International Journal of Smart Home, 10*(6), 155–162. doi:10.14257/ijsh.2016.10.6.16

Ahlgren, B., Hidell, M., & Ngai, E. C. (2016). Internet of Things for Smart Cities: Interoperability and Open Data. *IEEE Internet Computing, 20*(6), 52–56. doi:10.1109/MIC.2016.124

Ahmad. (2011). Smart Metering and Home Automation Solutions for the Next Decade. *International Conference on Emerging Trends in Networks and Computer Communications*, 200-204.

Ahmad, S., & Alam, N. (2018). Robust TFET SRAM cell for ultra-low power IoT applications. *AEU, 89*, 70–76.

Ahmad, S., Alam, N., & Hasan, M. (2018). Robust TFET SRAM cell for ultra-low power IoT applications. *Int. J. Electron. Commun., 89*, 70–76. doi:10.1016/j.aeue.2018.03.029

Aijaz, A., & Aghvami, A. (2015). Cognitive machine-to-machine communications for internet-of-things: A protocol stack perspective. *IEEE Internet of Things Journal, 2*(2), 103-112. Available at http://ieeexplore.ieee.org/xpl/articleDetails.jsp?tp=&arnumber=7006643

Air Navigation Services-India. (n.d.). *GAGAN-GPS-Aided Geo-Augmented Navigation*. Available at http://www.aaians.org/?q=buzz-words/gagan-gps-aided-geo-augmented-navigation

Al Ahasan, Hossain, Siddiquee, & Rahman. (2012). Obstacles invariant navigation of an autonomous robot based on GPS. *Proceedings of IEEE International Conference on Computer and Information Technology*, 524-532.

Ala Al, Guizani, Mohammadi, Aledhari, & Ayyash. (2015). Internet of Things: A Survey on Enabling Technologies, Protocols, and Applications. *IEEE Communication Surveys & Tutorials, 17*(4).

Alamr, Kausar, Kim, & Seo. (2016). A secure ECC-based RFID mutual authentication protocol for internet of things. *Journal of Super Computing, 1*.

Aldowah, H., Ul Rehman, S., Ghazal, S., & Naufal Umar, I. (2017). Internet of things in higher education: A study on future learning. *Journal of Physics: Conference Series*, *892*, 012017. doi:10.1088/1742-6596/892/1/012017

Alghamdi, A., & Shetty, S. (2016). Survey toward a smart campus using the internet of things. *Proceedings - 2016 IEEE 4th International Conference on Future Internet of Things and Cloud*, 235–239.

Ali & Ghazal. (2017). Real-time Heart Attack Mobile Detection Service (RHAMDS): An IoT Use Case for Software Defined Networks. *IEEE 30th Canadian Conference on Electrical and Computer Engineering (CCECE).*

Ali, Z. H., Ali, H. A., & Badawy, M. M. (2015, October). Internet of Things (IoT): Definitions, Challenges and Recent Research Directions. *International Journal of Computer Applications*, *128*(1), 37–47. doi:10.5120/ijca2015906430

Al-Kuwari, M., Ramadan, A., Ismael, Y., Al-Sughair, L., Gastli, A., & Benammar, M. (2018, April). Smart-home automation using IoT-based sensing and monitoring platform. In *2018 IEEE 12th International Conference on Compatibility, Power Electronics and Power Engineering (CPE-POWERENG 2018)* (pp. 1-6). IEEE. 10.1109/CPE.2018.8372548

Amaral, L., Tiburski, R., Matos, E., & Hessel, F. (2015). Cooperative middleware platform as a service for Internet of Things applications. *Proceedings of the 30th Annual ACM Symposium on Applied Computing*. 10.1145/2695664.2695799

Amatulla, P., Navnath, B., Bhoug, Y. P., & Zadbuke, A. S. (2017). IOT based water management system for smart city. *Journal of International Journal of Advance Research*, *3*(2).

Amir, M. F., Trivedi, A. R., & Mukhopadhyay, S. (2016). Exploration of Si/Ge tunnel FET bit cells for ultra-low power embedded memory. *IEEE J Emerg Sel Top Circuits Syst*, *6*(2), 185–197. doi:10.1109/JETCAS.2016.2547818

Andres, R. (2013). *M2M integration platforms enable complex IoT systems*. Received December 1, 2013, from http://www.embedded-computing.com/embedded-computing-design/m2m-integration-platforms-enable-complex-iot-systems

Anita, G., Rajesh, S., Mamta, M., Bhupendra, S., Ravindra, S., & Vikasgarg, G. (2018). *Hands on approach on Application of Internet of Things in Various Domain of Engineering*. Delhi: International Research Publication House.

Anita, G., Rajesh, S., Piyush, K., & Yadav, S. (2016). Wireless Personal Area Network and PSO-Based Home Security System. Academic Press. doi:10.1007/978-81-322-2523-2_24

Anita, G., Rajesh, S., Mamta, M., Bhupendra, S., Ravindra, S., & Vikasgarg, G. (2018). *Hands on approach on Application of Internet of Things in Various Domain of Engineering*. Delhi: International Research Publication House.

Ankit Sharma, A. (2015). *P-channel Tunneling Field Effect Transistor (TFET): Sub-10nm Technology Enablement by GaSb-InAs with Doped Source Underlap. In Device research conference (DRC)*. IEEE.

Anushka, K., & Mahadevasawamy, U.B. (2018). Automatic Plant Monitoring and Watering System using Raspberry Pi. *I.J. Engineering and Manufacturing,* (6), 55-67. Doi:10.5815/ijem.2018.06.05

Arduino UNO. (n.d.). Retrieved from https://datasheet.octopart.com/A000066-Arduino-datasheet-38879526.pdf

Arena, G. S. M. (n.d.). *Enhanced Data for Global Evolution.* Available at http://www.gsmarena.com/glossary.php3?term=edge

Arena, G. S. M. (n.d.). *General Packet Radio Service.* Available at http://www.gsmarena.com/glossary.php3?term=gprs

Ashton. (2009). Internet of things thing. *RFID Journal.* Available at Http://www.rfidjournal.com/article/print4986

Asthana & Pandey. (2018). Securing IoT: Threats and Vulnerabilities. IJCA Proceedings on International Information Security Conference, 2017(1), 12-16.

Athani, S., Tejeshwar, C.H., Patil, M., Patil, P., & Kulkarni, R. (2017). *Soil moisture monitoring using IoT enabled Arduino sensors with neural networks for improving soil management for farmers and predict seasonal rainfall for planning future harvest in North Karnataka-India.* Academic Press. doi:10.1109/I-SMAC.2017.8058385

Avci, U. E. (2012). Understanding the feasibility of scaled III–V TFET for logic by bridging atomistic simulations and experimental results. *Proc. VLSI Technol. (VLSIT) Symp.*, 183–184.

Ballazhi. (2012). *Wireless Indoor Positioning Techniques.* Comminications System Seminar.

Barakat, S. (2016). Education and the internet of everything. *International Business Management, 10*(18), 4301–4303.

Barcelona Digital City. (n.d.). *Digital transformation, 2018.* Retrieved from http://ajuntament.barcelona.cat/digital/en/digital-transformation.`

Barnaghi, P., & Sheth, A. (2016). On Searching the Internet of Things: Requirements and Challenges. *IEEE Intelligent Systems, 31*(6), 71–75. doi:10.1109/MIS.2016.102

Barnaghi, P., Wang, W., Henson, C., & Taylor, K. (2012). Semantics for the IOT: Early progress and back to the future. *Journal of Centre of Communication and System Research. International Journal on Semantic Web and Information Systems, 8*(1), 1–21. doi:10.4018/jswis.2012010101

Barrett, D. M., & Lloyd, B. (2012). Advanced preservation methods and nutrient retention in fruits and vegetables. *Journal of the Science of Food and Agriculture, 92*(1), 7–22. doi:10.1002/jsfa.4718 PMID:22102258

Beale, G. O., & Li, M. (1997). Robust temperature control for microwave heating of ceramics. *IEEE Transactions on Industrial Electronics, 44*(1), 124–131. doi:10.1109/41.557507

Bedi, G., Venayagamoorthy, G. K., & Singh, R. (2016). Navigating the challenges of Internet of Things (IoT) for power and energy systems. *Proceedings of 2016 Clemson University Power Systems Conference (PSC)*, 1-5. 10.1109/PSC.2016.7462853

Behera, A. R., Devi, J., & Mishra, D. S. (2015, October). A comparative study and implementation of real time home automation system. In *2015 International Conference on Energy Systems and Applications* (pp. 28-33). IEEE. 10.1109/ICESA.2015.7503308

Bekara, C. (2014). Security Issues and Challenges for the IoT-based Smart Grid. *Procedia Computer Science, 34*, 532–537. doi:10.1016/j.procs.2014.07.064

Bellavista, P., Cardone, G., Corradi, A., & Foschini, L. (2013). Convergence of MANET and WSN in IoT urban scenarios. *IEEE Sensors Journal, 13*(10), 3558–3567. doi:10.1109/JSEN.2013.2272099

Bendel, O. (2017). *Service Robots in Public Spaces.* Retrieved from https://www.heise.de/tp/features/Service-Robots-in-Public-Spaces-3754173.html?seite=all

Bernstein, T., & Chang, P. P. (1969). Calculation of Self-Heating Current for a Resistance Wire Temperature Sensor. *IEEE Transactions on Industrial Electronics and Control Instrumentation*, (1), 99-102.

Bezawada, B., Liu, A. X., Jayaraman, B., Wang, A. L., & Li, R. (2015). Privacy Preserving String Matching for Cloud Computing. *Proceedings of the 35th IEEE International Conference onDistributed Computing Systems, ICDCS '15*, 609–618. 10.1109/ICDCS.2015.68

Bhagat, K., Deshmukh, S., Dhonde, S., & Ghag, S. (2016). Obstacle avoidance robot. *International Journal of Science, Engineering and Technology Research*, *5*(2), 439–442.

Bhoomika & Muralidhara. (2015). Secured Smart Healthcare Monitoring System Based on IoT. *International Journal on Recent and Innovation Trends in Computing and Communication*, *3*(7), 4958-4961.

Bluetooth Module HC-05. (n.d.). Retrieved from https://www.electronicwings.com/sensors-modules/bluetooth-module-hc-05-

Boboc, R. G., Horaţiu, M., & Talabă, D. (2014). An Educational Humanoid Laboratory Tour Guide Robot. *Procedia: Social and Behavioral Sciences*, *141*, 424–430. doi:10.1016/j.sbspro.2014.05.074

Borgia, E. (2014). The internet of things vision: Key features, applications and open issues. *Computer Communications*, *54*, 1–31. doi:10.1016/j.comcom.2014.09.008

Bosona, T., & Gebresenbet, G. (2013). Food traceability as an integral part of logistics management in food and agricultural supply chain. *Food Control*, *33*(1), 32–48. doi:10.1016/j.foodcont.2013.02.004

Boujelben, M., Rekik, C., & Derbel, N. (2013). Hierarchical fuzzy controller to avoid mobile obstacle for a mobile robot. Proceedings of 10th International Multi-Conference on Systems, Signals & Devices (SSD), 1-8. doi:10.1109/SSD.2013.6564094

Boxall, J. (2014). *Tutorial - L298N Dual Motor Controller Module 2A and Arduino*. Retrieved from https://tronixlabs.com.au/news/tutorial-l298n-dual-motor-controller-module-2a-and-arduino/

Brandt, C. (2016). *The Internet of Things and Its Impact on Education*. University Herald.

Bugard, W. (2000). Experiences with an Interactive Museum Tour-Guide Robot. *Artificial Intelligence*, *114*(1/2).

Bui, N., & Zorzi, M. (2011). Health Care Applications: A Solution Based on The Internet of Things. Department of Information Engineering, University of Padova Via Gradenigo. doi:10.1145/2093698.2093829

Byrne, J., O'Sullivan, K., & Sullivan, K. (2017). An IoT and Wearable Technology Hackathon for Promoting Careers in Computer Science. *IEEE Transactions on Education*, *60*(1), 50–58. doi:10.1109/TE.2016.2626252

Calduwel Newton, P. (2009). A Quality of Service Strategy to Select Coding Schemes in General Packet Radio System. *International Journal of Advance Science and Technology*, *7*, 1–10.

Cambria, E. (2016). Affective Computing and Sentiment Analysis. *IEEE Intelligent Systems*, *31*(2), 102–107. doi:10.1109/MIS.2016.31

Camps-Mur, D., Garcia-Saavedra, A., & Serrano, P. (2013). Device-to-Device Communications with WiFi-Direct: Overview and Experimentation. *IEEE Wireless Communications*, *20*(3), 96–104. doi:10.1109/MWC.2013.6549288

Cao, Y., Jiang, T., & Han, Z. (2016). A Survey of Emerging M2M Systems: Context, Task, and Objective. *IEEE Internet of Things Journal*, *3*(6), 1246–1258. doi:10.1109/JIOT.2016.2582540

Caro De, N., & Colitti, W. (2013). *Comparison of two lightweight protocols for smartphone -based sensing*. Department of ETRO-IRIS Vrije University Brussels. doi:10.1109/SCVT.2013.6735994

Castaneda, D., Esparza, A., & Ghamari, M. (2018). A review on wearable photoplethysmography sensors and their potential future applications in health care. *Int J Biosen Bioelectron*, *4*(4), 195–202. doi:10.15406/ijbsbe.2018.04.00125 PMID:30906922

Castellani, A. P., Bui, N., Casari, P., Rossi, M., Shelby, Z., & Zorzi, M. (2010). *Architecture and Protocols for the Internet of Things: A Case Study*. PERCOM.

Cata, M. (2015). Smart university, a new concept in the Internet of Things. *2015 14th RoEduNet International Conference - Networking in Education and Research (RoEduNet NER)*, 195–197.

Cavalheiro, D., Moll, F., & Valtchev, S. (2017). TFET-Based Power Management Circuit for RF Energy Harvesting. *IEEE Journal of the Electron Devices Society*, *5*(1), 7–17. doi:10.1109/JEDS.2016.2619908

Chakrabarty, S., & Engels, D. W. (2016, January). A secure IoT architecture for Smart Cities. In 2016 13th IEEE annual consumer communications & networking conference (CCNC) (pp. 812-813). IEEE. doi:10.1109/CCNC.2016.7444889

Chang, C. H. (2011). Smart classroom roll caller system with IOT architecture. *Proc. - 2011 2nd Int. Conf. Innov. Bio-Inspired Comput. Appl. IBICA 2011*, 356–360. 10.1109/IBICA.2011.94

Chasta, R., Singh, R., Gehlot, A., Mishra, R., & Choudhury, S. (2016). A smart building automation system. *International Journal of Smart Home, 10*(9), 217-224. . doi:10.14257/ijsh.2016.10.9.20

Chen, Y. N. (2014). Ultra-low voltage mixed TFET-MOSFET 8T SRAM cell. *International symposium on low power electronics and design*, 255–8.

Chen, A., Gil-de-Castro, A., Palacios-García, E. J., Flores-Arias, J. M., & Bellido-Outeiriño, F. J. (2015, June). In-home data acquisition and control system based on BLE. In *2015 International Symposium on Consumer Electronics (ISCE)* (pp. 1-2). IEEE. 10.1109/ISCE.2015.7177780

Cheng, H., & Liao, W. (2012). *Establishing an lifelong learning environment using IOT and learning analytics*. Advanced Communication Technology.

Chen, Y. N., Fan, M.-L., Hu, V. P.-H., Su, P., & Chuang, C.-T. (2013). Design and analysis of robust tunneling FET SRAM. *IEEE Transactions on Electron Devices*, *60*(3), 1092–1098. doi:10.1109/TED.2013.2239297

Chen, Y. S., & Juang, J. H. (2009). Intelligent obstacle avoidance control strategy for wheeled mobile robot. *Proceedings of ICROS-SICE International Joint Conference*, 3199-3204.

Chen, Y., & Dong, X. (2013). The Development and Prospect of New Technology in Modern distance education. *International Conference Information Science Computer Application*, 40–44. 10.2991/isca-13.2013.7

Chew, C.B. (2015). Sensors-Enabled Smart Attendance Systems Using Nfc and Rfid Technologies. *Int. J. New Comput. Archit. their Appl., 5*(1), 19–28.

Chin, J., & Callaghan, V. (2013). Educational living labs: A novel internet-of-things based approach to teaching and research. *Proc. - 9th Int. Conf. Intell. Environ. IE 2013*, 92–99. 10.1109/IE.2013.48

Choudhury, S., Kuchhal, P., Singh, R., & Anita. (2015). ZigBee and Bluetooth network based sensory data acquisition system. *Procedia Computer Science*, *48*, 367–372. doi:10.1016/j.procs.2015.04.195

Chowdhury, A. R. (2017). *IoT and Robotics: A Synergy*. Peer J Preprints; doi:10.7287/peerj.preprints.2760v1

Cirillo, M., Karlsson, L., & Saffiotti, A. (2010). Human-aware task planning: an application to mobile robots. *ACM Trans Intell Syst Technol*, *1*(2), 15:1–15:26. Retrieved from https://dl.acm.org/citation.cfm?id=1869404

Cuff, D., Hansen, M., & Kang, J. (2008). Urban sensing: Out of the woods. *Communications of the ACM*, *51*(3), 24–33. doi:10.1145/1325555.1325562

Da Cruz, M. A. A., Rodrigues, J. J. P. C., Al-Muhtadi, J., Korotaev, V. V., & de Albuquerque, V. H. C. (2018). A Reference Model for Internet of Things Middleware. *IEEE Internet of Things Journal*, *5*(2), 871–883. doi:10.1109/JIOT.2018.2796561

Datta, S., Liu, H., & Narayanan, V. (2014). Tunnel FET technology: A reliability perspective. *Microelectronics and Reliability*, *54*(5), 861–874. doi:10.1016/j.microrel.2014.02.002

De Donno, D., Catarinucci, L., & Tarricone, L. (2014, July). RAMSES: RFID augmented module for smart environmental sensing. *IEEE Transactions on Instrumentation and Measurement*, *63*(7), 1701–1708. doi:10.1109/TIM.2014.2298692

De Ruyter, B., Saini, P., Markopoulos, P., & van Breemen, A. (2005). Assessing the Effects of Building Social Intelligence in a Robotic Interface for The Home. *INT COMPUT*, *17*(5), 522–541. doi:10.1016/j.intcom.2005.03.003

Decuir, J. (2010). *Bluetooth 4.0: Low Energy, Presentation slides*. Retrieved from http://chapters.comsoc.org/vancouver/BTLER3.pdf

Deepali. (2017). A Survey on Automatic Irrigation System Using Wireless Senor Network. *International Journal of Current Engineering And Scientific Research, 4*(8), 69-73.

Deepiga, T., & Sivasankari, A. (2015). Smart water monitoring system using wireless sensor network at home/office. *International Research Journal of Engineering and Technology*, *2*(4), 1305–1314.

Dempsey, P. (2017). The Teardown: Google Home personal assistant. *Engineering Technology*, *12*(3), 80–81. doi:10.1049/et.2017.0330

Dharani, S. (2018). IOT Based Smart Restaurant. *National Conference on Emerging Trends in Information, Management and Engineering Sciences.*

Digital Intelligence Blog. (2016). *Wearable's impact on cancer patient's quality of life to be tested.* Received July, 26 2016 From http://www.pmlive.com/blogs/digital_intelligence/archive/2016/july/wearables_impact_on_cancer_patients_quality_of_life_to_be_tested

Ding, C., Yang, L. J., & Wu, M. (2011). Security architecture and key technologies for IoT/CPS. *ZTE Technology Journal, 17*(1).

Dohler, M., Vilajosana, I., Vilajosana, X., & Llosa, J. (2011). Smart Cities: An action plan. *Proceedings Barcelona Smart Cities Congress*, 1–6.

Donno De, D., Catarinucci, L., & Tarricone, L. (2014, April). A battery-assisted sensor-enhanced RFID tag enabling heterogeneous wireless sensor networks. *IEEE Sensors Journal*, *14*(4), 1048–1055. doi:10.1109/JSEN.2013.2293177

Drakopoulou, S. (2010). A Movement of Experimentation: Spatial Practice and Representation of Space as narrative Elements in Location based Games. *Journal of Media Geography*, *5A*, 73–76.

Elangovan, R., Banu, A., Rozario, R., & Santhanakrishan, N. (2018). *Tome: A Plant Monitoring and Gardening System using IoT*. Academic Press.

El-Rabbany, A. (2002). *Introduction to GPS: The Global Positioning System*. Artech House Mobile Communication Series.

Endler, M. (2014). Microsoft Azure Intelligent Systems: 4 Facts. *Information Week, 906*, 174.

Evans, D. (2011). The internet of things how the next evolution of the internet is changing everything. *Advances in Internet of Things*, *6*(4).

Fan, M.-L., Chen, Y.-N., Su, P., & Chuang, C.-T. (2016). *Challenges and designs of tfet for digital applications. In Tunneling Field Effect Transistor Technology* (pp. 89–109). Springer.

Fan, X., Huang, H., Qi, S., Luo, X., Zeng, J., Xie, Q., & Xie, C. (2015). Sensing home: A cost-effective design for smart home via heterogeneous wireless networks. *Sensors (Basel)*, *15*(12), 30270–30292. doi:10.3390151229797 PMID:26633424

Farooq, M. U. (2015). A Review on Internet of Things (IoT). *International Journal of Computers and Applications*, *113*(1), 1–7. doi:10.5120/19787-1571

Firner, B., Moore, R. S., Howard, R., Martin, R. P., & Zhang, Y. (2011). Poster: Smart buildings, sensor networks, and the internet of things. *Proceedings of the 9th ACM Conference on Embedded Networked Sensor Systems*, 337–338.

Folea, S., Bordencea, D., Hotea, C., & Valean, H. (2012, May). Smart home automation system using Wi-Fi low power devices. In *Proceedings of 2012 IEEE International Conference on Automation, Quality and Testing, Robotics* (pp. 569-574). IEEE. 10.1109/AQTR.2012.6237775

Franklin, M. (2013). The Berkeley Data Analytic Stack. *IEEE International Conference on Big Data*.

Frontech, F. (2007). *Service Robot Enon from Fujitsu Acts as Visitors' Guide at Museum of Best-Selling Mystery Novelist*. Retrieved from http://www.frontech.fujitsu.com/en/release/20070920.html

Gaggl, L. (2011). *Education and the Internet of Things*. Retrieved from: http://www.brightcookie.com/wp-content/uploads/ 2011/11/IoT_in_Education-3.pdf

Gill, S. S., Chana, I., & Buyya, R. (2017). IoT based agriculture as a cloud and big data service: The beginning of digital India. *Journal of Organizational and End User Computing*, *29*(4), 1–23. doi:10.4018/JOEUC.2017100101

Gleason & Gziabher. (2009). *GNSS Applications and Methods*. Norwood: Artech House.

Global Sensors in Internet of Things (IoT) Devices Market, Analysis & Forecast: 2016 to 2022; (Focus on Pressure, Temperature, Light, Chemical, & Motion Sensors; and Applications in Healthcare, Manufacturing, Retail, & Transportation. (n.d.). Retrieved from https://www.marketresearch.com/BIS-Research-v4011/Global-Sensors-Internet-Things-IoT-10703194/

Goel, D., Chaudhury, S., & Ghosh, H. (2017). Smart water management - an ontology-driven context aware IOT application. *Journal of Pattern Recognition and Machine Intelligence*, 639-646.

Gomez, C., Oller, J., & Paradells, J. (2012). Overview and evaluation of Bluetooth low energy: An emerging low-power wireless technology. *Sensors (Basel)*, *12*(9), 11734–11753. doi:10.3390120911734

Gope, P., & Hwang, T. (2016). BSN-Care: A secure IoT-based modern healthcare system using body sensor network. *IEEE Sensors Journal*, *16*(5), 1368–1376. doi:10.1109/JSEN.2015.2502401

Gowrishankar, S., & Prachita, M. (2017). IoT based heart attack detection, heart rate and temperature monitor. *International Journal of Computers and Applications*, *170*(5), 26–30. doi:10.5120/ijca2017914840

Graf, B., Hans, M. & Schraft, R.D. (2004). *Care-O-bot II—Development of a Next Generation Robotic Home Assistant*. Academic Press.

Granjal, J., Monteiro, E., & Sa Silva, J. (2015). Security for the internet of things: A survey of existing protocols and open research issues. *IEEE Communications Surveys and Tutorials*, *17*(3), 1294–1312. doi:10.1109/COMST.2015.2388550

Gray, K. W. (2000). *Obstacle detection and avoidance for an autonomous farm tractor* (Master of Science thesis). Utah State University.

Guanshan, H. U. (2009). Neural network based on information fusion using for avoiding obstacle robot. *Proceedings of WASE International Conference on Information Engineering*, 565-568.

Guizzo, E. (2011). How Google's Self-Driving Car Works. *IEEE Spectrum.* Retrieved from https://spectrum.ieee.org/automaton/robotics/artificial-intelligence/how-google-self-driving-car-works

Gultom, J. H., Harsono, M., Khameswara, T. D., & Santoso, H. (2017). Smart IoT Water Sprinkle and Monitoring System for chili plant. *2017 International Conference on Electrical Engineering and Computer Science (ICECOS)*, 212-216. 10.1109/ICECOS.2017.8167136

Guo, S., Qiang, M., Luan, X., Xu, P., He, G., Yin, X., ... Fang, D. (2015). The application of the Internet of Things to animal ecology. *Integrative Zoology*, *10*(6), 572–578. doi:10.1111/1749-4877.12162 PMID:26338071

Gupta, N. (2016). Ultra-Low-Power Compact TFET Flip-Flop Design for High Performance Low-Voltage Applications. *IEEE Symposium on Quality Electronic Design.*

Gyeyoung & Yim. (2012). A Review of the Techniques for Indoor Location based Service. *International Journal of Grid and Distributed Computing*, *5*(1), 1–22.

Haghi, M., Thurow, K., & Stoll, R. (2017). Wearable Devices in Medical Internet of Things: Scientific Research and Commercially Available Devices. *Healthcare Informatics Research*, *23*(1), 4–15. doi:10.4258/hir.2017.23.1.4 PMID:28261526

Han, H., & Xu, H. (2012). Interactive Charging Strategy of Electric Vehicles connected in Smart Grids. *7th International Power Electronics and Motion Control Conference*, 2099-2103.

Han, J., Haihong, E., Le, G., & Du, J. (2011). Survey on NoSQL database. *6th International Conference on Pervasive Computing and Applications (ICPCA)*, 363-66.

Hanumante, V., Roy, S., & Maity, S. (2013). Low cost obstacle avoidance robot. *International Journal of Computing and Engineering*, *3*(4), 52–55.

Harper, C., & Siller, M. (2015). OpenAG: A Globally Distributed Network of Food Computing. *IEEE Pervasive Computing*, *14*(4), 24–27. doi:10.1109/MPRV.2015.72

Heidarsson, H. K., & Sukhatme, G. S. (2011). Obstacle detection and avoidance for an autonomous surface vehicle using a profiling sonar. *Proceedings of IEEE International Conference on Robotics and Automation*, 731-736. 10.1109/ICRA.2011.5980509

Henrik, I., & Christensen, G.D. (2016). Sensing and Estimation. In Springer Handbook of Robotics (2nd ed.; pp. 91-111). Springer International Publishing.

Hipa Journal. (n.d.). *Healthcare Data Security*. Retrieved from https://www.hipaajournal.com/category/healthcare-data-security/page/6/

How Internet of Things effects on our daily lifestyle. (2016). Retrieved from https://www.futurescope.co/internet-of-things-iot/

Hu, Chen, Fan, & Wang. (2017). An intelligent and secure health monitoring scheme using IoT sensor based on cloud computing. *Journal of Sensors*.

Huang, Y., & Li, G. (2010). *Descriptive Models for Internet of Thing. In IEEE International Conference on Intelligent Control and Information Processing* (pp. 483–486). ICICIP.

IEEE Trial-Use Standard for Wireless Access in Vehicular Environments – Security Services for Applications and Management Messages. (2006). IEEE Std 1609.2TM.

IETF. (n.d.). *IPv6 over Networks of Resource-constrained Nodes (6lo)*. Retrieved from https://datatracker.ietf.org/wg/6lo/documents/

Ikhlef, A., Kihel, M., Boukhezzar, B., Guerrouj, A., & Mansouri, N. (2014, November). Online temperature control system. In *2014 International Conference on Interactive Mobile Communication Technologies and Learning (IMCL2014)* (pp. 75-78). IEEE. 10.1109/IMCTL.2014.7011108

Imenabadi, Saremi, & Vandenberghe. (2017). A Novel PNPN-Like Z-Shaped Tunnel Field-Effect Transistor With Improved Ambipolar Behavior and RF Performance. *IEEE Transactions on Electron Devices*, 64.

Infosec. (n.d.). Internet of Things: How Much are We Exposed to Cyber Threats? Retrieved from http://resources.infosecinstitute.com/internet-thingsmuch-exposed-cyber-threats/

IoT. (2015). Retrieved from https://dzone.com/articles/the-internet-of-things-gateways-and-next-generation

iRobot.com. (2015). *iRobot Corporation: We Are The Robot Company*. Retrieved from http://www.irobot.com/

Isa, E., & Sklavos, N. (2017). Smart home automation: GSM Security Design & Implementation. *Journal of Engineering Science and Technology Review.*, *3*(10), 170–174. doi:10.25103/jestr.103.22

iScoop. (n.d.). *The Internet of Robotic Things (IoRT): Definition, Market and Examples*. Retrieved from https://www.i-scoop.eu/internet-of-things-guide/internet-robotic-things-iort/

Islam, S. M. R., Kwak, D., Kabir, M. H., Hossain, M., & Kwak, S. (2015). The Internet Of Things For Health Care: A Comprehensive Survey. *IEEE Access: Practical Innovations, Open Solutions*, *3*, 678–708. doi:10.1109/ACCESS.2015.2437951

ISRO. (2013). *Indian SBAS System-GAGAN*. GNSS. Asia India-Industry workshop.

Jagtap, A. M., Sikandar, B. S., Shivaji, S. S., Pramod, K. V., & Sarangdhar, N. K. (2016). Aquarius-smart IOT technology for water level monitoring system. *International Journal of Computational Science*, *5*(8).

Jakhete, M.D., & Mankar, P.C. (2015). Implementation of Smart Restaurant With The E-Menu Card. *International Journal of Computer Applications, 119*(21).

Jathavara, V., & Ashwath, P. (2017). Obstacle avoiding robot: a Lego EV3 robot. *Proceedings of International Conference on Recent Advances in Electronics and Communication Technology*, 43-49. 10.1109/ICRAECT.2017.61

Jayaraman, P.P., Yavari, A., Morshed, A., Zaslavsky, A., & Geograkopolous, D. (2016). *Internet of Things Platform for Smart Farming: Experiences and Lessons Learnt*. Academic Press. Doi:10.339016111884

Jedermann, R., Behrens, C., Westphal, D., & Lang, W. (2006). Applying autonomous sensor systems in logistics—Combining sensor networks, RFIDs and software agents. *Sensors and Actuators. A, Physical*, *132*(1), 370–375. doi:10.1016/j.sna.2006.02.008

Jeyanthi, N., & Thandeeswaran, R. (2017). *Security Breaches and Threat Prevention in the Internet of Things – common issues*. IGI Global. doi:10.4018/978-1-5225-2296-6

Kamienski, C., Soininen, J.P., Taumberger, M., Dantas, R., Toscano, A., Cinotti, T.S., … Neto, A.T. (2018). Smart water management platform: IOT-based precision irrigation for agriculture. *Journal of MDPI Sensors, 19*(276).

Kanase, P., & Gaikwad, S. (2016). Smart hospitals using internet of things (IOT). *International Research Journal of Engineering and Technology, 3,* 1-3.

Karagiannis, V., Chatzimisios, P., Vazquez-Gallego, F., & Alonso-Zarate, J. (2015). Asurvey on application layer protocols for the internet of things. *Transaction on IoT and Cloud Computing, 3*(1), 11-17. Available at https://jesusalonsozarate.files.wordpress.com/2015/01/2015-transaction-on-iot-and-cloudcomputing.pdf

Karl, H. (2004). *Data transmission in Mobile Communication Systems.* Location-Based Services. doi:10.1016/B978-155860929-7/50009-1

Kaur, P., & Katiyar, M. (2012). The energy-efficient hierarchical routing protocols for WSN: A review. *International Journal of Advanced Research in Computer Science and Software Engineering, 2*(11).

Keerthana, B., Nivetha, P., Boomika, M., & Mathivatahni, M. (2018). Literature Review on IOT Based Smart Security and Monitoring Devices for Agriculture. *International Journal on Future Revolution in Computer Science & Communication Engineering, 4*(2), 342-345.

Kehoe, B., Patil, S., Abbeel, P. & Goldberg, K. (2015). A Survey of Research on Cloud Robotics and Automation. *IEEE Transactions on Automation Science and Engineering, 12*(2), 398-409.

Khatami, Y., & Banerjee, K. (2009). Steep Subthreshold Slope n- and p-Type Tunnel-FET Devices for Low-Power and Energy-Efficient Digital Circuits. *IEEE Transactions on Electron Devices*, 56.

Khoi, M. N., & Ahlund, C. (2015). *IReHMo: An Efficient IoT Based Remote Health Monitoring System for Smart Regions.* Department of Computer Science, Electrical and Space Engineering Lulea University of Technology Skelleftea.

Kim, A., Hekland, F., Petersen, S., & Doyle, P. (2008). When HART goes wireless: Understanding and implementing the Wireless HART standard. *IEEE International Conference on Emerging Technologies and Factory Automation (ETFA 2008),* 899-907. Available at https://library.e.abb.com/public/eb20fe80a391ca8485257bc600667573/When%20HART%20Goes%20Wireless%20Understanding%20and%20Implementing%20the%20WirelessHART%20Standard.pdf

Kirtana, R. N., & Lokeswari, Y. V. (2017). An IoT based remote HRV monitoring system for hypertensive patients. *IEEE International Conference on Computer, Communication and Signal Processing*. 10.1109/ICCCSP.2017.7944086

Kodali, R. K., Jain, V., Bose, S., & Boppana, L. (2016, April). IoT based smart security and home automation system. In *2016 international conference on computing, communication and automation (ICCCA)* (pp. 1286-1289). IEEE.

Koo, D., Piratla, K., & Matthews, C. J. (2015). Towards sustainable water supply: Schematic development of big data collection using internet of things (IoT). *Procedia Engineering, 118*, 489–497. doi:10.1016/j.proeng.2015.08.465

Kosmatos, E.A., Tselikas, N.D., & Boucouvalas, A.C. (2011). Integrating RFIDs and Smart Objects into a Unified Internet of Things Architecture. *Advances in Internet of Things*: *Scientific Research, 1*, 5-12. Available at doi:10.4236/ait.2011.11002

Krebs. (n.d.). *Gunter's Space Page Site*. Available at http://space.skyrocket.de/doc_sdat/uragan-k1.htm

Krotov, V. (2017). The Internet of Things and new business opportunities. *Business Horizons, 60*(6), 831–841. doi:10.1016/j.bushor.2017.07.009

Kulkarni, P. V., & Joshi, M. S. (2016). An IOT based water supply monitoring and controlling system with theft identification. *Journal of Innovation Research in Science, 5*(9), 16153–16157.

Kumar, N., Kumar, P., Archana, B., & Nishant, G. (2017). Arduino based Automatic Plant Watering System with Internet of Things (IOT). *International Research Journal of Advanced Research in Electrical, Electronics and Instrumentation Engineering, 6*(3), 1-8. Doi:10.15662/IJAREEIE.2017.0603144

Kumar, J. (2014, March). A survey on Internet of Things: Security and Privacy Issues. *International Journal of Computers and Applications*, *90*(11).

Kumar, M. J., & Sahay, S. (2016). Controlling BTBT-induced parasitic BJTaction in junctionless FETs using a hybrid channel. *IEEE Transactions on Electron Devices*, *63*(8), 3350–3353. doi:10.1109/TED.2016.2577050

Kumar, P. (2017, February). Design and implementation of Smart Home control using LabVIEW. In *2017 Third International Conference on Advances in Electrical, Electronics, Information, Communication and Bio-Informatics (AEEICB)* (pp. 10-12). IEEE. 10.1109/AEEICB.2017.7972317

L. Zhou, N. Chen, Z. Chen, and C. Xing, "ROSCC: An Efficient Remote Sensing Observation-Sharing Method Based on Cloud Computing for Soil Moisture Mapping in Precision Agriculture," IEEE Journal of Selected Topics in Applied Earth Observations and Remote Sensing, vol. PP, no. 99, pp. 1–11, 2016.

Lade, P. S., Vyas, P., Walavalkar, V., Wankar, B., & Yadav, P. (2018). Water management system using IOT with WSN. *Journal of International Research of Engineering and Technology*, *5*(3), 3079.

Lahgere, A., & Kumar, M. J. (2017). A Tunnel Dielectric-Based Junctionless Transistor With Reduced Parasitic BJT Action. *IEEE Transactions on Electron Devices*, *64*(8), 3470–3475. doi:10.1109/TED.2017.2713423

Lardear, J. L. (1990). Control of self-regulating heating cable for use in pipeline heating applications. In *37th Annual Conference on Petroleum and Chemical Industry*. IEEE. 10.1109/PCICON.1990.218089

LaValle, S. M. (2006). Planning Algorithms. Cambridge University Press.

Lee, K. (2016). *Healthcare IoT security issues: Risks and what to do about them.* Retrieved from https://internetofthingsagenda.techtarget.com/

Lee, Y. (2017). *Robots at South Korea Airports to Help Passengers Finding Their Gates.* Retrieved from https://skift.com/2017/07/04/robots-at-south-korean-airports-to-help-passengers-find-their-gates/

Lee, J. (2013). SAP HANA Distributed In-Memory Database Systems: *Transaction, Session and Metadata Management. ICDE Conference*, 1165 – 1173.

Lee, Y., Yoon, D., Kim, Y., Blaauw, D., & Sylvester, D. (2013). Circuit and system design guidelines for ultra-low power sensor nodes. *IPSJ Trans Syst LSI Des Method*, *6*(2), 17–26. doi:10.2197/ipsjtsldm.6.17

Lennvall, T., Gidlund, M. & Akerberg, J. (2017). Challenges When Bringing IOT into Industrial Automation. *IEEE Africon,* 905–910.

Li, C. (n.d.). *Introduction and overview of Biosensors and Electrochemistry.* Retrieved from http://nanohub.org/resources/2261/download/nanobiotechnology_and_biosensors.pdf

Li, C.-T., Lee, C.-C., & Weng, C.-Y. (2016). A secure cloud-assisted wireless body area network in mobile emergency medical care system. Journal of Medical Systems, 40(5), 1–15. doi:10.100710916-016-0474-9

Liao, Y., & Leeson, S. M. (2016). *Analysis of In -to-Out Wireless Body Area Network Systems: Towards QoS-Aware Health Internet of Things Applications.* School of Engineering, University of Warwick.

Li, C., Hu, X., & Zhang, L. (2017). The IoT-based heart disease monitoring system for pervasive healthcare service. *Procedia Computer Science*, *112*, 2328–2334. doi:10.1016/j.procs.2017.08.265

Li, Liu, H., Wang, S., Chen, S., & Wang, Q. (2017). The Programming Optimization of Capacitorless 1T DRAM Based on the Dual-Gate TFET. *Nanoscale Research Letters*, *12*(1), 524. doi:10.118611671-017-2294-3 PMID:28875269

Li, M., Yu, S., Zheng, Y., Ren, K., & Lou, W. (2012). Scalable and secure sharing of personal health records in cloud computing using attributebased encryption. *IEEE Transactions on Parallel and Distributed Systems*, *24*(1), 131–143. doi:10.1109/TPDS.2012.97

Lin, Y.-B., Haung, Y.-R., Chen, Y.-K., & Chlamtac, I. (2001). Mobility Management: From GPRS to UMTS. *Wireless Communications and Mobile Computing*, *1*(4), 339–360. doi:10.1002/wcm.27

Liquid Crystal Display. (n.d.). Retrieved from https://www.elprocus.com

Liu, Li, Vaddi, Ma, Datta, & Narayanan. (2014). Tunnel FET RF Rectifier Design for Energy Harvesting Applications. *IEEE Journal on Emerging and selected topics in Circuits and Systems, 4*, 400 – 411.

Liu, F., & Li, T. (2018). A clustering k-anonymity privacy-preserving method for wearable IoT devices. *Security and Communication Networks*, *2018*, 1–8. doi:10.1155/2018/1635081

Liu, X., Yang, T., & Yan, B. (2015, November). Internet of Things for wildlife monitoring. In *2015 IEEE/CIC International Conference on Communications in China-Workshops (CIC/ICCC)* (pp. 62-66). IEEE.

Li, W., Liu, H., Wang, S., & Chen, S. (2017, March). Reduced Miller Capacitance in U-Shaped Channel Tunneling FET by Introducing Heterogeneous Gate Dielectric. *IEEE Electron Device Letters*, 38.

Li, W., Liu, H., Wang, S., Chen, S., & Yang, Z. (2017). Design of High Performance Si/SiGe Heterojunction Tunneling FETs with a T-Shaped Gate. *Nanoscale Research Letters.*

Li, X. (2014). RF-powered systems using steep-slope devices. *Proc. IEEE Int. New Circuits Syst. Conf.*

LoRa. (n.d.). In Wikipedia. Retrieved from https://en.wikipedia.org/wiki/LoRa

Louis, L. (2016). Working Principle of Arduino And Using It As A Tool For Study And Research. *International Journal of Control, Automation Communication and Systems*, *1*(2), 21. doi:10.5121/ijcacs.2016.1203

Lounis, A., Hadjidj, A., Bouabdallah, A., & Challal, Y. (2016). Healing on the cloud: Secure cloud architecture for medical wireless sensor networks. *Future Generation Computer Systems*, *55*, 266–277. doi:10.1016/j.future.2015.01.009

Lucio, C., Francesco, F., Gianluca, I., Longhi, S., & Giacomo, T. (2016). *IOT Based Indoor Personal Comfort Levels Monitoring*. Academic Press. . doi:10.1109/ICCE.2016.7430548

Lund, D., Turner, V., MacGillivray, C., & Morales, M. (2014). Worldwide and Regional Internet of Things (IoT) 2014-2020 Forecast: A Virtuous Circle of Proven Value and Demand. Framingham: International Data Corporation (IDC).

Luo, R. C., Yih, C. C., & Su, K. L. (2002). Multisensor fusion and integration: Approaches, applications, and future research directions. *IEEE Sensors Journal*, *2*(2), 107–119. doi:10.1109/JSEN.2002.1000251

Mahale, R. B., & Sonavane, S. S. (2016). Smart Poultry Farm Monitoring Using IOT and Wireless Sensor Networks. *International Journal of Advanced Research in Computer Science*, *7*(3).

Mahmoud, M. S., & Mohamad, A. A. H. (2016). A Study of Efficient Power Consumption Wireless Communication Techniques / Modules for Internet of Things (IoT) Applications. In Advances in Internet of Things (pp. 19–29). Academic Press.

Makosiej, A., Gupta, N., Vakul, N., Vladimirescu, A., Cotofana, S., Mahapatra, S., ... Anghel, C. (2016). Ultra-low leakage SRAM design with sub-32 nm tunnel FETs for low standby power applications. *Micro & Nano Letters*, *11*(12), 828–831. doi:10.1049/mnl.2016.0442

Malan, D. (2004). CodeBlue: An Ad hoc sensor Network Infrastructure for Emergency Medical Care. WIBSN'04.

Malche, I., & Maheshwary, P. (2017). Internet of things IOT based water level monitoring system for smart village. *Journal of INT International Conference on Communication and Networks*, *508*(12), 1–6.

Mallik, A., & Chattopadhyay, A. (2012). Tunnel field-effect transistors for analog/mixed-signal system-on-chip applications. *IEEE Transactions on Electron Devices*, *59*(4), 888–894. doi:10.1109/TED.2011.2181178

Mandula, K., Parupalli, R., Murty, C. A., Magesh, E., & Lunagariya, R. (2015, December). Mobile based home automation using Internet of Things (IoT). In *2015 International Conference on Control, Instrumentation, Communication and Computational Technologies (ICCICCT)* (pp. 340-343). IEEE.

Marc, B. (2015). *Industrial Internet of Things: Unleashing the Potential of Connected Products and Services*. Retrieved from http://www3.weforum.org/docs/WEFUSA_IndustrialInternet_Report2015.pdf

Marquez, J., Villanueva, J., Solarte, Z., & Garcia, A. (2016). *IoT in Education: Integration of Objects with Virtual Academic Communities. In New Advances in Information Systems and Technologies, no. 115* (pp. 201–212). Springer International Publishing.

Mathews, K. (2018). *6 Exciting IoT Use Cases in Healthcare*. Received May 03, 2018 From https://www.iotforall.com/exciting-iot-use-cases-in-healthcare/

McColl, D., Hong, A., Hatakeyama, N., Nejat, G., & Benhabib, B. (2016). A Survey of Autonomous Human Affect Detection Methods for Social Robots Engaged in Natural HRI. *Journal of Intelligent & Robotic Systems*, *82*(1), 101–133. doi:10.100710846-015-0259-2

McGrath, M. J., & Scanaill, C. N. (2013). Body-worn, ambient, and consumer sensing for health applications. In Sensor Technologies, (pp. 181–216). Springer.

Mehta, Y. (2015). *What Is The Internet Of Things Definition – IoT Definition | Details*. Retrieved from http://iotworm.com/what-is-internet-of-things-definition/

Mei-Ju, S., Hung-yu, W., & Guan-Yu, L. (2016). Two Paradigm in Cellular IOT Access for Energy-Harvesting M2M Devices: Push-Based Versus Pull-Based. *IET Wireless Sensor System.*, *6*. doi:10.1049/iet-wss.2015.0040

Miao, Y., Ma, J., Liu, X., Wei, F., Liu, Z., & Wang, X. A. (2016). m2-ABKS: Attribute-based multi-keyword search over encrypted personal health records in multi-owner setting. *Journal of Medical Systems*, *40*(11), 246. doi:10.100710916-016-0617-z PMID:27696175

Michal, S., Jiri, V., Jan, M., & Palvik, J. (2016). *Internet of Things (IoT) in Agriculture - Selected Aspects*. Academic Press. . doi:10.7160/aol.2016.080108

Mirjana. (2017). IOT concept application in educational sector using collaboration, series of Teaching. *Learning and Teacher Education*, *1*(2), 137–150.

Mishra, N. (2018). Automation in Restaurants: Ordering to Robots in Restaurant via Smart Ordering System. *International Journal of Technologies and Management.*

Mishra, N. (2018). Spine in Robot: Designing of Lightweight Robot with the Spine to Provide Balancing and Load Bearing Capability. *15th International Conference on Recent Trends in Engineering, Applied Science and Management (ICRTESM-18).*

Mishra, N. (2018). *Issues in Existing Robotic Service in Restaurants and Hotels. 3rd International Conference on Internet of Things and Connected Technologies.*

Misra, S., Maheswaran, M., & Hashmi, S. (2017). Case Studies of Selected IoT Deployments. In *Security Challenges and Approaches in Internet of Things*. Cham: Springer. doi:10.1007/978-3-319-44230-3_6

Mohamed, M. (2016). *Internet Of Things for Home Automation*. Academic Press. doi:10.13140/RG.2.2.15237.35046

Mohammeda & Ahmedb. (2017). *Internet of Things Applications, Challenges and Related Future Technologies*. Available at www.worldscientificnews.com/EISSN

Mohammed, J., Lung, C. H., Ocneanu, A., Thakral, A., Jones, C., & Adler, A. (2014). Internet of Things: Remote Patient Monitoring Using Web Services and Cloud Computing. *IEE International Conference on Internet of Things (iThings)*, 256-263. 10.1109/iThings.2014.45

Mohanraj, I., Ashokumar, K., & Naren, J. (2016). Field monitoring and automation using IOT in agriculture domain. *Procedia Computer Science*, *93*, 931–939. doi:10.1016/j.procs.2016.07.275

Mookerjea, S., Krishnan, R., Datta, S., & Narayanan, V. (2009). Effective capacitance and drive current for tunnel FET (TFET) CV/I estimation. *IEEE Transactions on Electron Devices*, *56*(9), 2092–2098. doi:10.1109/TED.2009.2026516

Moosavi, Gia, Rahmani, Nigussie, Virtanen, Isoaho, & Tenhunen. (2015). SEA: A Secure and Efficient Authentication and Authorization Architecture for IoT-Based Healthcare Using Smart Gateways. *Proceedings of 6th International Conference on Ambient Systems, Networks and Technologies (ANT 2015)*.

Mora, H., Gil, D., Terol, R. M., Azorín, M., & Szymanski, J. (2017). An IoT-Based Computational Framework for Healthcare Monitoring in Mobile Environments. *Sensors (Basel)*, *17*(10), 2302. doi:10.339017102302 PMID:28994743

Morris, D. H., Avci, U. E., & Young, I. A. (2015). Variation-tolerant dense TFET memory with low VMIN matching low-voltage TFET logic. *Symposium on VLSI technology*, T24–T25.

Mouly & Bernadette. (1992). *The GSM System for Mobile Communication*. Academic Press.

Moving from Uniform to Variable Fertilizer Rates on Iowa Corn. (1998). *Effects on Rates and Returns by Bruce A.* Babcock and Gregory R. Pautsch.

Mueller, R. (2007). Demo: A Generic Platform for Sensor Network Applications. MASS'07, 1–3. doi:10.1109/MOBHOC.2007.4428693

Mugisha, J. C., Munyazikwiye, B., & Karimi, H. R. (2015, November). Design of temperature control system using conventional PID and Intelligent Fuzzy Logic controller. In *2015 International Conference on Fuzzy Theory and Its Applications (iFUZZY)* (pp. 50-55). IEEE. 10.1109/iFUZZY.2015.7391893

Muragesh, S. K., & Rao, S. (2014). Automated internet of things for underground drainage and manhole monitoring system for metropolitan cities. *International Journal of Information & Computation Technology*.

Nasucha, M. (2015). Development of an obstacle avoiding robot. *Jurnal Sistem Komputer*, *5*(2), 55–63.

Nayyar, A. (2018). Smart Surveillance Robot for Real-Time Monitoring and Control System in Environment and Industrial Applications. Information Systems Design and Intelligent Applications, 229-243.

Nazeem Basha, S., Jilani, S.A.K., & Arun, S. (2016). An intelligent door system using raspberry pi and amazon web services IOT. *International Journal of Engineering Trends and Technology, 33*(2), 1-6.

Neun, M. (n.d.). *Techniques for LBS Cryptography*. Available at http://www.e-cartouche.ch/content_reg/cartouche/LBStech/en/text/LBStech.pdf

Newberry, S., & Gupta, G. S. (2015, April). Wireless sensor based home automation as an educational springboard. In *2015 IEEE Sensors Applications Symposium (SAS)* (pp. 1-6). IEEE. 10.1109/SAS.2015.7133610

Nirschl & Wang. (2004). *The tunneling field effect transistor (TFET): the temperature dependence, the simulation model, and its application*. IEEE.

Norbert. (2015). *Partialiternal models-techniques for integration into the standard formula*. Available at: https://www.munichre.com/site/corporate/get/documents_E352744188/mr/assetpool.shared/Documents/0_Corporate%20Website/1_The%20Group/Focus/Emerging%20Risks/CROF-ERI-2015-The-Smart-Factory.pdf

Nunberg, G. (2012). *The Advent of the Internet: 12th April*. Courses.

Object Management Group. (2015). *Data Distribution Service V1.4*. Retrieved from http://www.omg.org/spec/DDS/1.4

Occhiuzzi, C., Vallese, C., Amendola, S., Manzari, S., & Marrocco, G. (2014). NIGHT-care: A passive RFID system for remote monitoring and control of overnight living environment. *Procedia Computer Science*, *32*, 190–197. doi:10.1016/j.procs.2014.05.414

Okokpujie, K. O., John, S. N., Noma-Osaghae, E., Okokpujie, I. P., & Okonigene, R. E. (2019). A wireless sensor network based fire protection system with SMS alerts. *International Journal of Mechanical Engineering and Technology*, *10*(2), 44–52.

Open Geospatial Consortium. (n.d.). *Internet source*. Available at http://www.opengeospatial.org

Padmadevi, S., & Deepigha, V. (2016). Smart e-agriculture monitoring using internet of things. *International Conference on Emerging trends in Engineering and Technology*, 47-50.

Pappu, S., Vudatha, P., & Niharika, A. V. (2017). Intelligent IOT based water quality monitoring system. *International Journal of Applied Engineering and Research*, *12*(16), 5447–5454.

Park, J. K., & Han, K. S. (2018). Intensity Classification Background Model Based on the Tracing Scheme for Deep Learning Based CCTV Pedestrian Detection. 2018 IEEE 9th International Conference on Mechanical and Intelligent Manufacturing Technologies. doi:10.1109/ICMIMT.2018.8340453

Park, M. (2015). IEEE 802.11ah: sub-1-GHz license-exempt operation for the internet of things. *IEEE Communications Magazine, 53*(9), 145-151. Available at: http://ieeexplore.ieee.org/xpl/articleDetails.jsp?arnumber=7263359

Parra, J. M., Valdez, W., Guevara, A., Cedillo, P., & Ortiz-Segarra, J. (2017). Intelligent Pillbox: Automatic and Programmable Assistive Technology Device. *Proceedings of the IASTED International Conference Biomedical Engineering*, 74-81. 10.2316/P.2017.852-051

Patel, K.K. & Patel, S.M. (2016). Internet of Things-IOT: Definition, Characteristics, Architecture, Enabling Technologies, Application & Future Challenges. *International Journal of Engineering Science and Computing, 6*(5).

Patel, H. N. (2016). Optimizing SRAM bitcell reliability and energy for IoT applications. *17th international symposium on quality electronic design (ISQED)*, 12–17. 10.1109/ISQED.2016.7479149

Patil, V. C., Al-Gaadi, K. A., Biradar, D. P., & Rangaswamy, M. (2012). Internet of things (Iot) and cloud computing for agriculture: An overview. Proceedings of Agro-Informatics and Precision Agriculture (AIPA 2012), 292-296.

Patil, A. T. (2016). IOT Based Remote Access Human Control Robot Using MEMS Sensor. *International Journal of Computer Science and Mobile Computing*, *5*(3), 816–826.

Pavana, N. R., & Padma, M. C. (2016). Design of low cost system for real time monitoring of water quality parameters in IOT environment. *International Journal of Advanced Research in Computer Science and Management Studies*, *4*(5), 12–15.

Pavithra, D., & Balakrishnan, R. (2015, April). IoT based monitoring and control system for home automation. In *2015 global conference on communication technologies (GCCT)* (pp. 169-173). IEEE.

Penlikousis. (2005). Wireless Data Networks. *Internet Protocol Journal, 8*(1).

Peruma, T., Sulaiman, M. N., & Leong, C. Y. (2015). Internet of things enabled water monitoring system. *IEEE 4th Global Conference on Consumer Electronics,* (8), 42-43.

Pirbhulal, S., Zhang, H. E., Alahi, M., Ghayvat, H., Mukhopadhyay, S., Zhang, Y. T., & Wu, W. (2017). A novel secure IoT-based smart home automation system using a wireless sensor network. *Sensors (Basel), 17*(1), 69. PMID:28042831

Piyush, K., Bhupendra, S., Rajesh, S., & Anita, G. (2018). *Wireless Methods & Devices For Home Automation.* Delhi: International Research Publication House.

Poczter & Jankovic. (2014). The Google Car: Driving Toward A Better Future? *Journal of Business Case Studies, 10*(1).

Prachi. (2017). *IOT Fundamentals- networking technologies, Protocols and use cases for the Internet of Things.* ciscopress.com.

Pradeep, P., Narendran, S., & Ramesh, M. V. (2017). An internet of things (iot) based sustainable water management. *Journal of IEEE Global Humanitarian Technology Conference, 6*(2), 68–72.

Project Coordination Group. (2013). *The Third Generation Partnership Project Working Procedures* (Version 15.0). Available at http://www.3gpp.org

Pulse Sensor. (n.d.). Retrieved from https://components101.com/sensors/pulse-sensor

Pundir, Y., Sharma, N., & Singh, Y. (2016). Internet of things (IOT): Challenges & future directions. *International Journal of Advanced Research in Computer and Communication Engineering, 5*(3), 1-5.

PYMNTS. (2017, August 3). *Robot v. Food.* Retrieved from https://www.pymnts.com/intelligence-of-things/2017/miso-robotics-flippy-the-iot-restaurant robot/

Quigley, M., Conley, K., Gerkey, B. P., Faust, J., Foote, T., Leibs, J., ... Ng, A. Y. "ROS: an open-source Robot Operating System," in *ICRA Workshop on Open Source Software*, 2009.

Qu, T., Lei, S. P., Wang, Z. Z., Nie, D. X., Chen, X., & Huang, G. Q. (2016). IoT-based real-time production logistics synchronization system under smart cloud manufacturing. *International Journal of Advanced Manufacturing Technology, 84*(1-4), 147–164. doi:10.100700170-015-7220-1

Ragahv, A., Mishra, P., Verma, P., Randhawa, K. S., & Thakur, T. (2017). Obstacle avoiding robot. *International Journal of Advanced Research in Computer Science and Software Engineering, 7*(4), 415–417. doi:10.23956/ijarcsse/V7I4/0250

Rahmani, A. (2015). Smart e-Health Gateway: Bringing Intelligence to IoT-Based Ubiquitous Healthcare Systems. CCNC'15.

Rajesh, S., Anita, G., Bikarama, Y., & Bhupendera, S. (2018). *IOT Enabled Fire Safety and Security devices for building. In Pen2Print.* EduPedia Publication.

Ramesh, M. V. (2009, June). Real-time wireless sensor network for landslide detection. In *2009 Third International Conference on Sensor Technologies and Applications* (pp. 405-409). IEEE. 10.1109/SENSORCOMM.2009.67

Ram, K. S. S., & Gupta, A. N. P. S. (2016). IoT based Data Logger System for weather monitoring using Wireless sensor networks. *International Journal of Engineering Trends and Technology, 32*(2), 71–75. doi:10.14445/22315381/IJETT-V32P213

Rasouli, S. H. (2006). *Double Edge Triggered Feedback Flip-Flop in Sub-100nm Technology*. ASP-DAC.

Rata, G., & Rata, M. (2016, October). Temperature control solution with PLC. In *2016 International Conference and Exposition on Electrical and Power Engineering (EPE)* (pp. 571-575). IEEE. 10.1109/ICEPE.2016.7781405

Ratti, C., Frenchman, D., Pulselli, R. M., & Williams, S. (2006). Mobile Landscapes: Using Location Data from Cell Phones for Urban Analysis. *Environment and Planning. B, Planning & Design*, *33*(5), 727–748. doi:10.1068/b32047

Ravichandran, S. (2016). Liquid level monitoring system using IOT. *International Science Press*, *9*(31), 215–218.

Ray, P. P. (2016). Internet of Robotic Things: Concept, Technologies, and challenges. *IEEE Access: Practical Innovations, Open Solutions*, *4*, 9489–9500. doi:10.1109/ACCESS.2017.2647747

Raza, S., & Voigt, T. (2010). Interconnecting WirelessHART and legacy HART networks. *6th IEEE International Conference on Distributed Computing in Sensor Systems Workshops (DCOSSW)*, 1-8. Available at: http://ieeexplore.ieee.org/xpl/articleDetails.jsp?arnumber=5593285

Read, B. (2017). *Rise of the Airport Robots*. Retrieved from https://www.aerosociety.com/news/rise-of-the-airport-robots/

Reclus, F., & Drouard, K. (2009). Geofencing for Fleet and Freight Management. 9th International Conference on Intelligent Transport Systems Telecommunications (ITST), 353-356. doi:10.1109/ITST.2009.5399328

Reddy, P. S. N., Reddy, K. T. K., Reddy, P. A. K., Ramaiah, G. K., & Kishor, S. N. (2016, October). An IoT based home automation using android application. In *2016 International Conference on Signal Processing, Communication, Power and Embedded System (SCOPES)* (pp. 285-290). IEEE. 10.1109/SCOPES.2016.7955836

Redondi, A., Chirico, M., Borsani, L., Cesana, M., & Tagliasacchi, M. (2013). An integrated system based on wireless sensor networks for patient monitoring, localization, and tracking. *Ad Hoc Networks*, *11*(1), 39–53. doi:10.1016/j.adhoc.2012.04.006

Renka. (n.d.). Location based Services on Mobile Ecommerce. *International Journal of Computer Science and Information Technologies, 3*(1), 3147-3151.

Rescorla, E., & Modadugu, N. (2012). Datagram Transport Layer Security (DTLS) Version 1.2. RFC 5238.

RFC 7452. (2015). Architectural Considerations in Smart Object Networking. Retrieved from https://tools.ietf.org/html/rfc7452

Robles, T., Alcarria, R., Mart'ın, D., Navarro, M., Calero, R., Iglesias, S., & Lopez, M. (2013). An IOT based reference architecture for smart water management processes. *Journal of Wireless Networks*, *6*(1), 4–23.

Robotics+IoT=The Internet of Robotic Things. (2017, November 20). Retrieved from https://blog.techdesign.com/robotics-internet-of-robotic-things/

Rohde & Schwarz. (2007). Headed Towards Next Mobile Radio Generation with UMTS LTE. *News on Mobile Radio Signal Generators/Signal Analyzers, 193*. Available at www.rohde-schwarz.com

Roots, W. K., & Gonenc, G. (1969). Temperature control in industrial processes. *IEEE Transactions on Industrial Electronics and Control Instrumentation*, (2), 147-156.

Rosi, A., Berti, M., Bicocchi, N., Castelli, G., Corsini, A., Mamei, M., & Zambonelli, F. (2011). Landslide monitoring with sensor networks: Experiences and lessons learnt from a real-world deployment. *International Journal of Sensor Networks*, *10*(3), 111–122. doi:10.1504/IJSNET.2011.042195

Rossi, A., Vila, Y., Lusiani, F., Barsotti, L., Sani, L., Ceccarelli, P., & Lanzetta, M. (2019). Embedded smart sensor device in construction site machinery. *Computers in Industry*, *108*, 12–20. doi:10.1016/j.compind.2019.02.008

Roth, J. (2004). Data collection. In Location-Based Services. Morgan Kaufmann.

Roth. (2004). Data collection. In *Location-Based Services*. Morgan Kaufmann.

Rouse, M. (2018). *IoT analytics guide: Understanding Internet of Things Data*. Received May 2018, from https://internetofthingsagenda.techtarget.com/definition/machine-to-machine-M2M

Ruiz-Garcia, L., Barreiro, P., & Robla, J. I. (2008). Performance of ZigBee-based wireless sensor nodes for real-time monitoring of fruit logistics. *Journal of Food Engineering*, *87*(3), 405–415. doi:10.1016/j.jfoodeng.2007.12.033

Rusu, R. B., Gerkey, B., & Beetz, M. (2008). Robots In The Kitchen: Exploiting Ubiquitous Sensing and Actuation. *Robotics and Autonomous Systems*, *56*(10), 844–856. doi:10.1016/j.robot.2008.06.010

Ryther, C. A., & Madsen, O. B. (2009). *Obstacle detection and avoidance for mobile robots* (Bachelor's thesis). Technical University of Denmark.

Rytivaara, A. (2012). Collaborative classroom management in a co-taught primary school classroom. Academic Press.

Sadasivan, S., Vadhri, V., & Ramesh, S. (2015). *Smart Plant Monitoring System*. Academic Press. Doi:10.19140/RG.2.1.2915.6563/1

Sagar, V. R., & Kumar, P. S. (2010). Recent advances in drying and dehydration of fruits and vegetables: A review. *Journal of Food Science and Technology*, *47*(1), 15–26. doi:10.100713197-010-0010-8 PMID:23572596

Sahay & Kumar. (2016b). Realizing efficient volume depletion in SOI junctionless FETs. *IEEE J. Electron Devices Soc.*, *4*.

Sahay, S., & Kumar, M. J. (2016a). Controlling L-BTBT and volume depletionin nanowire JLFETs using core–shell architecture. *IEEE Transactions on Electron Devices*, *63*(9), 3790–3794. doi:10.1109/TED.2016.2591588

Sahoo, P. M. (n.d.). *Introduction to Geographic Information System (GIS) – IASRI*. Retrieved from www.iasri.res.in/ebook/gis_ta/m3_1_intogis.pdf

Saint-Andre, P. (2011). *Extensible messaging and presence protocol (XMPP): Core*. IETFRFC 6120. Available at: https://tools.ietf.org/html/rfc6120

Salunke, P., & Nerkar, R. (2017). IoT Driven Healthcare System for Remote Monitoring of Patients. Department of E&TC, SITRC.

Santoso, F. K., & Vun, N. C. (2015, June). Securing IoT for smart home system. In *2015 International Symposium on Consumer Electronics (ISCE)* (pp. 1-2). IEEE.

Saquib, N., Papon, M. T. I., Ahmad, I., & Rahman, A. (2015). Measurement of heart rate using photoplethysmography. *Networking Systems and Security (NSysS) 2015 International Conference*, 1-6.

Saripalli, V. (2011). Variation-tolerant ultra-low-power heterojunction tunnel FET SRAM design. *Proceedings IEEE/ACM international symposium on nanoscale architectures*, 45–52. 10.1109/NANOARCH.2011.5941482

Saurabh & Kumar. (2012). Novel Attributes of a Dual Material Gate Nanoscale Tunnel Field-Effect Transistor. *IEEE Transactions on Electron Devices*, *58*(February), 1023–1029.

Schiller & Voisard. (2004). *Location based Services*. Morgan Kaufmann Publishers.

Sebastian, S. (2018). *How to protect against IoT security threats in healthcare*. Received December 13 2018 from https://www.cabotsolutions.com/how-to-protect-against-iot-security-threats-in-healthcare www.cabotsolutions.com

Sethi, P., & Sarangi, S. R. (2017). Internet of Things: Architectures, Protocols, and Applications. *Journal of Electrical and Computer Engineering*. doi:10.1155/2017/9324035

Sezer, O. B., Dogdu, E., & Ozbayoglu, A. M. (2018). Context-Aware Computing, Learning, and Big Data in Internet of Things: A Survey. *IEEE Internet of Things Journal*, *5*(1), 1–27. doi:10.1109/JIOT.2017.2773600

Shahdib, Ullah, Hasan, & Mahmud. (2013). Obstacle detection and object size measurement for autonomous mobile robot using sensor. *International Journal of Computers and Applications*, *66*(9), 28–33.

Sheehan, B., Murphy, F., Mullins, M., & Ryan, C. (2018). (in press). Connected and autonomous vehicles: A cyber-risk classification framework. *Transportation Research Part A, Policy and Practice*.

Shen, W. (2011). Smart Border Routers for eHealthCare Wireless Sensor Networks. WiCOM'11, 1–4. doi:10.1109/wicom.2011.6040606

Shingala, K., & Patel, J. (2017). Automatic home appliances and security of smart home with RFID, SMS, Email and real time algorithm based on IOT. *International Research Journal of Engineering and Technology, 4,* 1-7.

Shirehjini, A. A. N., Yassine, A., & Shirmohammadi, S. (2012, November). Equipment location in hospitals using RFID-based positioning system. *IEEE Transactions on Information Technology in Biomedicine*, *16*(6), 1058–1069. doi:10.1109/TITB.2012.2204896 PMID:24218700

Shi, W., Cao, J., Zhang, Q., Li, Y., & Xu, L. (2016). Edge Computing: Vision and Challenges. *IEEE Internet Of Things Journal*, *3*(5), 637–646. doi:10.1109/JIOT.2016.2579198

Shvachko, K., & Kuang, H. (2010). The Hadoop Distributed File System. *26th IEEE Symposium on Mass Storage Systems and Technologies*, 1-10.

Siddagangaiah, S. (2016). A Novel Approach to IoT Based Plant Health Monitoring System. *International Research Journal of Engineering and Technology, 3*(11), 880-884.

Simic, K., Despotovic-Zrakic, M., Đuric, I., Milic, A., & Bogdanovic, N. (2015). A Model of Smart Environment for E-Learning Based on Crowdsourcing, RUO. *Rev. za Univerzalno Odlicnost*, *4*(1), A1–A10.

Simoens, P. (2018). The Internet of Robotic Things: A review of concept, added value and applications. *International Journal of Advanced Robotic Systems*, 1–11.

Singh, M., Rajan, M., Shivraj, V., & Balamuralidhar, P. (2015). Secure MQTT for Internet ofThings (IoT). *Fifth International Conference on Communication Systems and NetworkTechnologies (CSNT 2015),* 746-751. Available at http://ieeexplore.ieee.org/xpl/articleDetails.jsp?arnumber=7280018

Singh, R., Kuchhal, P., Choudhury, S., & Gehlot, A. (2015). Wireless controlled intelligent heating system using hpso. *Procedia Computer Science*, *48*, 600–605. doi:10.1016/j.procs.2015.04.141

Singh, S., & Singh, N. (2015). Internet of Things (IoT): Security challenges, business opportunities & reference architecture for E-commerce. *Proceedings of 2015 International Conference on Green Computing and Internet of Things (ICGCIoT)*, 1577-1581. 10.1109/ICGCIoT.2015.7380718

Sinha, S. (2018). *Introduction to Internet of Things: IoT Tutorial with IoT Application*. Retrieved December 24, 2018, from https://www.edureka.co/blog/iot-tutorial/

Siva Kumar, P., Jilani, S.A.K., & Venkata Raman, B. (2015). Design and Development of Indore Plant Monitoring System. *International Journal of Advanced Research in Electronics and Communication Engineering, 4*(10), 2531-2535.

Sivagami, S., & Revathy, D. (2016). Smart Healhcare System implemented using IoT. *International Journal of Contemporary Research in Computer Science and Technology*, *2*(3), 641–646.

Smart Nation and Digital Government Office. (2016). *Self-Driving Vehicles (SDVs): Future of Mobility in Singapore*. Retrieved from https://www.smartnation.sg/initiatives/Mobility /self-driving-vehicles-sdvs--future-of-mobility-in-singapore

Smart Nation and Digital Government Office. (n.d.). *Self-Driving Vehicles (SDVs): Future of Mobility in Singapore, 2016*. Retrieved from https://www.smartnation.sg/initiatives/Mobility /self -driving-vehicles-sdvs--future-of-mobility-in-singapore

Somayya, Ramaswamy, & Siddharth. (2015). Internet of things (IOT): A literature review. *Journal of Computer and Communications, 3*, 164-173.

Somayya, M., Ramaswamy, R., & Siddharth, T. (2015). Internet of Things (IoT): A Literature review. *Journal of Computer and Communication*, *3*(05), 164–173. doi:10.4236/jcc.2015.35021

Sonar, K., & Upadhyay, H. (2016). *An Approach to Secure Internet of Things Against DDoS*. Singapore: Springer Singapore.

Soo-Jin, K. (2017). *Room Service is Now Delivered by Robots at These Hotels in Singapore*. Retrieved from http://www.scmp.com/magazines/style/travel-food/article/2119123/room-service-now-delivered-robots-singapores-hotel-jen

Stankovic, J. (2014). Research directions for the internet of things. *Internet of Things Journal, IEEE*, *1*(1), 3–9. doi:10.1109/JIOT.2014.2312291

Steinfield, C. (2004). The Development of Location based Services in Mobile Commerce. E-Life after Dot Com Bust, 4, 177-197.

Steiniger, S. (2012). *Foundation of LBS*. Lecture notes on LBS, Version 1.0. Available at http://www.e-cartouche.ch/content_reg/cartouche/LBSbasics/en/text/LBSbasics.pdf

Steven. (2011). Restaurant Staffed By Dancing Robot Samurai Waiters. *Sora News 24*. Retrieved from https://soranews24.com/2011/06/06/restauraunt-staffed-by-dancing-robot-samurai-waiters/

Stojmenovic, I. (2019). Fog computing: a cloud to the ground support for smart things and machine-to-machine networks. *Proceedings of the Australasian Telecommunication Networks and Applications Conference (ATNAC '14)*, 117–122.

Strangio, S., Palestri, P., Esseni, D. A. V. I. D., Selmi, L., Crupi, F., Richter, S., ... Mantl, S. (2015). Impact of TFET unidirectionality and ambipolarity on the performance of 6T SRAM cells. *IEEE J Electron Devices Soc*, *3*(3), 223–232. doi:10.1109/JEDS.2015.2392793

Sudha, Reddi, Vasanthi, & Venkatesh. (2018). IOT Based Smart Security and Smart Home Automation. *International Journal of Engineering Research and Technology*, *7*(4).

Sukriti, G. S., & Indumathy, K. (2016). IOT based smart irrigation and tank monitoring system. *International Journal of Innovative Research in Computer and Communication Engineering*, *4*(9), 15718–15722.

Sun, E., Zhang, X., & Li, Z. (2012). The internet of things (IOT) and cloud computing (CC) based tailings dam monitoring and pre-alarm system in mines. *Safety Science*, *50*(4), 811–815. doi:10.1016/j.ssci.2011.08.028

Sun, W., Cai, Z., Li, Y., Liu, F., Fang, S., & Wang, G. (2018). Security and Privacy in the Medical Internet of Things: A Review. *Security and Communication Networks*, *2018*, 1–9.

Sushabhan, C., Bhupendra, S., Rajesh, S., & Anita, G. (2018). *Internet of Things [IoT] Enabled Automation in Agriculture.* NIPA Publication House.

Symphony Link vs. LoRaWAN. (2017). Link Labs. Available: https://www.link-labs.com/whitepapersymphony-link-vs-lorawan

Tabassum, Lopa, Tarek, & Ferdosi. (2017). Obstacle avoiding robot. *Global Journal of Researches in Engineering: H Robotics & Nano-Tech, 17*(1), 19-24.

Taheri, S., & Yuan, J.-S. (2017). Security Analysis of Tunnel Field-Effect Transistor for Low Power Hardware. *International Journal of Computer Science and Information Technologies, 8*(2), 271–275.

Taur, Y., & Ning, T. H. (2013). *Fundamentals of modern VLSI devices.* Cambridge University Press.

Thamaraimanalan, T., Vivekk, S. P., Satheeshkumar, G., & Saravanan, P. (2018). Smart Garden Monitoring System using IOT. *Asian Journal of Applied Science and Technology, 2*(2), 186-192.

The Internet of Things with ESP8266. (n.d.). Retrieved from http://esp8266.net/

The Things Network. (2017). Ground breaking world record! LoRaWAN packet received at 702 km (436 miles) distance. Retrieved from: https://www.thethingsnetwork.org/article/ground-breaking-world-record-lorawan-packet-received-at-702-km-436-miles-distance"

Thrun, S. (1999). MINERVA: A second-generation museum tour-guide robot. *Proc. of the International Conference on Robotics and Automation (ICRA' 99).*

Thrun, S., & Leonard, J. J. (2008). Simultaneous localization and mapping. In B. Siciliano & O. Khatib (Eds.), *Springer handbook of robotics* (pp. 871–889). Cham: Springer. doi:10.1007/978-3-540-30301-5_38

Tjiharjadi, S., & Fajar, A. (2017). Human Heart Rate Detection Application. *International Conference on Soft Computing, Intelligent System and Information Technology (ICSIIT)*, 167-172.

Todica, M. (2016). *Controlling Arduino board with smartphone and Blynk via internet.* Academic Press. . doi:10.13140/RG.2.2.23956.30080

Todica, M. (2017). *Home Security alarm notification with smartphone and Blynk.* Academic Press. . doi:10.13140/RG.2.2.32529.43369

Tongke, F. (2013). *Smart Agriculture Based on Cloud Computing and IOT.* Academic Press.

Tripathi, B. K., & Anuradha, J. (2018). *Internet of Things (IoT) Technologies, Applications, Challenges, and Solutions.* CRC Press.

Tsang, Y. P., Choy, K. L., Wu, C. H., Ho, G. T. S., Lam, H. Y., & Koo, P. S. (2017). An IoT-based cargo monitoring system for enhancing operational effectiveness under a cold chain environment. *International Journal of Engineering Business Management, 9*, 1847979017749063. doi:10.1177/1847979017749063

Turcu, C., Turcu, C., & Gaitan, V. (2012). *Merging the Internet of Things and Robotics.* Recent Researches in Circuits and Systems.

Ungurean, O., & Brezulianu, A. (2017). An Internet of Things Framework for Remote Monitoring of the HealthCare Parameters. *Advances in Electrical and Computer Engineering, 17*(2), 11–16. doi:10.4316/AECE.2017.02002

Uygar, E. (2014). *Tunnel Field-Effect Transistors: Prospects and Challenges. Journal of Electron Device Society.*

Vaghela, A., Patel, J., & Vaghela, S. (2016). Obstacle avoidance robotic vehicle using ultrasonic sensor, android, and Bluetooth for obstacle detection. *International Research Journal of Engineering and Technology*, *3*(2), 339–348.

Vandna & Ravi. (2016). A Review paper on "IOT" & it's Smart Applications. *International Journal of Science, Emerging and Technology Research, 5*(2).

Verma, P., Rajnish, R., & Fatima, S. (2017). Challenges: Wearable Computing for Internet of Things. *International Journal of Science and Research, 6*(4), 1306-1311.

Vidya, B., Poonam, K., Priyanka, G., Gaurav, D., & Chandgude, A. S. (2016). Water level monitoring system in real time mode using WSN. *International Journal of Emerging Technology and Advanced Engineering*, *6*(9), 212–214.

Viswanath, S., Belcastro, M., Barton, J., O'Flynn, B., Holmes, N., & Dixon, P. (2015). Low-power wireless liquid monitoring system using ultrasonic sensors. *International Journal on Smart Sensing and Intelligent Systems*, *8*(1), 26–44. doi:10.21307/ijssis-2017-747

W3C Standards. (n.d.). Available at http://www.w3.org/standards/

Wahab, M. H. A. (2016, September). IoT-based home automation system for people with disabilities. In *2016 5th International Conference on Reliability, Infocom Technologies and Optimization (Trends and Future Directions) (ICRITO)* (pp. 51-51). IEEE. 10.1109/ICRITO.2016.7784923

Walravens, N., & Ballon, P. (2013). Platform business models for smart cities: From control and value to governance and public value. *IEEE Communications Magazine*, *51*(6), 72–79. doi:10.1109/MCOM.2013.6525598

Wang, Y., & Dasgupta, P. (2016, April). Intelligent and adaptive temperature control for large-scale buildings and homes. In *2016 IEEE 13th International Conference on Networking, Sensing, and Control (ICNSC)* (pp. 1-6). IEEE. 10.1109/ICNSC.2016.7478971

Wang, W. (2014, January). Design of U-Shape Channel Tunnel FETs WithSiGe Source Regions. *IEEE Transactions on Electron Devices*, 61.

Want, R., Hopper, A., Falcão, V., & Gibbons, J. (1992). The Active Badge Location Based System. *ACM Transactions on Information Systems*, *10*(1), 91–102. doi:10.1145/128756.128759

Western PA Healthcare News Team. (2018). *The Market vs. technologists: Treating type one Diabetes*. Received November 21, 2018, from https://www.wphealthcarenews.com/?s=OpenAps

Whittenberger, R. T. (1951). Changes in specific gravity, starch content, and sloughing of potatoes during storage. *American Journal of Potato Research*, *28*(10), 738–747. doi:10.1007/BF02851330

Wikipedia. (n.d.). *Enhanced Data Rate for GSM Evolution*. Available at http://en.wikipedia.org/wiki/Enhanced_Data_Rates_for_GSM_Evolution

Wikipedia. (n.d.). *High Speed Downlink Packet Access*. Available at http://en.wikipedia.org/wiki/High-Speed_Downlink_Packet_Access

Wikipedia. (n.d.). *Radio Frequency Identification*. Available at http://en.wikipedia.org/wiki/Radio-frequency_identification

Wikipedia. (n.d.). *Radio Network Controller*. Available at http://en.wikipedia.org/wiki/Radio_Network_Controller#RNC_Roles

Wu, C., Gu, W., Jiang, P., Li, Z., Cai, H., & Li, B. (2018). Combined economic dispatch considering the time-delay of district heating network and multi-regional indoor temperature control. *IEEE Transactions on Sustainable Energy*, *9*(1), 118–127. doi:10.1109/TSTE.2017.2718031

Wu, Q., Ding, G., Xu, Y., Feng, S., Du, Z., Wang, J., & Long, K. (2014). Cognitive Internet of Things: A New Paradigm beyond Connection. *IEEE Internet of Things Journal*, *1*(2), 129–143. doi:10.1109/JIOT.2014.2311513

Wu, S., Rendall, J. B., Smith, M. J., Zhu, S., Xu, J., Wang, H., & Qin, P. (2017). Survey on Prediction Algorithms in Smart Homes. *IEEE Internet of Things Journal*, *4*(3), 636–644. doi:10.1109/JIOT.2017.2668061

Wu, T. F., Tsai, P. S., Hu, N. T., & Chen, J. Y. (2014). Use of ultrasonic sensors to enable wheeled mobile robots to avoid obstacles. *Proceedings of 10th International Conference on Intelligent Information Hiding and Multimedia Signal Processing*, 958-961. 10.1109/IIH-MSP.2014.240

Xiaojun, C., Xianpeng, L., & Peng, X. (2015, January). IOT-based air pollution monitoring and forecasting system. In *2015 International Conference on Computer and Computational Sciences (ICCCS)* (pp. 257-260). IEEE. 10.1109/ICCACS.2015.7361361

Xiong, Y., Xiao, S., & Zha, F. (2011). An avoiding obstacle algorithm of mobile robot under complex unknown environments. *Proceedings of 2nd International Conference on Mechanic Automation and Control Engineering*, 1019-1022. 10.1109/MACE.2011.5987106

Yamada, T., Sa, Y. L., & Ohya, A. (2013). Moving obstacle avoidance for mobile robot moving on designated path. *Proceedings of 10th International Conference on Ubiquitous Robots and Ambient Intelligence*, 509-514. 10.1109/URAI.2013.6677324

Yang, D., Xia, J., & Xiang, L. (2010). The research on intelligent mobile robot's avoiding obstacle by ultrasound. *Proceedings of International Conference on Artificial Intelligence and Computational Intelligence*, 77-79. 10.1109/AICI.2010.23

Yang, Y., Tong, X., Yang, L.-T., Guo, P.-F., Fan, L., & Yeo, Y.-C. (2010). Tunneling field-effect transistor: Capacitance components and modeling. *IEEE Electron Device Letters*, *31*(7), 752–754. doi:10.1109/LED.2010.2047240

Yang, Y., Wu, L., Yin, G., Li, L., & Zhao, H. (2017). A Survey on Security and Privacy Issues in Internet-of-Things. *IEEE Internet of Things Journal*, *4*(5), 1250–1258. doi:10.1109/JIOT.2017.2694844

Yang, Z. (2016, July). Tunnel Field-Effect Transistor Withand L-Shaped Gate. *IEEE Electron Device Letters*, 37.

Yinghua, Z., Guanghua, F., Zhigang, Z., Zhian, H., Hongchen, L., & Jixing, Y. (2012). Discussion on application of IOT technology in coal mine safety supervision. *Procedia Engineering*, *43*, 233–237. doi:10.1016/j.proeng.2012.08.040

Yue, Y., Xu, Q., Luo, A., Guo, P., He, Z., & Li, Y. (2018). Analysis and control of tundish induction heating power supply using modular multilevel converter. *IET Generation, Transmission & Distribution*, *12*(14), 3452–3460. doi:10.1049/iet-gtd.2017.1274

Zanella, A., Bui, N., Castellani, A., Vangelista, L., & Zorzi, M. (2014). Internet of Things for Smart Cities. *IEEE Internet of Things Journal*, *1*(1), 22–32. doi:10.1109/JIOT.2014.2306328

Zeinab, Aldein, & Ali. (2017). Internet of Things Applications, Challenges and Related Future Technolofies. *World Scientific News*, *67*(2), 126–148.

Zhang, Y., Chen, B., & Lu, X. (2011, August). Intelligent monitoring system on refrigerator trucks based on the internet of things. In *International Conference on Wireless Communications and Applications* (pp. 201-206). Springer.

Zhao, Y., Chai, X., Gao, F., & Qi, C. (2018). Obstacle avoidance and motion planning scheme for a hexapod robot Octopus-III. *Robotics and Autonomous Systems*, *103*, 199–212. doi:10.1016/j.robot.2018.01.007

Zhiqiang & Junming. (2011). The Application of Internet of Things in Education and Its Trend of Development. *Modern Distance Education Research, 2*, 19.

Zhou, Z., Yao, B., Xing, R., Shu, L., & Bu, S. (2015). E-CARP: An energy efficient routing protocol for UWSNs in the internet of underwater things. *IEEE Sensors Journal*. Available at: http://ieeexplore.ieee.org/xpl/articleDetails.jsp?arnumber=7113774

Zhou, L., & Chao, H. C. (2011). Multimedia traffic security architecture for the IOT. *Journal of IEEE Networks*, *25*(3), 35–40. doi:10.1109/MNET.2011.5772059

ZigBee Standards Organization. (2008). *ZigBee Specification*. Document 053474r17. Available at http://home.deib.polimi.it/cesana/teaching/IoT/papers/ZigBee/ZigBeeSpec.pdf

About the Contributors

Vishal Jain is currently working as Associate Professor in Bharati Vidyapeeth's Institute of Computer Applications and Management (BVICAM), Affiliated to GGSIPU and Accredited by AICTE, from August, 2010 to till date. Before joining BVICAM, I have worked four years in Guru Presmsukh Memorial College of Engineering, Affiliated to GGSIPU and Accredited by AICTE, from July 2004 to July, 2008. I have completed Ph.D (Computer Science and Engineering) from Lingaya's University, Faridabad, Haryana, completed M.Tech (Computer Science and Engineering) from University School of Information Technology (USIT), Guru Gobind Singh Indraprastha University, MBA (HR) from Shobhit University, Meerut, MCA from Sikkim Manipal University, Sikkim. In additional qualification I have obtained DOEACC 'A' Level and DOEACC 'O' Level, Post Graduate Diploma in Computer Software Training from A.M Informatics, Advance Diploma in Computer Software Training from ET&T, Delhi, Diploma in Business Management from All India Institute of Management Studies, Chennai, Diploma in Programming from Oxford Computer Education, Delhi, MCP and CCNA.

* * *

Akanksha Aggarwal is currently working toward Ph.D. degree from the Department of Electrical Engineering, Deenbandhu Chhotu Ram University of Science & Technology, Sonepat, Haryana, India. She received B. Tech degree in Electrical and Electronics Engineering from GGSIP University, Delhi in 2012 and M. Tech degree in Instrumentation and Control Engineering from Deenbandhu Chhotu Ram University of Science & Technology, Sonepat, India in 2016. Her research interests include signal processing, condition monitoring and pattern recognition.

Jyoti Batra Arora is PhD in computer science from Banasthali University Rajsthan.She has completed her doctorate in area of mobile transaction. She has 8 years of enriched experience n teaching graduate and post graduate students. she has authored books on C, C++. She has publications in international and national journal. Her area of Interest is mobile commerce, computer network software engineering.

Shubham Asthana, Scholar, at Amity Institute of Information Technology, Amity University Uttar Pradesh, Lucknow. He has a varied experience in project development and research in the field of IoT, Android and Artificial Intelligence in our daily lives. He has published papers and articles in journals of repute, about the same too.

Neelendra Badal is Professor in the Department of Computer Science and Engineering at Kamla Nehru Institute of Technology (KNIT), at Sultanpur (U.P.). He pursued Ph.D. in 2009 from MNNIT, Allahabad in Computer Science and Engineering. He is Chartered Engineering (CE) from Institution of Engineers (IE), India. He has published more than 65 papers in International/National Journals, Conferences and Seminars. He is a Life member of IE, IETE, ISTE, and CSI India. His research interests are Distributed System, Parallel Processing, GIS, Data Warehouse and Data Mining, Software Engineering, and Networking.

Shaveta Bhatia is working as an Associate Professor and Head of Department in Manav Rachna International Institute of Research & Studies, Faridabad. She has publish more than 35 papers in various reputed International/National Journals and Conferences. She is also guested as Session chair in various International Conference organised by IEEE & ACM.She has experience of 17 years of Teaching. She is member of renowned professional organisations such as ACM, IEEE and CSI.

Parth Brahmbhatt received the Master of Science (C.A.& I.T.). degree in Computer Science from the Hemchndracharya North Gujarat University, India, in 2011, and currently pursuing Ph.D. degrees in Computer Science from the Sakarchand Patel University, Visnagar, Gujarat, India. In 2011, he joined the Gujarat International Finance Tec-City Ltd. (India's First Smart City) as an Engineer, and in 2016 became a Manager (ICT). His primary responsibility is taking care of the City Command and Control Center. His current research interests include the Internet of Things (IoT) with a different kind of IoT architecture and IoT data storage system. Mr. Parth Brahmbhatt is a Fellow of the Gujarat Forensic Science University, member of Computer Society of India and working as a volunteer to Suraksha Setu Society – a unit of CID Crime.

Ahmed A. Elngar currently is an Assistant Professor at Faculty of Computers & Artificial Intelligence, Department of Computer Science, Beni Suef University. Their current project is 'Biometrics, AI-based Security, Image processing, AI-based Control, AI-based Smart Grids' . Internet of Things Security. He is a Founder and Chair of the Scientific Innovation Research Group (SIRG). Also he is a Managing Editor of Journal of Cybersecurity and Information Management (JCIM).

Shahnaz Fatima is MCA and Ph.D. She has completed her Ph.D from Integral University. Her research area is Human computer interaction. Currently she is working as an Asst. Professor in Amity University. Dr. Shahnaz Fatima has published many of the valuable research papers in various national and international conferences and journals.

Dinesh Goyal is Professor of Cloud Computing and Security and Academic Dean of Suresh Gyan Vihar University, holding Ph.D. degree from Suresh Gyan Vihar University in 2014, also has 16 years of rich experience in teaching with 7 years in pure R&D sector of Image Processing, cloud computing/information security and over 90 research papers in international publications followed by 21 papers presented in conferences, out of above 8 papers are SCI/Scopus/Web of Sciences/ Thomson Reuter Indexed Journals. Also, in order to have inquisitiveness for the current trends in technology, I have conducted and organized 4 conferences and 5 workshops followed by 5 National paper publications. In addition to this, I have strived hard to guide mentor over 35 research scholars during my career followed by 8 doctoral guidance with 5 as supervisor and 3 as Co-supervisor.

Pragya Gupta is pursuing her B.Tech in Computer Science from SRM Institute of Science and Technology, Chennai. She believes in Project based learning and working enthusiastically to grasp required application oriented skills.

Narinder Singh Jassal has been working with CSIR-CSIO since last 15 years and has vast academic, research and administrative experience in the field of Manufacturing Technologies and Computer Integrated Manufacturing and Automation Systems. He is life member of Indian Institute of Industrial Engineering.

Simranjeet Kaur is pursuing Business Administration-Accounting from Reynolds Community Collage, Richmond, Virginia, U.S.A.

Suruchi Kaushik is M.Sc in Physics (with specialization in Electronics) from Delhi University. She has qualified GATE twice in 2012 and 2014.She has 9 years of enriched experience in Teaching. She has also guided many students at graduate level in their projects.

Brijesh Khandelwal has more than 25 years of encyclopedic academic experience in Computer Science, Management and Insurance domain. Currently, he is working as Associate Professor and HOD-Department of Computer Science, Amity School of Engineering and Technology, Amity University Chhattisgarh, Raipur. Dr. Khandelwal has rich and; diverse experience in academia and has several publications in International/ National Journals and Conferences. Dr. Khandelwal did Masters in Computer Applications (MCA) from University of Lucknow, Lucknow in year 1994. In 2001, he became Sun Certified Programmer with Sun Microsystems. In 2007 he was awarded Doctorate in Philosophy in Applied Economics from University of Lucknow, Lucknow. He did Masters in Business Administration (MBA) in 2010 from Punjab Technical University. In 2010 he became licentiate in Life Insurance from Insurance Institute of India, Mumbai. He also got awarded Ph.D. in Computer Science from Sri Venkateshwara University, Gajurala, UP in 2017.

Ankur Kohli is pursuing Bachelor's of Technology in Mechatronics Engineering from University of Petroleum & Energy Studies. Internship in Internet of Things with projects like Dam Monitoring System, Smart Movable Bin Monitoring System, Smart Garden Monitoring, and Robot Controlled through Voice. Another internship at ONGC Dehradun in AC Plant System with project Industrial Air Conditioning. Other Projects like Research in Bubbly Flow Feedback Control System, Smart Path follower using AI, Design of Terrain Rover: The Shrimp for multipurpose like security purpose, climatic condition purpose etc and controlled through voice. Completed the training on Structure Design and Analysis. Currently working under Artificial Intelligence, Machine Learning and currently pursuing internship from Bosch Rexroth, Ahmedabad (India).

Rohit Kohli is currently working under Parkland Fuel Corporation as an Assistant Manager in Canada. Completed his undergraduate degree in Bachelor's of Computer Application from Graphic Era University, Dehradun (India) and completed postgraduate degree in Software Enterprises and Resource Planning from Selkrik College, Castlegar (Canada).

Kriti received her BTech in Electronics and Communication Engineering from Maharishi Markandeshwar Engineering College, Maharishi Markandeshwar University, Ambala in 2013 and MTech in Electronics and Communication Engineering from Jaypee University of Information Technology, Waknaghat, Solan, Himachal Pradesh, India in 2015.Currently she is a PhD student at Thapar Institute of Engineering and Technology, Patiala, Punjab, India.

Sunil Kumar is currently works as an Assistant Professor in Computer Science and Engineering Department of Amity School of Engineering & Technology, Amity University, Noida, India. He has rich experience in teaching & training the undergraduate and postgraduate classes. He has published 3 text books and 20 research papers publications in different International Journals and in Proceedings of International Conferences of repute. Dr. Sunil Kumar completed his Ph.D. in Computer Science & Engineering from Amity University, Noida, India. He Completed his M.Tech (Computer Engineering) in 2011 and B.Tech (Information Technology) in 2007 from Uttar Pradesh Technical University, Lucknow. He has 11 years of teaching, training and research experience with leading institutions. His area of interest includes computer networks, distributed systems, wireless sensor networks and network security. He filed 2 international patents in the field of networks. He is Cisco certified instructor for CCNA routing & switching course and Tech Mahindra certified instructor for networking Course. He is global CCNA & CCNP certified by Cisco.

Rashima Mahajan has more than 14 years of experience with strong professional record in teaching, research and administration at universities of repute and dedicated research centers like NBRC (National Brain Research Centre) Manesar. She is working as Associate professor with the Department of Computer Science and Engineering Manav Rachna International Institute of Research and Studies, Faridabad. Prior to joining this institute, she has worked with G D Goenka University and Apeejay Stya University, Gurgaon. She did her Ph.D in Signal Processing; M.Tech and B.Tech in ECE. She has published more than 35 research papers/book chapters in peer reviewed International Journals and Conferences. She has authored a book with Elsevier's: Title- EEG Based Brain Computer Interfaces: Cognitive Analysis and Control applications, March 2019. She is a reviewer of International journals of repute including IEEE Transactions on Biomedical engineering, IEEE Journal of Biomedical and Health Informatics, IET Science, Measurement & Technology, International Journal of Information Technology (Springer Publishers). She has chaired many technical sessions and is a member Technical Program Committee of many International Conferences.

Neelima Mishra is pursuing Ph.D. in Computer Science from Suresh Gyan Vihar University, Jaipur. She is having teaching experience of 3.5 years and industrial experience of 4 years. She is an entrepreneur and her research interests include Cloud Computing, Internet of Robotics Things and Robotics. She has attended five national and international conferences and published more than ten research papers with prestigious publishers and earned two provisional patents in robotics.

Rajiv Pandey is a Faculty at Amity Institute of Information Technology, Amity University, Uttar Pradesh, Lucknow Campus India. He possesses a diverse back ground experience of around 30 years to include 15 years of Industry and 13 years of academic. His research interests include the contemporary technologies as Semantic Web Provenance, Cloud and Big Data, and Data Analytics. He has published more than 30 research papers in Scopus, and other science indexed journals of repute. He has been Session chairs, Technical committee member for various IEEE and Elsevier conference He has been on technical committees of various government and private universities. He is intellectually involved in supervising Doctorate Research Scholars and Post graduate Students. He is also an active contributor in professional bodies like IEEE, IET and Lucknow Management Association. He is also a member of Machine Intelligence Research Labs.

Bhushan Patil has done Master of Engineering in Electronics and Telecommunication. He has worked as an Automation engineer. He has various International Conference Paper Publications published under his name. He is an active Research Scholar. He is currently writing various chapters for different publication houses and he is also a Reviewer and Editorial Review Board Member of more than 30 International Journals. He is also a co-editor of a book which shall be published next year.

Nimisha Patel is currently working as an Associate Professor in the Computer Engineering department and Head in CE/IT department of Sankalchand Patel College of Engineering, Visnagar, Gujarat, India. She has more than 15 years of teaching experience in the field of Computer Engineering and Information Technology. She is acting as Dean, Faculty of Computer Science, Sankalchand Patel University, Visnagar. She has many professional memberships like CSI, ISTE, Sr. member UACEE, Editorial Board member, IAENG, etc. She is the Chairman, Board of Study of Computer Engineering and Information Technology and member of Information Communication Technology, under the Faculty of Engineering and Technology, Sankalchand Patel University. Dr. Nimisha Patel obtained her Ph.D. in the domain of Cloud Computing in Computer Science and Engineering. She completed her MTech in Computer Engineering from S. V. National Institute of Technology (NIT), Surat and BE in IT from HNGU, Patan. She has more than 37 publications in National and International Conferences (including IEEE, ACM, Science Direct, Springer, Elsevier, etc.) and Journals (UGC, Scopus, eSCI indexing). She participated and presented the more than 09 National/International indexing conferences. She has delivered talks, acted as session chair, session expert in seminar, project fair, International/National level conferences. She has participated in many one-week/two-week FDPs/IFDP/STTPs/Workshop/Seminar. Her area of domain is Cloud Computing, Networking, Information Security and IoT. She organized and coordinated many workshop/talk/seminar/STTP at Institute and University level. She received the Honored Certificate in the Technical Communication workshop from IIT, Bombay. She guided the more than 09 PG students and 50 groups of UG students. She is recognized Ph.D. research supervisor at Sankalchand Patel University, Visnagar. She acted as an Editorial board member/Reviewer/TPC/Advisory member of IEEE, Springer and other International/National conferences and Journals.

Rajan Patel is currently working as an Associate Professor in the Computer Engineering department of Gandhinagar Institute of Technology (GIT, Gandhinagar). He has worked as Faculty (Associate Professor, Assistant Professor, Sr. Lecturer and Lecturer) in Computer Department of Sankalchand Patel College of Engineering (Faculty of Engineering and Technology), Sankalchand Patel University, Visnagar. He has more than 14 years of teaching experience in the field of Computer Science and Engineering and also associated with the IT industry. Dr. Patel has more than 13 several professional academic and administrative memberships like CSI, ISTE, Senior Member UACEE, editorial board member, BoS member etc. He obtained his M.Tech. in Computer Engineering from S. V. National Institute of Technology (NIT), Surat in 2009 and B.E. in Computer Engineering from Saurashtra University in 2004. He completed his Ph.D. in Computer Engineering in the domain of MANET Security from RK University in March 2017. Dr. Patel has more than 36 publications in various National and International journals and conferences. He presented 13 articles in National/International conferences including IEEE, Science Direct, Springer,

Elsevier, etc. More than 11 articles are UGC indexing and more than 6 articles are Scopus indexing of his publications. Dr. Patel received the appreciation from the Dewang Mehta Innovation IT award and Honored in the Technical Communication workshop from IIT, Bombay. He also acted as session chair, session expert and delivered talks in seminar, project fair, International/National conferences, etc. Dr. Patel coordinated and organized many STTP/Seminar/Talk at Institute and University level. He also received the certificate of excellence from IIT Bombay for coordinating the 2-Week FDP. He also participated in more than 29 STTP/Seminar/Workshop/Talks etc. He guided the more than 13 PG students and 55 groups of UG students. He is the recognized Ph.D. research supervisor at Sankalchand Patel University, Visnagar and guiding the three Ph.D. scholars. Dr. Patel is acted as an Editorial board member /Reviewer/TPC/Advisory member of IEEE, Springer and other International/National conferences and Journals such as IET Network, Taylor & Francis. Dr. Patel also worked for ISEAP sponsored MHRD project during his Post graduation period at NIT, Surat. His main area of interest is Internet of Things, Information Security, and Networking.

Manish Kumar Saini did his B.E., M.E. in Electrical Engineering from Maharishi Dayanand University, Rohtak with Honors. He completed Ph.D. in Electronics and Communication Engineering from Delhi College of Engineering, New Delhi (INDIA). He is currently associated with Department of Electrical Engineering, Deenbandhu Chhotu Ram University of Science and Technology, Sonepat, Haryana, India. His field of interest includes digital signal processing, power quality, image processing, automation and power systems. He has been the author of more than 80 papers in different journals and conference in the area of interest.

Sunita Saini received her B.A. in Arts, M.A. in English and M.B.A. in Human Resource Management from Maharshi Dayanand University, Rohtak, Haryana in 2011. She is currently pursuing her Ph.D. degree from the Department of Management Studies, Deenbandhu Chhotu Ram University of Science & Technology, Sonepat, Haryana, India. She is the author of around 30 papers in international journals and conferences. She is currently working on service quality management in power sector and retail business.

Sahil Sandhu is a student of Electronics (2016-117) at Indo-Swiss Training Centre Chandigarh. Department of Mechatronics and Industrial Automation.

Sudhir Kumar Sharma is currently working as a professor in the Department of Computer Science, Institute of Information Technology & Management affiliated to GGSIPU, New Delhi, India. He has a rich experience of more than 19 years in the field of Computer Science and Engineering. He obtained his Ph.D. degree in Information Technology from USICT, GGSIPU, New Delhi, India. He obtained the M. Tech degree in Computer Science & Engineering in 1999 from the Guru Jambheshwar University, Hisar, India and M.Sc. degree in Physics from the University of Roorkee (now IIT Roorkee), Roorkee, in 1997. He is interested in the Machine Learning, Data Mining, and Security. He has published several research papers in various International Journals and International Conferences of repute. He is a life member of CSI and IETE. He is an Associate Editor for International Journal of End-User Computing and Development (IJEUCD), IGI Global, USA. He is a convener of ICETIT-2019.

Bhupendra Singh is Founder & Managing Director of Schematics Microelectronics, Dehradun (India). CCNA and MCSE Licenses & Certificate issued by Cisco and Microsoft. Having patent product and publication related to Arduino, Atmel AVR, IOT, Sensor and Wireless Technology.

Jasjit Singh is pursuing Bachelors of Technology in Mechatronics from University of Petroleum & Energy Studies. Internship in Internet of Things , projects - home automation using blynk , air pollution monitoring via Things speak server , smart parking , smart bin , Remote robot , high energy trapping solar panel , line follower robot .Pursuing Research and Development Internship at Plateforme ensmm (Handispoert Bourgogne Franche-Cmte) (France) on project Instrumental Athleticism gloves for handisport.

Kavita Srivastava is currently working as an Associate Professor in the Department of Computer Science, Institute of Information Technology & Management affiliated to GGSIPU, New Delhi, India. She obtained the M.Tech degree in IT in 2007 from USICT, GGSIPU, Delhi, India. She has an interest in Machine Learning, Wireless Sensor Networks, Internet of Things and Image Processing.

Suman Lata Tripathi has completed her Ph.D in the area of low power micro-electronic devices from MNNIT, Allahabad. She did her M.Tech in Electronics Engineering from UP Technical University, Lucknow and B.Tech in Electrical Engineering from Purvanchal University, Jaunpur. She is associated with Lovely Professional University as Professor with more than sixteen years of experience in academics. She has published more than 35 research papers in referred journals and conference. She has organized a number of workshops, summer internships and expert lectures for students. She has worked as session chair, reviewer and editorial board member in several National/International conferences and Journals.

Parul Verma is working as an Assistant Professor in Information Technology department in Amity University, Lucknow. She had completed her Ph.D. in Computer Science in the year 2012 and has 12+ years of teaching experience. There are more than 35 papers to her credit in the International and National journals and conferences. Her research interests are Natural Language Processing, Web Mining, Word Sense Disambiguation, Semantic Web and IoT. She is a member of Review Board of several International Journals. She is nominated as a member of Technical Program Committee and Organizing Committee of many International Conferences. She is also a member of many International and National bodies like IAENG, IACSIT, Internet Society, ACM, and CSI.

Jitendra Virmani received his BTech (Hons) in Instrumentation Engineering from Sant Longowal Institute of Engineering and Technology, Punjab in 1999 and MTech in Electrical Engineering with specialisation in Measurement and Instrumentation from Indian Institute of Technology, Roorkee in 2006. He received his PhD on Analysis and Classification of B-Mode Liver Ultrasound Images from Indian Institute of Technology, Roorkee in 2014. He served in academia for 12 years before joining the ISTC, CSIR-CSIO, Chandigarh, India.

Manisha Vohra has done Master of Engineering in Electronics and Telecommunication. She has various International Conference Paper Publications published under her name. She is an active Research Scholar. She is currently writing various chapters for different publication houses and she is also a Reviewer and Editorial Review Board Member of more than 25 International journals. She is also a co-editor of a book which shall be published next year.

Neeta Yadav completed her B.Tech from UNSIET, VBS Purvanchal University in 2016 in Information Technology. She is completed M.Tech in Computer Science and Engineering from Kamla Nehru Institute of Technology, Sultanpur. She has published three papers in International journals. Her area of interest is Data mining and Warehousing, Software Testing, Cryptography and Network Security and Computer Network.

Index

H

I

L

M

N

P

Q

R

S

T

U

V

W

www.ingramcontent.com/pod-product-compliance
Lightning Source LLC
LaVergne TN
LVHW060358300926
849162LV00006B/1296